任福尧

数学论文选

任福尧◎著

復旦大學出版社

任福尧教授

大学时的任福尧

任福尧、何又清夫妇(1957 年)

全家合影(1980 年)

东北人民大学数学系毕业留影，第三排右二为任福尧（1954年7月）

和导师陈建功的儿子陈翰林（前）合影（1954年）

和吴卓人(左)、欧阳光中(中)合影(1958年)

和函数论同行及学生合影，何成奇(左三)、李忠(左四)、吴卓人(左五)、夏道祥(左六)、王键(左七)、任福尧(右三)、陈纪修(右二)

西安单叶函数会议期间同西北大学刘书琴教授(右二)等合影(1982 年)

和大学时的老师江泽坚教授以及陈建功先生门人合影(后排左起:张锦豪、何成奇、龚升、江泽坚、程民德、龙瑞麟、严绍宗、任福尧;前排左起:邱维元、陈纪修、黄立丰、肖建斌、高福昌。复旦大学)

和函数论同行合影（左起：金亿丹、杨维奇、任福尧、姚碧云、张顺燕。浙江大学）

和原东北人民大学老同学李岳生（右二）、陈惠国（左一）在广州中山大学合影（2007 年）

访问日本大阪大学时和大阪大学 Owa 教授（左）合影（1997 年）

邀请加拿大维多利亚大学 Srivastava 教授来家作客(左起：Srivastava，任夫人何又清，任福尧)

纽约州立大学 Kra 教授夫妇访问复旦大学时合影(中排左起：陈有根、张锦豪、李立康、Kra、Kra 夫人、任福尧、何成奇、蒋尔雄)

美国 Miller 教授(右)访问复旦大学时合影

美国 Keen 教授夫妇访问复旦大学时合影(前排左起:Keen、Keen 丈夫、任福尧、陈正;后排左起:陈红斌、黄立丰、邱维元、马建国)

复旦同仁合影(左起:邱维元、陈纪修、张锦豪、任福尧、何成奇、朱洪)(1998年)

任福尧教授和他的研究生的合影(左起:梁金荣、肖建斌、王键、乔建永、任福尧、喻祖国、张文俊、尹永成、邱维元、龚志民、李万社)(1998年10月)

美国 Humboldt 大学留影(1989 年)

同夫人在美国休斯顿的 Rice 大学合影(2000 年)

参观香港艺术馆留影(1998 年)

访问深圳大学数学研究所留影(2007 年)

在杭州大学作学术报告(1993 年)

在学术会议上作报告

在家中书房(1992 年)

老年心静,养生长寿

祝贺任福尧、何成奇教授 80 华诞合影(2009 年)

在杭州和亲戚一起过
80 大寿

80 寿辰宴会上切蛋糕

和大儿子任铁工及孙子在复旦校门口合影

和小儿子任铁军在原数学系大楼前合影(2019 年)

前　言

任福尧先生是复旦大学数学科学学院退休教授，我国知名的复分析专家，在复分析的单叶函数、拟共形映射、H_p 空间、复解析动力系统、多复变几何函数论等基础研究领域，以及复分析在流体力学中的应用等方面得到了许多重要成果．除复分析外，任先生还涉足分形几何、分数次微积分及其在统计力学和期权定价理论中的应用等领域，也做出了许多重要贡献．今年是任先生 90 华诞，我们将任先生发表的部分代表性学术论文整理成册，献给我们的导师任福尧先生．

任先生 1930 年 9 月 14 日出生于浙江省东阳县泉塘村，父亲是忠厚老实、勤恳劳动的农民，母亲是勤俭持家的家庭妇女．因家庭困难，任先生的初等教育是断断续续完成的．他先后在家乡的崇正小学、东阳中学、衢州高级中学，以及江西玉山的浙赣铁路子弟中学等完成中小学教育．1950 年，任先生高中毕业后考入上海复旦大学数理系，但因交不起学费没有入学．之后他辗转进入沈阳东北工学院数理系，1952 年院系调整转入长春东北人民大学（现吉林大学）数学系，在那里接受了良好的数学基础训练．1954 年，任先生大学毕业后进入复旦大学数学系，跟随著名数学家陈建功先生攻读单叶函数论方向研究生．1958 年研究生毕业留校任教，之后一直在复旦大学数学系（现数学科学学院）工作，历任助教、讲师、副教授、教授、博士生导师，直到 1996 年退休，时间长达 38 年．任先生酷爱数学，退休后仍坚持指导青年教师及学生并继续从事研究工作，直到 2017 年 88 岁高龄时才告别他挚爱的数学研究．

任先生的主干研究领域是复分析，他的突出特点是具有敏锐的洞察力，思维开阔，与时俱进，总能抓住最前沿、最有生命力的研究方向去开拓研究和培养人才．从理论到应用、从单叶函数到复动力系统、从复分析到分形几何、从统计物理到期权定价等，都体现了他的这一特点．

任先生早年的学术研究是单叶函数论，这是复分析中一个非常热门的研究方向.在那个年代,单叶函数论最著名的问题是关于单位圆上规范化单叶函数幂级数展开系数估计的 Bieberbach 猜想,与此相关还有一系列系数估计和偏差估计问题,任先生在这些方面都得到了重要成果.例如,在第三项系数限制下的 Bieberbach 猜想,他给出了当时最好的结果;关于单叶半纯函数逆函数的系数估计,他先后证明了 $k=6,7,8$ 时的 Springer 猜想,也是当时最好的成果.1984 年,美国数学家德布兰奇(Louis de Brange)证明了 Bieberbach 猜想,标志着单叶函数理论的研究达到了顶峰.之后,任先生一方面迅速转向 H_p 空间理论,将单复变函数 H_p 空间理论的结果系统地推广到多维有界对称域上;另一方面他在龚升教授工作的启发下,依靠他在单叶函数理论方面深厚的研究功底和研究经验,及时转向多复变几何函数论的研究,将 Lowner 从属链理论等应用到多复变双全纯星型映射的刻画等问题,都取得了许多有意义的成果.1985 年任先生获得原国家教委优秀科技成果奖.

另外,任先生早年还在拟共形映射理论研究中做出了重要贡献,他在 1962 年就给出了一种具有无界特征(局部伸缩商)的拟共形映射的存在性定理,这可能是关于无界特征拟共形映射的第一个工作,是一个重要创新,具有重要应用价值.由于任先生的工作是用中文发表的,国外并不知道任先生的工作,直到 1968 年,芬兰数学家 Lehto 获得了类似的成果,并产生了很大影响.1988 年,法国数学家 David 进一步发展了相关理论,并在复解析动力系统等领域的研究中发挥了重要作用.

在 1970 年代和 80 年代初,基于当时倡导理论联系实际的背景,任先生和复旦大学力学系柳兆荣教授一起研究流体管路传输理论,同时和上海地质处等单位合作,进行冬灌夏用控制地面沉降的理论研究,取得了一系列非常有价值的成果.后来,任先生又将在应用研究中用到的复分析工具,结合基本复变函数理论,总结编写成专著兼教材《应用复分析》.

1980 年代,复解析动力系统成为复分析在国际上的一个主流研究方向.1985 年末,杨乐院士介绍了复动力系统的进展后,任先生率先在国内开展了这一方向的研究并最早招收研究生,在多项式、有理函数、整函数、亚纯函数、代数体函数以及多复变函数动力系统理论中取得了较为丰富的成果.特别是结合复解析动力系统和当时分形几何迭代函数系统的思想,任先生创造性地提出了有理函数和超越函数的随机迭代动力系统概念,并在 1992 年,和学生周维民合作建立了随机迭代动力系统的基本理论,而后和学生龚志民合作将随机迭代动力系统推广到无穷生成情形.国际上,直到 1996 年,才由 Hinkkanen 和 Martin 给出了类似的理论(只是名称不同,他们称为有理函数半群动力系统,简称有理半群).任先生和他学生建立的随机迭代动力系统理论在国际上产生了较大影响,被后续工作广泛引用,日本京都大学数学家 Sumi 在其论文中多次指出他所研究的对象是由任福尧及其团队最先提出的.以随机迭代动力系统为主要内容的项目“随机复动力系统与分形几何中的若干问题”获得了教育部高等学校科学技术奖自然科学二等奖.任先生及其学生在复解析动力系统研究中做出的主要创新工作都汇编在他的专著《复解析动力系统》中.

1987 年,任先生邀请了分形几何专家文志英教授到复旦大学介绍分形几何的发展情况,随后便开展了分形几何的研究,并在理论上取得了重要进展.例如,他和许友合作证明了关于 Cartesian 乘积集的填充测度的 Taylor 猜想,在广义 Moran 分形的 Hausdorff 维数和测度的研究中也取得了较好的成果.

1990 年代,任先生注意到了在凝聚态物理和统计物理中出现的奇异核和分形介质的关系,于是就带领学生开展这个方面的研究,并取得了丰富成果.在统计物理中,有大量不符合标准扩散方程的奇异扩散过程,包括次扩散过程和超扩散过程等,任先生发现这些奇异扩散过程可以解释为分形介质上的扩散过程.基于这个假设,任先生从理论上推导出了扩散过程的奇异核,从而利用分数次导数和分数次积分推导出各种奇异扩散过程的分数次 Forker-Planck 方程并求得其渐近解,并证明这些奇异扩散过程满足广义 Einstein 关系,工作引起了统计物理学家的关注.在这个方面,任先生还解决了 Barkai, Klafter 和 Klafter, Metzler 等提出的两个公开问题.作为上述研究的数学基础,任先生建立了网分形上的微积分,这个成果已被研究相关问题的数学家和统计物理学家作为基本工具大量引用.在 Metzler 的一篇综述中,作者肯定了任福尧教授团队所做的贡献.值得指出的是,任先生的上述工作,有很大部分是在他退休以后独自完成的.

随着数学在金融经济中发挥的作用越来越重要,任先生还指导学生研究期权定价理论,研究了以非 Brown 运动驱动的期权定价问题,包括分式 Brown 运动、Lévy 过程,以及一些复合过程情形.鉴于期权交易实际是有交易费的,任先生还首次给出了具有交易费的 Black-Scholes 期权定价公式.

除学术研究外,任先生在人才培养方面花费了很大心血,先后指导了 17 位博士研究生和 22 位硕士研究生.如今,他们大多在各自的行业和研究领域发挥着重要作用.任先生指导学生继承了他的导师陈建功先生对学生学术上严格要求、培养学生独立工作能力的风格,并且,对学生无比关怀,全身心为学生的学术前程而付出,不求回报.在编辑本论文集时,任先生只同意收集以他为第一作者或者他本人执笔的论文，只是在我们学生的强力要求下，才勉强同意收录几篇体现任先生重要学术思想的而他不是第一作者的论文. 任先生高风亮节的品格影响着学生们，也必定会在他的子弟中传承下去.

值得一提的是，任先生对祖国和中国共产党有一颗真挚的心. 任先生常说，按他过去的家庭情况，没有新中国他是不可能成为一位复旦大学教授的. 因此，任先生对党和新中国有一种特别的感情，非常感谢党和国家对他的培养. 1959 年，任先生光荣地加入了中国共产党. 任先生对党组织和领导安排的任务总是不折不扣地完成，1958 年毕业留校后，他服从组织分配，长时间负责学生工作，并且不计较个人在教学和科研上的得失，全身心地投入学生工作中去，和学生同吃同住同劳动，直到 1964 年才回到教研室工作，专心投入教学科研. 这种专心致志、认真做好每一件事的精神也是值得我们学习的.

任福尧教授全体学生

目　　录

开拓戈鲁辛和夏尔绳斯基的几个定理*

任福尧

1. S 表示 $|z|<1$ 中正则且单叶的函数 $f(z)=z+a_2z^2+\cdots$ 的全体所成之族. Σ 表示在区域 $|\zeta|>1$ 中半纯且单叶的函数 $F(\zeta)=\zeta+\alpha_0+\dfrac{\alpha_1}{\zeta}+\cdots$ 的全体所成之族.

设 $f(z)/f'(0)\in S$，且当 $|z|<1$ 时 $|f(z)|<1$. 当 $f'(0)\geqslant T$ $(0<T<1)$ 时，记这种函数的全体为 F_T.

设 $f(z)\in F_T$，θ 是实数，记 $e^{i\theta}f(z)$ 之全体所成之族为 F_T^*.

设函数 $W=F(\zeta)=\alpha_{-1}\zeta+\alpha_0+\dfrac{\alpha_1}{\zeta}+\cdots$ 在区域 $|\zeta|>1$ 上是正则、单叶的，而且 $|F(\zeta)|>1$. 当 $|\alpha_{-1}|\leqslant\dfrac{1}{T}$ $(0<T<1)$ 时，记这种函数的全体为 Σ_T.

设函数 $H=H(x_2,\cdots,x_N,y_2,\cdots,y_N)$ 是 $2N-2$ 个实变数的实函数，它在足够大的区域中有意义，且具有一次的连续偏导数. 又设这些偏导数在任何一点绝不同时为 0. 我们简称具有这些性质的 H 为夏尔绳斯基函数.

戈鲁辛利用单叶函数的变分，研究函数族 S 的系数之绝对和相对极值问题[1]，以及函数族 Σ 的函数模之极值问题[1]，本篇第一部分的目的是要开拓戈鲁辛的结果.

夏尔绳斯基曾用变分法研究 F_T 的系数函数 H 之绝对极值问题[2]，得到一个"基本定理". 夏道行给夏尔绳斯基定理以一个简短的证明[3]，本篇第二部分的目的是采用戈鲁辛所展开的方法考察函数族 F_T 的系数之相对极值问题及函数族 Σ_T 的函数模的极值问题.

---

* 原载《复旦学报(自然科学版)》，1957 年，第 1 期，117 - 133.

本篇中还证明:对于函数族 S 的函数 $f(z)=z+a_2z^2+\cdots$,成立着准确的不等式:

$$\left|a_5-2a_2a_4+4a_2^2a_3-\frac{3}{2}a_3^2-\frac{3}{2}a_2^4\right|\leqslant\frac{1}{2}.$$

2. 当 $f(z)\in S$ 时,令 $x_n=\operatorname{Re}\{a_{\alpha_n}\}$, $y_n=\operatorname{Im}\{a_{\alpha_n}\}$, $n=2,\cdots,N$, $\alpha_2,\cdots,\alpha_N$ 是一组各不相同的自然数,得到 H 的一个数值 $H_f(\alpha_2,\cdots,\alpha_N)$,此时 H 是 S 上的一个泛函数.

定理 $H(\alpha_2,\cdots,\alpha_N)$对于 S 中的一切函数 $f(z)=z+\sum\limits_{\nu=2}^{\infty}a_\nu z^\nu$ 的最大值只被如下的函数 $W=f(z)$ 所达到:它映照 $|z|<1$ 于 W 平面除去有限根解析的若当弧线的区域,而且每个极值函数在 $|z|<1$ 中满足微分方程

$$\left(\frac{zf'(z)}{f(z)}\right)^2M(f(z))=N(z),\tag{2.1}$$

此处

$$M(w)=\sum_{n=2}^{N}H_n\left(\sum_{\nu=1}^{\alpha_n-1}\frac{\{f(\xi)^{\nu+1}\}_{\alpha_n}}{w^\nu}\right),\tag{2.2}$$

$$N(z)=\sum_{n=2}^{N}\left\{H_n(\alpha_n-1)a_{\alpha_n}+\sum_{\nu=1}^{\alpha_n-1}(\alpha_n-\nu)\left(\frac{H_na_{\alpha_n-\nu}}{z^\nu}+\overline{H_na_{\alpha_n-\nu}}\right)\right\},\tag{2.3}$$

$$H_n=\frac{\partial}{\partial x_n}H_f-\mathrm{i}\frac{\partial}{\partial y_n}H_f,\quad n=2,3,\cdots,N,\tag{2.4}$$

$\{g(z)\}_n$ 表示 $g(z)$的幂级数展开式中 z^n 的系数,其所对应的境界曲线的参数表示 $W=W(t)$——取适当参数 t,乃是微分方程

$$[W'(t)]^2\cdot\frac{M(W(t))}{(W(t))^2}+1=0\tag{2.5}$$

的积分曲线. 函数 $N(z)$在圆周 $|z|=1$ 上取实值而不取负值且至少有一个二重根. 设 $P(z_1,\cdots,z_{N-1})$是 $z_1,\cdots,z_{N-1}$ 的多项式,

$$H_f(2,3,\cdots,N)=\operatorname{Re}\{P(a_2,a_3,\cdots,a_N)\},$$

那么 $H_n=\dfrac{\partial}{\partial a_n}P(a_2,\cdots,a_N)(n=2,\cdots,N)$.

证明 由函数族 S 的致密性,极值函数是存在的. 设 $f(z)$是一极值函数,考虑变分函数[1]

$$f_*(z)=f'(z)+\lambda\frac{Af(z)^2}{f(z)-f(\zeta)}-\lambda f(z)A\left(\frac{f(\zeta)}{\zeta f(\zeta)}\right)^2$$
$$-\lambda zf'(z)A\left(\frac{f(\zeta)}{\zeta f'(\zeta)}\right)^2\frac{\zeta}{z-\zeta}+\lambda\bar{A}zf'(z)\overline{\left(\frac{f(\zeta)}{\zeta f'(\zeta)}\right)^2}\frac{\bar{\zeta}z}{1-\bar{\zeta}z}+O(\lambda^2),$$

A 是任意的, $|\zeta|<1$. 经过平易的计算,可以得到

$$H_{f*}-H_f=\lambda\,\mathrm{Re}\left\{A\left[-M(f(\zeta))+N(\zeta)\left(\frac{f(\zeta)}{\zeta f'(\zeta)}\right)^2\right]\right\}+O(\lambda^2).$$

由于 $f(z)$的极值性及 A 的任意性,得到

$$\left(\frac{\zeta f'(z)}{f(\zeta)}\right)^2M(f(\zeta))=N(\zeta),\quad |\zeta|<1.$$

首先证明极值区域 $f(|z|<1)$ 无外点:假如它有一个外点 ξ,那么考虑变分函数

$$f_*(z)=f(z)+\lambda A\frac{f(z)^2}{f(z)-\xi},$$

当 λ 甚小时, $f_*(z)\in S$,

$$H_{f*}-H_f=-\lambda\,\mathrm{Re}\{AM(\xi)\}.$$

由于 $f(z)$之极值性及 A 的任意性,得到

$$M(\xi)\equiv 0.$$

这是不可能的,因为如果 $M(\xi)\equiv 0$ 的话, $\frac{1}{\xi_{a_2-1}}$, $\frac{1}{\xi_{a_3-1}}$, …, $\frac{1}{\xi_{a_n-1}}$ 之各项系数都是 0,就是说,我们会有如下的方程组:

$$\begin{cases}H_2+(*)H_3+\cdots+(*)H_N=0,\\ \qquad H_3+\cdots+(*)H_N=0,\\ \qquad\cdots\cdots\\ \qquad\qquad H_N=0.\end{cases}$$

由于 H_2, H_3, …, H_N 不同时为零,所以上面方程组之系数行列式为 0,但是事

实上行列式之值等于 1,所以 $f(|z|<1)$ 无外点.

从微分方程的解析理论,知道极值区域的境界由有限根解析的若当弧线组成(参见[1]),因此必有一根解析弧线,其一端 W^* 是有限的单纯境界点. 设 $W^*=f(\xi_0)$, $|\xi_0|=1$,易知 $f'(\xi_0)=0$,因此 $N(\xi_0)=0$. 变分函数

$$f_*(z)=f(z)+t\left[k\left(f(z)-zf'(z)\frac{\xi_1+z}{\xi_1-z}\right)-\left(f(z)-zf(z)\frac{\xi_0+z}{\xi_0-z}\right)\right]+O(t^2),$$

$t>0$, $0<k<1$, ξ_1 是 $|z|=1$ 上的任意点,当 t 充分小时, $f_*(z)\in S$,经过平易计算,获得

$$H_{f*}-H_f=t\mathrm{Re}\{N(\xi_0)-kN(\xi_1)\}.$$

由 $f(z)$ 的极值性, $\mathrm{Re}\{N(\xi_0)\}\leqslant\mathrm{Re}\{N(\xi_1)\}$. 另一方面,对于极值函数 $f(z)=z+\sum_{n=2}^{\infty}a_nz^n\in S$ 成立着关系式[3]

$$\mathrm{Im}\{\sum_{n=2}^{N}a_{\alpha_n}(\alpha_n-1)H_n\}=0.$$

因此

$$N(\xi_1)\geqslant N(\xi_0),\quad |\xi_1|=1,$$

但 $N(\xi_0)=0$, 所以 $N(\xi)$ 在 $|\xi|=1$ 上是非负的,因而 $|\xi|=1$ 上一切根都是重根.

由(2.1)不难推导出(2.5).

最后,从关系式

$$\mathrm{Re}\{\sum_{n=2}^{N}H_n\cdot\Delta a_n\}=\mathrm{Re}\left\{\sum_{n=2}^{N}\frac{\partial}{\partial a_n}P(a_2,\cdots,a_N)\right\}+O(|\Delta a|^2)$$

知道: $\mathrm{Re}\left\{\mathrm{e}^{i\theta}\left(H_n-\frac{\partial}{\partial a_n}P(a_2,\cdots,a_N)\right)\right\}\equiv 0\ (0\leqslant\theta\leqslant 2\pi)$.

因此

$$H_n=\frac{\partial}{\partial a_n}P(a_2,\cdots,a_N)\ (n=2,\cdots,N).$$

应用上述定理,能彻底地解决某些系数不等式的估值问题,例如,我们能证明:对于 S 中的一切函数 $f(z)=z+\sum_{n=2}^{\infty}a_nz^n$, 成立着估计式:

$$\left|a_5-2a_2a_4+4a_2^2a_3-\frac{3}{2}a_3^2-\frac{3}{2}a_2^4\right|\leqslant\frac{1}{2}. \tag{2.6}$$

为此，考察极值问题

$$\operatorname{Re}\left\{a_5-2a_2a_4+4a_2^2a_3-\frac{3}{2}a_3^2-\frac{3}{2}a_2^4\right\}=\text{最大值} \tag{2.7}$$

就行了.从(2.5)即知(2.7)的极值区域的境界曲线满足微分方程

$$\left(\frac{\mathrm{d}W}{W}\right)\left(\frac{1}{W^2}+\frac{a_2}{W}\right)\leqslant 0, \tag{2.8}$$

由是得到

$$\operatorname{Re}\left\{\frac{1+2a_2w}{w^2}\right\}=0. \tag{2.9}$$

从此关系式和许瓦兹对称原则得到极值函数 $f(z)$ 的表达式

$$f(z)=\frac{a_2z^2+z\sqrt{1+(a_2^2\pm 2\mathrm{i})z^2-z^4}}{1\pm 2\mathrm{i}z^2-z^4}. \tag{2.10}$$

由实际计算，知(2.7)之最大值等于$\frac{1}{2}$.函数

$$W=\frac{z}{(1-\mathrm{e}^{\mathrm{i}\theta}z)^2},\quad W=\frac{z}{1\pm\mathrm{i}z^2}$$

是(2.7)的极值函数，

从(2.6)又得如下的事实：对于 Σ 中任何函数 $F(\zeta)=\zeta+\alpha_0+\frac{\alpha_1}{\zeta}+\cdots$ 在 $|\zeta|>1$ 中没有零点的话，成立着

$$|2\alpha_3+\alpha_1^2|\leqslant 1^{①}. \tag{2.11}$$

3. 设有 $m+1$ 个夏尔绳斯基函数 $H^{(k)}=H^{(k)}(x_2^{(k)},\cdots,x_{N_k}^{(k)},y_2^{(k)},\cdots,y_{N_k}^{(k)})$，$k=0,1,2,\cdots,m$. 我们引入下面的记号：函数族 S 中的函数满足下述条件时，记它们的全体为

① 后来发觉 Z. Nesari 已证明了此不等式.

$$S(a_1^{(0)}, a_2^{(0)}, \cdots, a_m^{(0)}; r_1, r_2, \cdots, r_m);$$
$$|H_f^{(k)} - a_k^{(0)}| \leqslant r_k, \quad k=1, 2, \cdots, m,$$

$H_f^{(k)}$ 表示 $H^{(k)}(x_2^{(k)}, \cdots, x_{N_k}^{(k)}, y_2^{(k)}, \cdots, y_{N_k}^{(k)})$，当

$$x_\nu^{(k)} = \mathrm{Re}\{a_{n_{k\nu}}\}, \ y_\nu^{(k)} = \mathrm{Im}\{a_{n_{k\nu}}\} \quad (\nu=2, \cdots, N_k)$$

时的数值.

同样我们可以定义函数族 F_T 的子族 $F_T(a_1^0, \cdots, a_n^0; r_1, \cdots, r_m)$，但此时 $x_\nu^{(k)} = \mathrm{Re}\left\{\frac{a_{n_{k\nu}}}{a_1}\right\}, \ y_\nu^{(k)} = \mathrm{Im}\left\{\frac{a_{n_{k\nu}}}{a_1}\right\}$.

下述引理是我们研究系数的相对极值问题时的基本工具之一.

引理 假如函数 $L_0(\zeta), L_1(\zeta), \cdots, L_m(\zeta)$ $(m \geqslant 1)$ 在某一区域 B 中都是正则的，并且对于 B 的任何 m 个点 $\zeta_1, \cdots, \zeta_m$ 和任何复数 $A_1, \cdots, A_m$，$m+1$ 个不等式

$$\mathrm{Re}\{A_1 L_k(\zeta_1) + A_2 L_k(\zeta_2) + \cdots + A_m L_k(\zeta_m)\} > 0 \ (k=0, 1, \cdots, m)$$

绝不都能成立，那么，$L_0(\zeta), \cdots, L_m(\zeta)$在 B 中是线性相倚的[1].

援用戈鲁辛在研究系数的相对极值问题时所展开的方法，不难证明下述诸定理.

定理 1 假如函数族 $S(a_1^0, \cdots, a_m^0; r_1, \cdots, r_m)$不是空的，那么此函数族中一切函数 $f(\zeta) = \sum_{\nu=1}^{\infty} a_\nu \zeta^\nu$ 给 $H_f^0 = H^0(x_2^{(0)}, \cdots, x_{N_0}^{(0)}, y_2^{(0)}, \cdots, y_{N_0}^{(0)})$——$x_\nu^{(0)} = \mathrm{Re}\{a_{n_{0\nu}}\}, y_\nu^{(0)} = \mathrm{Im}\{a_{n_{0\nu}}\}, \nu=2, \cdots, N_0$——以最大值的必属于函数族 S'(像无外点)，并且在 $|\zeta|>1$ 中满足微分方程

$$\sum_{k=0}^{m} \lambda_k L_{n_k}^{(s)}(\zeta) = 0, \tag{3.1}$$

此处

$$L_{n_k}^{(s)}(\zeta) = \varepsilon_k \left(\frac{\zeta f'(\zeta)}{f(\zeta)}\right)^2 \sum_{\nu=2}^{N_k} H_\nu^{(k)} \left(\sum_{l=1}^{n_{k\nu}-1} \frac{\{f(\xi)^{l+1}\}_{n_{k\nu}}}{f(\zeta)^l}\right)$$
$$-\sum_{\nu=2}^{N_k}\left[H_\nu^{(k)} a_{n_{k\nu}}(n_{k\nu}-1)\varepsilon_k + \sum_{l=1}^{n_{k\nu}-1}(n_{k\nu}-l)\left(\frac{\varepsilon_k H_\nu^{(k)} a_{n_{k\nu}-l}}{\zeta^l} + \overline{\varepsilon_k H_\nu^{(k)} a_{n_{k\nu}-l} \zeta^l}\right)\right]$$
$$(k=0, 2, \cdots, m), \tag{3.2}$$

常数 λ_k 不同时等于 0，$\varepsilon_k=\pm 1$.

定理 2 假如函数族 $S(a_1^0, \cdots, a_m^0; 0, \cdots, 0)$ 不是空的，那么此函数族中一切函数 $f(\zeta)=\sum_{\nu=1}^{\infty} a_\nu \zeta^\nu$ 给 $H_f^{(0)}=H^{(0)}(x_2^{(0)}, \cdots, x_{N_0}^{(0)}, y_2^{(0)}, \cdots, y_{N_0}^{(0)})$——$x_\nu^{(0)}=\mathrm{Re}\{a_{n_{0\nu}}\}$，$y_\nu^{(0)}=\mathrm{Im}\{a_{n_{0\nu}}\}$，$\nu=2, \cdots, N_0$——以最大值的必属于 S'（像无外点），并且在 $|\zeta|<1$ 中满足微分方程(3.1)，其中 $L_{n_k}(\zeta)$ 是由公式(3.2)决定的，常数 λ_k 不都等于 0，常数 ε_k 之模等于 1.

定理 3 假如函数族 $S(a_1^0, \cdots, a_m^0; 0, \cdots, 0)$ 不是空的，那么族中函数 $f(\zeta)$ 给 $H_f^{(0)}=H^{(0)}(x_2^{(0)}, \cdots, x_{N_0}^{(0)}, y_2^{(0)}, \cdots, y_{N_0}^{(0)})$——$x_\nu^{(0)}=\mathrm{Re}\{f(\zeta_\nu)\}$，$y_\nu^{(0)}=\mathrm{Im}\{f(\zeta_\nu)\}$；$\zeta_2, \zeta_3, \cdots, \zeta_{N_0}$ 是区域 $|\zeta|<1$ 的定点——以最大值的必属于 S'，并且在 $|\zeta|<1$ 中满足微分方程

$$\lambda_0 L_0(\zeta)+\sum_{k=1}^{m} \lambda_k L_k^{(s)}(\zeta)=0,$$

此处

$$L_0(\zeta)=\left(\frac{\zeta f'(\zeta)}{f(\zeta)}\right)^2\left(\sum_{\nu=2}^{N_0} H_\nu^{(0)} \frac{f^2(\zeta_\nu)}{f(\zeta_\nu)-f(\zeta)}\right)-\sum_{\nu=2}^{N_0} H_\nu^{(0)} f(\zeta_\nu)\left[1+\frac{\zeta_\nu f'(\zeta_\nu)}{f(\zeta_\nu)} \frac{\zeta}{\zeta_\nu-\zeta}\right]$$
$$+\sum_{\nu=2}^{N_0} \overline{H_\nu^{(0)} \zeta_\nu f'(\zeta_\nu)} \frac{\zeta \bar{\zeta}_\nu}{1-\zeta \bar{\zeta}_\nu}. \tag{3.3}$$

而 $L_k^{(0)}(\zeta)$ $(k=1, \cdots, m)$ 由公式(3.2)决定，ε_k，λ_k 都是常数，$\varepsilon_k=\pm 1$，λ_k 不都等于 0.

定理 4 假使函数族 $F_T(a_1^0, a_2^0, \cdots, a_m^0; r_1, r_2, \cdots r_m)$ 不是空的，那么族中一切函数 $f(\zeta)=\sum_{\nu=1}^{\infty} a_\nu \zeta^\nu$ 给 $H_f^{(0)}=H^{(0)}(x_2^{(0)}, \cdots, x_{N_0}^{(0)}, y_2^{(0)}, \cdots, y_{N_0}^{(0)})$——$x_\nu^{(0)}=\mathrm{Re}\left\{\frac{a_{n_{0\nu}}}{a_1}\right\}$，$y_\nu^{(0)}=\mathrm{Im}\left\{\frac{a_{n_{0\nu}}}{a_1}\right\}$ $(\nu=2, \cdots, N_0)$——以最大值的必属于 F_T'（像无外点），并且在 $|\zeta|<1$ 中满足微分方程

$$\sum_{k=0}^{m+1} \lambda_k L_{n_k}^{(F_T)}(\zeta)=0, \tag{3.4}$$

此处

$$L_{n_k}^{(F_T)}(\zeta)=\varepsilon_k\left(\frac{\zeta f'(\zeta)}{f(\zeta)}\right)^2\sum_{\nu=2}^{N_k}\left[H_\nu^{(k)}\left(\sum_{l=1}^{n_{k\nu}-1}\frac{\{f(\xi)^{l+1}\}_{n_{k\nu}}}{f(\zeta)^l}\right)+H_\nu^{(k)}\left(\sum_{l=1}^{n_{k\nu}-1}\{f(\xi)^{l+1}\}_{n_{k\nu}}\overline{f(\zeta)^l}\right)\right]$$
$$-\sum_{\nu=2}^{N_k}\left[H_\nu^{(k)}a_{n_{k\nu}}(n_{k\nu}-1)\varepsilon_k+\sum_{l=1}^{n_{k\nu}-1}(n_{k\nu}-l)\left(\frac{\varepsilon_k H_\nu^{(k)}a_{n_{k\nu}-1}}{\zeta^l}+\overline{\varepsilon_k H_\nu^{(k)}a_{n_{k\nu}-l}}\right)\right]$$
$(k=0,1,\cdots,m)$,
$$L_{n_{m+1}}^{(F_T)}(\zeta)=\left(\frac{\zeta f'(\zeta)}{f(\zeta)}\right)^2-1. \tag{3.5}$$

常数 λ_k 不同时为 0，$\varepsilon_k=\pm1$.

证明 极值函数的存在是显然的. 设 $f(\zeta)=\sum_{\nu=1}^{\infty}a_\nu\zeta^\nu$ 是一个极值函数，$z=f(\zeta)$ 映照圆 $|\zeta|<1$ 于区域 D，我们要证明 D 无外点.

假如 D 有外点，那么从其外点 $z_0, z_1, \cdots, z_m$ 可以作变分函数[3]

$$f_*(\zeta)=f(\zeta)+\lambda\sum_{\nu=0}^{m}\left[A_\nu f(\zeta)\frac{f(\zeta)+z_\nu}{f(\zeta)-\bar{z}_\nu}-\bar{A}_\nu f(\zeta)\frac{1+f(\zeta)\bar{z}_\nu}{1-f(\zeta)\bar{z}_\nu}\right]+O(\lambda^2),$$

这个函数当 $|\lambda|$ 甚小时在单位圆中是正则、单叶的，而且当 $\lambda\operatorname{Re}\left(\sum_{\nu=0}^{m}A_\nu\right)\leqslant 0$ 时 $f_*(\zeta)\in F_T$. 对于这函数，我们有

$$H_{f_*}^{(k)}-H_f^{(k)}=-\frac{2\lambda}{a_1}\operatorname{Re}\left\{\sum_{\nu=0}^{m}A_\nu K_{n_k}(z_\nu)\right\},\ k=0,1,\cdots,m,$$

此处

$$K_{n_k}(z_\nu)=\sum_{\nu=2}^{N_k}H_\nu^{(k)}\left(\sum_{l=1}^{n_{k\nu}-1}\frac{\{f(\xi)^{l+1}\}_{n_{k\nu}}}{z^l}\right)+\overline{\sum_{\nu=2}^{N_k}H_\nu^{(k)}\left(\sum_{l=1}^{n_{k\nu}-1}\{f(\xi)^{l+1}\}_{n_{k\nu}}\bar{z}^l\right)}.$$

显然，行列式

$$\Delta=\begin{vmatrix}K_{n_1}(z_0) & \cdots & K_{n_1}(z_m)\\ K_{n_2}(z_0) & \cdots & K_{n_2}(z_m)\\ \vdots & & \vdots\\ K_{n_m}(z_0) & \cdots & K_{n_m}(z_m)\\ K_{n_0}(z_0) & \cdots & K_{n_0}(z_m)\end{vmatrix}$$

不恒等于 0,我们总可以取外点 $z_0, z_1, \cdots, z_m$ 使此行列式不等于 0. 此时 $m+1$ 个方程

$$A_0K_{n_k}(z_0)+\cdots+A_mK_{n_k}(z_m)=0 \quad (k=1, 2, \cdots, m),$$
$$A_0K_{n_0}(z_0)+\cdots+A_mK_{n_0}(z_m)=x+\mathrm{i}y$$

关于 A_ν 有满足 $\lambda\,\mathrm{Re}\left\{\sum_{\nu=0}^{m}A_\nu\right\}\leqslant 0$ 及 $\lambda\,\mathrm{Re}\left\{\sum_{\nu=0}^{m}A_\nu K_{n_0}(z_\nu)\right\}<0$ 的解. 事实上,记

$$\bar{\Delta}=\begin{vmatrix} K_{n_1}(z_0) & \cdots & K_{n_1}(z_m) \\ \vdots & & \vdots \\ K_{n_m}(z_0) & \cdots & K_{n_m}(z_m) \\ 1 & \cdots & 1 \end{vmatrix},$$

我们不妨假设

$$\Delta'=\begin{vmatrix} K_{n_1}(z_1) & \cdots & K_{n_1}(z_m) \\ \vdots & & \vdots \\ K_{n_m}(z_1) & \cdots & K_{n_m}(z_m) \end{vmatrix}\neq 0.$$

那么关于 A_ν 所欲满足的条件可写成

$$\lambda\,\mathrm{Re}\left\{\frac{A_0}{\Delta'}\bar{\Delta}\right\}\leqslant 0, \quad \lambda\,\mathrm{Re}\left\{\frac{A_0}{\Delta'}\Delta\right\}<0.$$

因此,适当地取 $\arg A_0$ 可使上面两式都成立.

记这组方程的解为 A_ν. 当 $|\lambda|$ 甚小时,函数 $f_*(\zeta)$ 属于 $F_T(a_1^0, a_2^0, \cdots, a_m^0; r_1, r_2, \cdots, r_m)$,而

$$H_{f_*}^{(0)}=H_f^{(0)}-\frac{2\lambda}{a_1}\mathrm{Re}\left\{\sum_{\nu=0}^{m}A_\nu K_{n_0}(z_\nu)\right\}>H_f^{(0)},$$

这是有悖于 $f(\zeta)$ 的极值性的. 因此区域 D 绝无外点.

关于极值函数 $f(\zeta)$ 满足方程(3.4)的证明,我们考虑变分函数[3]

$$f_*(\zeta)=f(\zeta)+\lambda\left\{\sum_{\nu=1}^{m_1+1}\left[A_\nu f(\zeta)\frac{f(\zeta)+f(\zeta_\nu)}{f(\zeta)-f(\zeta_\nu)}-\bar{A}_\nu f(\zeta)\frac{1+f(\zeta)\overline{f(\zeta_n)}}{1-f(\zeta)\overline{f(\zeta_\nu)}}\right.\right.$$
$$\left.\left.-A_\nu\zeta f'(\zeta)\frac{2f(\zeta_\nu)^2}{\zeta_\nu f'(\zeta_\nu)^2}\frac{1}{\zeta-\zeta_\nu}+\bar{A}_\nu\zeta f'(\zeta)\left(\overline{\frac{2f(\zeta_\nu)^2}{\zeta_\nu f'(\zeta_\nu)^2}}\right)\frac{\zeta}{1-\zeta\bar{\zeta}_\nu}\right]\right\}+O(\lambda^2),$$

此处，$1 \leqslant m_1 \leqslant m$，当$|\lambda|$充分小并且

$$\lambda \operatorname{Re}\left\{\sum_{\nu=1}^{m_1+1} A_\nu \left[1-\frac{f(\zeta_\nu)^2}{\zeta_\nu^2 f'(\zeta_\nu)^2}\right]\right\} < 0$$

时，$f_*|\zeta|$是属于F_T^*的. 往后，我们参照戈鲁辛的证法，知道结论是真实的. 定理证毕.

定理 5 假如函数族$F_T(a_1^0, a_2^0, \cdots, a_m^0; 0, 0, \cdots, 0)$非空，那么族中一切函数$f(\zeta)=\sum_{\nu=1}^{\infty} a_\nu \zeta^\nu$给$H_f^{(0)}=H^{(0)}(x_2^{(0)}, \cdots, x_{N_0}^{(0)}, y_2^{(0)}, \cdots, y_{N_0}^{(0)})$——$x_\nu^{(0)}=\operatorname{Re}\left\{\frac{a_{n_{0\nu}}}{a_1}\right\}$，$y_\nu^{(0)}=\operatorname{Im}\left\{\frac{a_{n_{0\nu}}}{a_1}\right\}$ $(\nu=2, \cdots, N_0)$——以最大值的必属于F_T'(像无外点)，并且在$|\zeta|<1$中满足微分方程(3.4)，其中$L_{n_k}^{(F_T)}(\zeta)$ $(k=0, 1, \cdots, m+1)$由公式(3.5)决定，ε_k，λ_k都是常数，$\varepsilon_k=\pm 1$，λ_k不同时为0.

用类似上文的讨论，我们也能证明下述定理.

定理 6 假如函数族$F_T(a_1^0, a_2^0, \cdots, a_m^0; 0, 0, \cdots, 0)$不是空的，那么在$\zeta_0$，$|\zeta_0|<1$，族中函数取最大模$|f(\zeta_0)|$的，必属于函数族$F_T'$(像无外点)，此极值函数$f(\zeta)$满足微分方程

$$\sum_{k=0}^{m+1} \lambda_k L_{n_k}^{(F_T)}(\zeta)=0 \quad (|\zeta|<1),$$

此处

$$L_{n_0}(\zeta)=\left(\frac{\zeta f'(\zeta)}{f(\zeta)}\right)^2\left[\frac{f(\zeta_0)+f(\zeta)}{f(\zeta_0)-f(\zeta)}-\frac{1+\overline{f(\zeta_0)}f(\zeta)}{1-\overline{f(\zeta_0)}f(\zeta)}-2\frac{\zeta_0 f'(\zeta_0)}{f(\zeta_0)}\frac{\zeta}{\zeta_0-\zeta}\right.$$
$$\left.+2\frac{\zeta_0 f'(\zeta_0)}{f(\zeta_0)}\frac{\bar{\zeta}_0\zeta}{1-\bar{\zeta}_0\zeta}\right],$$

而$L_{n_k}^{(F_T)}(\zeta)$ $(k=1, \cdots, m+1)$由公式(3.5)定义，λ_k，ε_k都是常数，$\varepsilon_k=\pm 1$，λ_k不同时为0.

设$f(\zeta)\in F_T$，当θ是实数时，函数$e^{i\theta}f(\zeta)$的全体记为F_T^*，显然$H_f^{(k)}=H_{e^{i\theta}f}^{(k)}(x_2^{(k)}, \cdots, x_{N_k}^{(k)}, y_2^{(k)}, \cdots, y_{N_k}^{(k)})$——$x_\nu^{(k)}=\operatorname{Re}\left\{\frac{a_{n_{k\nu}}}{a_1}\right\}$，$y_\nu^{(k)}=\operatorname{Im}\left\{\frac{a_{n_{k\nu}}}{a_1}\right\}$ $(\nu=2, \cdots, N_k)$；对于函数族F_T^*，同样可定义$F_T^*(H^{(1)}, a_{n_1}^0, r_1; \cdots; H^{(m)}, a_{n_m}^0, r_m)$，特别是$F_T^*(H^{(1)}, a_{n_1}^0, \cdots, H^{(m)}, a_{n_m})$，我们能证明下述定理.

定理7 假设函数族 $F_T^*(a_1^0, a_2^0, \cdots, a_n^0; 0, 0, \cdots, 0)$ 不是空的，那么族中函数 $f(\zeta)$ 给 $H_f^{(0)}=H^{(0)}(x_2^{(0)}, \cdots, x_{N_0}^{(0)}, y_2^{(0)}, \cdots, y_{N_0}^{(0)})$——$x_\nu^{(0)}=\mathrm{Re}\{f(\zeta_\nu)\}$，$y_\nu^{(0)}=\mathrm{Im}\{f(\zeta_\nu)\}$；$\zeta_2, \cdots, \zeta_{N_0}$ 是区域 $|\zeta|<1$ 的定点——以最大值的必属于 F_T^*（像无外点），且在 $|\zeta|<1$ 中满足微分方程

$$\lambda_0 L_{n_0}^{(F_T^*)}(\zeta)+\sum_{k=1}^{m+1}\lambda_k L_{n_k}^{(F_T^*)}(\zeta)=0,$$

此处

$$L_{n_0}^{(F_T^*)}(\zeta)=\left(\frac{\zeta f'(\zeta)}{f(\zeta)}\right)^2\left\{\sum_{\nu=2}^{N_0}H_\nu^{(0)}f(\zeta_\nu)\left[\frac{f(\zeta_\nu)+f(\zeta)}{f(\zeta_\nu)-f(\zeta)}-\frac{1+\overline{f(\zeta_\nu)}f(\zeta)}{1-\overline{f(\zeta_\nu)}f(\zeta)}\right]\right\}$$
$$-2\sum_{\nu=2}^{N_0}H_\nu^{(0)}f(\zeta_\nu)\left[\frac{\zeta_\nu f'(\zeta_\nu)}{f'(\zeta_\nu)}\cdot\frac{\zeta}{\zeta_\nu-\zeta}-\frac{\overline{\zeta_\nu f'(\zeta_\nu)}}{f(\zeta_\nu)}\,\frac{\overline{\zeta}_\nu\zeta}{1-\overline{\zeta}_\nu\zeta}\right],$$

而 $L_{n_k}^{(F_T^*)}(\zeta)\ (k=1, 2, \cdots, m+1)$ 由公式(3.5)决定，ε_k，λ_k 都是常数，$\varepsilon_k=\pm1$，λ_k 不同时为 0.

戈鲁辛定义函数族 S 中满足下述诸条件的函数 $f(\zeta)=\zeta+\sum_{\nu=2}^{\infty}a_\nu\zeta^\nu$ 之全体为

$$S(a_n^0;\ a_{n_2}^0, \cdots, a_{n_m}^0),$$

$a_{n_1}^0, \cdots, a_{n_m}^0$ 都是定值，$m\geqslant 1$；$n_1, \cdots, n_m\geqslant 2$；

$$|a_{n_k}^0|=|a_{n_k}|,\quad k=1, \cdots, m.$$

同样我们可定义 $F_T(a_{n_1}^0, a_{n_2}^0, \cdots, a_{n_m}^0)$，不过此时的条件是

$$\left|\frac{a_{n_k}^0}{a_1}\right|=\left|\frac{a_{n_k}}{a_1}\right|.$$

对于此两函数族可考虑较广泛的系数之相对极值问题，即给 $H_f=H(x_2, \cdots, x_N;\ y_2, \cdots, y_N)$——$x_\nu=\mathrm{Re}\left\{\frac{a_{\alpha_\nu}}{a_1}\right\}$，$y_\nu=\mathrm{Im}\left\{\frac{a_{\alpha_\nu}}{a_1}\right\}$，$\nu=2, \cdots, N$ 是一组正整数，它们都不同于 $n_1, \cdots, n_m$ 中之任何一个——以最大值者. 对于此类极值问题，显然我们能获得和戈鲁辛所得的相类似的定理.

4. 在本节里我们将考察函数族 Σ 和 Σ_T 中的一类极值问题.

设 $\gamma_{\nu,\nu'}^{(2)}, \gamma_{\nu,\nu'}^{(3)}, \cdots, \gamma_{\nu,\nu'}^{(N)}(\nu, \nu'=1, 2, \cdots, m)$ 都是固定的复数，且 $\{\gamma_{\nu,\nu'}^{(n)}+\gamma_{\nu',\nu}^{(n)}\}$ 都不同时为 0，而 $\zeta_\nu^{(2)}, \zeta_\nu^{(3)}, \cdots, \zeta_\nu^{(N)}, \nu=1, \cdots, m$ 是区域 $|\zeta|>1$ 中的 N 组定点；记

$$a_n=\sum_{\nu,\nu'=1}^{m}\gamma_{\nu,\nu'}^{(n)}\log\frac{F(\zeta_\nu^{(n)})-F(\zeta_{\nu'}^{(n)})}{\zeta_\nu^{(n)}-\zeta_{\nu'}^{(n)}}=\sum_{\nu,\nu'=1}^{m}\gamma_{\nu,\nu'}^{(n)}\log\frac{F_\nu^{(n)}-F_{\nu'}^{(n)}}{\zeta_\nu^{(n)}-\zeta_{\nu'}^{(n)}}.$$

函数 $H=H(x_2, \cdots, x_N, y_2, \cdots, y_N)$ 是夏尔绳斯基函数，当 $F(\zeta)\in\Sigma$ 或 Σ_T 时，令 $x_n=\mathrm{Re}\{a_n\}$, $y_n=\mathrm{Im}\{a_n\}$，得到 H 的一个函数值 H_F. 因此 H 可以看作函数 Σ 或 Σ_T 的一个泛函.

定理 8 在 Σ 中的一切函数 $z=F(\zeta)$，能使数量

$$H_F=H(x_2, \cdots, x_N, y_2, \cdots, y_N)$$

取最大值的，它必映照于 z 平面除去有限根解析曲线的区域，每一极值函数 $F(\zeta)$ 在 $|\zeta|>1$ 中满足微分方程

$$\left(\frac{\zeta F'(\zeta)}{F(\zeta)}\right)^2 M_\Sigma(F(\zeta))=N_\Sigma(\zeta), \tag{4.1}$$

此处

$$M_\Sigma(w)=\sum_{n=2}^{N}H_n\left[\sum_{\nu,\nu'=1}^{m}\gamma_{\nu,\nu'}^{(n)}\frac{w^2}{(F_\nu^{(n)}-w)(F_{\nu'}^{(n)}-w)}\right],$$

$$N_\Sigma(\zeta)=\sum_{n=2}^{N}H_n\left\{\sum_{\nu,\nu'=1}^{m}\gamma_{\nu,\nu'}^{(n)}+\sum_{\nu,\nu'=1}^{m}\gamma_{\nu,\nu'}^{(n)}\left[\frac{\dfrac{\zeta_\nu^{(n)2}F_\nu^{(n)'}}{\zeta-\zeta_\nu^{(n)}}-\dfrac{\zeta_{\nu'}^{(n)2}F_{\nu'}^{(n)'}}{\zeta-\zeta_{\nu'}^{(n)}}}{F_\nu^{(n)}-F_{\nu'}^{(n)}}\right]\right\}$$

$$-\sum_{n=2}^{N}H_n\sum_{\nu,\nu'=1}^{m}\gamma_{\nu,\nu'}^{(m)}\left[\frac{\dfrac{\zeta_\nu^{(n)}F_\nu^{(n)'}}{\bar{\zeta}\zeta_\nu^{(n)}-1}-\dfrac{\zeta_{\nu'}^{(n)}F_{\nu'}^{(n)'}}{\bar{\zeta}\zeta_{\nu'}^{(n)}-1}}{F_\nu^{(n)}-F_{\nu'}^{(n)}}\right]. \tag{4.2}$$

函数 $N_\Sigma(\zeta)$ 在 $|\zeta|=1$ 上取实值而不取负值，且具有至少一个二重根.

这是戈鲁辛定理的开拓.

证明 我们不妨假设 $F(\zeta)$ 在 $|\zeta|>1$ 中无零点，设 $F(\zeta)$ 是一极值函数，作变分函数[1]

$$F_*(\zeta)=F(\zeta)+\lambda A\frac{F(\zeta_0)F(\zeta)}{F(\zeta)-F(\zeta_0)}+\lambda AF(\zeta)\left(\frac{F(\zeta_0)}{\zeta_0F'(\zeta_0)}\right)^2$$

$$-\lambda A\zeta F'(\zeta)\left(\frac{F(\zeta_0)}{\zeta_0 F'(\zeta_0)}\right)^2\frac{\zeta}{\zeta-\zeta_0}$$
$$+\lambda\bar{A}\zeta F'(\zeta)\left(\frac{F(\zeta_0)}{\zeta_0 F'(\zeta_0)}\right)^2\frac{1}{1-\bar{\zeta}_0\zeta}+O(\lambda^2).$$

经过平易的计算，我们获得

$$H_{F_*}-H_F=\lambda\,\mathrm{Re}\left\{A\left[M_\Sigma(F_0)-\left(\frac{F_0}{\zeta_0 F_0'}\right)^2 N_\Sigma(\zeta_0)\right]\right\}+O(\lambda^2).$$

由 $F(\zeta)$的极值性和 A 的任意性，得到

$$\left(\frac{\zeta_0 F'(\zeta_0)}{F(\zeta_0)}\right)^2 M_\Sigma(F(\zeta_0))=N_\Sigma(\zeta_0),\quad |\zeta_0|>1.$$

由是可知极值函数满足微分方程(4.1)，

极值区域必无外点，假使它有外点 z_1，那么作变分函数

$$F_*(\zeta)=F(\zeta)+\lambda A\,\frac{z_1 F(\zeta)}{F(\zeta)-z_1}+O(\lambda^2),\quad \lambda>0$$

时，获得

$$H_{F_*}-H_F=-\lambda\,\mathrm{Re}\{AM_\Sigma(z_1)\}\leqslant 0.$$

因此

$$M_\Sigma(z_1)=0.$$

但 $M_\Sigma(z_1)$是 z_1 的正则函数(所有奇点均为可去奇点)，因此上式在整个平面上成立. 所以

$$\sum_{n=2}^{N}H_n\left[\sum_{\nu,\nu'=1}^{m}\gamma_{\nu,\nu'}^{(n)}\,\frac{\dfrac{\zeta_\nu^{(n)2}F_\nu^{(n)'}}{\zeta-\zeta_\nu^{(n)}}-\dfrac{\zeta_{\nu'}^{(n)2}F_{\nu'}^{(n)'}}{\zeta-\zeta_{\nu'}^{(n)}}}{F_\nu^{(n)}-F_{\nu'}^{(n)}}\right]$$

在 $|\zeta|>1$ 内是正则的，特别在 $\zeta=\zeta_\nu^{(n)}$ $(n=2,\cdots,N;\ \nu=1,\cdots,m)$ 处是正则的，因此

$$H_n(\gamma_{\nu,\nu'}^{(n)}+\gamma_{\nu',\nu}^{(n)})=0\quad(\nu,\nu'=1,2,\cdots,m).$$

因为 $H_n(n=2,\cdots,N)$ 不同时为 0，所以必有 n_0，$2\leqslant n_0\leqslant N$，使得 $H_{n_0}\neq 0$，因此

$$\gamma_{\nu,\nu'}^{(n_0)}+\gamma_{\nu',\nu}^{(n_0)}=0 \quad (\nu, \nu'=1, 2, \cdots, m).$$

此与假设矛盾,所以极值区域没有外点.

根据微分方程的解析理论,从(4.2)知道,在圆周 $|\zeta|=1$ 上除有限个代数支点外, $F(\zeta)$是正则的,因此,极值区域的境界是由有限支解析曲线所构成.

下面证明 $N_\Sigma(\zeta)$在 $|\zeta|=1$ 上$\geqslant 0$. 作变分函数[1]

$$F_*(\zeta)=F(\zeta)-t\left(F(\zeta)+\zeta F'(\zeta)\frac{\zeta_0+\zeta}{\zeta_0-\zeta}\right)+O(t^2),\ t>0,\ |\zeta_0|=1,$$

则得

$$H_{F_*}-H_F=-t\operatorname{Re}\{N_\Sigma(\zeta_0)\}\leqslant 0,$$

所以

$$\operatorname{Re}\{N(\zeta_0)\}\geqslant 0,\ |\zeta_0|=1.$$

但

$$\operatorname{Im}\{N(\zeta)\}=\operatorname{Im}\left\{\sum_{n=2}^{N}H_n\left[\sum_{n-2}^{N}\gamma_{\nu,\nu'}^{(n)}\left(1-\frac{\zeta_\nu^{(n)}F_\nu^{(n)'}-\zeta_{\nu'}^{(n)}F_{\nu'}^{(n)'}}{F_\nu^{(n)}-F_{\nu'}^{(n)}}\right)\right]\right\}.$$

函数 $F_\alpha(\zeta)=\mathrm{e}^{-\mathrm{i}\alpha}F(\zeta\mathrm{e}^{\mathrm{i}\alpha})\in\Sigma$, 由于 $F(\zeta)$是极值函数,所以

$$\left.\frac{\partial H_{F_\alpha}}{\partial\alpha}\right|_{\alpha=1}=0,$$

即

$$\operatorname{Im}\left\{\sum_{n=2}^{N}H_n\left[\sum_{\nu,\nu'=1}^{m}\gamma_{\nu,\nu'}^{(n)}\left(1-\frac{\zeta_\nu^{(n)}F_\nu^{(n)'}-\zeta_{\nu'}^{(n)}F_{\nu'}^{(n)'}}{F_\nu^{(n)}-F_{\nu'}^{(n)}}\right)\right]\right\}=0,$$

因此 $N(\zeta)$在 $|\zeta|=1$ 上$\geqslant 0$. 由于极值区域的境界曲线中必有一单纯境界点,对此点而言, $N(\zeta)=0$. 所以 $N(\zeta)$在 $|\zeta|=1$ 至少具有一个二重根.

定理 9 在 Σ_T 中的一切函数 $w=F(\zeta)$, 能使

$$H_F=H(x_2, \cdots, x_N, y_2, \cdots, y_N)$$

取最大极值的,它必映照 $|\zeta|>1$ 于单位圆外($|w|>1$) 具有有限支解析割线的区域,每个极值函数 $F(\zeta)$在 $|\zeta|>1$ 中满足微分方程

$$\left(\frac{\zeta F'(\zeta)}{F(\zeta)}\right)^2 P(F(\zeta))=Q(\zeta), \tag{4.3}$$

此处

$$P(w)=A(w)+\overline{A\left(\frac{1}{\bar{w}}\right)}-2p,$$

$$Q(\zeta)=B(\zeta)-2p,$$

$$p=\min_{|w|=1}\mathrm{Re}\{A(w)\},\tag{4.4}$$

$$A(w)=\sum_{n=2}^{N}H_n\left[\sum_{\nu,\nu'=1}^{m}\gamma_{\nu,\nu'}^{(n)}\frac{w^2+w(F_\nu^{(n)}+F_{\nu'}^{(n)})-F_\nu^{(n)}F_{\nu'}^{(n)}}{(F_\nu^{(n)}-w)(F_{\nu'}^{(n)}-w)}\right],$$

$$B(\zeta)=-2\sum_{n=2}^{N}H_n\left[\sum_{\nu,\nu'=1}^{m}\gamma_{\nu,\nu'}^{(n)}\frac{\dfrac{\zeta_\nu^{(n)2}F_\nu^{(n)'}}{\zeta_\nu^{(n)}-\zeta}-\dfrac{\zeta_{\nu'}^{(n)2}F_{\nu'}^{(n)'}}{\zeta_{\nu'}^{(n)}-\zeta}}{F_\nu^{(n)}-F_{\nu'}^{(n)}}\right]$$

$$+2\sum_{n=2}^{N}H_n\left[\sum_{\nu,\nu'=1}^{m}\gamma_{\nu,\nu'}^{(n)}\frac{\dfrac{\zeta_\nu^{(n)}F_\nu^{(n)'}}{1-\bar{\zeta}\zeta_\nu^{(n)}}-\dfrac{\zeta_{\nu'}^{(n)}F_{\nu'}^{(n)'}}{1-\bar{\zeta}\zeta_{\nu'}^{(n)}}}{F_\nu^{(n)}-F_{\nu'}^{(n)}}\right].$$

函数 $P(w)$和 $Q(\zeta)$分别在 $|w|=1$ 和$|\zeta|=1$ 上取实值而不取负值,两者各具一个二重点.

证明 我们首先导出微分方程(4.3),作变分函数[3]

$$F_*(\zeta)=F(\zeta)+\lambda AF(\zeta)\frac{F(\zeta)+F(\zeta_0)}{F(\zeta)-F(\zeta_0)}-\lambda\bar{A}F(\zeta)\frac{1+\overline{F(\zeta_0)}F(\zeta)}{1-\overline{F(\zeta_0)}F(\zeta)}$$

$$-\lambda A\zeta F'(\zeta)2\left(\frac{F(\zeta_0)}{\zeta_0F'(\zeta_0)}\right)^2\frac{\zeta}{\zeta-\zeta_0}$$

$$+2\lambda\bar{A}\zeta F'(\zeta)\left(\frac{F(\zeta_0)}{\zeta_0F'(\zeta_0)}\right)^2\frac{1}{1+\bar{\zeta}_0\zeta}+O(\lambda^2).$$

当 $\lambda\,\mathrm{Re}\left\{2A\left[1-\left(\frac{F(\zeta_0)}{\zeta_0F'(\zeta_0)}\right)^2\right]\right\}<0$, 且 λ 充分小时, $F_*(\zeta)\in\Sigma_T$, 其中 ζ_0 是 $|\zeta_0|>1$ 中的任意点,经过平易的计算我们得到

$$H_{F_*}-H_F=-\lambda\,\mathrm{Re}\left\{A\left[A(F(\zeta_0))+\overline{A\left(\frac{1}{\overline{F}(\zeta_0)}\right)}-\left(\frac{F(\zeta_0)}{\zeta_0F'(\zeta_0)}\right)^2B(\zeta_0)\right]\right\}$$

$$=-\lambda\,\mathrm{Re}\{A\Phi(\zeta_0)\}\leqslant 0,$$

即

$$\lambda \operatorname{Re}\{A\Phi(\zeta_0)\} \geqslant 0.$$

因此，假如$\dfrac{F(\zeta)}{\zeta}$不是常数，那么

$$\lambda \operatorname{Re}\left\{A\left[1-\left(\frac{F(\zeta_0)}{\zeta_0 F'(\zeta_0)}\right)^2\right]\right\}<0$$

含有 $\lambda \operatorname{Re}\{A\Phi(\zeta_0)\} \geqslant 0$，这就必须对任一 ζ_0，

$$\Phi(\zeta_0)\Big/\left[1-\left(\frac{F(\zeta_0)}{\zeta_0 F'(\zeta_0)}\right)^2\right]\leqslant 0, \quad |\zeta_0|>1.$$

上式左边是 ζ_0 的解析函数，因此必须是一常数. 所以有如下的 θ：

$$\Phi(\zeta_0)+\theta\left[1-\left(\frac{F_0}{\zeta_0 F'_0}\right)^2\right]=0, \quad \theta \geqslant 0.$$

写开来就是

$$\left(\frac{\zeta F'(\zeta)}{F(\zeta)}\right)^2\left[A(F(\zeta))+\overline{A\left(\frac{1}{\overline{F(\zeta)}}\right)}+\theta\right]=B(\zeta)+\theta. \tag{4.5}$$

其次证明极值区域必无外点. 设 $w=F(\zeta)$ 将 $|\zeta|>1$ 映照于 G，若 G 有外点 ξ，$|\xi|>1$，则作变分函数[3]

$$F_*(\zeta)=F(\zeta)+\lambda\left[AF(\zeta)\frac{F(\zeta)+\xi}{F(\zeta)-\xi}-\overline{A}F(\zeta)\frac{1+\bar{\xi}F(\zeta)}{1-\bar{\xi}F(\zeta)}\right]+O(\lambda^2).$$

当 $\lambda \operatorname{Re}\{A\} \leqslant 0$，且 λ 充分小时，$F_*(\zeta) \in \Sigma_T$，经过计算，得到

$$H_{F_*}-H_F=-\lambda \operatorname{Re}\left\{A\left[A(\xi)+\overline{A\left(\frac{1}{\bar{\xi}}\right)}\right]\right\}.$$

因此 $\lambda \operatorname{Re}\{A\} \leqslant 0$ 含有 $\lambda \operatorname{Re}\left\{A\left[A(\xi)+\overline{A\left(\frac{1}{\bar{\xi}}\right)}\right]\right\} \geqslant 0$. 所以存在一常数 C，使得当 ξ 是 G 在单位圆外的外点时，

$$A(\xi)+\overline{A\left(\frac{1}{\bar{\xi}}\right)} \equiv C.$$

左边是 ξ 的解析函数，因此上式在整个平面成立，所以(4.5)右端在 $|\xi|>1$ 中是解析的，从而函数

$$\sum_{n=2}^{N} H_n \left[\sum_{\nu, \nu'=1}^{m} \gamma_{\nu, \nu'}^{(n)} \left(\frac{\dfrac{\xi_\nu^{(n)2} F_\nu^{(n)'}}{\xi_\nu^{(n)} - \zeta} - \dfrac{\zeta_{\nu'}^{(n)2} F_{\nu'}^{(n)'}}{\zeta_{\nu'}^{(n)} - \zeta}}{F_\nu^{(n)} - F_{\nu'}^{(n)}} \right) \right]$$

在 $|\zeta| > 1$ 中是解析的,特别地,当 $\zeta = \zeta_\nu^{(n)}$ 时应该是解析的. 因此

$$H_n(\gamma_{\nu,\nu'}^{(n)} + \gamma_{\nu',\nu}^{(n)}) = 0 \quad (\nu, \nu' = 1, 2, \cdots, m).$$

由假设必有 n_0, $2 \leqslant n_0 \leqslant N$,使 $H_{n_0} \neq 0$, 所以

$$\gamma_{\nu,\nu'}^{(n_0)} + \gamma_{\nu',\nu}^{(n_0)} = 0 \quad (\nu, \nu' = 1, 2, \cdots, m).$$

这是与假设相矛盾的,故极值区域无外点.

记(4.5)为

$$\left(\frac{\zeta F'(\zeta)}{F(\zeta)} \right)^2 P[F(\zeta)] = Q(\zeta), \quad |\zeta| > 1. \tag{4.5$'$}$$

利用微分方程的解析理论,知道极值区域 G 的境界是由有限根解析若当弧线组成的,但 G 无外点,所以 G 是由 $|w| > 1$ 中割去有限根解析若当弧线而成的. 因此 G 的境界中必有一弧线 γ,它除其一端点 w_0 在 $|w_0| = 1$ 外都在 $|w| > 1$ 中,利用上述方法可知 $P(w_0) = 0$. 同样,G 的境界上有一弧线 γ^*,其一端点 w^* 为单纯境界点,且在 $|w| > 1$ 中. 设 $F(\zeta_0) = w^*$, $|\zeta_0| = 1$,易知 $F'(\zeta_0) = 0$. 因此 $Q(\zeta_0) = 0$.

现在要证明 $|w_1| = 1$ 时 $\mathrm{Re}\{A(w_1)\} \geqslant \mathrm{Re}\{A(w_0)\}$, 不然的话,变分函数[3]

$$F(\zeta, \tau) = F(\zeta) + \tau F(\zeta) \frac{F(\zeta) + w_0}{F(\zeta) - w_0} + k\tau F(\zeta) \frac{1 + \bar{w}_1 F(\zeta)}{1 - \bar{w}_1 F(\zeta)} + O(\tau^2), \ \tau < 0,$$

当 $k < 1$ 时,它属于 Σ_T,对此函数,有

$$H_{F(\zeta, \tau)} - H_F = -\tau \mathrm{Re}\{A(w_0) - kA(w_1)\}.$$

因为 $\mathrm{Re}\{A(w_1)\} < \mathrm{Re}\{A(w_0)\}$, 所以必然有 k 适合于

$$1 > k > 0, \quad \mathrm{Re}\{A(\varphi_0) - kA(w_1)\} > 0,$$

因此 $H_{F(\zeta, \tau)} > H_F$, 这和 $F(\zeta)$的极值性矛盾. 设

$$p = \min_{|w|=1} \mathrm{Re}\{A(w)\} = \mathrm{Re}\{A(w_0)\}.$$

但 $p(w_0)=0$,因此 $2p+\theta=0$,由是 $P(w)=M_{\Sigma^*_{F_T}}(w)$ 在 $|w|=1$ 上 $\geqslant 0$,所以每一根都是二重根,因此 w_0 是一个二重根.

$F(\zeta_0)$是 G 的境界在 $|w|>1$ 中的单纯境界点,ξ 是 $|\zeta|=1$ 上任意点. 当 $k<1$ 时,变分函数[3]

$$F_*(\zeta,\tau)=F(\zeta)+\zeta F'(\zeta)\left(\tau\frac{\zeta_0+\zeta}{\zeta_0-\zeta}-k\tau\frac{1+\zeta\bar{\xi}}{1-\zeta\bar{\xi}}\right)+O(\tau^2),\ \tau>0$$

属于 Σ_T,但

$$H_{F_*}-H_F=\frac{\tau}{2}\mathrm{Re}\{B(\zeta_0)-kB(\xi)\}\leqslant 0.$$

由此同样可证得

$$\mathrm{Re}\{B(\xi)\}\geqslant\mathrm{Re}\{B(\zeta_0)\}.$$

因

$$\mathrm{Im}\{B(\zeta_0)\}=\mathrm{Im}\{B(\xi)\}=-2\mathrm{Im}\left\{\sum_{n=2}^{N}H_n\left(\sum_{\nu,\nu'=1}^{m}\gamma_{\nu,\nu'}^{(n)}\frac{\zeta_\nu^{(n)}F_\nu^{(n)'}-\zeta_{\nu'}^{(n)}F_{\nu'}^{(n)'}}{F_\nu^{(n)}-F_{\nu'}^{(n)}}\right)\right\},$$

所以 $Q(\xi)-Q(\zeta_0)=\mathrm{Re}\{B(\xi)-B(\zeta_0)\}\geqslant 0$,但 $Q(\zeta_0)=0$,所以 $Q(\xi)\geqslant 0$ $(|\xi|=1)$. 这说明 $N(\zeta)$ 在 $|\zeta|=1$ 上是非负的,因而在 $|\zeta|=1$ 上的一切根都是二重根,因此 ζ_0 是一个二重根,即 $N(\zeta)$在 $|\zeta|=1$ 上至少有一个二重根,定理证毕.

◇参考文献◇

[1] Г. М. Голузин:(1) Метод вариаций в конформном отображении, I, Матем. сборник 19, 203-236, 1946.

(2) Некоторые вопросы теории одяолистных функций, Труды Матем. ИН Имени Стеклова, 1949(有陈建功教授的中译本).

[2] Z. Charzinski: Sur les fonctions univalentes bornées, Rozprawy Matem. II, 1953, 1-51, Warszawa.

[3] 夏道行:Z. 夏尔绳斯基定理的较短的证明,数学进展,第 2 卷第 4 期(1956).

Extensions of Some Theorems of Golusin and Charzinski

By Jen Fu-Yao

Abstract

Let S be the class of functions $f(z)=z+a_2z^2+\cdots$ regular and schlicht in the unit circle $|z|<1$. Let Σ be the class of functions $F(\zeta)=\zeta+\alpha_0+\dfrac{\alpha_1}{\zeta}+\cdots$ regular and schlicht in the domain $1<|\zeta|<\infty$. Let F_T, $0<T<1$, be the class of functions $f(z)$ such that $f(z)/f'(0)\in S$, that $f'(0)\geqslant T$, and that $|f(z)|<1$ for $|z|<1$. Let F_T^* be the class of functions $e^{i\theta}f(z)$ with real θ and $f(z)\in F_T$. Let Σ_T be the class of functions $F(\zeta)=\alpha_{-1}\zeta+\alpha_0+\dfrac{\alpha_1}{\zeta}+\cdots$ regular and schlicht in the domain $1<|\zeta|<\infty$ such that $|F(\zeta)|>1$ for $|\zeta|>1$ and that $|\alpha_{-1}|\leqslant\dfrac{1}{T}$, $0<T<1$.

Any functions $H=H(x_2,\cdots,x_N;y_2,\cdots,y_N)$ of $2N-2$ real variables is said to be a Charzinski's function, if it is continuously differentiable with respect to each of its arguments in some sufficiently large "sphere" $x_2^2+\cdots+y_N^2<R^2$ such that the first partial derivatives do not vanish simultaneously at any point in the sphere.

The following theorems can be established by the method of vasriation.

Theorem I *Let $f(z)=z+a_2z^2+\cdots$ and let $H=H_f(x_2,\cdots,x_N;y_2,\cdots,y_N)$ be a Charzinski's function with $a_n=x_n+iy_n$. If the function $W=f(z)$ maximizes H_f, then it maps the unit circle $|z|<1$ onto the W-plane with a finite number of slits which are analytic Jordan arcs Γ satisfying the differential eqution*

$$\left(\frac{zf'(z)}{f(z)}\right)^2 M(f(z))=N(z),$$

with

$$M(w)=\sum_{n=2}^{N}H_n\cdot\left(\sum_{\nu=1}^{\alpha_n-1}\frac{\{f(\xi)^{\nu+1}\}_{\alpha_n}}{w^{\nu}}\right),$$

$$N(z)=\sum_{n=2}^{N}\left\{H_n(\alpha_n-1)a_{\alpha_n}+\sum_{\nu=1}^{\alpha_n-1}(\alpha_n-\nu)\left(\frac{H_n a_{\alpha_n-\nu}}{z^{\nu}}+\overline{H_n a_{\alpha_n-\nu}z^{\nu}}\right)\right\},$$

$$H_n=\frac{\partial}{\partial x_n}H_f-\mathrm{i}\frac{\partial}{\partial y_n}H_f,\quad n=2,3,\cdots,N,$$

Where $\{g(z)\}_n$ *being the coefficients of* z^n *in the Maclaurin expansion of* $g(z)$. *On the circle*, $|z|=1$, *the function* $N(z)$ *is real and non-negative and possesses at least one double zero. If* $P(z_1, z_2, \cdots, z_{N-1})$ *is a polynomial and* $H_f=\mathrm{Re}\{P(a_2,\cdots,a_N)\}$, *then*

$$H_n=\frac{\partial}{\partial a_n}P(a_2,a_3,\cdots,a_N)\quad(n=2,3,\cdots,N).$$

Theorem II *Let* $\gamma_{\nu,\nu'}^{(2)}, \gamma_{\nu,\nu'}^{(3)}, \cdots, \gamma_{\nu,\nu'}^{(N)}$ $(\nu,\nu'=1,2,\cdots,m)$ *be a set of complex numbers such that the sums* $\{\gamma_{\nu,\nu'}^{(n)}+\gamma_{\nu',\nu}^{(n)}\}$ *do not vanish simultaneously for each n, and let* $\zeta_\nu^{(2)}, \zeta_\nu^{(3)}, \cdots, \zeta_\nu^{(N)}$ $(\nu=1,2,\cdots,m)$ *be N fixed points in* $|\zeta|>1$. *If the function* $z=F(\zeta)\in\Sigma_T$ *maximizes* $H_F=H(x_2,\cdots,x_N;\ y_2,\cdots,y_N)$ *with* $x_n=\mathrm{Re}\left(\sum_{\nu,\nu'=1}^{m}\gamma_{\nu,\nu'}^{(n)}\log\frac{F(\zeta_\nu^{(n)})-F(\zeta_{\nu'}^{(n)})}{\zeta_\nu^{(n)}-\zeta_{\nu'}^{(n)}}\right)$ *and* $y_n=\mathrm{Im}\left(\sum_{\nu,\nu'=1}^{m}\gamma_{\nu,\nu'}^{(n)}\log\frac{F(\zeta_\nu^{(n)})-F(\zeta_{\nu'}^{(n)})}{\zeta_\nu^{(n)}-\zeta_{\nu'}^{(n)}}\right)$, *then it maps the domain* $|\zeta|>1$ *onto the z-plane with a finite number of cuts which are analytic Jordan curves satisfying the diffirential equation*

$$\left(\frac{\zeta F'(\zeta)}{F(\zeta)}\right)^2 P(F(\zeta))=Q(\zeta),$$

where

$$P(w)=A(w)+\overline{A(1/\overline{w})}-2p,$$

$$Q(\zeta)=B(\zeta)-2p,$$

$$A(w)=\sum_{n=2}^{N}H_n\left[\sum_{\nu,\nu'=1}^{m}\gamma_{\nu,\nu'}^{(n)}\frac{w^2+(F_\nu^{(n)}+F_{\nu'}^{(n)})w-F_\nu^{(n)}\cdot F_{\nu'}^{(n)}}{(F_\nu^{(n)}-w)(F_{\nu'}^{(n)}-w)}\right],$$

$$B(\zeta)=-2\sum_{n=2}^{N}H_n\cdot\left[\sum_{\nu,\nu'=1}^{m}\gamma_{\nu,\nu'}^{(n)}\frac{F_\nu^{(n)'}\zeta_\nu^{(n)^2}/(\zeta_\nu^{(n)}-\zeta)-\zeta_{\nu'}^{(n)^2}F_{\nu'}^{(n)'}/(\zeta_{\nu'}^{(n)}-\zeta)}{F_\nu^{(n)}-F_{\nu'}^{(n)}}\right]$$

$$+2\sum_{n=2}^{N} H_n \left[\sum_{\nu,\nu'=1}^{m} \gamma_{\nu,\nu'}^{(n)} \frac{\zeta_\nu^{(n)} F_\nu^{(n)'}/(1-\bar{\zeta}\zeta_\nu^{(n)}) - \zeta_{\nu'}^{(n)} F_{\nu'}^{(n)'}/(1-\bar{\zeta}\zeta_{\nu'}^{(n)})}{F_\nu^{(n)} - F_{\nu'}^{(n)}} \right],$$

$$p = \min \operatorname{Re}\{A(w)\}.$$

The functions $P(w)$ and $Q(\zeta)$ are real and non-negative, and possess at least one double zero on the circles $|w|=1$ and $|\zeta|=1$ respectively.

For the functions $f(z)=z+a_2z^2+\cdots$ of S, we prove the relation

$$\left| a_5 - 2a_2a_4 + 4a_2^2a_3 - \frac{3}{2}a_3^2 - \frac{3}{2}a_2^4 \right| \leqslant \frac{1}{2},$$

the estimation is precise.

On the Functions of Bieberbach and of Lebedev-Milin*

Jen Fu-yao
Fu-Tan University

(*Communicated by Prof. K.K. Chen, Member of Academia Sinica*)

1. Let the function $W = f(z) = \sum_{n=1}^{\infty} a_n z^n$ be regular in $|z| < 1$. We denote by B the class of functions $f(z)$ which satisfy the condition

$$f(z_1) \cdot f(z_2) \neq 1, \quad |z_1| < 1, \ |z_2| < 1,$$

and by L the class of functions $f(z)$ which satisfy the condition

$$f(z_1) \cdot \overline{f(z_2)} \neq -1, \quad |z_1| < 1, \ |z_2| < 1.$$

Jenkins[1] and Shah[5] have considered the maximum of $|f(z)|$ on $|z| = r$ for the functions of B and L. For the class B, Jenkins[3] expresses $\max_\theta |f(re^{i\theta})|$ and $\min_\theta |f(re^{i\theta})|$ in terms of $|a_1|$. The purpose of this paper is to discuss the maximum and the minimum of $|f(r_2)|$ for the univalent functions of B and L by fixing $|f(-r_1)|$, where $0 < r_1 < 1$, $1 < r_2 < 1$.

2. We shall construct a family of functions which provide the extremal in our problem. Let us consider the function

$$\zeta = \int^z \left[\frac{(z-p)(z-p^{-1})}{z(z+r_1)(1+r_1 z)(r_2 - z)(1 - r_2 z)} \right]^{1/2} \mathrm{d}z$$

on the semicircle $|z| < 1$, $\mathrm{Im}(z) \geqslant 0$, where $-1 \leqslant p \leqslant 1$. It is understood that this function is defined by continuity in such cases that p assumes either

* Originally published in *Science Record*, *New Ser.*, Vol. II, No. 4, (1958), 117 - 125.

one of the values of 0, r_2, $-r_1$. The radicals are to have their positive determinations on the real axis to the right of 0. The lower limit of the integral may be taken at any suitable point. Let the images of 0, p, r_2, 1, -1 and $-r_1$ be respectively denoted by A, B, C, D, E and F. We shall assume r_1, r_2 to be fixed and describe the images of the upper semicircle in various cases, depending on the value of p. We obtain the following canonical domains:

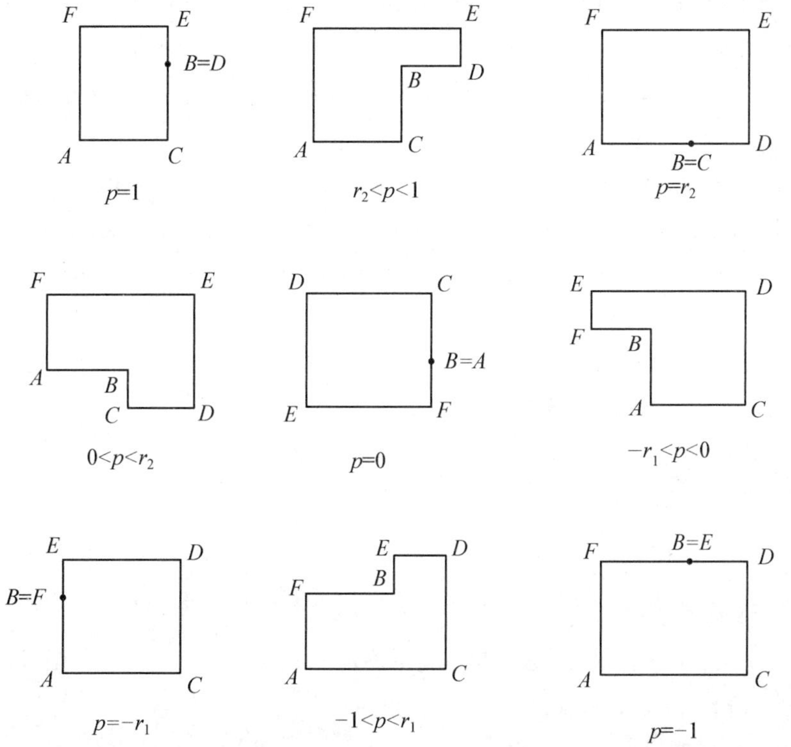

In each case, we rotate the domain so obtained through 180° about the midpoint P of DE, denoting the images of A, B, C, D, E and F by A', B', C', D', E' and F'. Identifying the point of DE and $D'E'$ which coincide with each other in pairs, we obtain a new domain. The latter domain we map conformally on the left-hand halfplane $\mathrm{Re}(W)<0$ in the W-plane in such a way that A goes into $W=0$, A' goes into $W=\infty$, and P goes into $W=-1$.

Let the images of B, C, D, F be ib, ic, id, $-if$. The quadratic differential on W-sphere can then be written as

$$d\zeta^2 = K\frac{(w-ib)(w+ib^{-1})}{iw(w+if)(w-if^{-1})(w-ic)(w+ic^{-1})}dw^2 = Q(w)dw^2$$

with a suitable positive constant K.

Let us consider the above combined mapping which maps the upper semicircle in z-plane into the w-plane. Reflecting the upper unit semicircle and its image in w-plane under above combined mapping across respectively the segment joining 1 and -1 in z-plane and $-i/d$ and id in w-plane, we obtain a function $f(z, r_1, r_2, p)$ regular and schlicht in $|z|<1$. It maps $|z|<1$ on a domain $D(r_1, r_2, p)$ in w-plane. We see that $f(z, r_1, r_2, p)$ belongs to B as well as L and that $f(z, r_1, r_2, r_2) \equiv i(1-r_1^2)^{1/2}z/(1+r_1z)$[3, 5].

For the case $p=-1$ $(p=1)$, we rotate the domain so obtained through $180°$ about any point Q lying between P and $D(E)$. The remainder of the construction is the same except that Q goes into $W=-1$ instead of P. We denote the continuum of functions so obtained by $f(z, r_1, r_2, \mp 1, \lambda)$ where λ runs over $0<\lambda<1$ as Q runs over the open segment $PD(PE)$. As before, $f(z, r_1, r_2, \mp 1, \lambda)$ belongs to both B and L.

3. **Theorem 1** *Let $f(z)$ belong to B and be univalent in $|z|<1$. If $|f(-r_1)|=|f(-r_1, r_1, r_2, p)|$ $(|f(-r_1, r_1, r_2, +1, \lambda)|)$, $r_2 \leqslant p \leqslant 1$, $1<\lambda<1$, then*

$$|f(r_2)| \geqslant |f(r_2, r_1, r_2, p)| (|f(r_2, r_1, r_2, 1, \lambda)|).$$

Equality can be attained only by the corresponding function $\pm f(z, r_1, r_2, p)$ $(\pm f(z, r_1, r_2, 1, \lambda))$.

Theorem 2 *Let $f(z)$ belong to B and be univalent in $|z|<1$. If $|f(-r_1)|=|f(-r_1, r_1, r_2, p)| (|f(-r_1, r_1, r_2, -1, \lambda)|)$, then*

$$|f(r_2)| \leqslant |f(r_2, r_1, r_2, p)| (|f(r_2, r_1, r_2, -1, \lambda)|),$$

provided either that $-1<p<-r_1$, $0<\lambda<1$, or that $-r_1<p<r_2$ and $f(z)$ is purely imaginary on the real axis.

Equality can be attained respectively only by the function $\pm f(z, r_1, r_2, p)$ $(\pm f(z, r_1, r_2, -1, \lambda))$.

We shall give here a brief proof for the case $r_2<p<1$. We consider the trajectories on which $d\zeta^2 = Q(w)dw^2 < 0$. In this case $b > c$ and there emerge from ib three arcs of trajectories. One of these proceeds along the imaginary axis to ic. We denote this segment by T_1. Joining the other two, we obtain a Jordan curve T_2. Let l be the trajectory in the neighborhood of $W = 0$. T_1, T_2 and l describe a doubly connected domain D_1. The region between the boundary of $D(r_1, r_2, p)$ and T_2 is also a doubly connected domain D_2.

Let us consider the image of $D(r_1, r_2, p)$ by $w' = f\{f^{-1}(w, r_1, r_2, p)\}$. Denote by D_i', T_i' and l' the images of D_i, T_i and l respectively. Applying to the doubly connected domain D_i', the circle symmetrization with respect to the negative axis, we obtain a new doubly connected domain $\widetilde{D}_i$. The boundary of $\widetilde{D}_1$, $\widetilde{T}_1$, $\widetilde{T}_2$ and $\widetilde{l}$ correspond respectively to D_1', T_1', T_2' and l'. $\widetilde{T}_1$ and $\widetilde{l}$ lie on the imaginary axis in the opposite sides of $W = 0$. Let M_i, M_i' and $\widetilde{M}_i$ be respectively the modules of D_i, D_i' and $\widetilde{D}_i$ $(i = 1, 2)$, then we have

$$M_i = M_i' \leqslant \widetilde{M}_i \quad (i = 1, 2).$$

Let l^* be the image of l under the transformation $W^* = 1/W$. Let $\mathscr{D}$ be the doubly connected domain bounded by l and l^*. Let X be a point on the positive imaginary axis in $\mathscr{D}$, and X^*, its image under the transformation $W^* = 1/W$. In $\mathscr{D}$, let us consider the following problem: Denote by C_1 the class of all the rectifiable Jordan curves lying in $\mathscr{D}$ and separating l from X, X^* and l^*; and by C_2 the class of all the rectifiable Jordan curves lying in $\mathscr{D}$, separating l and X from l^* and X^*, and being further restricted to be homotopic to such a curve symmetric with respect to the imaginary axis. Finally let C_3 denote the class of all the rectifiable Jordan curves lying in $\mathscr{D}$ and separating l^* from l, X and X^*. Let ρ be a non-negative real function of integrable square over $\mathscr{D}$ and such that for $\gamma_i \in C_i$ $(i = 1, 2, 3)$, $\int_{\gamma_i} \rho |dw|$ exists and that

$$\int_{\gamma_1} \rho |dw| \geqslant \alpha, \quad \int_{\gamma_2} \rho |dw| \geqslant \beta, \quad \int_{\gamma_3} \rho |dw| \geqslant \alpha,$$

α and β being respectively the distances from $\overline{AC}$ and $\overline{BD}$ to $\overline{EF}$. Denote by M

(α, β, l, X) the greatest lower bound of $\iint_{\mathscr{D}} \rho^2 \mathrm{d}\xi \mathrm{d}\eta$ $(w=\xi+\mathrm{i}\eta)$ for all such functions ρ.

Now let X coincide with the point $f(r_2, r_1, r_2, p)$; we then verify at once that $|Q(w)|^{1/2}$ provides the extremal metric in the corresponding module problem and

$$M(\alpha, \beta, l, X)=2(\alpha^2 M_1+\beta^2 M_2).$$

Now suppose that we had, contrary to the statement of our theorem, $|f(r_2)|<|f(r_2, r_1, r_2, p)|$; then in the configuration obtained from the above symmetrization, the end point $\widetilde{X}$ of $\widetilde{T}_1$ not on $\widetilde{T}_2$ would have an affix $\mathrm{i}k$, $k<|f(r_2, r_1, r_2, p)|$, and we have $M(\alpha, \beta, l, X) \geqslant M(\alpha, \beta, \widetilde{l}, \widetilde{X})+d$ where α, β are as above and $d>0$ dependent only on $|f(r_2)|$. On the other hand, let $\widetilde{D}_i^*$ be the image of $\widetilde{D}_i$ $(i=1, 2)$ under the transformation $W^*=1/W$. We have

$$M(\alpha, \beta, \widetilde{l}, \widetilde{X}) \geqslant 2(\alpha^2 M_1+\beta^2 M_2).$$

Combining this with the foregoing imequalities, we obtain

$$M(\alpha, \beta, \widetilde{l}, \widetilde{X})+d \leqslant M(\alpha, \beta, \widetilde{l}, \widetilde{X}),$$

which is however impossible. Therefore we must have

$$|f(r_2)| \geqslant |f(r_2, r_1, r_2, p)|.$$

If $f(z)$ is an extremal function, then $\widetilde{X}$ and $\widetilde{l}$ coincide with X and l respectively and $M(\alpha, \beta, l, X)=M(\alpha, \beta, \widetilde{l}, \widetilde{X})$. Hence, $M_i'=\widetilde{M}_i$ $(i=1, 2)$. Applying the principle of uniqueness in the theory of symmetrization[1], we have $D_i'=\pm\widetilde{D}_i$ $(i=1, 2)$. Thus we obtain $f(z) \equiv \pm f(z, r_1, r_2, p)$.

In a similar manner we can prove the corresponding theorems for the univalent functions of L.

It follows from the principle of uniqueness that no two values of $|f(-r_1, r_1, r_2, p)|$ $(-1 \leqslant p \leqslant r_1)$ and $|f(-r_1, r_1, r_2, -1, \lambda)|$ $(0<\lambda<1)$ can be equal. Also by the theory of normal families, we see that

$|f(-r_1, r_1, r_2, -1, \lambda)|$ tends to zero as λ approaches 1. Further since $|f(-r_1, r_1, r_2, r_2)| = r_1/(1-r_1^2)^{1/2}$, it follows by continuity that for fixing r_1, r_2 and $0 < a \leqslant r_1^{1/2}/(1-r_1^2)^{1/2}$ there exists in the above set just one function whose modulus at $z = -r_1$ is equal to a. Similar results hold true for the families $f(z, r_1, r_2, p)$ $(r_2 \leqslant p \leqslant 1)$ and $f(z, r_1, r_2, 1, \lambda)$ $(0 < \lambda < 1)$.

◇ References ◇

[1] Jenkins, J. A. 1949 *Trans. Amer. Math. Soc.* **67** 327 - 350.

[2] ———1953 *Amer. J. Math.* **75** 510 - 522.

[3] ———1954 *Trans. Amer. Math. Soc.* **76** 389 - 396.

[4] ———1954 *Annals of Math.* **61** 106 - 115.

[5] Shah Tao-shing 1955 *Acta Math. Sinica.* **5** 454.

具有无界型特征的拟似共形映照的存在性定理*

任福尧

设 D 是 z 平面上的一个区域，$z \in D$，$p(z)$和 $\theta(z)$是如下的两个函数：

$$p = p(z) \geqslant 1, \quad 0 \leqslant \theta(z) = \theta \leqslant \pi \quad (p \neq 1),$$

$p(z)$在 D 上是连续的，当 $p(z) \neq 1$ 时，$\theta(z)$有意义并且也是连续的，在这种情况下，我们说特征 $p(z)$，$\theta(z)$连续分布在 D 上.

设特征 $p(z)$，$\theta(z)$连续分布在 D 上，假如 D 上的连续单叶函数 $w = f(z)$ 对 D 中任一点 z，映照微小椭圆 $E_h(p(z), \theta(z); z)$于微小圆，即 $z' \in E_h(p(z), \theta(z); z)$，

$$\lim_{h \to 0} \frac{\max|f(z') - f(z)|}{\min|f(z') - f(z)|} = 1,$$

并且保持着方向，那么，我们说 $f(z)$为在 D 上以 $p(z)$，$\theta(z)$为其特征的拟似共形映照，其中 $E_h(p(z), \theta(z); z)$表示 z 平面上以点 z 为中心的椭圆，它的长轴与实轴成 $\theta(z)$的角，长轴与短轴的长分别等于 $2p(z)h$ 和 $2h$，$h > 0$.

拉夫连捷也夫首先提出并证明这样的基本定理[1]：

对于平面区域 D 上任何事先给定的有界的连续分布特征 p，θ，必有以 p，θ 为特征的拟似共形映照，这种映照，除开一个共形映照外，是完全确定的. 后来，别林斯基利用拉夫连捷也夫所展开的方法，证明即使有第一种不连续点，只要 $p(z)$有界，存在定理仍然是成立的[2]. 本文的目的在于证明具有无界型特征的存在性定理.

定理 1 设 $G(\gamma_i, \zeta_i)$表示单位圆 $|z| < 1$ 内除去有限条彼此不相交的光滑曲线 $\gamma_1, \gamma_2, \cdots, \gamma_k$ 和有限个孤立点 $\zeta_1, \zeta_2, \cdots, \zeta_l$ 所成的区域. 设函数 $p(z)$

* 原载《复旦大学数学系数学论文集》，1962，427－432.

和 $\theta(z)$ 为区域 $G(\gamma_i, \zeta_i)$ 上的连续分布特征，而在 $G(\gamma_i, \zeta_i)$ 的每个境界点 ζ 上，积分

$$\int_0^c \frac{\mathrm{d}r}{rq(r, \zeta)}$$

是发散的，其中 $q(r, \zeta) = \sup_z\{p(z)\}$, $z \in G(\gamma_i, \zeta_i) \cdot (|z-\zeta|=r)$, c 是某一常数，则存在从 $|z| \leqslant 1$ 到 $|w| \leqslant 1$ 的拓扑映照 $w=f(z)$, $f(0)=0$，它在 $G(\gamma_i, \zeta_i)$ 上是以 $p(z)$, $\theta(z)$ 为特征的拟似共形映照.

证 作以一组光滑曲线为境界的包含在区域 $G(\gamma_i, \zeta_i)$ 内部的 $\partial(k+l)$ 连的区域序列 G_n，且在卡拉泰屋独利的意义下[4]，以区域 $G(\gamma_i, \zeta_i)$ 为其核. 不难作连续特征函数序列 $p_n(z)$ 和 $\theta_n(z)$，使得

(1) $p_n(z)$ 在 $|z| \leqslant 1$ 上是连续的，且在区域 $G(\gamma_i, \zeta_i)$ 上, $p(z) \geqslant p_n(z) \geqslant 1$, $\theta_n(z)$ 在 $|z| \leqslant 1$ 上 $p_n(z) \neq 1$ 的地方也是连续的.

(2) 在区域 G_n 上, $p_n(z) \equiv p(z)$, $\theta_n(z) \equiv \theta(z)$.

根据拉夫连捷也夫的存在定理 3[1]，存在以 $p_n(z)$ 和 $\theta_n(z)$ 为特征的拟似共形映照 $w=f_n(z)$, $f_n(0)=0$，它将 $|z| \leqslant 1$ 拓扑地映照成 $|w| \leqslant 1$，这样就得到映照函数序列 $\{f_n(z)\}$. 采用对角线法，即知存在子序列 $\{f_{n_m}(z)\}$，在区域 $G(\gamma_i, \zeta_i)$ 上内闭地匀敛于某一函数 $w=f(z)$. 由于当 m 充分大时，在 $G(\gamma_i, \zeta_i)$ 上内闭地成立着

$$p_{n_m}(z) \equiv p(z), \quad \theta_{n_m}(z) \equiv \theta(z),$$

根据拉夫连捷也夫的致密性定理 4[1]，此极限函数 $w=f(z)$ 在区域 $G(\gamma_i, \zeta_i)$ 上恰以 $p(z)$ 和 $\theta(z)$ 为其特征. 设 $\tilde{z}=F(z)$ 为区域 $G(\gamma_i, \zeta_i)$ 的内闭子区域 E 上以 $p(z)$ 和 $\theta(z)$ 为特征的拟似共形映照，它映照 E 于 $\tilde{z}$ 平面上的单叶区域 $\tilde{E}$，则当 m 充分大时, $f_{n_m}(F^{-1}(\tilde{z}))$ 为区域 $\tilde{E}$ 上的正则单叶函数[1]，根据正则单叶函数的极限函数仍然是正则单叶函数[4]，从而知道此极限函数 $w=f(z)$ 在区域 $G(\gamma_i, \zeta_i)$ 上是拓扑映照，故 $f(z)$ 是在区域 $G(\gamma_i, \zeta_i)$ 上的以 $p(z)$ 和 $\theta(z)$ 为特征的拟似共形映照.

现在，我们来证明此极限函数 $w=f(z)$ 的角状境界值是处处存在的，而且也是连续的. 事实上，仿引理 1[1] 的证明，即得

$$|f_{n_m}(z_1) - f_{n_m}(z_2)| < K' \Big/ \sqrt{\int_{|z_i-\zeta|=r}^c \frac{\mathrm{d}r}{rq_m(r, \zeta)}},$$

其中 z_1, $z_2 \in G(\gamma_i, \zeta_i)$, $|\zeta| \neq 1$, $q_m(r, \zeta) = \sup_z\{p_{n_m}(z)\}$, $z \in G(\gamma_i, \zeta_i) \cdot (|z-\zeta|=r)$, K', c 是与 m 和 r 无关的常数. 根据 $f_{n_m}(z)$和 $p_n(z)$的假设及法都引理知道

$$\lim_{m\to\infty} q_m(r, \zeta) = q(r, \zeta)$$

及

$$|f(z_1) - f(z_2)| \leqslant K' \Big/ \sqrt{\int_{|z_i-\zeta|=r}^{c} \frac{\mathrm{d}r}{rq(r, \zeta)}} = K'\mu(r).$$

根据定理的假设, $\lim\limits_{r\to 0}\mu(r) = 0$, 因此 $f(z)$的角状境界值是处处存在的. 定义此角状境界值为 $w = f(z)$ 在 $z = \zeta$ 的函数值,则 $w = f(z)$ 在 $|z| \leqslant 1$ 内是连续的.

现在我们来证明 $w = f(z)$ 在单位圆周 $|\zeta| = 1$ 上也处处存在着角状境界值. 若不然,设有 ζ_0, $|\zeta_0| = 1$ 及区域 $G(\gamma_i, \zeta_i)$内以 ζ_0 为端点的两条半曲线 τ_1 和 τ_2,当 z 沿 τ_1, τ_2 移动而达到点 ζ_0 时,对应的点沿两曲线 $f(\tau_1)$, $f(\tau_2)$而终止于相异两点 w_1, w_2, $|w_1| \leqslant 1$, $|w_2| \leqslant 1$. 但在点 ζ_0 的任何环境 $O(\zeta_0)$ 中 τ_1 和 τ_2 可用 $G(\gamma_i, \zeta_i)$ 中介于 τ_1 和 τ_2 之间的连续曲线来连接. 当 $O(\zeta_0)$ 收缩成点 ζ_0 时,圆 $|w| < 1$ 中的对应的若当曲线的两端分别收敛于 w_1, w_2. 在这些曲线上, $f^{-1}(w)$ 匀敛于 ζ_0. 另一方面,我们总可作一扇形 $\pi_0(\zeta_0)$:

$$|z - \zeta_0| \leqslant \rho, \quad |\arg(z - \zeta_0)| < \frac{\pi}{2m},$$

使得此扇形包含在区域 $G(\gamma_i, \zeta_i)$中介于 τ_1 和 τ_2 之间, m 为自然数. 将扇形 $\pi_0(\zeta_0)$按正(或负)方向回转,回转角是 $\dfrac{2\pi}{m}$, $\dfrac{4\pi}{m}$, $\cdots$, $\dfrac{2(m-1)\pi}{m}$,得到扇形 $\pi_1(\zeta_0)$, $\pi_2(\zeta_0)$, $\cdots$, $\pi_{m-1}(\zeta_0)$. 显然, $\pi_0(\zeta_0)$, $\pi_1(\zeta_0)$, $\cdots$, $\pi_{m-1}(\zeta_0)$的和集恰好构成圆 $|z - \zeta_0| \leqslant \rho$. 我们来构造圆环 $0 < |z - \zeta_0| \leqslant \rho$ 上的连续函数 $p^*(z)$和 $\theta^*(z)$. 当 $z \in \pi_0(\zeta_0)$ 时,我们令 $p^*(z) \equiv p(z)$, $\theta^*(z) \equiv \theta(z)$;当 $z \in \pi_1(\zeta_0)$ 时,我们定义 $p^*(z)$和 $\theta^*(z)$的函数值,分别等于 $p(z)$和 $\theta(z)$在 $\pi_0(\zeta_0)$中关于 $\pi_0(\zeta_0)$和 $\pi_1(\zeta_0)$的交线 l_1 与 z 对称的点的函数值,显然, $p^*(z)$和 $\theta^*(z)$在 l_1 上是连续的. 用同样的方法,从 $\pi_1(\zeta_0)$的 $p^*(z)$和 $\theta^*(z)$的函数值出发,定义 $p^*(z)$和 $\theta^*(z)$在 $\pi_2(\zeta_0)$的函数值,显然,它们在 $\pi_2(\zeta_0)$和 $\pi_1(\zeta_0)$的交线 l_2 上也是连续的. 依此类推,我们就构造好了圆环 $0 < |z - \zeta_0| \leqslant \rho$ 上

的连续函数. 根据 $p^*(z)$和 $\theta^*(z)$的作法,知道

$$q^*(r,\zeta_0)=\sup_{|z-\zeta_0|=r<\rho}\{p^*(z)\}=\sup_{\substack{|z-\zeta_0|=r\\ z\in\pi_0(\zeta_0)}}\{p(z)\}\leqslant q(r,\zeta_0),$$

因此,积分

$$\int_0^c \frac{\mathrm{d}r}{rq^*(r,\zeta_0)}$$

是发散的. 据拉夫连捷也夫的定理 7[1],在圆环 $0<|z-\zeta_0|\leqslant\rho$ 上存在以 $p^*(z)$和 $\theta^*(z)$为其特征的拟似共形映照 $w=\varphi(z)$,它将圆$|z-\zeta_0|\leqslant\rho$ 拓扑地映照成 $|\omega|\leqslant\rho^*$, $\varphi(\zeta_0)=0$. 在 $\omega=\varphi(z)$ 的映照下,扇形 $\pi_0(\zeta_0)$拓扑地映照成曲三角形区域. 由于 $\omega=\varphi[f^{-1}(w)]$ 是正则单叶的,且当 w 沿任何途径趋近连接 w_1 和 w_2 的边界时, $\varphi[f^{-1}(w)]$均趋于极限值零. 据寇勃引理 4,则 $\varphi[f^{-1}(w)]\equiv 0$, 这与 $f(z)$和 $\varphi(z)$的性质矛盾,故 $w=f(z)$ 在$|\zeta|=1$ 上处处存在着角状境界值. 定义此境界值为 $f(z)$在 $z=\zeta(|\zeta|=1)$ 的函数值,则得 $w=f(z)$ 在 $|z|\leqslant 1$ 上是连续的.

至于 $w=f(z)$ 在区域 $G(\gamma_i,\zeta_i)$的境界上的单叶性的证明,可以这样来考虑:设有同一境界线段上的相异两点 ζ_1, ζ_2,而 $f(\zeta_1)\equiv f(\zeta_2)$, 则在区域 $G(\gamma_i,\zeta_i)$内作两条半曲线 λ_1 和 λ_2,分别以 ζ_1 和 ζ_2 为一端,它们在 $G(\gamma_i,\zeta_i)$内恰有一个交点(不妨假定就是它们的另一端),并且在由 λ_1, λ_2 和介于 ζ_1, ζ_2 之间的境界线段 $\widehat{\zeta_1\zeta_2}$ 所围成的单连区域 g 内再不含有其他的境界点;在区域 g 内充分接近 $\widehat{\zeta_1\zeta_2}$ 的地方存在一点 z_0,圆 $|z-z_0|=r$ 有弧长$\geqslant\frac{2\pi r}{m}$之一圆弧($m$ 是一整数)露出在 g 的外部,且当 g 中的点 z 从圆 $|z-z_0|<r$ 趋近于 g 的境界 $\widehat{\zeta_1\zeta_2}$ 时, $f(z)$的境界值为常数 $f(\zeta_1)$,我们将证明 $f(z)$必恒等于 $f(\zeta_1)$. 为使证明简洁,不妨设 $f(\zeta_1)=0$, $z_0=0$, 若不然,只须就 $f(z)-f(\zeta_1)$ 和 $z-z_0$ 来考虑就行了. 在 $z=0$ 的周围,将 g 回转,回转角是$\frac{2\pi}{m}$, $\frac{4\pi}{m}$, …, $\frac{2(m-1)\pi}{m}$,得到区域 g', g'', …, $g^{(m-1)}$. 通集

$$g_1=g'\cdot g''\cdot\cdots\cdot g^{(m-1)}\cdot g$$

含有 $z=0$ 而不含有$|z|=r$ 上的任何点,因此, g_1 完全在圆 $|z|<r$ 的内部. 函数

$$F(z)=f(z)\cdot f(\eta z)\cdot\cdots\cdot f(\eta^{m-1}z),\ \eta=\exp\frac{2\pi}{m}$$

在 g_1 中是连续单叶的. 当 z 从 g_1 的内部趋近于 g_1 的境界时,

$$|F(z)|=0,$$

从而上式在 g_1 的内部也必定成立. 这是因为若 $|F(z)|$ 在 g_1 上的上确界 M 大于0,那么 g_1 中必有点列 $z_k(k=1, 2, \cdots)$, $|F(z_k)|\to M$. 设 $z_k\to z_0$, z_0 绝非 g_1 的境界点,因为在境界上, $|F(z)|=0$, 而 $|F(z_0)|$ 必大于 0. 但是由最大模定理, $|F(z)|$ 在 g_1 中取到最大值也是不可能的,所以 $|F(z)|$ 在 g_1 的内部必须恒等于零,由此, $f(z)$ 在 g_1 内部必须恒等于零,但这是不可能的. 故 $w=f(z)$ 在 $G(\gamma_i, \zeta_i)$ 的境界上是单叶的.

其余就不难证明了.

定理 2 设 $G(\gamma_i, \zeta_i)$ 即为定理 1 中所述的区域,函数 $p(z, w)$ 和 $\theta(z, w)$ 为区域 $G(\gamma_i, \zeta_i)$ 上的连续分布特征,而在区域 $G(\gamma_i, \zeta_i)$ 的每个境界点 ζ 上,积分

$$\int_0^c \frac{\mathrm{d}r}{rq(r, \zeta)}$$

是发散的,其中 $q(r, \zeta)=\sup\limits_{z, w}\{p(z, w)\}$, $z\in G(\gamma_i, \zeta_i)\cdot(|z-\zeta|=1)$, $|w|\leqslant 1$, c 是某一常数,则存在 $|z|\leqslant 1$ 到 $|w|\leqslant 1$ 的拓扑映照 $w=f(z)$, $f(0)=0$, 它在 $G(\gamma_i, \zeta_i)$ 上以 $p(z, w)$ 和 $\theta(z, w)$ 为其特征.

证 令 $p_1(z)=p(z, z)$, $\theta_1(z)=\theta(z, z)$, 则 $p_1(z)$ 和 $\theta_1(z)$ 满足定理 1 中所有的条件.

设

$$q_1(p, \zeta)=\sup_z\{p_1(z)\},\quad z\in G(\gamma_i, \zeta_i)\cdot(|z-\zeta|=r),$$

则显然有

$$q_1(r, \zeta)\leqslant q(r, \zeta),$$

故积分

$$\int_0^c \frac{\mathrm{d}r}{rq_1(r, \zeta)}$$

是发散的. 因此根据定理 1,存在函数 $w=f_1(z)$, 它拓扑地映照 $|z|\leqslant 1$ 于

$|w| \leqslant 1$,在区域 $G(\gamma_i, \zeta_i)$上以 $p_1(z)$和 $\theta_1(z)$为其特征.

置 $p_2(z)=p_1(z, f_1(z))$, $\theta_2(z)=\theta_1(z, f_1(z))$, $p_2(z)$和 $\theta_2(z)$同样也满足定理 1 中所有的条件,因此,存在一个函数 $w=f_2(z)$,它将 $|z| \leqslant 1$ 拓扑地映照成 $|w| \leqslant 1$,在区域 $G(\gamma_i, \zeta_i)$上以 $p_2(z)$和 $\theta_2(z)$为其特征. 如此经过 $n-1$ 次后,我们置

$$p_n(z)=p(z, f_{n-1}(z)), \quad \theta_n(z)=\theta(z, f_{n-1}(z)),$$

存在映照函数 $w=f_n(z)$,它将 $|z| \leqslant 1$ 拓扑地映照成 $|w| \leqslant 1$,在区域 $G(\gamma_i, \zeta_i)$上以 $p_n(z)$和 $\theta_n(z)$为其特征. 如此无限制地继续下去,我们得到映照函数序列$\{f_n(z)\}$和特征函数序列$\{p_n(z)\}$和$\{\theta_n(z)\}$.

首先证明$\{f_n(z)\}$在区域 $G(\gamma_i, \zeta_i)$上存在内闭匀敛的子函数序列. 事实上, $p_n(z)$在区域 $G(\gamma_i, \zeta_i)$上内闭均匀有界,根据有界型拟似共形映照的致密性定理[1]和对角线过程法,存在子函数序列$\{f_{n_m}(z)\}$,它内闭匀敛于某一函数 $w=f(z)$. 由于 $p(z, w)$和 $\theta(z, w)$是 $G(\gamma_i, \zeta_i)$和 $|w| \leqslant 1$ 上的双变数连续函数,因此, $p_n(z)$和 $\theta_n(z)$在区域 $G(\gamma_i, \zeta_i)$上分别内闭匀敛于 $p(z)$和 $\theta(z)$. 根据推广了的致密性定理[1],[4],在区域 $G(\gamma_i, \zeta_i)$上函数 $w=f(z)$ 恰以 $p(z)$和 $\theta(z)$为其特征,而

$$p(z)=p(z, f(z)), \quad \theta(z)=\theta(z, f(z)).$$

我们不难证明 $w=f(z)$ 在区域 $G(\gamma_i, \zeta_i)$上是单叶的. 若不然,在区域 $G(\gamma_i, \zeta_i)$上必有 z_1 和 z_2,使得

$$f(z_1)=f(z_2).$$

设区域 D 为包含在 $G(\gamma_i, \zeta_i)$中含有 z_1 和 z_2 的单连区域,根据引理 8 - 2[5], $f(z)$在此区域 D 上应该是单叶的,这是有悖于假设的,故 $f(z)$在 $G(\gamma_i, \zeta_i)$上是单叶的.

现在我们来证明 $w=f(z)$ 的角状境界值是处处存在的,而且是连续的. 设 ζ 为区域 $G(\gamma_i, \zeta_i)$的境界点,且 $|\zeta| \neq 1$,则

$$\begin{aligned}|f_{n_m}(z_1)-f_{n_m}(z_2)| &< K' \Big/ \sqrt{\int_{|z_i-\zeta|=r}^{c} \frac{\mathrm{d}r}{rq_m(r, \zeta)}} \\ &< K' \Big/ \sqrt{\int_{|z_i-\zeta|=r}^{c} \frac{\mathrm{d}r}{rq(r, \zeta)}}.\end{aligned}$$

先令 $m\to\infty$,然后令 $r\to 0$,立即知道 $f(z)$ 在 $z=\zeta$ 处的境界值是存在的.若定义此境界值为 $f(z)$ 在 $z=\zeta$ 处的函数值,则不难知道 $f(z)$ 在 $|z|<1$ 内是连续的.

仿效定理 1 中的后一段的证明,就不难证明定理中所欲证明的事实是真实的.定理证毕.

◇参考文献◇

[1] M. A. Lavrentieff: Sur une classe de repesentations continues, Матем. сб. (1935), 407-424.

[2] П. П. Белинский: Теорема существования и единственности квазиконформных отображении), У. М. Н. VI 2(42) (1951), 145.

[3] Г. М. 戈鲁辛:复变函数的几何理论(陈建功译).

[4] Л. И. Волковыский: Квазиконформные отображения.

无界的广义拟共形映照的孤立奇点*

任福尧

设 $w=w(z)=u(x, y)+\mathrm{i}v(x, y)(z=x+\mathrm{i}y)$ 是区域 D 上连续可微的连续映照,其偏导数 $w_x(z)$ 和 $w_y(z)$ 在 D 上满足如下的不等式:

$$|\nabla w(z)|^2 \leqslant RJ(z)+p_1+p_2|\nabla w(z)|^\delta, \tag{1}$$

$$|\nabla w(z)|^2=|w_x(z)|^2+|w_y(z)|^2,\ J(z)=u_x v_y-u_y v_x,$$
$$R>1,\ p_1 \geqslant 0,\ p_2 \geqslant 0,\ 0 \leqslant \delta \leqslant 1,$$

则称 $w(z)$是区域 D 上有界的广义拟共形映照.关于这种映照的赫尔特连续性和可去性孤立奇点的研究,紧密地联系着椭圆型偏微分方程组和非线性椭圆型偏微分方程解的连续性. R. 费姆和 J. 塞林[1]及 P. 哈德门[2]曾研究过这类问题.本文的目的在于开拓他们的结果,使它至少适用于某种非一致椭圆型的情况.

1. 定义和预备引理.

设 $R=(a, b; c, d)$ 表示闭的矩形:

$$a \leqslant x \leqslant b,\ c \leqslant y \leqslant d.$$

如果 R 上的连续函数 $w(z)$具有如下的性质:几乎对所有的 $x_0(a \leqslant x_0 \leqslant b)$, $w(z)$作为 y 的函数在区间(c, d)上是绝对连续的,并且,几乎对所有的 $y_0(c \leqslant y_0 \leqslant d)$, $w(z)$作为 x 的函数在区间(a, b)上是绝对连续的,则我们说函数 $w(z)$在矩形 $R=(a, b; c, d)$ 上具有 ACL 性质,或说在截线上是绝对连续的.如果区域 D 上的连续函数 $w(z)$在 D 中的任一矩形R $(\bar{R} \subset D)$ 上具有 ACL 性质,则我们说 $w(z)$在区域 D 上具有 ACL 性质.

设 $S(z_0; r_1, r_2)$表示闭环:

* 原载《复旦大学学报(自然科学版)》,1963 年,第 8 卷,第 2 期,237 - 242.

$$r_1 \leqslant |z - z_0| \leqslant r_2.$$

如果环 S 上的连续函数 $w(z)$ 具有如下的性质：几乎对所有的 $r_0(r_1 \leqslant r_0 \leqslant r_2)$，$w(r_0 e^{i\theta})$ 作为 θ 的函数在区间 $[0, 2\pi]$ 上是绝对连续的，那么我们说函数 $w(z)$ 在环 $S(z_0; r_1, r_2)$ 上具有 ACC 性质，如果区域 D 上的连续函数 $w(z)$ 在 D 中的任一圆环 S $(\bar{S} \subset D)$ 上具有 ACC 性质，则我们说 $w(z)$ 在区域 D 上具有 ACC 性质.

如果函数 $w(z)$ 在区域 D 上具有 ACL 性质，并且 $|w_x(z)|^2$ 和 $|w_y(z)|^2$ 在 D 内内闭 L-可积，则称 $w(z)$ 属于函数族 $\mathscr{R}_2$.

下面我们为了完整起见，不加证明地援引一个预备引理[3]：

设函数 $w = w(z) = u(x, y) + iv(x, y)$ 在区域 D 上属于函数族 $\mathscr{R}_2$，并且 $J(z) = u_x v_y - u_y v_x \geqslant 0$，则对于 D 中的每个可测集 E 有可测的映像 E'，并且

$$m(E') = \iint_E J(z) \mathrm{d}x \mathrm{d}y.$$

这引理的证明主要基于毛雷的引理七[4].

2. 赫尔特连续性和可去性孤立奇点.

引理 1 设 $w = w(z)$ 是区域 $D - (z_0)(z_0 \in D)$ 上连续的内部映照，它将 $D - (z_0)$ 映照到有有限面积的有界区域 Σ，它在 $D - (z_0)$ 上属于函数族 $\mathscr{R}_2$ 并且具有 ACC 性质，此外，

(i) $|\nabla w(z)|^2 \leqslant R(z)J(z) + |z - z_0|^{-\lambda}(p_1 + p_2 |\nabla w(z)|)$, (2)

$$p_1 \geqslant 0,\ p_2 \geqslant 0,\ 0 \leqslant \lambda < 1.$$

(ii) $J(z) \geqslant 0$.

(iii) $R(z)$ 在 $D - (z_0)$ 上是内闭有界可测的，并且大于 1，那么，存在常数 M，使得当 $r = |z - z_0| < c$ 时，成立着

$$|w(z_0 + re^{i\theta'}) - w(z_0 + re^{i\theta''})| < M \Big/ \sqrt{\int_r^c \frac{\mathrm{d}r}{rR(r)}}. \tag{3}$$

若 $w(z_0)$ 是有意义的，则成立着

$$|w(z_0 + re^{i\theta}) - w(z_0)| < M \Big/ \sqrt{\int_r^c \frac{\mathrm{d}r}{rR(r)}}, \tag{4}$$

其中
$$R(r) = \sup_{|z - z_0| = r} R(z),$$

而$M=M(p_1, p_2, \lambda, m(\Sigma))$是仅和$p_1$, p_2, λ及Σ的面积$m(\Sigma)$有关的常数.

证明 为了方便起见,不妨假设$z_0=0$. 取c'充分地小,使得圆盘$|z|\leqslant c'$包含在D内. 设圆环$r\leqslant |z|\leqslant c'$在$w(z)$映照下,其映像的面积记为$A(r, c')$,根据2.的预备引理,则

$$A(r, c')=\int_r^{c'}\int_0^{2\pi} J(\rho\cos\theta, \rho\sin\theta)\rho\,\mathrm{d}\rho\,\mathrm{d}\theta. \tag{5}$$

于是

$$\frac{\mathrm{d}A(r, c')}{\mathrm{d}r}=-\int_0^{2\pi} J(r\cos\theta, r\sin\theta)r\,\mathrm{d}\theta. \tag{6}$$

从条件(i)得到

$$\begin{aligned} J(z) &\geqslant \frac{|\nabla w(z)|^2}{R(z)}-\frac{|z|^{-\lambda}}{R(z)}(p_1+p_2|\nabla w(z)|) \\ &> \frac{|\nabla w(z)|^2}{R(r)}-|z|^{-\lambda}(p_1+p_2|\nabla w(z)|). \end{aligned} \tag{7}$$

但另一方面,从条件(i)得到

$$\begin{aligned} |\nabla w(z)| &\leqslant \frac{p_2}{2}|z|^{-\lambda}\left[1+\sqrt{1+\frac{4p_1}{p_2^2}|z|^{\lambda}+\frac{4}{p_2^2}|z|^{2\lambda}R(z)J(z)}\right] \\ &\leqslant \frac{p_2}{2}|z|^{-\lambda}\left[2+\frac{4p_1}{p_2^2}|z|^{\lambda}+\frac{4}{p_2^2}|z|^{2\lambda}R(z)J(z)\right], \end{aligned}$$

即

$$|\nabla w(z)|\leqslant p_2|z|^{-\lambda}+\frac{2p_1}{p_2}+\frac{2}{p_2}|z|^{\lambda}R(z)J(z). \tag{8}$$

结合(7)和(8)得到

$$3J(z)>\frac{|\nabla w|^2}{R(r)}-3p_1|z|^{-\lambda}-p_2^2|z|^{-2\lambda}. \tag{9}$$

将(9)式代入(6)式,得到

$$\begin{aligned} 3\frac{\mathrm{d}A(r, c')}{\mathrm{d}r} &< 6\pi p_1 r^{1-\lambda}+2\pi p_2^2 r^{1-2\lambda}-\frac{1}{R(r)}\int_{|z|=r}|\nabla w|^2\mathrm{d}s \\ &< c_1 r^{1-\lambda}+c_2 r^{1-2\lambda}-\frac{1}{R(r)}\int_{|z|=r}\left|\frac{\partial w}{\partial s}\right|^2\mathrm{d}s. \end{aligned}$$

利用许瓦茨不等式,

$$\left(\int_{|z|=r}\left|\frac{\partial w}{\partial s}\right|\mathrm{d}s\right)^2\leqslant 2\pi r\int_{|z|=r}\left|\frac{\partial w}{\partial s}\right|^2\mathrm{d}s.$$

此外,由于 $w(z)$在 $D-(z_0)$ 上具有 ACC 性质,所以几乎对所有的 $0<r<c'$ 有

$$l(c_r)=\int_{|z|=r}\left|\frac{\partial w}{\partial s}\right|\mathrm{d}s,$$

此处,$l(c_r)$表示 $|z|=r$ 在映照 $w(z)$下像曲线的长度.由是,获得

$$\frac{[l(c_r)]^2}{2\pi rR(r)}<c_1r^{1-\lambda}+c_2r^{1-2\lambda}-3\frac{\mathrm{d}A(r,c')}{\mathrm{d}r}. \tag{10}$$

置 $$\delta(c_r)=\max_{|z'|=|z''|=r}|w(z')-w(z'')|,$$

显然 $$2\delta(c_r)<l(c_r).$$

因此

$$\frac{2[\delta(c_r)]^2}{\pi rR(r)}<c_1r^{1-\lambda}+c_2r^{1-2\lambda}-3\frac{\mathrm{d}A(r,c')}{\mathrm{d}r}. \tag{11}$$

上式两端对 r 从固定的 r 到 $c(<c')$积分,由于 $w(z)$是正向的内映照,得到

$$c'_1c^{2-\lambda}+c'_2c^{2(1-\lambda)}+3A(r,c')>2\int_r^c\frac{[\delta(c_r)]^2}{\pi rR(r)}\mathrm{d}r>\frac{2}{\pi}[\delta(c_r)]^2\int_r^c\frac{\mathrm{d}r}{rR(r)}, \tag{12}$$

由于 $$\delta(c_r)\geqslant|w(re^{i\theta'})-w(re^{i\theta''})|,\quad 0\leqslant\theta',\theta''\leqslant 2\pi,$$
$$A(r,c')<m(\Sigma),$$

故存在常数 $M(p_1,p_2,\lambda,m(\Sigma))$,使得

$$|w(re^{i\theta'})-w(re^{i\theta''})|<M\Big/\sqrt{\int_r^c\frac{\mathrm{d}r}{rR(r)}}.$$

若 $w(0)$有意义,则有

$$|w(re^{i\theta})-w(0)|<M\Big/\sqrt{\int_r^c\frac{\mathrm{d}r}{rR(r)}}.$$

这样,证明了我们的引理.

引理 2 在引理 1 的诸条件中若把(iii)代替以条件

(iii)′ $R(z)$在区域 $D-(z_1)$ 中的 $|z-z_0|=r\ (0<r)$ 上按勒贝格意义是可积的,并且大于 1,那么,存在常数 M,使当 $r=|z-z_0|<c$ 时,成立着①

$$|w(z_0+re^{i\theta'})-w(z_0+re^{i\theta''})|<M\Big/\sqrt{\int_r^c\frac{dr}{r\int_0^{2\pi}R(z_0+re^{i\theta})d\theta}}.\tag{13}$$

若 $w(z_0)$是有意义的,则成立着

$$|w(z_0+re^{i\theta})-w(z_0)|<M\Big/\sqrt{\int_r^c\frac{dr}{r\int_0^{2\pi}R(z_0+re^{i\theta})d\theta}},\tag{14}$$

$M=M(p_1,\ p_2,\ \lambda,\ m(\Sigma))$.

证明 从不等式(9)和关系式(6)得到

$$\int_{|z|=r}\frac{\left|\dfrac{\partial w}{\partial s}\right|^2}{R(z)}ds<c_1r^{1-\lambda}+c_2r^{1-2\lambda}-3\frac{dA(r,\ c')}{dr}.\tag{15}$$

根据许瓦茨不等式,有

$$\left(\int_{|z|=r}\left|\frac{\partial w}{\partial s}\right|ds\right)^2\leqslant\int_{|z|=r}\frac{\left|\dfrac{\partial w}{\partial s}\right|^2}{R(z)}ds\cdot\int_{|z|=r}R(z)ds.\tag{16}$$

由是,从(15)式得到

$$\frac{\left(\int_{|z|=r}\left|\dfrac{\partial w}{\partial s}\right|ds\right)^2}{\int_{|z|=r}R(z)ds}<c_1r^{1-\lambda}+c_2r^{1-2\lambda}-3\frac{dA(r,\ c')}{dr}.\tag{17}$$

从引理 1 的证明,可见

$$[\delta(c_r)]^2\int_r^c\frac{dr}{r\int_0^{2\pi}R(re^{i\theta})d\theta}<M^2,$$

① 当 $p_1=p_2=0$ 时,在较强的条件下,首先由何成奇获得这样的结果.

其中，M 仅依赖于 p_1，p_2，λ 和 Σ 的面积. 由是即得不等式(13)和(14). 引理证毕.

这些引理推广了 M. A. 拉夫连捷也夫的引理 1[5]、R. 费姆和 J. 塞林及 P. 哈德门等相应的结果. 这引理在研究拟共形映照的同等连续性及其孤立奇点时是一基本的工具.

作为上述引理的直接的推论，立即可获得下述定理.

定理 设 S 是区域 D 中之一孤立点集，函数 $w=w(z)$ 是区域 $D-S$ 上连续的内部映照，它将 $D-S$ 映照到有有限面积的有界区域 Σ，它在 $D-S$ 上属于函数族 $\mathscr{R}_2$，并且具有 ACC 性质，此外，适合下列诸条件：

(i) 对 S 中的每个点 ζ，成立着

$$|\nabla w(z)|^2 \leqslant R(z)J(z) + |z-\zeta|^{-\lambda}(p_1 + p_2|\nabla w(z)|),$$
$$p_1 \geqslant 0,\ p_2 \geqslant 0,\ 0 \leqslant \lambda < 1.$$

(ii) $J(z) \geqslant 0$.

(iii) $R(z)$ 是 $D-S$ 上大于 1 的有界可测函数，而对于 S 中的每个点 ζ，成立着

$$\int_0^c \frac{\mathrm{d}r}{rR(r,\zeta)} = +\infty,$$

其中 $r=|z-\zeta|$，$R(r,\zeta)=\sup\limits_{|z-\zeta|=r} R(z)$，$c$ 是任一充分小的正数，或者

(iii)′ $R(z)$ 是 $D-S$ 上大于 1 的上半连续函数，而对于 S 中的每个点 ζ，成立着

$$\int_0^c \frac{\mathrm{d}r}{r\int_0^{2\pi} R(\zeta + r\mathrm{e}^{\mathrm{i}\theta})\mathrm{d}\theta} = +\infty.$$

那么，将 $w(z)$ 在 S 上适当定义后，可使得 $w(z)$ 成为 D 上的内部映照.

这定理推广了 M. A. 拉夫连捷也夫[5]、R. 费姆和 J. 塞林的结果[1].

证明 设 ζ 是 S 中任意一个定点. 我们证明区域 $D-S$ 中任一以 ζ 为极限点的点列 z_n，数列 $w(z_n)$ 必趋向同一的极限. 事实上，若不然，存在两点列 $z_n'\to\zeta$，$z_n''\to\zeta$，而

$$\lim_{n\to\infty} w(z'_n) = w_1,\ \lim_{n\to\infty} w(z''_n) = w_2,\ w_1 \neq w_2.$$

记 $z_n'-\zeta = r_n'\mathrm{e}^{\mathrm{i}\theta_n'}$，$z_n''-\zeta = r_n''\mathrm{e}^{\mathrm{i}\theta_n''}$. 不妨假设 r_n' 和 r_n'' 单调减少地趋于零，根据引理

1～2,当 $n\to\infty$ 时, $|z-\zeta|=r'_n$ 的像曲线 $c_{r'_n}=w(|z-\zeta|=r'_n)$ 收缩成点 w_1,而 $|z-\zeta|=r''_n$ 的像曲线 $c_{r''_n}=w(|z-\zeta|=r''_n)$ 收缩成点 w_2.

另一方面,不妨假设

$$r'_1>r''_1>r'_2>r''_2>r'_3>r''_3>\cdots,$$

因此,当 $n\to\infty$ 时, $|z-\zeta|=r'_n$ 的像曲线 $c_{r'_n}$ 又必须收缩成点 w_2,这是有悖于 $w_1\neq w_2$ 的假设的. 故当 z 不管以任何方式趋向于 ζ 时, $w(z)$ 趋向同一极限. 若定义此极限值为 $w(z)$ 在 $z=\zeta$ 的函数值,则如此定义的 $w(z)$ 在 D 上是连续的. 定理证毕.

◇参考文献◇

[1] R. Finn, J. Serrin, On the Hölder continuity of quasiconformal and elliptic mappings. Trans. Amer. Math. Soc. 89, (1958) 1-15.

[2] P. Hartman, Hölder continuity and non-linear elliptic partial differential equations. Duke Math. J. 25, (1958) 57-65.

[3] F. W. Gehring, The definitions and exceptional sets for quasiconformal mappings. Ann. Acad. Sci. Fenn. A I 281, 1960.

[4] J. R. Morrey, On the solutions of quasi-linear elliptic partial differential equations. Trans. Amer. Math. Soc. 43,(1938) 126-166.

[5] M. A. Lavrentief, Sur une classe de representations continues. Matem. сб. 42,(1935) 407-424.

On the Hölder Continuity and Isolated Singularities of Quasiconformal Mappings with Unbounded Character

Jen Fu-Yao

In this note we improve the results of R. Finn and J. Serrin on the Hölder continuity and isolated singularities of quasiconformal mappings with almost everywhere bounded character.

流体瞬态理论在双风机并联运行管系中的应用*

任福尧　柳兆荣

一、引　言

铁矿石必须经过烧结，脱去矿石中的硫等有害物质，增强还原性后，再送入高炉冶炼成铁. 烧结需要高温、大风. 其中风是第一位的，风助火势，有了大风才能促进燃烧，产生高温. 大风需要大功率的巨型抽风机，在实际应用中，往往采用两台大型抽风机并联来代替一台巨型抽风机.

然而，正如梅山工程指挥部烧结厂双机并联时曾遇到的那样，抽风机并联运行后，若不加注意，管路中会立即出现严重的抢风现象：两台抽风机交替出现时而超载，时而空载. 此时穿过料层抽进管路的气流忽而被这台风机全盘抢过来，继而又被另一台风机统统夺过去，犹如脱缰野马，东奔西撞，不可驾驭. 伴随着气流的这种异乎寻常的运动状态，管路中将发出撼人心弦的吼叫声，机组也将产生强烈的振动.

实践出真知. 为制止抢风，保证并联风机正常运行，早投产、多出铁，梅山工程指挥部烧结厂广大工人与革命科技人员发扬了大无畏的革命精神，以辩证唯物主义为指导，认真总结经验，终于制定出一套防止抢风发生的简便易行的操作规程.

我们在梅山工人制止抢风、保证风机正常运行的实践经验基础上，利用流体管路的瞬态理论，分析了两台风机并联运行时可能出现抢风或失风的条件. 并在此基础上，提出保证并联风机正常运行、防止产生抢风的方法. 本文所得结果与梅山工程指挥部烧结厂双机并联时曾出现过的抢风现象，以及后来为防止抢风所采取的有效措施一致.

---

* 原载《复旦大学学报(自然科学版)》，1977 年，第 3 期，85 - 95.

二、频域上分叉管道流的压力和流量表达式

在双风机并联的情况下,其传输管道是如图 1 所示的“二进一出”的 Y 形分支管道,其中 A_1, A_2, A_e 分别表示 3 条均匀圆管的横截面积, x, y 和 w 分别表示距离各管始端的距离, L_1, L_2 和 L_e 分别表示各管道的长度. 支管(Ⅰ)和(Ⅱ)的始端分别连接着抽风机 1 与 2,其终端与汇合管(e)的始端连接,烧结炉料层作为汇合管(e)的终端.

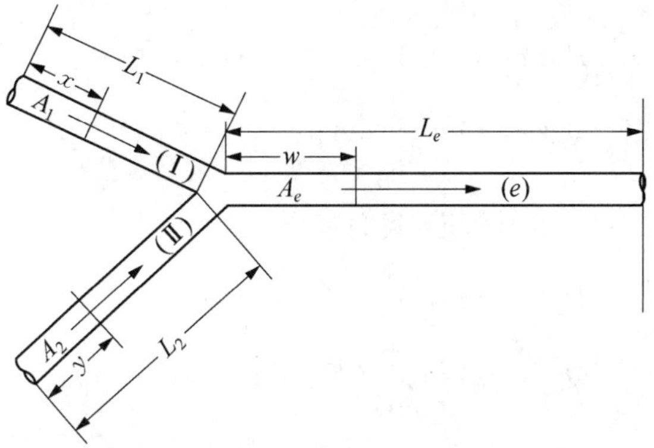

图 1 双机并联分叉管道流的示意图

对于理想流体的均匀圆管,其传输方程(参阅本刊 1975 年第 2 期第 42 页)为

$$\begin{cases} \dfrac{\partial P(x,\ s)}{\partial x} = -slQ(x,\ s), \\ \dfrac{\partial Q(x,\ s)}{\partial x} = -scP(x,\ s), \end{cases} \tag{1}$$

其中

$$P(x,\ s) = \int_0^\infty e^{-st} p(x,\ t)\mathrm{d}t \doteq p(x,\ t),$$

$$Q(x,\ s) = \int_0^\infty e^{-st} q(x,\ t)\mathrm{d}t \doteq q(x,\ t).$$

$p(x,\ t)$, $q(x,\ t)$分别表示在 x 处 t 时刻的压力和流量, $l = \dfrac{1}{A}$ 表示单位长度管

路的气感，$c=A/a^2$ 表示单位长度管路的气容，a 是声速. 我们熟知方程(1)的通解为

$$\begin{cases} P(x,\ s)=A_1\mathrm{e}^{-\gamma x}+A_2\mathrm{e}^{\gamma x}, \\ Q(x,\ s)=\dfrac{1}{Z_c}(A_1\mathrm{e}^{-\gamma x}-A_2\mathrm{e}^{\gamma x}), \end{cases} \tag{2}$$

其中 $\gamma=\sqrt{lc}\cdot s=s/a$ 称为传播因子，$Z_c=\sqrt{l/c}=a/A$ 称为所讨论管道的特性阻抗.

于是，对Ⅰ号管道，我们有

$$\begin{cases} P_1(x,\ s)=A_{11}\mathrm{e}^{-sx/a}+A_{12}\mathrm{e}^{sx/a}, \\ Q_1(x,\ s)=\dfrac{1}{Z_{c_1}}(A_{11}\mathrm{e}^{-sx/a}-A_{12}\mathrm{e}^{sx/a}), \end{cases} \tag{3}$$

其中 $Z_{c_1}=a/A_1$. 对于Ⅱ号管道，我们有

$$\begin{cases} P_2(y,\ s)=A_{21}\mathrm{e}^{-sy/a}+A_{22}\mathrm{e}^{sy/a}, \\ Q_2(y,\ s)=\dfrac{1}{Z_{c_1}}(A_{21}\mathrm{e}^{-sy/a}-A_{22}\mathrm{e}^{sy/a}), \end{cases} \tag{4}$$

其中 $Z_{c_2}=a/A_2$. 对于汇合管道，我们有

$$\begin{cases} P_e(w,\ s)=A_{e1}\mathrm{e}^{-ws/a}+A_{e2}\mathrm{e}^{sw/a}, \\ Q_e(w,\ s)=\dfrac{1}{Z_{c_0}}(A_{e1}\mathrm{e}^{-ws/a}-A_{e2}\mathrm{e}^{sw/a}), \end{cases} \tag{5}$$

其中 $Z_{ce}=a/A_e$.

上述方程组(3)，(4)和(5)就是各管道在频域上的压力和流量的表达式.

下面我们通过边界条件来确定包含于上述基本解中的 6 个待定系数 A_{11}，A_{12}；A_{21}，A_{22}；A_{e1}，A_{e2}.

对于Ⅰ，Ⅱ管的始端，即 $x=0$，$y=0$，其边界条件为

$$P_1(0,\ s)=P_1(s)-Z_{01}Q_1(0,\ s), \tag{6}$$

$$P_2(0,\ s)=P_2(s)-Z_{02}Q_2(0,\ s), \tag{7}$$

其中 Z_{01}，Z_{02} 分别表示Ⅰ，Ⅱ管始端的输入阻抗，$P_1(s)$，$P_2(s)$分别表示Ⅰ，Ⅱ管在始端输入的压力信号 $p_{01}(t)$，$p_{02}(t)$的拉普拉斯变换的像，即

$$P_1(s)=\int_0^{\infty} e^{-st} p_{01}(t)\mathrm{d}t \doteq p_{01}(t),$$

$$P_2(s)=\int_0^{\infty} e^{-st} p_{02}(t)\mathrm{d}t \doteq p_{02}(t).$$

对于汇合管道的终端 ($w=L_e$)，其边界条件为

$$P_e(L_e, s)=Z_{L_e}Q_e(L_e, s), \tag{8}$$

其中 Z_{L_e} 是汇合管终端 ($w=L_e$) 的负载阻抗.

在 3 根管道的汇合处,其边界条件为

$$p_1(L_1, t)=p_2(L_2, t)=p_{0e}(t), \tag{9}$$

$$q_1(L_1, t)+q_2(L_2, t)=q_e(0, t). \tag{10}$$

因而,其在频域上对应的边界条件为

$$P_1(L_1, s)=P_2(L_2, s)=P_e(s)=P_e(0, s)+Z_{0e}Q_e(0, s), \tag{11}$$

$$Q_1(L_1, s)+Q_2(L_2, s)=Q_e(0, s), \tag{12}$$

其中 Z_{0e} 是汇合管道始端 ($w=0$) 的输入阻抗.

利用上述边界条件(6),(7),(8),(11)和(12),恰好能确定 6 个待定系数. 其具体步骤如下:

将方程组(3)代入边界条件(6),得

$$A_{11}+A_{12}=P_1(s)-Z_{01}(A_{11}-A_{12})/Z_{c_1},$$

因而

$$(1+Z_{01}/Z_{c_1})A_{11}+(1-Z_{01}/Z_{c_1})A_{12}=P_1(s).$$

于是,得

$$A_{11}=\frac{P_1(s)}{1+Z_{01}/Z_{c_1}}+\delta_{01}A_{12}, \tag{13}$$

其中, $\delta_{01}=\dfrac{Z_{01}-Z_{c_1}}{Z_{01}+Z_{c_1}}$ 称为 I 管始端的反射系数.

同理,将方程组(4)代入边界条件(7),则得

$$A_{21}=\frac{P_2(s)}{1+Z_{02}/Z_{c_2}}+\delta_{02}A_{22}, \tag{14}$$

其中，$\delta_{02}=\dfrac{Z_{02}-Z_{c_2}}{Z_{02}+Z_{c_2}}$ 称为Ⅱ管始端的反射系数.

将方程组(5)代入边界条件(8)，则得

$$A_{e1}e^{-sL_e/a}+A_{e2}e^{sL_e/a}=\frac{Z_{L_e}}{Z_{c_e}}(A_{e1}e^{-sL_e/a}-A_{e2}e^{sL_e/a}),$$

因而

$$e^{-sL_e/a}(1-Z_{0e}/Z_{c_0})A_{e1}=-(1+Z_{0e}/Z_{c_e})e^{sL_e/a}A_{e2},$$

于是，有

$$A_{e2}=\delta_{L_e}e^{-2sL_e/a}\cdot A_{e1},\tag{15}$$

其于 $\delta_{L_e}=\dfrac{Z_{L_e}-Z_{c_e}}{Z_{L_e}+Z_{c_e}}$ 称为汇合管终端的反射系数.

将方程组(3)，(4)和(5)的第一个方程式代入边界条件(11)，则

$$A_{11}e^{-sL_1/a}+A_{12}e^{sL_1/a}=A_{21}e^{-sL_s/a}+A_{22}e^{sL_s/a}=(A_{e1}+A_{e2})+Z_{0e}(A_{e1}-A_{e2})/Z_{c_e}.$$

将(13)，(14)和(15)式代入上式，则得

$$\begin{aligned}&\frac{P_1(s)e^{-sL_1/a}}{1+Z_{01}/Z_{c_1}}+(1+\delta_{01}e^{-2sL_1/a})e^{sL_1/a}A_{12}\\
=&\frac{P_2(s)e^{-sL_2/a}}{1+Z_{02}/Z_{c_2}}+(1+\delta_{02}e^{-2sL_1/a})\cdot e^{sL_1/a}\cdot A_{22}\\
=&(1+Z_{0e}/Z_{c_e})A_{e1}-(Z_{0e}/Z_{c_e}-1)A_{e2}\\
=&(1+Z_{0e}/Z_{c_e})(1-\delta_{0e}\cdot\delta_{L_e}e^{-2sL_e/a})A_{e1}.\end{aligned}$$

于是，我们获得

$$A_{12}e^{sL_1/a}=\frac{1}{(1+\delta_{01}e^{-2sL_1/a})}\left[\left(1+\frac{Z_{0e}}{Z_{c_e}}\right)(1-\delta_{0e}\cdot\delta_{L_e}e^{-2sL_e/a})A_{e1}-\frac{P_1(s)e^{-sL_1/a}}{1+Z_{01}/Z_{c_1}}\right],\tag{16}$$

$$A_{22}e^{sL_2/a}=\frac{1}{(1+\delta_{02}e^{-2sL_2/a})}\left[\left(1+\frac{Z_{0e}}{Z_{c_e}}\right)(1-\delta_{0e}\cdot\delta_{L_e}e^{-2sL_e/a})A_{e1}-\frac{P_2(s)e^{-sL_2/a}}{1+Z_{02}/Z_{c_2}}\right],\tag{17}$$

其中，$\delta_{0e}=\dfrac{Z_{0e}-Z_{c_e}}{Z_{0e}+Z_{c_e}}$ 称为汇合管的始端反射系数.

将方程组(3)，(4)和(5)的第二个方程式代入边界条件(12)，则得

$$\frac{1}{Z_{c_1}}(A_{11}e^{-sL_1/a}-A_{12}e^{sL_1/a})+\frac{1}{Z_{c_2}}(A_{21}e^{-sL_1/a}-A_{22}e^{sL_1/a})=\frac{1}{Z_{c_e}}(A_{e1}-A_{e2}),$$

将关系式(13)，(14)和(15)代入上式，则得

$$\frac{1}{Z_{c_1}}\left[\frac{P_1(s)e^{-sL_1/a}}{1+Z_{01}/Z_{c_1}}-(1-\delta_{01}e^{-2sL_1/a})A_{12}e^{sL_1/a}\right]$$
$$+\frac{1}{Z_{c_2}}\left[\frac{P_2(s)e^{-sL_2/a}}{1+Z_{02}/Z_{c_2}}-(1-\delta_{02}e^{-2sL_2/a})e^{sL_2/a}A_{22}\right]=\frac{1}{Z_{c_e}}(1-\delta_{L_e}e^{-2sL_e/a})A_{e1}.$$

将(16)，(17)式代入上式，则得

$$\frac{1}{Z_{c_1}}\left\{\frac{P_1(s)e^{-sL_1/a}}{1+Z_{01}/Z_{c_1}}-\left[\left(1+\frac{Z_{0e}}{Z_{c_e}}\right)(1-\delta_{0e}\cdot\delta_{L_e}e^{-2sL_e/a})A_{e1}\right.\right.$$
$$\left.\left.-\frac{P_1(s)e^{-sL_1/a}}{1+Z_{01}/Z_{c_1}}\right]\left(\frac{1-\delta_{01}e^{-2sL_1/a}}{1+\delta_{01}e^{-2sL_1/a}}\right)\right\}+\frac{1}{Z_{c_2}}\left\{\frac{P_2(s)e^{-sL_2/a}}{1+Z_{02}/Z_{c_2}}\right.$$
$$\left.-\left[\left(1+\frac{Z_{0e}}{Z_{c_e}}\right)(1-\delta_{0e}\cdot\delta_{L_e}e^{-2sL_e/a})A_{e1}-\frac{P_2(s)e^{-sL_2/a}}{1+Z_{02}/Z_{c_2}}\right]\left(\frac{1-\delta_{02}e^{-2sL_2/a}}{1+\delta_{02}e^{-2sL_2/a}}\right)\right\}$$
$$=\frac{1}{Z_{c_e}}(1-\delta_{L_e}e^{-2sL_e/a})A_{e1.}$$

于是，我们获得

$$A_{e1}=\frac{1}{(1+Z_{0e}/Z_{c_e})(1-\delta_{0e}\delta_{L_e}e^{-2sL_e/a})}\frac{A(s)}{B(s)},\tag{18}$$

$$A(s)=\frac{2P_1(s)e^{-sL_1/a}}{(Z_{01}+Z_{c_1})(1+\delta_{01}e^{-2sL_1/a})}+\frac{2P_2(s)e^{-sL_2/a}}{(Z_{02}+Z_{c_2})(1+\delta_{02}e^{-2sL_2/a})},\tag{19}$$

$$B(s)=\frac{1}{Z_{c_1}}\left(\frac{1-\delta_{01}e^{-2sL_1/a}}{1+\delta_{01}e^{-2sL_1/a}}\right)+\frac{1}{Z_{c_2}}\left(\frac{1-\delta_{02}e^{-2sL_2/a}}{1+\delta_{02}e^{-2sL_2/a}}\right)+\frac{1}{Z_{0e}+Z_{c_e}}\left(\frac{1-\delta_{L_e}e^{-2sL_e/a}}{1-\delta_{0e}\delta_{L_e}e^{-2sL_e/a}}\right).\tag{20}$$

将关系式(18)代入(16),(17),我们获得

$$A_{12}=\frac{e^{-sL_1/a}}{1+\delta_{01}e^{-2sL_1/a}}\left[\frac{A(s)}{B(s)}-\frac{P_1(s)e^{-sL_1/a}}{1+Z_{01}/Z_{c_1}}\right], \tag{21}$$

$$A_{22}=\frac{e^{-sL_2/a}}{1+\delta_{02}e^{-2sL_2/a}}\left[\frac{A(s)}{B(s)}-\frac{P_2(s)e^{-sL_2/a}}{1+Z_{02}/Z_{c_2}}\right], \tag{22}$$

将它们代入关系式(13),(14)和(15),我们获得

$$A_{11}=\frac{P_1(s)}{1+Z_{01}/Z_{c_1}}+\frac{\delta_{01}e^{-sL_1/a}}{(1+\delta_{01}e^{-2sL_1/a})}\left[\frac{A(s)}{B(s)}-\frac{P_1(s)e^{-sL_1/a}}{1+Z_{01}/Z_{c_1}}\right], \tag{23}$$

$$A_{21}=\frac{P_2(s)}{1+Z_{02}/Z_{c_2}}+\frac{\delta_{02}e^{-sL_2/a}}{(1+\delta_{02}e^{-2sL_2/a})}\left[\frac{A(s)}{B(s)}-\frac{P_1(s)e^{-sL_2/a}}{1+Z_{02}/Z_{c_2}}\right], \tag{24}$$

$$A_{e2}=\frac{\delta_{L_e}e^{-2sL_e/a}}{(1+Z_{0e}/Z_{c_e})(1-\delta_{0e}\delta_{L_e}e^{-2sL_e/a})}\cdot\frac{A(s)}{B(s)}. \tag{25}$$

这里特别要指出的是,在我们所讨论的情况下,汇合管(e)终端阻抗 Z_{L_e} 是由烧结炉料层的透风性所决定的. 当烧结料层太厚、太湿或温度太高使料层透风性十分差时,$Z_{L_e}\gg Z_{c_e}$,即 $\delta_{L_e}\doteq 1$; 当料层又薄又干,透风性十分好,甚至跑空料而风箱伐又没关上时,则 $Z_{L_e}\ll Z_{c_e}$,即 $\delta_{L_e}\doteq -1$.

三、对双风机并联时抢风条件的分析

1. 造成抢风的充要条件

在双风机并联工作过程中,有时双风机会出现振荡性抢风现象. 在这里,我们将研究造成抢风的充要条件及避免抢风的途径.

我们知道,在双风机并联的工作过程中,出现抢风时,在管道的汇合处,抢到风的那一根管道抢走了汇合管道的全部流量,而被抢走风的另一根管道的流量为零或几乎为零. 这意味着这根被抢走风的管道的终端动态流阻为无限大. 不妨假设Ⅰ号管道被抢走了风. 因此,由于

$$Z_{L_1}=\frac{P_1(L_1,\ s)}{Q_1(L_1,\ s)}=\infty$$

等价于

$$Q_1(L_1, s)=0 \text{ 或 } O(\varepsilon), \tag{1}$$

所以,关系式(1)是Ⅰ号管道被抢走风的充要条件. 此处, $O(\varepsilon)$表示其绝对值与正的小量 ε 同一量级.

但是,根据前一节的(3),(21)和(23)式,则

$$\begin{aligned} Z_{c_1}Q_1(L_1, s) &= A_{11}e^{-sL_1/a}-A_{12}e^{sL_1/a}=\frac{P_1(s)e^{-sL_1/a}}{1+Z_{01}Z_{c_1}}+\frac{\delta_{01}e^{-2sL_1/a}}{1+\delta_{01}e^{-2sL_1/a}}\left[\frac{A(s)}{B(s)}\right.\\ &\left.-\frac{P_1(s)e^{-sL_1/a}}{1+Z_{01}/Z_{c_1}}\right]-\frac{1}{1+\delta_{01}e^{-2sL_1/a}}\left[\frac{A(s)}{B(s)}-\frac{P_1(s)e^{-sL_1/a}}{1+Z_{01}/Z_{c_1}}\right]\\ &=\frac{2P_1(s)e^{-sL_1/a}}{(1+Z_{01}/Z_{c_1})(1+\delta_{01}e^{-2sL_1/a})}-\frac{A(s)}{B(s)}\left(\frac{1-\delta_{01}e^{-2sL_1/a}}{1+\delta_{01}e^{-2sL_1/a}}\right), \end{aligned}$$

因此, $Q_1(L_1, s)=O(\varepsilon)$ 等价于

$$2P_1(s)e^{-sL_1/a}\cdot B(s)/(1+Z_{01}/Z_{c_1})=(1-\delta_{01}e^{-2sL_1/a})A(s)+O(\varepsilon). \tag{2}$$

将第二节中 $A(s)$, $B(s)$的关系式(19),(20)代入上式,则

$$\begin{aligned} &\frac{2P_1(s)e^{-sL_1/a}}{1+Z_{01}/Z_{c_1}}\left[\frac{1}{Z_{c_1}}\left(\frac{1-\delta_{01}e^{-2sL_1/a}}{1+\delta_{01}e^{-2sL_1/a}}\right)+\frac{1}{Z_{c_2}}\left(\frac{1-\delta_{02}e^{-2sL_2/a}}{1+\delta_{02}e^{-2sL_2/a}}\right)+\right.\\ &\left.\frac{1}{Z_{0e}+Z_{c_e}}\left(\frac{1-\delta_{L_e}e^{-2sL_e/a}}{1-\delta_{0e}\delta_{L_e}e^{-2sL_e/a}}\right)\right]=(1-\delta_{01}e^{-2sL_1/a})\left[\frac{2P_1(s)e^{-sL_1/a}}{(Z_{01}+Z_{c_1})(1+\delta_{01}e^{-2sL_1/a})}\right.\\ &\left.+\frac{2P_2(s)e^{-sL_1/a}}{(Z_{02}+Z_{c_2})(1+\delta_{02}e^{-2sL_2/a})}\right]+O(\varepsilon), \end{aligned}$$

亦即

$$\begin{aligned} &\frac{2P_1(s)e^{-sL_1/a}}{1+Z_{01}/Z_{c_1}}\left[\frac{1}{Z_{c_2}}\left(\frac{1-\delta_{02}e^{-2sL_2/a}}{1+\delta_{02}e^{-2sL_2/a}}\right)+\frac{1}{Z_{0e}+Z_{c_e}}\left(\frac{1-\delta_{Le}e^{-2sL_e/a}}{1-\delta_{0e}\delta_{L_e}e^{-2sL_e/a}}\right)\right]\\ &=\frac{2P_2(s)e^{-sL_2/a}}{Z_{02}+Z_{c_2}}\left(\frac{1-\delta_{01}e^{-2sL_1/a}}{1+\delta_{02}e^{-2sL_2/a}}\right)+O(\varepsilon). \end{aligned} \tag{3}$$

当出现抢风时,根据风机的特性曲线,我们知道,此时成立着(设Ⅰ号管失风,Ⅱ号管抢到风)

$$Z_{01}/Z_{02} \ll 1. \tag{4}$$

同时,2 号风机超载,有

$$Z_{02} \doteq \infty, \ \delta_{02} \doteq 1. \tag{5}$$

因而,由关系式(3),则得

$$P_1(s)\mathrm{e}^{-sL_1/a}\left[\frac{1}{Z_{c_2}}\left(\frac{1-\mathrm{e}^{-2sL_2/a}}{1+\mathrm{e}^{-2sL_2/a}}\right)+\frac{1}{Z_{0e}+Z_{c_e}}\left(\frac{1-\delta_{L_e}\mathrm{e}^{-2sL_e/a}}{1-\delta_{0e}\delta_{L_e}\mathrm{e}^{-2sL_e/a}}\right)\right]=O(\varepsilon). \tag{6}$$

由于 $P_1(s)\neq 0$, 因此,上述(6)式等价于

$$\begin{aligned}&(Z_{0e}+Z_{c_e})(1-\mathrm{e}^{-2sL_2/a})(1-\delta_{0e}\delta_{L_e}\mathrm{e}^{-2sL_e/a})\\&+Z_{c_2}(1+\mathrm{e}^{-2sL_2/a})(1-\delta_{L_e}\mathrm{e}^{-2sL_e/a})=O(\varepsilon).\end{aligned} \tag{7}$$

我们不妨假设

$$A_1=A_2=A_e, \ Z_{0e}=0. \tag{8}$$

因而

$$Z_{c_1}=Z_{c_2}=Z_{c_e}, \quad \delta_{0e}=-1.$$

于是,关系式(7)即等价于

$$(1+\delta_{L_e}\mathrm{e}^{-2sL_e/a})(1-\mathrm{e}^{-2sL_2/a})+(1-\delta_{L_e}\mathrm{e}^{-2sL_e/a})(1+\mathrm{e}^{-2sL_2/a})=O(\varepsilon),$$

亦即

$$1-\delta_{L_e}\mathrm{e}^{-2s(L_e+L_2)/a}=O(\varepsilon). \tag{9}$$

可见,当传输管道为无限长或其终端阻抗恰好等于其管道的特性阻抗时,风机不可能出现抢风现象.

事实上,当 $L_e=+\infty$ 时,则 $\mathrm{e}^{-2sL_e/a}=0$;当 $Z_{L_e}=Z_{c_e}$ 时,则 $\delta_{L_e}=0$. 因而,若 $Q_1(L_1, s)=0$ 或 $O(\varepsilon)$ 时,则应成立

$$1=0 \text{ 或 } O(\varepsilon),$$

但这是不可能的,这说明假设 $Q_1(L_1, s)=0$ 或 $\mathrm{O}(\varepsilon)$ 是不成立的. 故当 $L_e=+\infty$ 或 $Z_{L_e}=Z_{c_e}$ 时,风机不可能出现抢风现象.

当汇合传输管道为有限长时,我们知道

$$s=\mathrm{j}\omega=2\pi f\mathrm{j},\quad \mathrm{j}=\sqrt{-1},\tag{10}$$

f 是分支管路的固有频率，因而，关系式(9)等价于

$$1-\delta_{L_e}\mathrm{e}^{-\mathrm{j}4\pi f(L_e+L_2)/a}=O(\varepsilon).\tag{11}$$

它等价于

$$\begin{cases}\sin[4\pi f(L_e+L_2)/a]=O(\varepsilon),\\ 1-\delta_{L_e}\cdot\cos[4\pi f(L_e+L_2)/a]=O(\varepsilon).\end{cases}\tag{12}$$

于是，当 $f\doteq\dfrac{a}{4(L_e+L_2)}\cdot 2m\ (m=1,\ 2,\ \cdots)$ 时，则得

$$1-\delta_{L_e}=O(\varepsilon)，即\ \delta_{L_e}=1-O(\varepsilon);\tag{13}$$

当 $f\doteq\dfrac{a}{4(L_e+L_2)}(2m+1),\ m=1,\ 2,\ 3,\ \cdots$ 时，则得

$$1+\delta_{L_e}=O(\varepsilon)，即\ \delta_{L_e}=-1+O(\varepsilon).\tag{14}$$

由于风机对管内的扰动频率是可变的，因此，$Q_1(L_1,\ s)=0$ 或 $O(\varepsilon)$ 的充要条件为

$$\delta_{L_e}=1-O(\varepsilon)，即\ Z_{c_e}/Z_{L_e}=O(\varepsilon),\tag{15}$$

或者

$$\delta_{L_e}=-1+O(\varepsilon)，即\ Z_{L_e}/Z_{c_e}=O(\varepsilon).\tag{16}$$

由此可见，汇合传输管道的终端阻抗 Z_{L_e} 相对于其管道的特性阻抗 Z_{c_e} 来说太大或太小都会造成抢风现象. 所以，避免抢风的关键在于使传输管道的终端阻抗不要太大或太小. 因而当 $\delta_{L_e}\doteq 1$ 即 $Z_{c_e}/Z_{L_e}\doteq 0$ 时，使Ⅱ号管抢到风、Ⅰ号管失风的频率为

$$f\doteq\frac{a}{4(L_e+L_2)}\cdot 2m\quad (m=1,\ 2,\ \cdots);\tag{17}$$

当 $\delta_{L_e}\doteq -1$ 即 $Z_{c_e}/Z_{L_e}=\infty$ 时，使Ⅱ号管抢到风、Ⅰ号管失风的频率为

$$f\doteq\frac{a}{4(L_e+L_2)}\cdot(2m+1)\quad (m=1,\ 2,\ 3,\ \cdots).\tag{18}$$

由于风机转速并非绝对恒定，特别是刚起动时，风机从转速为零逐步增大到

额定转速，此时风机对管内气流的扰动频率完全可能落入关系式(17)，(18)所指出的抢风频率范围之内.这样，会否出现抢风，关键是看汇合管(e)的终端反射系数 δ_{L_e} 是否满足关系式(15)，(16).显然，料层透风性十分差，以致有 $Z_{L_e} \gg Z_{ce}$，或料层透风性十分好又未关上风箱伐，以致有 $Z_{L_e} \ll Z_{ce}$，均可能出现抢风.

2. 抢风的切换问题

前面我们假设是Ⅰ号管被抢走了风，那怎么又会造成抢风的切换呢？即怎么会从 $Q_1(L_1, s)=0$ 或 $O(\varepsilon)$ 转化成 $Q_2(L_2, s)=0$ 或 $O(\varepsilon)$ 呢？

根据第二节中的方程组(4)和关系式(22)，(23)，则

$$\begin{aligned} Z_{c_2}Q_2(L_2, s) &= \frac{2P_2(s)e^{-sL_2/a}}{(1+Z_{02}/Z_{c_2})(1+\delta_{02}e^{-2sL_2/a})} - \frac{A(s)}{B(s)}\left(\frac{1-\delta_{02}e^{-2sL_2/a}}{1+\delta_{02}e^{-2sL_2/a}}\right) \\ &= \frac{2P_2(s)e^{-sL_2/a}}{(1+Z_{02}/Z_{c_2})(1+\delta_{02}e^{-2sL_2/a})B(s)}\left\{\frac{1}{Z_{c_1}}\left(\frac{1-\delta_{01}e^{-2sL_1/a}}{1+\delta_{01}e^{-2sL_1/a}}\right)\right. \\ &\quad + \frac{1}{Z_{c_2}}\left(\frac{1-\delta_{02}e^{-2sL_2/a}}{1+\delta_{02}e^{-2sL_2/a}}\right) + \frac{1}{Z_{0e}+Z_{c_e}}\left(\frac{1-\delta_{L_e}e^{-2sL_e/a}}{1-\delta_{0e}\delta_{L_e}e^{-2sL_e/a}}\right) \\ &\quad - \frac{(1+Z_{02}/Z_{c_2})(1-\delta_{02}e^{-2sL_2/a})}{2P_2(s)e^{-sL_2/a}}\left[\frac{2P_1(s)e^{-sL_1/a}}{(Z_{01}+Z_{c_1})(1+\delta_{01}e^{-2sL_1/a})}\right. \\ &\quad \left.\left. + \frac{2P_2(s)e^{-sL_2/a}}{(Z_{02}+Z_{c_2})(1+\delta_{02}e^{-2sL_2/a})}\right]\right\}. \end{aligned} \tag{19}$$

由于 $Z_{c_1}=Z_{c_2}=Z_{c_e}$，$L_1=L_2$，$Z_{0e}=0$，于是

$$\begin{aligned} Z_{c_2}Q_2(L_2, s) &= \frac{2P_2(s)e^{-sL_2/a}}{Z_{c_1}(Z_{02}+Z_{c_2})(1+\delta_{02}e^{-2sL_2/a})B(s)}\left\{\frac{1-\delta_{01}e^{-2sL_2/a}}{1+\delta_{01}e^{-2sL_2/a}}\right. \\ &\quad \left. + \frac{1-\delta_{L_e}e^{-2sL_e/a}}{1+\delta_{L_e}e^{-2sL_e/a}} - \frac{P_1(s)}{P_2(s)}\left(\frac{Z_{02}+Z_{c_2}}{Z_{01}+Z_{c1}}\right)\left(\frac{1-\delta_{02}e^{-2sL_2/a}}{1+\delta_{01}e^{-2sL_2/a}}\right)\right\}. \end{aligned} \tag{20}$$

由于当Ⅰ号管被抢走风时，成立

$$\begin{cases} \dfrac{P_1(s)}{P_2(s)} = \dfrac{P_1}{P_2} = \left(\dfrac{P_1\cdot Z_{02}}{P_2\cdot Z_{01}}\right)\left(\dfrac{Z_{01}}{Z_{02}}\right) = \left(\dfrac{Q_1}{Q_2}\right)\left(\dfrac{Z_{01}}{Z_{02}}\right), \\ Q_1/Q_2 \ll 1, \end{cases} \tag{21}$$

因而，$Q_2(L_2, s)=0$ 或 $O(\varepsilon)$ 等价于

$$\frac{1-\delta_{01}\mathrm{e}^{-2sL_2/a}}{1+\delta_{01}\mathrm{e}^{-2sL_2/a}}+\frac{1-\delta_{L_e}\mathrm{e}^{-2sL_e/a}}{1+\delta_{L_e}\mathrm{e}^{-2sL_e/a}}=0 \text{ 或 } O(\varepsilon). \tag{22}$$

由于随着Ⅰ号管被抢走风的时间的增长，1 号风机的空载程度越来越大，甚至完全空载，这意味着Ⅰ号管的始端动态流阻越来越小，以致 $Z_{01}=0$，即 $\delta_{01}=-1$，因此，关系式(22)即为

$$\frac{1+\mathrm{e}^{-2sL_2/a}}{1-\mathrm{e}^{-2sL_2/a}}+\frac{1-\delta_{L_e}\mathrm{e}^{-2sL_e/a}}{1+\delta_{L_e}\mathrm{e}^{-2sL_e/a}}=0 \text{ 或 } O(\varepsilon). \tag{23}$$

于是得 $Q_2(L_2, s)=0$ 或 $O(\varepsilon)$，等价于

$$1+\delta_{L_e}\mathrm{e}^{-2s(L_e+L_2)/a}=0 \text{ 或 } O(\varepsilon). \tag{24}$$

它等价于

$$\begin{cases}\sin[4\pi f(L_e+L_2)/a]\doteq 0,\\ 1+\delta_{L_e}\cdot\cos[4\pi f(L_e+L_2)/a]\doteq 0.\end{cases} \tag{25}$$

由此可见，如果 $\delta_{L_e}\doteq 1$，则当

$$f\doteq\frac{a}{4(L_e+L_2)}(2m+1)\quad(m=1, 2, 3, \cdots) \tag{26}$$

时，关系式(24)成立；如果 $\delta_{L_e}\doteq -1$，则当

$$f\doteq\frac{a}{4(L_e+L_2)}\cdot 2m\quad(m=1, 2, 3, \cdots) \tag{27}$$

时，关系式(24)式成立，因而 $Q_2(L_2, s)=0$ 或 $O(\varepsilon)$.

由此可见，随着Ⅰ号管被抢走风的时间的增长，以致 1 号风机完全空载即达到 $Z_{01}=0$ 时，如果 $\delta_{L_e}\doteq 1$，即 $Z_{L_e}/Z_{c_e}\doteq\infty$，则当风机的扰动频率 f 满足关系式(26)时，抢风就会出现切换，即从 2 号风机抢到风、1 号风机被抢走了风转化成 1 号风机抢到风、而 2 号风机被抢走了风；如果 $\delta_{L_e}\doteq -1$，即 $Z_{L_e}/Z_{c_e}\doteq 0$，则当风机的扰动频率 f 满足关系式(27)时，抢风就出现切换.

综上所述，如果 $Z_{L_e}/Z_{c_e}\doteq\infty$，若 1 号风机失风、2 号风机得风的频率为

$$f\doteq\frac{a}{4(L_e+L_2)}\cdot 2m\quad(m=1, 2, 3, \cdots), \tag{28}$$

则 2 号风机失风、1 号风机得风的频率为

$$f \doteq \frac{a}{4(L_e + L_2)}(2m+1) \quad (m=1,\ 2,\ 3,\ \cdots), \tag{29}$$

反之亦然. 如果 $Z_{L_e}/Z_{c_e} \doteq 0$，若 1 号风机失风、2 号风机得风的频率 f 为

$$f \doteq \frac{a}{4(L_e + L_2)}(2m+1) \quad (m=1,\ 2,\ \cdots), \tag{30}$$

则 2 号风机失风、1 号风机得风的频率 f 为

$$f \doteq \frac{a}{4(L_e + L_2)} \cdot 2m \quad (m=1,\ 2,\ \cdots), \tag{31}$$

反之亦然.

我们不难证明上述结论对单风机来说也是真实的. 不过此时的抢风表征为风机时而空载、时而超载，如此交替进行. 事实上，在第二节中的(19)，(20)，(21)和(23)式先令 $A_2=0$，即 $Z_{c_2}=\infty$，$Z_{0e}=0$，则得

$$Z_{L_1'} = P_1(L_1',\ s)/Q_1(L_1',\ s) = (1+\delta_{L_e'}\mathrm{e}^{-2sL_e'/a})/(1-\delta_{L_e'}\mathrm{e}^{-2sL_e'/a}). \tag{32}$$

由于风机空载等价于 $Z_{L_1'}=\infty$，即 $1-\delta_{L_e'}\mathrm{e}^{-2sL_e'/a}=0$，风机超载等价于 $Z_{L_1'}=0$，即 $1+\delta_{L_e'}\mathrm{e}^{-2sL_e'/a}=0$，又由于 $L_2'=L_1'=0$，因此，$L_e'=L_e+L_1$，因而风机空载等价于 $1-\delta_{L_e}\mathrm{e}^{-2sL_e/a}=0$，风机超载等价于 $1+\delta_{L_e}\mathrm{e}^{-2sL_e}=0$. 故上述结论对单风机的抢风也是真实的.

四、抢风现象的讨论

从抢风与抢风切换条件的分析中不难发现，在不改动机组性能的情况下，会否出现抢风将由汇合管(e)终端料层透风性与风机对气流的扰动频率两方面因素所决定.

从料层情况看，当料层湿度、温度与厚度调整得恰当，使料层透风性恰巧达到或接近阻抗匹配，即使得汇合管(e)终端流阻 Z_{L_e} 满足

$$\xi = Z_{L_e}/Z_{ce} \approx 1,$$

这时，风机对管内气流的扰动波到达管(e)终端后没有反射波存在，因而绝对不会出现抢风，管内气流的脉动很小. 自然，作为检测管中气流压力或流量的仪表

指针摆动也必将很小.

当料层状况改变,使管(e)终端不匹配,此时,随着比值$\xi=Z_{L_e}/Z_{ce}$偏离1越大,风机对管内气流的扰动波到达管(e)终端后,其反射波强度也越大,因而,管路中驻波成分越来越大,管内气流的脉动也越来越大,此时检测仪表的指针摆动也越来越大.

当料层透风性使得管(e)终端达到严重不匹配时,即比值$\xi=Z_{L_e}/Z_{ce}$趋于零或趋于无限大时,此时风机对管内的扰动波到达管(e)终端后几乎全部反射回到管中,反射流量波与入射流量波位相恰巧相反,在管内形成驻波. 管内某些点(特别是波腹上),压力或流量有大幅度的起落变化,反映在测试仪表上,是指针大幅度来回摆动,伴随而来的是管路中将发出撼人心弦的吼叫声.

此时,若风机对气流的扰动频率恰巧扫过抢风条件或切换条件所示的频率时,上述驻波将表征为两台风机的来回抢风. 时而风机2超载,风机1空载;继而却是风机1超载,风机2空载. 超载风机所对应的支管其始端是流量波腹,空载风机所对应支管的始端却是流量波节. 波腹、波节交替出现. 这样,风机将发生喘振,并使机组强烈振动. 同时,它又将反过来对管内气流产生进一步的扰动,从而加剧了两风机的抢风. 只要条件(28),(30)与(29),(31)满足,这将往复不止,永远继续下去.

由此可见,在造成抢风的两方面因素中,反映管(e)终端透风性的δ_{L_e}是决定性的. 当管(e)终端并非极端不匹配(即$|\delta_{L_e}|$并不接近1)时,即使风机扰动频率等于条件(28),(30)与(29),(31)所指的频率,管内也绝不会出现抢风. 反之,即使原来风机正常运行,对管内的扰动频率不满足条件(28),(30)与(29),(31)所指的频率,此时管(e)终端开始极端不匹配,管内几乎全部是驻波,这就可能导致风机喘振,机组振动. 它将反过来对管内气流产生进一步的扰动. 风机喘振、机组振动对气流的扰动频率完全可能与抢风频率及失风频率一致,这时管内气流的谐振将以抢风的形式表征出来.

五、避免抢风的方法

上面关于抢风及其切换条件的分析,不但可以用来解释双机并联时的抢风现象,更重要的是可以从中寻找到避免抢风、保证风机正常运行的方法.

既然抢风会否发生是由管(e)终端反射系数δ_{L_e}与风机对气流的扰动频率两个因素所决定的,因而,制止抢风自然可以从这两方面入手.

调整料层透风性，不使管(e)终端处于极不匹配的情况下(即 $|\delta_{L_e}| \doteq 1$) 工作，就可避免抢风. 梅山工人从实践中总结出来的如下几条防止抢风的措施与上面所得结论是一致的：

“跑空料时一定要关上风箱伐.”这原来是为了防止冷空气进入，影响烧结而制定的措施，但它却有效地避开管(e)终端因透风太好，导致 $\delta_{L_e} \doteq -1$ 而抢风的可能. 因此，在烧结过程中，双机并联若出现抢风只能是料层供风不足，导致 $\delta_{L_e} \doteq 1$ 所引起的.

“合理调整料层厚度、湿度、温度，使料层透风性适当，避免供风不足.”这是关键的措施. 有了上述两条，则完全避免了管(e)终端处于极不匹配(即 $|\delta_{L_e}| \doteq 1$) 情况下工作，因而也避免了抢风的可能.

“风机空载启动，即关好风机伐启动，待风机运转正常后再开风机伐.”这个措施原来是为防止启动电流过大制定的，但它在一定程度上也减少了风机启动时，转速从低到高，极易扫过 $\dfrac{a}{4(L_e+L_2)}$ 的整数倍的某些频率而可能产生抢风的机会.

“若出现抢风征兆，立即关掉一台风机，待料层调整适当后，再开这台风机.”既然在烧结时抢风必是供风不足导致 $\delta_{L_e} \doteq 1$ 而引起的，那么此时立刻关掉一台风机，管内流量显著减少. 由于管(e)终端流阻的非线性，此时所对应的流阻 Z_{L_e} 也将显著减小. 这就是说，即使料层状态没来得及改变，但由于关一台风机，流量显著下降，会使管(e)终端 Z_{L_e} 下降. 从而关系式 $\delta_{L_e} \doteq 1$(或 $Z_{L_e} \gg Z_{ce}$) 不成立，抢风即可制止. 待料层调整适当，供风充足时，再开这台风机，就可避免抢风.

必须指出，正如第三节中所证明的那样，上面对双机并联时的抢风所作的论述也完全适用于单机工作的情况. 双机并联时，管内谐振表征为双风机的交替抢风. 对单机来说，若料层透风性使得管(e)终端极端不匹配，管内也会产生谐振，风机喘振，这时这台风机将时而超载，继而空载，往复交替. 其中当 $\delta_{L_e} \doteq 1$，即 $Z_{L_e}/Z_{ce} \doteq \infty$ 时，风机扰动频率是 $a/4L$(这里 L 为总管长)的偶数倍时，风机表现为空载；当扰动频率是 $a/4L$ 的奇数倍时，风机表现为超载. 当 $\delta_{L_e} \doteq -1$，即 $Z_{L_e}/Z_{ce} \doteq 0$ 时，则相反. 显然，我们同样可以通过调整料层的透风性，使管终端避免在极端不匹配情况下工作，即可防止这种现象产生.

梅山工程指挥部烧结厂技术组对本文工作曾给予热情支持，特此致谢.

单叶亚纯函数的系数*

任福尧

1. 设 Σ 表示在区域 $1<|\zeta|<\infty$ 上正则且单叶的函数

$$F(\zeta)=\zeta+b_0+b_1\zeta^{-1}+b_2\zeta^{-2}+\cdots$$

的全体所组成的函数族，Σ' 表示函数

$$g(\zeta)=\zeta+b_1\zeta^{-1}+b_2\zeta^{-2}+\cdots\in\Sigma$$

的全体所组成的子族. 设 $g(\zeta)\in\Sigma'$，令

$$G(w)=w+c_1w^{-1}+c_2w^{-2}+\cdots$$

是 $g(\zeta)$ 的逆函数. 已知 $|c_1|=|b_1|\leqslant 1$，$|c_2|=|b_2|\leqslant 2/3$，Springer[3] 证明了 $|c_3|\leqslant 1$ 并猜测

$$|c_{2k-1}|\leqslant (2k-2)!\ /k!\ (k-1)!\qquad (k=1,\ 2,\ 3,\ \cdots).$$

1977 年 Kubota[1] 证明了 $k=3$，4 和 5 时，Springer 猜测是真实的.

本文证明了下列定理.

定理 1 设 $g(\zeta)=\zeta+b_1\zeta^{-1}+b_2\zeta^{-2}+\cdots\in\Sigma'$，函数 $G(w)=w+\sum_{n=1}^{\infty}c_nw^n$ 是 $g(\zeta)$ 的逆函数，则当 $k=6$，7 时上述 Springer 猜测是成立的，即

$$|c_{11}|\leqslant 42,\quad |c_{13}|\leqslant 132.$$

同时重新简单地证明了 $k=1$，2，3，4 和 5 时，Springer 猜想是真的，等号对 $\zeta+$

* 原载《复旦大学学报(自然科学版)》，1978 年，第 4 期，93 - 96.

ζ^{-1} 的逆函数 $(1/2)w(1+\sqrt{1-4/w^2})$ 成立①.

定理 2 在定理 1 的假设下,又若 $b_2=0$,则当 $k=8, 9$ 时 Springer 猜想

$$|c_{15}|\leqslant 429, \quad |c_{17}|\leqslant 1\,430$$

是真实的,等号对 $\zeta+\zeta^{-1}$ 的逆函数成立.

定理 3 设 $g(\zeta)=\zeta+\sum_{k=1}^{\infty}b_{2k-1}\zeta^{-(2k-1)}\in\Sigma'_{\text{odd}}$,令 $G(w)=w+\sum_{k=1}^{\infty}c_{2k-1}w^{-(2k-1)}$ 是 $g(\zeta)$ 的逆函数,则对一切正整数 k,成立着

$$|c_{2k-1}|\leqslant(2k-2)!\,/k!\,(k-1)!,$$

等号对 $\zeta+\zeta^{-1}$ 的逆函数成立①.

2. 定理 2 的证明.

引理 1(Fitzgerald)[2] 设 $f(z)=z+a_2z^2+a_3z^3+\cdots$ 是在 $|z|<1$ 内正则且单叶的函数(即 $f\in S$),则对一切复数 $\alpha_1, \alpha_2, \cdots, \alpha_m (m=1, 2, \cdots)$ 及单位圆内的点列 $z_1, z_2, \cdots, z_m$ 成立着

$$\left|\sum_{\nu,\mu=1}^{m}\alpha_\nu\alpha_\mu\frac{f(z_\nu)f(z_\mu)}{z_\nu z_\mu}\frac{z_\nu-z_\mu}{f(z_\nu)-f(z_\mu)}\right|\leqslant\sum_{\nu,\mu=1}^{m}\alpha_\nu\bar{\alpha}_\mu\frac{1}{1-z_\nu\bar{z}_\mu}. \tag{1}$$

引理 2(面积定理) 设函数 $F(\zeta)=\zeta+\sum_{n=1}^{\infty}b_n\zeta^{-n}\in\Sigma$,则

$$\sum_{n=1}^{\infty}n|b_n|^2\leqslant 1. \tag{2}$$

现在,我们来证明定理 2. 对 $g(\zeta)\in\Sigma'$,函数 $f(z)=1/g(\zeta)\in S(z=1/\zeta)$,根据引理 1,得

$$\left|\sum_{\nu,\mu=1}^{m}\alpha_\nu\alpha_\mu\frac{\zeta_\nu-\zeta_\mu}{g(\zeta_\nu)-g(\zeta_\mu)}\right|\leqslant\sum_{\nu,\mu=1}^{m}\alpha_\nu\bar{\alpha}_\mu\frac{1}{1-(\zeta_\nu\bar{\zeta}_\mu)^{-1}} \tag{3}$$

对任何自然数 m 和复数列 $\{\alpha_\nu\}$ 及点列 $\{\zeta_\nu\}(|\zeta_\nu|>1;\ \nu=1, 2, \cdots, m)$ 成立.

因而,对 $g\in\Sigma'$ 的逆函数 $G(w)=w-\sum_{n=1}^{\infty}c'_nw^{-n}(c'_n=-c_n)$,

① 本文另稿投《中国科学》后,发现上述定理 1 和 3 与不久前发表的 Schober 的结果[4]相重,但作者不知道他的结果,本文是独立完成的.

$$\left|\sum_{\nu,\mu=1}^{m}\alpha_\nu\alpha_\mu\frac{G(w_\nu)-G(w_\mu)}{w_\nu-w_\mu}\right|\leqslant\sum_{\nu,\mu=1}^{m}\alpha_\nu\bar{\alpha}_\mu\frac{1}{1-(G(w_\nu)\overline{G(w_\mu)})^{-1}} \tag{4}$$

对任何自然数 m 和复数 $\{\alpha_\nu\}$ 及点列 $\{w_\nu=g(\zeta_\nu)\in g(|\zeta|>1)$ 成立. 将 $G(w)$ 的幂级数展开式代入上式,则得

$$\begin{aligned}&\Big|\sum_{\nu,\mu=1}^{m}\alpha_\nu\alpha_\mu\{1+c'_1(w_\nu w_\mu)^{-1}+c'_2(w_\nu+w_\mu)(w_\nu w_\mu)^{-2}+c'_3(w_\nu^2+w_\nu w_\mu\\&+w_\mu^2)(w_\nu w_\mu)^{-3}+c'_4(w_\nu^3+w_\nu^2w_\mu+w_\nu w_\mu^2+w_\mu^3)(w_\nu w_\mu)^{-4}+\cdots\\&+c'_{2k-1}(w_\nu^{2k-2}+w_\nu^{2k-3}w_\mu+\cdots+w_\nu^{k-1}w_\mu^{k-1}+\cdots+w_\nu w_\mu^{2k-3}+w_\mu^{2k-2})(w_\nu w_\mu)^{-(2k-1)}\\&+c'_{2k}(w_\nu^{2k-1}+w_\nu^{2k-2}w_\mu+\cdots+w_\nu^{2k-1-l}w_\mu^{l}+\cdots+w_\mu^{2k-1})(w_\nu w_\mu)^{-2k}+\cdots\}\Big|\\&\leqslant\sum_{\nu,\mu=1}^{m}\alpha_\nu\bar{\alpha}_\mu\Big\{1+\sum_{k=1}^{n}D_k\cdot(w_\nu\bar{w}_\mu)^{-k}+\sum_{\substack{s\neq t,\\ t\geqslant 2}}O(w_\nu^s\bar{w}_\mu^t)\Big\},\end{aligned} \tag{5}$$

其中,

$$\begin{aligned}&D_1=1,\ D_2=1,\ D_3=1+|c'_1|^2,\ D_4=|c'_2|^2+4|c'_1|^2+1,\\&D_5=|c'_3|^2+2\mathrm{Re}\{c'^2_1\bar{c}'_3\}+|c'_1|^4+4|c'_2|^2+9|c'_1|^2+1,\\&D_6=|c'_4|^2+4\mathrm{Re}\{c'_1c'_2\bar{c}'_4\}+4|c'_1c'_2|^2+4|c'_3|^2+12\mathrm{Re}\{c'^2_1\bar{c}'_3\}+9|c'_1|^4\\&\qquad+9|c'_2|^2+16|c'_1|^2+1,\end{aligned} \tag{6}$$

$$\begin{aligned}D_7=&|c'_5|^2+2\mathrm{Re}\{(c'^2_2+2c'_1c'_3)\bar{c}'_5\}+2\mathrm{Re}\{c'^3_1\bar{c}'_5\}+|c'^2_2+2c'_1c^1_3|^2\\&+2\mathrm{Re}\{(c'^2_2+2c'_1c'_3)\bar{c}'^3_1\}+|c'_1|^6+4|c'_4|^2+24\mathrm{Re}\{c'_1c'_2\bar{c}'_4\}\\&+36|c'_1c'_2|^2+9|c'_3|^2+36\mathrm{Re}\{c'^2_1\bar{c}'_3\}+36|c'_1|^4+16|c'_2|^2+25|c'_1|^2\\&+1,\end{aligned} \tag{7}$$

$$\begin{aligned}D_8=&|c'_6|^2+4\mathrm{Re}\{(c'_1c'_4+c'_2c'_3)\bar{c}'_6\}+6\mathrm{Re}\{c'^2_1c'_1\bar{c}'_6\}+4|c'_1c'_4+c'_2c'_3|^2\\&+12\mathrm{Re}\{\overline{c'^2_1c'_2(c'_1c'_4}+\bar{c}'_2\bar{c}'_3)+9|c'^2_1c'_2|^2+4|c'_5|^2+12\mathrm{Re}\{(c'^2_2\\&+2c'_1c'_3)\bar{c}'_5\}+9|c'^2_2+2c'_1c'_3|^2+16\mathrm{Re}\{c'^8_1\bar{c}'_5\}+24\mathrm{Re}\{\bar{c}^3_1(c'^2_2\\&+2c'_1c'_5)\}+16|c'_1|^6+9|c'_4|^2+72\mathrm{Re}\{c'_1c'_2\bar{c}'_4\}+144|c'_1c'_2|^2\\&+16|c'_3|^2+80\mathrm{Re}\{c'^2_1c'_3\}+100|c'_1|^4+25|c'_2|^2+36|c'_1|^2+1,\end{aligned} \tag{8}$$

$$\begin{aligned}D_9=&|c'_7|^2+2\mathrm{Re}\{(c'^2_3+2c'_1c'_5+2c'_2c'_4\}\bar{c}'_7\}+6\mathrm{Re}\{(c'^2_1c'_3+c'_1c'^2_2)\bar{c}'_7\}\\&+|c'^2_3+2c'_1c'_5+2c'_2c'_4|^2+6\mathrm{Re}\{(c'^2_1c'_3+c'_1c'^2_2)\overline{(c'^2_3+2c'_1c'_5+2c'_2c'_4)}\}\end{aligned}$$

$$
\begin{aligned}
&+2\mathrm{Re}\{c_1'^4\bar{c}_7'\}+9|c_1'^2c_3'+c_1'c_2'^2|^2+2\mathrm{Re}\{\bar{c}_1'^4(c_3'^2+2c_1'c_5'+2c_2'c_4')\}\\
&+6\mathrm{Re}\{\bar{c}_1'^4(c_1'^2c_3'+c_1'c_2'^2)\}+|c_1'|^8+4|c_6'|^2+24\mathrm{Re}\{(c_1'c_4'+c_2'c_8')\bar{c}_6'\}\\
&+36|c_1'c_4'+c_2'c_3'|^2+48\mathrm{Re}\{c_1'^2c_2'\bar{c}_6'\}+144\mathrm{Re}\{c_1'^2c_2'\overline{(c_1'c_4'+c_2'c_3')}\}\\
&+144|c_1'^2c_2'|^2+9|c_5'|^2+36\mathrm{Re}(\{c_2'^2+2c_1'c_2')\bar{c}_5'\}+36|c_2'^2+2c_1'c_3'|^2\\
&+60\mathrm{Re}\{c_1'^3\bar{c}_5'\}+120\mathrm{Re}\{\bar{c}_1'^3(c_2'^2+2c_1'c_3')\}+100|c_1'|^6+16|c_4'|^2\\
&+160\mathrm{Re}\{c_1'c_2'\bar{c}_4'\}+400|c_1'c_2'|^2+25|c_3'|^2+150\mathrm{Re}\{c_1'^2\bar{c}_3'\}\\
&+225|c_1'|^4+36|c_2|^2+49|c_1'|^2+1. \qquad (9)
\end{aligned}
$$

对给定的任一自然数 k，令

$$\alpha_\nu=\rho^k\beta_\nu,\ w_\nu=\rho x_\nu^{-1}(\nu=1,\ 2,\ \cdots,\ m;\ m>k), \qquad (10)$$

这里，$\rho>1$，$\{x_\nu\}$是位于正实数轴上紧靠零点且互不相同的 m 个点. 选取实数列 $\beta_1,\ \beta_2,\ \cdots,\ \beta_m$，使得对每个从零到 $m-1$ 的正整数 s 成立着

$$\sum_{\nu=1}^{m}\beta_\nu x_\nu^s=\begin{cases}1,\ s=k,\\0,\ s\neq k.\end{cases} \qquad (11)$$

根据 Vandermonde 行列式的性质，大家熟知，这是做得到的. 于是，从不等式(5)得

$$\left|c_{2k-1}\sum_{\nu,\ \mu=1}^{m}\alpha_\nu\alpha_\mu(w_\nu w_\mu)^{-k}+O(\rho^{-2})\right|\leqslant D_k\cdot\left[\sum_{\nu,\ \mu=1}^{m}\alpha_\nu\alpha_\mu(w_\nu w_\mu)^{-k}\right]+O(\rho^{-2}).$$

由于 $\sum\limits_{\nu,\ \mu=1}^{m}\alpha_\nu\alpha_\mu(w_\nu w_\mu)^{-k}=\left(\sum\limits_{\nu=1}^{m}\beta_\nu x_\nu^k\right)^2=1$，于上不等式中令 $\rho\to+\infty$便得

$$|c_{2k-1}|\leqslant D_k. \qquad (12)$$

当 $k=8$ 时，由于 $|c_2'|=|b_2|=0$，所以，由不等式(12)和 D_8 的表示式(8)，我们有

$$
\begin{aligned}
|c_{15}|\leqslant{}&|c_6'|^2+4\mathrm{Re}\{c_1'c_4'\bar{c}_6'\}+9|c_4'|^2+4|c_1'c_4'|^2+4|c_5'|^2+24\mathrm{Re}\{c_1'c_3'\bar{c}_5'\}\\
&+36|c_1'c_3'|^2+16\mathrm{Re}\{c_1'^3\bar{c}_5'\}+48\mathrm{Re}\{\bar{c}_1'^3c_1'c_3'\}+16|c_1'|^6+16|c_3'|^2\\
&+80\mathrm{Re}\{c_1'^2c_3'\}+100|c_1'|^4+36|c_1'|^2+1. \qquad (13)
\end{aligned}
$$

但是，由于 $|c_1'|\leqslant 1$，$|c_3'|\leqslant 1$，$|c_5'|\leqslant 2$，所以

$$|c_{15}|\leqslant 81+348|b_1|^2+13|c_4|^2+4|c_4c_6|+|c_6|^2. \qquad (14)$$

我们知道

$$c_4=-(b_4+3b_1b_2),\ c_6=-(b_6+5b_1b_4+10b_1^2b_2+5b_2b_3),$$

因而,在 $b_2=0$ 的条件下,利用不等式 $ab\leqslant\frac{1}{2}(a^2+b^2)$,便得

$$|c_4|=|b_4|,\ |c_6|^2=|b_6+5b_1b_4|^2\leqslant 6|b_6|^2+30|b_4|^2, \tag{15}$$

$$4|c_4c_6|\leqslant 4|b_4b_8|+20|b_4|^2\leqslant 2|b_6|^2+22|b_4|^2, \tag{16}$$

于是,由不等式(14)和引理 2,我们获得

$$\begin{aligned}|c_{15}|&\leqslant 81+348|b_1|^2+65|b_4|^2+8|b_6|^2\\&\leqslant 81+348(|b_1|^2+4|b_4|^2+6|b_6|^2)\\&\leqslant 429,\end{aligned}$$

这说明当 $k=8$ 时定理 2 的结论成立.

当 $k=9$ 时,在 $b_2=0$ 的条件下,由基本不等式(12)和 D_9 的表示式(9),我们有

$$\begin{aligned}|c_{17}|\leqslant{}&|c_7'|^2+2\mathrm{Re}\{(c_3'^2+2c_1'c_5')\bar{c}_1'\}+6\mathrm{Re}\{c_1'^2c_3'\bar{c}_7'\}+|c_3'^2+2c_1'c_5'|^2\\&+6\mathrm{Re}\{c_1'^2c_3'\overline{(c_3'^2+2c_1'c_5')}\}+2\mathrm{Re}\{c_1'^4\bar{c}_7'\}+9|c_1'^2c_3'|^2\\&+2\mathrm{Re}\{\bar{c}_1'^4(c_3'^2+2c_1'c_5')\}+6\mathrm{Re}\{\bar{c}_1'^4c_1'^2c_3'\}+|c_1'|^3+4|c_6'|^2\\&+24\mathrm{Re}\{c_1'c_4'\bar{c}_6'\}+36|c_1'c_4'|^2+9|c_5'|^2+72\mathrm{Re}\{c_1'c_3'\bar{c}_5'\}\\&+144|c_1'c_3'|^2+60\mathrm{Re}\{c_1'^3\bar{c}_5'\}+240\mathrm{Re}\{\bar{c}_1'^3c_1'c_3'\}+100|c_1'|^6\\&+16|c_4'|^2+25|c_3'|^2+150\mathrm{Re}\{c_1'^2\bar{c}_3'\}+255|c_1'|^4+49|c_1'|^2+1.\end{aligned}$$

由于 $|c_1'|\leqslant 1$, $|c_3'|\leqslant 1$, $|c_5'|\leqslant 2$, $|c_7'|\leqslant 5$, 则得

$$|c_{17}|\leqslant 1\,205+225|b_1|^2+4|c_6|^2+24|c_4c_6|+52|c_4|^2.$$

利用不等式(15),(16)和引理 2,我们获得

$$|c_{17}|\leqslant 1\,205+225|b_1|^2+304|b_4|^2+36|b_6|^2\leqslant 1\,430.$$

这说明当 $k=9$ 时定理 2 的结论成立. 定理 2 证毕.

同理,利用基本不等式(12)和引理 2 及 D_k 的表达式,容易知道,当 $k\leqslant 7$ 时 Springer 猜想是成立的.

◇参考文献◇

[1] Y. Kubota, *Kōdai Math. Sem. Rep.*, **28**, 253-261 (1977).

[2] C. H. Fitzgerald, *Arch. Rational Mech. Anal.*, **46**, 356-368 (1972).

[3] G. Springer, *Trans. Amer. Math. Soc.*, **70**, 421-450 (1951).

[4] G. Schober, *Proc. Amer. Mach. Soc.*, **67**, 111-116 (1977).

在第三项系数限制下单叶函数的 Bieberbach 猜测*

任福尧

摘　要　本文证明：若 $f(z)=z+\sum_{n=2}^{\infty}a_n z^n \in S$，则存在一单调上升数列 $\{K_n\}$ $(n\geqslant 7)$，且 $\lim\limits_{n\to\infty}K_n=K_\infty>2.449$，使得对一切自然数 $n\geqslant 7$ 和 $|a_3|\leqslant K_n$，

$$|a_k|<k,\quad k=n,\ n+1,\ \cdots.$$

特别是 $K_7=1.71$，$K_8=1.74$，$K_9=1.772$，$K_{20}=1.995$，对充分大的 N，它仅与常数 2.449 有关，而与 f 无关，$K_N>2.449$. 这就改进了目前所知的最好的估计[1].

一、引　言

1916 年，Bieberbach 猜想：设 S 是由在 $|z|<1$ 内单叶且解析的函数

$$f(z)=z+a_2z^2+a_3z^3+\cdots$$

的全体所成的函数族. 若 $f\in S$，则 $|a_n|\leqslant n$ 对一切 $n=2,\ 3,\ \cdots$ 成立；对所有 n，等号仅当 Koebe 函数 $K(z)=z/(1-z)^2$ 及其旋转成立. 我们已经知道，当 $n\leqslant 6$ 时，Bieberbach 猜想是成立的. 1974 年，G. Ehrig[1] 证明：若 $f\in S$，则存在一单调上升数列 $\{M_n\}$ $(n\geqslant 9)$，且 $\lim\limits_{n\to\infty}M_n=2.434$，使得对一切 $n\geqslant 9$ 和 $f\in S$，若 $|a_3|\leqslant M_n$，则

$$|a_n|<n.$$

* 原载《中国科学：数学专辑(I)》，1979 年，275－280.

特别地，当 $|a_3|\leqslant 1.032$ 时，$|a_n|<n$ 对 $n\geqslant 9$ 成立；当 $|a_3|\leqslant 1.703$ 时，$|a_n|<n$ 对 $n\geqslant 20$ 成立；当 $|a_3|\leqslant 2.434$ 时，$|a_n|<n$ 对 $n\geqslant N(f)$ 成立.

本文的目的在于改进这一最好的结果. 我们证明了下述定理.

定理 若 $f\in S$，则存在一单调上升的数列 $\{K_n\}$ $(n\geqslant 7)$，且 $\lim\limits_{n\to\infty}K_n>2.449$，使得对一切正整数 n 和 $f\in S$，当 $|a_3|\leqslant K_n$ 时，对一切 $k\geqslant n$ 成立

$$|a_k|<k, \tag{1}$$

其中 K_n 是下列方程式的解：

$$\psi_n(|a_3|)=\frac{(9-|a_3|^2-A_n)^2}{28+2|a_3|+3|a_3|^2+2D(|a_3|)-|a_3|^4}=B_n(\rho_0), \tag{2}$$

此处，$\rho_0=1.0657$，

$$B_n(\rho_0)=\frac{3\rho_0^2-1}{12}+(\rho_0^2-1)\left(\frac{1}{2n}-\frac{441}{n^4}\right)+\frac{3\rho_0^2-5}{12n^2}, \tag{3}$$

$$A_n=[\alpha_{n+1}(\rho_0)+4\alpha_n(\rho_0)+\alpha_{n-1}(\rho_0)]/n^2, \tag{4}$$

$$\alpha_{n+1}(\rho_0)\geqslant 0,\ \alpha_n(\rho_0)\geqslant 0,\ \alpha_{n-1}(\rho_0)\geqslant 0, \tag{5}$$

且

$$\begin{aligned}\alpha_{n+1}^2(\rho_0)=&(368-364\rho_0^2)+\frac{4}{3}\rho_0^2n^3+2(\rho_0^2+1)n^2+\frac{2}{3}(\rho_0^2+6)n\\&+\Big[(441.10-364\rho_0^2)+\frac{4}{3}(\rho_0^2+7)n^3+(2\rho_0^2+30)n^2\\&+\frac{2}{3}(\rho_0^2+142.84)n\Big]^{1/2},\end{aligned} \tag{6}$$

$$\begin{aligned}\alpha_n^2(\rho_0)=&(366-364\rho_0^2)+\frac{4}{3}\rho_0^2n^3+\frac{2}{3}n+\Big[(366.54-364\rho_0^2)\\&+\frac{4}{3}(\rho_0^2+7)n^3+2(\rho_0^2-1)n^2+\frac{2}{3}(\rho_0^2+94.84)n\Big]^{1/2},\end{aligned} \tag{7}$$

$$\begin{aligned}\alpha_{n-1}^2(\rho_0)=&366(1-\rho_0^2)+\frac{2}{3}\rho_0^2(2n^3-6n^2+7n)+\Big[(291.98-364\rho_0^2)\\&+\frac{4}{3}(\rho_0^2+7)n^3+2(\rho_0^2-17)n^2+\frac{2}{3}(\rho_0^2+142.84)n\Big]^{1/2},\end{aligned} \tag{8}$$

$$D(|a_3|)=\frac{16}{15}+\frac{452(1+|a_3|)}{180}+\frac{184(1+|a_3|)^2}{144}-\frac{140(1+|a_3|)^3}{576}. \quad (9)$$

特别地，当 $|a_3|\leqslant 1.71$ 时，$|a_k|<k$ 对一切 $k\geqslant 7$ 成立；当 $|a_3|\leqslant 1.74$ 时，$|a_k|<k$ 对一切 $k\geqslant 8$ 成立；当 $|a_3|\leqslant 1.772$ 时，$|a_k|<k$ 对一切 $k\geqslant 9$ 成立；当 $|a_3|\leqslant 1.995$ 时，$|a_k|<k$ 对一切 $k\geqslant 20$ 成立；当 $|a_3|\leqslant 2.449$ 时，$|a_k|<k$ 对一切 $k\geqslant N(f)$ 成立.

上述结果也改进了作者自己前不久获得的结果[2]：$K_7>1.6525$，$K_8>1.71$，$K_9>1.770$.

二、主要引理

引理 1(Fitzgerald[3]) 设 $f\in S$，则对任何自然数 l 和任何实数 x_1，x_2，…，x_l，我们有

$$\left(\sum_{p=1}^{l}|a_p|^2x_p\right)^2\leqslant\sum_{p=1}^{l}\sum_{q=1}^{l}\left(\sum_{k=1}^{p+q-1}\beta_k(p,q)|a_k|^2\right)x_px_q. \quad (10)$$

此处，$\beta_k(p,q)=\beta_k(q,p)$，且对 $p\leqslant q$，

$$\beta_k(p,q)=\begin{cases}p-q+k, & 当\ q-p+1\leqslant k\leqslant q,\\ p+q-k, & 当\ q+1\leqslant k\leqslant p+q-1,\\ 0, & 否则.\end{cases} \quad (11)$$

它等价于下述二次形式：

$$\sum_{p,q=1}^{l}a_{pq}x_px_q\geqslant 0, \quad (12)$$

此处，

$$a_{pq}=\sum_{k=1}^{p+q-1}\beta_k(p,q)|a_k|^2-|a_p|^2\cdot|a_q|^2. \quad (13)$$

引理 2(Ehrig[1]) 设 $f\in S$，$|a_k|=b_k$，则对一切自然数 n 成立

$$b_n^4\left[1+\frac{(q-b_3^2+T_n/b_n^2)^2}{1+2b_2^2+3b_3^2+2b_4^2+b_5^2-b_3^4}\right]\leqslant\sum_{k=1}^{n}kb_k^2+\sum_{k=n+1}^{2n-1}(2n-k)b_k^2, \quad (14)$$

其中,

$$T_n = b_{n+2}^2 + 2b_{n+1}^2 - 6b_n^2 + 2b_{n-1}^2 + b_{n-2}^2. \tag{15}$$

引理 3(Bishouty[4]) 设 $f \in S$, $t_n = b_{n+1}^2 - 2b_n^2 + b_{n-1}^2$, 则我们有

$$t_n^2 \leqslant 2 + 4\sum_{k=1}^{n-1} b_k^2 + 2b_n^2 + t_{2n}. \tag{16}$$

引理 4 设 $f \in S$,若 $7 \leqslant i \leqslant m$ 时, $b_i < \rho_0 j\ (\rho_0 > 1)$;当 $k \geqslant m+1$ 时, $b_k < k$, 则

$$t_{m:=k}^2 < 366 + \frac{2}{3}\rho_0^2(2k^3 + k - 546) + \Big[366 + \frac{2}{3}\rho_0^2(16k^3 + 2k - 546) + 62.56k + 0.54 - \frac{2}{3}(\rho_0^2 - 1)(14k^3 - 3k^2 + k)\Big]^{1/2} = \tilde{\alpha}_m^2(\rho_0), \tag{17}$$

$$t_{m+1:=k}^2 < 366 + \frac{2}{3}\rho_0^2(2k^3 + k - 546) - 2(\rho_0^2 - 1)k^2 + \Big[366 + \frac{2}{3}\rho_0^2(16k^3 + 2k - 546) + 62.56k + 0.54 - \frac{2}{3}(\rho_0^2 - 1)(14k^3 + 3k^2 + k)\Big]^{1/2} = \tilde{\alpha}_{m+1}^2(\rho_0), \tag{18}$$

$$t_{m-1:=k}^2 < 366 + \frac{2}{3}\rho_0^2(2k^3 + k - 546) + \Big[366 + \frac{2}{3}\rho_0^2(16k^3 + 2k - 546) + 62.56k + 0.54 - \frac{2}{3}(\rho_0^2 - 1)(14k^3 - 9k^2 - 11k - 6)\Big]^{1/2} = \tilde{\alpha}_{m-1}^2(\rho_0). \tag{19}$$

证 据引理 3,则有

$$t_m^2 \leqslant 2 + 4\sum_{k=1}^{m-1} b_k^2 + 2b_m^2 + t_{2m} < 2 + 4\sum_{k=1}^{6} k^2 + 4\rho_0^2 \sum_{k=7}^{m-1} k^2 + 2\rho_0^2 m^2 + t_{2m} = 366 + \frac{2}{3}(2m^3 + m - 546) + t_{2m}, \tag{20}$$

$$t_{2m}^2 \leqslant 2 + 4\sum_{k=1}^{2m-1} b_k^2 + 2b_{2m}^2 + t_{4m} < 2 + 4\sum_{k=1}^{6} k^2 + 4\rho_0^2 \sum_{k=7}^{m} k^2 + 4\sum_{k=m+1}^{2m-1} k^2 + 2(2m)^2 + t_{4m} = 366 + \frac{2}{3}\rho_0^2(2m^3 + 3m^2 + m - 546) + \frac{2}{3}(14m^3 - 3m^2 + m) + t_{4m} = 366 + \frac{2}{3}\rho_0^2(16m^3 + 2m - 546)$$

$$-\frac{2}{3}(\rho_0^2-1)(14m^3-3m^2+m)+t_{4m}. \tag{21}$$

根据 Pommerenke 的结果[5]，对一切 $k=2, 3, \cdots$,

$$-3.64<b_k-b_{k-1}<4.18,$$

则得

$$t_{4k}<62.56k+0.54. \tag{22}$$

于是，将(21)，(22)式代入(20)式，便得到不等式(17).

同理，据引理 3 和引理 4 的条件，我们有

$$t_{m+1}^2 \leqslant 2+4\sum_{j=1}^{m} b_j^2+2b_{m+1}^2+t_{2(m+1)}$$

$$<366+\frac{2}{3}\rho_0^2[m(m+1)(2m+1)-546]+2(m+1)^2+t_{2(m+1)},$$

$$t_{2m+2}^2 \leqslant 2+4\sum_{j=1}^{2m+1} b_j^2+2b_{2m+2}^2+t_{4(m+1)}$$

$$<366+\frac{2}{3}\rho_0^2[m(m+1)(2m+1)-546]+\frac{2}{3}[(2m+1)(2m+2)(4m+3)-m(m+1)(2m+1)]+8(m+1)^2+t_{4(m+1)},$$

于上二不等式中，令 $m+1=k$，便得

$$t_{m+1:=k}^2<366+\frac{2}{3}\rho_0^2(2k^3+k-546)-2(\rho_0^2-1)k^2+t_{2k},$$

$$t_{2(m+1)}^2<366+\frac{2}{3}\rho_0^2(2k^3-3k^2+k-546)+\frac{2}{3}(14k^3+3k^2+k)+t_{4k}$$

$$=366+\frac{2}{3}\rho_0^2(16k^3+2k-546)-\frac{2}{3}(\rho_0^2-1)(14k^3+3k^2+k)+t_{4k}.$$

于是，利用不等式(22)，由上二不等式便得不等式(18).

同理，据引理 3，且令 $m-1=k$，便得

$$t_{m-1}^2<366+\frac{2}{3}\rho_0^2[(m-2)(m-1)(2m-3)-546]+2\rho_0^2(m-1)^2$$

$$+t_{2(m-1)}=366+\frac{2}{3}\rho_0^2(2k^3+k-546)+t_{2(m-1)},$$

$$t_{2(m-1)}^2<366+\frac{2}{3}\rho_0^2[m(m+1)(2m+1)-546]+\frac{2}{3}[(2m-3)(2m-2)(4m-5)-m(m+1)(2m+1)]+8(m-1)^2+t_{4(m-1)}$$

$$=366+\frac{2}{3}\rho_0^2(16k^3+2k-546)-\frac{2}{3}(\rho_0^2-1)(14k^3-9k^2-11k-6)+t_{4k},$$

可见,不等式(19)是成立的.引理4得证.

引理5 设 $f\in S$, $T_m=b_{m+2}^2+2b_{m+1}^2-6b_m^2+2b_{m-1}^2+b_{m-2}^2$,又若当 $7\leqslant j\leqslant m$ 时, $b_j<\rho_0 j\ (\rho_0>1)$;当 $j\geqslant m+1$ 时, $b_j<j$,则我们有

$$T_m<\alpha_{m+1}(\rho_0)+4\alpha_m(\rho_0)+\alpha_{m-1}(\rho_0), \tag{23}$$

其中, $\alpha_{m+1}(\rho_0)\geqslant 0$, $\alpha_m(\rho_0)\geqslant 0$, $\alpha_{m-1}(\rho_0)\geqslant 0$,且分别由(6),(7)和(8)式所定义.

证 我们知道, $t_k=b_{k+1}^2-2b_k^2+b_{k-1}^2$,可见

$$T_m=t_{m+1}+4t_m+t_{m-1}.$$

从而,由引理4,我们获得

$$t_{m:=k}^2<\tilde{\alpha}_m^2(\rho_0),\ t_{m+1:=k}^2<\tilde{\alpha}_{m+1}^2(\rho_0),\ t_{m-1:=k}^2<\tilde{\alpha}_{m-1}^2(\rho_0).$$

但是, $\tilde{\alpha}_m^2(\rho_0)=\alpha_m^2(\rho_0)$, $\tilde{\alpha}_{m+1}(\rho_0)=\alpha_{m+1}^2(\rho_0)$, $\tilde{\alpha}_{m-1}^2(\rho_0)=\alpha_{m-1}^2(\rho_0)$,故引理5是真实的.

三、定理的证明

若定理不真,则存在一个函数 $f\in S$ 和自然数 m,使得 $m\geqslant n$,且 $b_m>m$. Hayman[6]证明了

$$\lim_{p\to\infty} b_p/p=\alpha\leqslant 1, \tag{24}$$

且等号仅对 Koebe 函数成立.于是,不妨假设

$$b_m>m;\text{对 } k=m+1,\ \cdots,\ b_k<k. \tag{25}$$

此外, D. Horowitz 证明了[7]:

$$\begin{aligned} b_n&\leqslant(1\,659\,164\,137/681\,080\,400)^{1/14}\cdot n\\ &<1.065\,7\cdot n\quad(n=2,\ 3,\ \cdots). \end{aligned} \tag{26}$$

因此, $b_j<\rho_0 j$, $\rho_0=1.065\,7$,对 $7\leqslant j\leqslant m$ 是成立的.

根据引理2,对 $f\in S$,我们有

$$b_m^4\left[1+\frac{(9-b_3^2+T_m/b_m^2)^2}{1+2b_2^2+3b_3^2+2b_4^2+b_5^2-b_3^4}\right]\leqslant\sum_{k=1}^{m}kb_k^2+\sum_{k=m+1}^{2m-1}(2m-k)b_k^2$$
$$<\frac{3\rho_0^2+11}{12}m^4+\frac{1}{2}(\rho_0^2-1)m^3+\frac{1}{12}(3\rho_0^2-5)m^2 \tag{27}$$
$$-441(\rho_0^2-1)=\theta_m(\rho_0).$$

根据引理 5 及 $b_m>m$，我们有

$$m^4\left[1+\frac{(9-b_3^2-A_m)^2}{1+2b_2^2+3b_3^2+2b_4^2+b_5^2-b_3^4}\right]<\theta_m(\rho_0), \tag{28}$$

倘若 $9-b_3^2-A_m>0$. 但是,这是容易证明的. 事实上,由于当 $m>6$ 时, $\alpha_{m+1}(\rho_0)/m^2$，$\alpha_m(\rho_0)/m^2$ 和 $\alpha_{m-1}(\rho_0)/m^2$ 是关于 m 单调下降的,因而,$A_m(\rho_0)$ 是关于 m 单调下降的. 因此,当 $m\geqslant 9$ 和 $b_3<2.546$ 时，$9-b_3^2-A_m>0$ 是成立的,当 $m=7, 8$ 且 $b_3\leqslant K_m$ 时，$9-b_3^2-A_m>0$ 是显然的.

由(28)式,显然有

$$\frac{(9-b_3^2-A_m)^2}{1+2b_2^2+3b_3^2+2b_4^2+b_5^2-b_3^4}<\theta_m(\rho_0)-1=B_m(\rho_0). \tag{29}$$

但是，Ehrig[1]证明了

$$1+2b_2^2+3b_3^2+2b_4^2+b_5^2-b_3^4\leqslant 28+2b_3+3b_3^2+2D(b_3)-b_3^4, \tag{30}$$

其中 $D(b_3)$ 由(3)式所定义. 因而,当 $b_3\leqslant 2.449$ 时,我们有

$$\psi_m(b_3)=\frac{(9-b_3^2-A_m)^2}{28+2b_3+3b_3^2+2D(b_3)-b_3^4}<B_m(\rho_0). \tag{31}$$

但是,根据 K_m 的定义，$K_m<2.449$，且

$$\psi_m(K_m)=B_m(\rho_0),$$

这是与(31)式矛盾的,故不等式(1)是真实的.

关于 K_m 的单调性,首先,我们证明:当 $m\geqslant 17$ 时，K_m 关于 m 是严格单调上升的. 事实上,设 $m_1<m_2$,由于 B_m 当 $m\geqslant 17$ 时关于 m 是严格单调下降的,因而，

$$B_{m_1}>B_{m_2}. \tag{32}$$

但是,我们知道，A_m 关于 m 是单调下降的,因而，

$$\psi_{m_1}(K_{m_2}) < \psi_{m_2}(K_{m_2}).$$

又由于 $\psi_m(b_3)$ 在 $b_3 \leqslant K_m$ 时，关于 b_3 是严格单调下降的[1]，因而，若 $K_{m_1} \geqslant K_{m_2}$，则

$$\psi_{m_1}(K_{m_1}) \leqslant \psi_{m_1}(K_{m_2}),$$

于是，若 $K_{m_1} \geqslant K_{m_2}$，则得

$$B_{m_1} = \psi_{m_1}(K_{m_1}) < \psi_{m_2}(K_{m_2}) = B_{m_2},$$

这是与(32)式矛盾的，故上述结论是真实的.

其次，我们证明，当 $7 \leqslant m \leqslant 17$ 时，K_m 关于 m 也是严格单调上升的. 事实上，根据 K_m 的定义，它是方程(2)的解，直接由数字计算，我们有

$$K_7 \doteq 1.71,\ K_8 \doteq 1.74,\ K_9 \doteq 1.772,\ K_{10} \doteq 1.803,\ 1.818 < K_{11} < 1.820,$$
$$1.856 < K_{12} < 1.86,\ 1.882 < K_{13} < 1.885,$$
$$1.89 < K_{14} < 1.921 < K_{15} < 1.93 < K_{16} < 1.945 < K_{17}.$$

此外，我们获得

$$K_{20} \doteq 1.995,\ \lim_{n\to\infty} K_n = K_\infty > 2.449.$$

故定理证完.

◇参考文献◇

[1] Ehrig, G., Coefficient estimates concerning the Bieberbach conjecture, *Math. Z.*, **140** (1974), 111 - 126.

[2] 任福尧，在第三项系数限制下单叶函数的 Bieberbach 猜想，自然杂志，1979，1.

[3] Fitzgerald, C. H., Quadratic inequalities and coefficient estimates for schlicht functions, *Arch. Rat. Mech. Anal.*, **46**(1972), 356 - 368.

[4] Bishouty, D. H., The Bieberbach conjecture for univalent functions with small second coefficients, *Math. Z.*, **149**(1976), 183 - 187.

[5] Pommerenke, Ch., *Univalent funchions*, Göttingen: Vandenhoeck & Ruprecht, 1975.

[6] Hayman, W. K., The asymptotic behaviour of p-valent functions, *Proc. London Math. Soc.* III., **5**(1955), 257 - 284.

[7] Horowitz, D., A further refinement for coefficient estimates of univalent functions, *Proc. Amer. Math. Soc.*, **71**(1978), 217 - 222.

The Bieberbach Conjecture for Univalent Functions with Restricted Third Coefficients

Ren Fuyao (Jen Fuyao 任福尧)

Abstract

In this paper, we prove that if $f(z) = z + \sum_{n=2}^{\infty} a_n z^n \in S$, then there exists a monotonic increasing sequence of real number $\{K_n\}$ $(n \geqslant 7)$ and $\lim\limits_{n \to \infty} K_n > 2.449$ such that for all $n \geqslant 7$ and $|a_3| \leqslant K_n$, we have

$$|a_k| < k \quad \text{for } k = n,\ n+1,\ \cdots,$$

particularly $K_7 = 1.71$, $K_8 = 1.74$, $K_9 = 1.772$, $K_{20} = 1.995$, for sufficient large N, where N depends only on 2.449 and not on f, $K_N > 2.449$. This improves the best estimate that has ever known in [1].

压力阶跃在复杂管系中的传输*

复旦大学　任福尧　柳兆荣

本文利用理想流体瞬态理论的线化模型，对实际中经常遇到的压力阶跃在分散流管系中的传输特性进行了分析. 对于如图 1 所示的分支管路，假设构成分支管路的 3 根管均为均匀的圆管，截面积分别为 A_1，A_2 与 A_3，管长分别为 L_1，L_2 与 L_3，并以 x_1，x_2 与 x_3 分别表示距其始端的距离. 在均匀管段内理想流体的传输方程为[1]

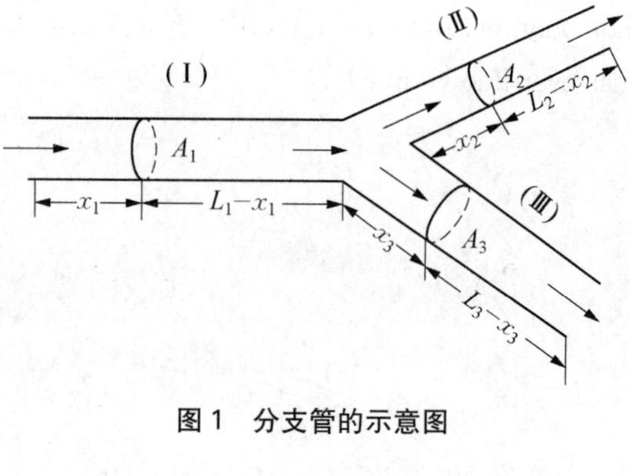

图 1　分支管的示意图

$$
\begin{aligned}
\partial P(x,\ s)/\partial x &= -slQ(x,\ s), \\
\partial Q(x,\ s)/\partial x &= -scP(x,\ s),
\end{aligned}
\tag{1}
$$

其中，

$$
P(x,\ s) = \int_0^s \mathrm{e}^{-st} p(x,\ t)\,\mathrm{d}t \approx p(x,\ t),
$$

---

* 原载《力学学报》，1979 年，第 4 期，380 - 384.

$$Q(x, s)=\int_0^s e^{-st} q(x, t) dt \approx q(x, t),$$

$p(x, t)$，$q(x, t)$为 x 处 t 时刻的压力与流量，$l=1/A$ 为单位管长的流感，$c=A/a^2$ 为单位管长的流容，a 为声速. 对于管段(Ⅰ)，(Ⅱ)与(Ⅲ)，方程(1)的通解为

$$\begin{cases} P_i(x_i, s)=A_{i1} e^{-\gamma_i x_i}+A_{i2} e^{\gamma_i x_i}, \\ Q_i(x_i, s)=\dfrac{A_{i1}}{Z_{c_i}} e^{-\gamma_i x_i}-\dfrac{A_{i2}}{Z_{c_i}} e^{\gamma_i x_i}, \end{cases} \tag{2}$$

其中，

$$l_i=1/A_i, \ c_i=A_i/a^2, \ \gamma_i=\sqrt{l_i c_i}\, s=s/a, \ Z_{c_i}=a/A_i \quad (i=1, 2, 3).$$

为了利用方程(2)来讨论分支管的瞬态特性，在对其进行 Laplace 反变换之前必须通过如下给定的边界条件来确定其待定系数 A_{i1} 与 $A_{i2}(i=1, 2, 3)$：在各支管始端阻抗不为零的一般情况下，图 1 所示分支管系所对应的边界条件为

$$P_1(0, s)=P_{01}(s)-Z_{01}Q_1(0, s), \tag{3}$$

$$P_2(L_2, s)=Z_{l_2}Q_2(L_2, s), \tag{4}$$

$$P_3(L_3, s)=Z_{l_3}Q_3(L_3, s), \tag{5}$$

$$Q_1(L_1, s)=Q_2(0, s)+Q_3(0, s), \tag{6}$$

$$P_2(0, s)=P_1(L_1, s)-Z_{02}Q_2(0, s), \tag{7}$$

$$P_3(0, s)=P_1(L_1, s)-Z_{03}Q_3(0, s), \tag{8}$$

式中 $P_{01}(s)=\int_0^s e^{-st} p_{01}(t) dt \approx p_{01}(t)$；$p_{01}(t)$是输入端 $(x_1=0)$ 输入的压力信号；Z_{01}，Z_{02} 与 Z_{03} 分别为(Ⅰ)，(Ⅱ)与(Ⅲ)管的始端阻抗；Z_{L_1}，Z_{L_2} 与 Z_{L_3} 分别为(Ⅰ)，(Ⅱ)与(Ⅲ)管的终端阻抗.

由边界条件(3)～(8)足以定出方程(2)的 6 个待定系数 A_{i1} 与 $A_{i2}(i=1, 2, 3)$，从而可定出对应于 3 个管段上某一点的压力与流量 $P_i(x_i, s)$ 与 $Q_i(x_i, s)(i=1, 2, 3)$. 特别地，当在(Ⅰ)管始端 $(x_1=0)$ 输入幅值为 P_0 的压力阶跃，即当 $P_{01}=P_0/s$ 时，可进一步对这些频域上的 $P_i(x_i, s)$ 与

$Q_i(x_i, s)$进行 Laplace 反变换,最后得对应于 3 个管段的压力与流量的瞬态表达式为

$$\begin{cases} p_1(x_1, t)=\dfrac{P_0}{1+(Z_{01}/Z_{c_1})}\Big\{U\Big(t-\dfrac{x_1}{a}\Big)+\eta_1 U\Big(t-\dfrac{2L_1-x_1}{a}\Big) \\ \quad +\delta_{01}\eta_1 U\Big(t-\dfrac{2L_1+x_1}{a}\Big)+\xi_{21}\xi_{12}\delta_{L_2}U\Big(t-\dfrac{2L_1+2L_2-x_1}{a}\Big) \\ \quad +\xi_{31}\xi_{13}\delta_{L_3}U\Big(t-\dfrac{2L_1+2L_3-x_1}{a}\Big)+\eta_1^2\delta_{01}U\Big(t-\dfrac{4L_1-x_1}{a}\Big) \\ \quad +\delta_{01}\xi_{21}\xi_{12}\delta_{L_2}U\Big(t-\dfrac{2L_1+2L_2+x_1}{a}\Big)+\cdots\Big\}, \\ q_1(x_1, t)=\dfrac{P_0}{Z_{01}+Z_{c_1}}\Big\{U\Big(t-\dfrac{x_1}{a}\Big)-\eta_1 U\Big(t-\dfrac{2L_1-x_1}{a}\Big) \\ \quad +\delta_{01}\eta_1 U\Big(t-\dfrac{2L_1+x_1}{a}\Big)-\xi_{21}\xi_{12}\delta_{L_2}U\Big(t-\dfrac{2L_1+2L_2-x_1}{a}\Big) \\ \quad -\xi_{31}\xi_{13}\delta_{L_3}U\Big(t-\dfrac{2L_1+2L_3-x_1}{a}\Big)-\eta_1^2\delta_{01}U\Big(t-\dfrac{4L_1-x_1}{a}\Big) \\ \quad +\delta_{01}\xi_{21}\xi_{12}\delta_{L_2}U\Big(t-\dfrac{2L_1+2L_2+x_1}{a}\Big)+\cdots\Big\}, \end{cases} \tag{9}$$

$$\begin{cases} p_2(x_2, t)=\dfrac{p_0}{1+(Z_{01}/Z_{c_1})}\Big\{\xi_{12}U\Big(t-\dfrac{L_1+x_2}{a}\Big)+\xi_{12}\delta_{L_2}U\Big(t-\dfrac{L_1+2L_2-x_2}{a}\Big) \\ \quad +\eta_2\xi_{12}\delta_{L_2}U\Big(t-\dfrac{L_1+2L_2+x_2}{a}\Big)+\xi_{32}\xi_{13}\delta_{L_3}U\Big(t-\dfrac{L_1+2L_3+x_2}{a}\Big) \\ \quad +\xi_{12}\eta_1\delta_0 U\Big(t-\dfrac{3L_1+x_2}{a}\Big)+\eta_2\xi_{12}\delta_{L_2}^2 U\Big(t-\dfrac{L_1+4L_2-x_2}{a}\Big)+\cdots\Big\}, \\ q_2(x_2, t)=\dfrac{p_0}{Z_{01}+Z_{c_1}}\Big\{\xi_{12}U\Big(t-\dfrac{L_1+x_2}{a}\Big)-\xi_{12}\delta_{L_2}U\Big(t-\dfrac{L_1+2L_2-x_2}{a}\Big) \\ \quad +\eta_2\xi_{12}\delta_{L_2}U\Big(t-\dfrac{L_1+2L_2+x_2}{a}\Big)+\xi_{32}\xi_{13}\delta_{L_3}U\Big(t-\dfrac{L_1+2L_3+x_2}{a}\Big) \\ \quad +\xi_{12}\eta_1\delta_0 U\Big(t-\dfrac{3L_1+x_2}{a}\Big)-\eta_2\xi_{12}\delta_{L_2}^2 U\Big(t-\dfrac{L_1+4L_2-x_2}{a}\Big)+\cdots\Big\}, \end{cases} \tag{10}$$

$$\begin{cases} p_3(x_3, t) = \dfrac{p_0}{1+(Z_{01}/Z_{c_1})}\Big\{\xi_{13}U\Big(t-\dfrac{L_1+x_3}{a}\Big)+\xi_{13}\delta_{L_3}U\Big(t-\dfrac{L_1+2L_3-x_3}{a}\Big) \\ \qquad +\xi_{23}\xi_{12}\delta_{L_2}U\Big(t-\dfrac{L_1+2L_2+x_3}{a}\Big)+\eta_3\xi_{13}\delta_{L_3}U\Big(t-\dfrac{L_1+2L_3+x_3}{a}\Big) \\ \qquad +\xi_{13}\eta_1\delta_0 U\Big(t-\dfrac{3L_1+x_3}{a}\Big)+\eta_3\xi_{13}\delta_{L_3}^2 U\Big(t-\dfrac{L_1+4L_3-x_3}{a}\Big)+\cdots\Big\}, \\ q_3(x_3, t) = \dfrac{p_0}{Z_{01}+Z_{c_1}}\Big\{\xi_{13}U\Big(t-\dfrac{L_1+x_3}{a}\Big)-\xi_{13}\delta_{L_3}U\Big(t-\dfrac{L_1+2L_3-x_3}{a}\Big) \\ \qquad +\xi_{23}\xi_{12}\delta_{L_2}U\Big(t-\dfrac{L_1+2L_2+x_3}{a}\Big)+\eta_3\xi_{13}\delta_{L_3}U\Big(t-\dfrac{L_1+2L_3+x_3}{a}\Big) \\ \qquad +\xi_{13}\eta_1\delta_0 U\Big(t-\dfrac{3L_1+x_3}{a}\Big)-\xi_{13}\eta_3\delta_{L_3}^2 U\Big(t-\dfrac{L_1+4L_3-x_3}{a}\Big)+\cdots\Big\}, \end{cases} \tag{11}$$

其中，

$$\eta_1 = \frac{(Z_{02}+Z_{c_2})(Z_{03}+Z_{c_3})-Z_{c_1}[(Z_{02}+Z_{c_2})+(Z_{03}+Z_{c_3})]}{(Z_{02}+Z_{c_2})(Z_{03}+Z_{c_3})+Z_{c_1}[(Z_{02}+Z_{c_2})+(Z_{03}+Z_{c_3})]}$$ 为管(Ⅰ)在分叉点的反射系数，

$$\eta_2 = \frac{Z_{c_1}(Z_{03}+Z_{c_3})+(Z_{02}-Z_{c_2})[((Z_{03}+Z_{c_3})+z_{c_1}]}{Z_{c_1}(Z_{03}+Z_{c_3})+(Z_{02}+Z_{c_2})[((Z_{03}+Z_{c_3})+z_{c_1}]}$$ 为管(Ⅱ)在分叉点的反射系数，

$$\xi_{12} = \frac{1}{1+(Z_{02}/Z_{c_2})}\left\{\frac{2(Z_{02}+Z_{c_2})(Z_{03}+Z_{c_3})}{(Z_{02}+Z_{c_2})(Z_{03}+Z_{c_3})+Z_{c_1}(Z_{02}+Z_{c_2}+Z_{03}+Z_{c_3})}\right\}$$

为管(Ⅰ)到管(Ⅱ)的传输系数，

$$\xi_{13} = \frac{1}{1+(Z_{03}/Z_{c_3})}\left\{\frac{2(Z_{02}+Z_{c_2})(Z_{03}+Z_{c_3})}{(Z_{02}+Z_{c_2})(Z_{03}+Z_{c_3})+Z_{c_1}(Z_{02}+Z_{c_2}+Z_{03}+Z_{c_3})}\right\}$$

为管(Ⅰ)到管(Ⅲ)的传输系数，

$$\xi_{32} = \frac{2Z_{c_1}Z_{c_2}}{Z_{c_1}(Z_{02}+Z_{c_2})+(Z_{03}+Z_{c_3})(Z_{02}+Z_{c_2}+Z_{c_1})}$$ 为管(Ⅲ)到管(Ⅱ)的传输系数，

$$\xi_{23} = \frac{2Z_{c_1}Z_{c_3}}{Z_{c_1}(Z_{03}+Z_{c_3})+(Z_{02}+Z_{c_2})(Z_{03}+Z_{c_3}+Z_{c_1})}$$ 为管(Ⅱ)到管(Ⅲ)的传输系数，

$$\xi_{21}=\frac{2Z_{c_1}(Z_{03}+Z_{c_3})}{Z_{c_1}(Z_{03}+Z_{c_3})+(Z_{02}+Z_{c_2})(Z_{03}+Z_{c_3}+Z_{c_1})}$$ 为管(Ⅱ)到管(Ⅰ)的传输系数，

$$\xi_{31}=\frac{2Z_{c_1}(Z_{02}+Z_{c_2})}{Z_{c_1}(Z_{02}+Z_{c_2})+(Z_{03}+Z_{c_3})(Z_{02}+Z_{c_2}+Z_{c_1})}$$ 为管(Ⅲ)到管(Ⅰ)的传输系数.

$$U(t-\tau)=\begin{cases}1, & \text{当 } t\geqslant\tau \text{ 时},\\ 0, & \text{当 } t<\tau \text{ 时}.\end{cases}$$

由式(9)～(11)不难看出，对于管路端点是纯流阻的情况，反射系数 δ_{01}，δ_{L_2}，δ_{L_3}，η_1，η_2 与 η_3，传输系数 ξ_{12}，ξ_{13}，ξ_{21}，ξ_{23}，ξ_{31} 与 ξ_{32} 等是纯实数. 因而管(Ⅰ)，(Ⅱ)和(Ⅲ)中某一点的压力与流量的瞬态波形将由一系列阶梯波形经适当延迟后叠加而成. 这一列阶梯波形随时间的推迟逐渐趋于平滑稳态波形. 显然，压力与流量的相应稳态值为

$$\begin{cases}
p_1(x_1,\infty)=P_0\Big/\left[1+Z_{01}\left(\dfrac{1}{Z_{02}+Z_{L_2}}+\dfrac{1}{Z_{03}+Z_{L_3}}\right)\right],\\
q_1(x_1,\infty)=\left[P_0\left(\dfrac{1}{Z_{02}+Z_{L_2}}+\dfrac{1}{Z_{03}+Z_{L_3}}\right)\right]\Big/\left[1+Z_{01}\left(\dfrac{1}{Z_{02}+Z_{L_2}}+\dfrac{1}{Z_{03}+Z_{L_3}}\right)\right],\\
p_2(x_2,\infty)=P_0Z_{L_2}\Big/\left[Z_{01}(Z_{02}+Z_{L_2})\left(\dfrac{1}{Z_{01}}+\dfrac{1}{Z_{02}+Z_{L_2}}+\dfrac{1}{Z_{03}+Z_{L_3}}\right)\right],\\
q_2(x_2,\infty)=P_0\Big/\left[Z_{01}(Z_{02}+Z_{L_2})\left(\dfrac{1}{Z_{01}}+\dfrac{1}{Z_{02}+Z_{L_2}}+\dfrac{1}{Z_{03}+Z_{L_3}}\right)\right],\\
p_3(x_3,\infty)=P_0Z_{L_3}\Big/\left[Z_{01}(Z_{03}+Z_{L_3})\left(\dfrac{1}{Z_{01}}+\dfrac{1}{Z_{02}+Z_{L_2}}+\dfrac{1}{Z_{03}+Z_{L_3}}\right)\right],\\
q_3(x_3,\infty)=P_0\Big/\left[Z_{01}(Z_{03}+Z_{L_3})\left(\dfrac{1}{Z_{01}}+\dfrac{1}{Z_{02}+Z_{L_2}}+\dfrac{1}{Z_{03}+Z_{L_3}}\right)\right].
\end{cases}\tag{12}$$

其次，从式(9)～(11)还可看出，压力与流量的瞬态波形不但取决于端点与分叉点的反射系数，而且还取决于管长 L_1，L_2 与 L_3 以及观察点位置 x_1，x_2 与 x_3. 例如选取 $L_1>L_2>L_3$，$x_1<L_1$，$x_2<L_2$，$x_3<L_3$，$L_3<x_3+L_2$ 时，瞬态波形的先后次序将如式(9)～(11)所示.

还可知道，压力阶跃自管(Ⅰ)始端 $(x_1=0)$ 输入到达分叉点后，其一部分将

以反射系数 η_1 反射回到管(Ⅰ)的始端,另一部分则以传输系数 ξ_{12} 穿过交叉点传输入管(Ⅱ),以传输系数 ξ_{13} 穿过交叉点传输入管(Ⅲ). 这些信号到达端点后还将继续反射,如此下去. 因而式(9)~(11)中各项的物理意义是十分清楚的.

最后,利用式(9)~(11)还可将端点的流阻用瞬态压力波前面若干个阶梯的高度表出:

$$
\begin{cases}
Z_{01} = Z_{c_1}[(\Delta h_{12}/\Delta h_{13})+1]/[(\Delta h_{12}/\Delta h_{13})-1], \\
Z_{L_1} = Z_{c_2}[(\Delta h_{21}/\Delta h_{22})+1]/[(\Delta h_{21}/\Delta h_{22})-1], \\
Z_{L_3} = Z_{c_3}[(\Delta h_{31}/\Delta h_{32})+1]/[(\Delta h_{31}/\Delta h_{32})-1], \\
Z_{03} = Z_{c_1}\left[1+\dfrac{Z_{c_3}}{Z_{c_2}}\dfrac{\Delta h_{21}}{\Delta h_{31}}\right]\Big/\left[\dfrac{(\Delta h_{11}/\Delta h_{12})-1}{(\Delta h_{11}/\Delta h_{12})+1}\right]-Z_{c_3}, \\
Z_{02} = Z_{c_1}\left[1+\dfrac{Z_{c_2}}{Z_{c_3}}\dfrac{\Delta h_{31}}{\Delta h_{21}}\right]\Big/\left[\dfrac{(\Delta h_{11}/\Delta h_{12})-1}{(\Delta h_{11}/\Delta h_{12})+1}\right]-Z_{c_2},
\end{cases}
\tag{13}
$$

式中 Δh_{ij} 表示第 i 管的压力波形中第 j 个相对高度. 这就是说,利用瞬态压力波形前面若干个阶梯波可定出管系各端点的瞬态流阻,从而进一步算出分支点处的传输系数与反射系数,因而使这些本来难以确定的参量可以有一个较简便的确定方法.

顺便指出,若图 1 中管(Ⅱ)或(Ⅲ)消失,则从式(9)~(11)可得弯管与异径管的特例. 若管(Ⅰ)与(Ⅱ)同截面,管(Ⅲ)细而短,则管(Ⅲ)终端感受到的压力波形将与文献[1]的均匀管的结果一样. 这就是说,当用旁侧分支管来传感某管路的压力时,只要这分支管与待测管相比是短而细的,就能保证从旁侧分支管终端传感到的压力就是待测管在待测点的压力.

◇参考文献◇

[1] 佘鎏,复旦学报,2(1975),42-57.

Pressure-step Transmission in A Complex Piping System

Ren Fu-yao Liu Zhao-rong

(*Fudan University*)

单叶半纯函数之逆函数的系数*

任福尧

1. 引　言

设 Σ' 表示在区域 $1 < |\zeta| < \infty$ 中由单叶函数

$$g(\zeta) = \zeta + \sum_{n=1}^{\infty} b_n \zeta^{-n}$$

所组成的函数族. 若 G 是 $g \in \Sigma'$ 的逆函数,那么, G 在 $w = \infty$ 附近可展成

$$G(w) = w + \sum_{n=1}^{\infty} c_n w^{-n}.$$

我们知道, $|c_1| = |b_1| \leqslant 1$, $|c_2| = |b_2| \leqslant \frac{2}{3}$, Springer[1]证明了 $|c_3| \leqslant 1$, 并猜想

$$|c_{2k-1}| \leqslant \frac{(2k-2)!}{k!\,(k-1)!}, \quad k = 2, 3, \cdots. \tag{1}$$

Y. Kubota[2]证明了 $k = 3, 4$,和 5 时 Springer 猜想是成立的, G. Schober[3]和作者[4]证明了 $k = 6, 7$ 时 Springer 猜想是成立的.

本文的目的在于证明 $k = 8$ 时, Springer 猜想是成立的. 我们证明了下述定理.

定理　若 $g(\zeta) \in \Sigma'$, $G(w) = g^{-1}(w) = w + \sum_{n=1}^{\infty} c_n w^{-n}$, 则

* 原载《科学通报》,1980 年,第 5 期,193 - 195.

$$|c_{15}| \leqslant 429, \tag{2}$$

等号当且仅当 $g(\zeta)=\zeta+\dfrac{\eta}{\zeta}$，$|\eta|=1$ 时成立.

2. 引　理

引理　若 $g(\zeta)\in\Sigma'$，$G(w)=g^{-1}(w)=w+\sum\limits_{n=1}^{\infty}c_n w^{-n}$，则

$$|c_{2k-1}| \leqslant D_{kk},\quad k=1,2,3,\cdots, \tag{3}$$

这里

$$\frac{1}{1-1/G(w)G(t)}=\sum_{p,q=0}^{\infty} D_{p,q} w^{-p}(\bar{t})^{-q}.$$

特别地，若记 $c'_n=-c_n$，则

$$\begin{aligned}
&D_{11}=D_{22}=1,\quad D_{33}=|c'_1|^2+1,\quad D_{44}=|c'_2|^2+4|c_1|^2+1,\\
&D_{55}=|c'_3+c'^2_1|^2+4|c'_2|^2+9|c'_1|^2+1,\\
&D_{66}=|c'_4+2c'_1c'_2|^2+|2c'_3+3c'^2_1|^2+9|c'_2|^2+16|c'_1|^2+1,\\
&D_{77}=|c'_5+(c'^2_2+2c'_1c'_3)+c'^3_1|^2+|2c'_4+6c'_1c'_2|^2+|3c'_3+6c'^2_1|^2\\
&\qquad +16|c'_2|^2+25|c'_1|^2+1,\\
&D_{88}=|c'_6+2(c'_1c'_4+c'_2c'_3)+3c'^2_1c'_2|^2+|2c'_5+3(c'^2_2+2c'_1c'_2)+4c'^3_1|^2+\\
&\qquad |3c'_4+12c'_1c'_2|^2+|4c'_3+10c'^2_1|^2+25|c'_2|^2+36|c'_1|^2+1.
\end{aligned}$$

证　对 $g(\zeta)\in\Sigma'$，根据夏道行不等式[5]①，有

$$\left|\sum_{\nu,\mu=1}^{m}\alpha_\nu\alpha_\mu\frac{\zeta_\nu-\zeta_\mu}{g(\zeta_\nu)-g(\zeta_\mu)}\right| \leqslant \sum_{\nu,\mu=1}^{m}\alpha_\nu\bar{\alpha}_\mu\frac{1}{1-1/\zeta_\nu\bar{\zeta}_\mu} \tag{4}$$

对任何自然数 m 和复数 $\{\alpha_\nu\}_1^m$ 及点列 $\{\zeta_\nu\}_1^m$ ($|\zeta_n|>1$) 成立. 因而，对

$$G(w)=g^{-1}(w)=w-\sum_{n=1}^{\infty}c'_n w^{-n},\ c'_n=-c_n,$$

① 1972 年 Fitzgerald 重新得到了这不等式[6].

$$\left|\sum_{\nu,\,\mu=1}^{m}\alpha_\nu\alpha_\mu\ \frac{G(w_\nu-G(w_\mu)}{w_\nu-w_\mu}\right|\leqslant\sum_{\nu,\,\mu=1}^{m}\alpha_\nu\bar{\alpha}_\mu\left(\frac{1}{1-1/G(w_\nu)\overline{G(w_\mu)}}\right)$$

对任何自然数 m 和复数$\{\alpha_\nu\}_1^m$ 及点列 $\{w_\nu=g(\zeta_\nu)\}_1^m$ 成立. 将 $G(w)$的幂级数展开式代入上不等式,则得

$$\left|\sum_{\nu,\,\mu=1}^{m}\alpha_\nu\alpha_\mu\Big\{1+c'_1(w_\nu w_\mu)^{-1}+c'_3(w_\nu w_\mu)^{-2}+\cdots+c'_{2k-1}(w_\nu w_\mu)^{-k}+\cdots\right.$$
$$\left.+\sum_{s\neq i,\,s,\,t\geqslant 2}O(w_\nu^{-s}w_\mu^{-t})\Big\}\right|\leqslant\sum_{\nu,\,\mu=1}^{m}\alpha_\nu\bar{\alpha}_\mu\Big(\sum_{p,\,q=0}^{\infty}D_{pq}w_\nu^{-p}\overline{w_\mu^{-q}}\Big).\qquad(5)$$

对给定的任一自然数 k,令

$$\alpha_\nu=\rho^k\beta_\nu,\ w_\nu=\rho x_\nu^{-t}\quad(\nu=1,\,2,\,\cdots,\,m,\,m>k),$$

其中 $\rho>1$, $\{x_\nu\}$ 是位于正实数轴上紧靠原点且不相同的 m 个点,选取 β_1, β_2, $\cdots$, β_m,使得对每个从零到 $m-1$ 的自然数 m 成立着

$$\sum_{\nu=1}^{m}\beta_\nu x_\nu^s=\begin{cases}1, & s=k,\\ 0, & \text{其他}.\end{cases}$$

根据 Vandermonde 行列式的性质,大家熟知,这是做得到的. 于是,从不等式(5)得

$$\left|c_{2k-1}\sum_{\nu,\,\mu=1}^{m}\alpha_\nu\alpha_\mu(w_\nu w_\mu)^{-k}+O(\rho^{-2})\right|$$
$$\leqslant D_{kk}\Big[\sum_{\nu,\,\mu=1}^{m}\alpha_\nu\alpha_\mu(w_\nu w_\mu)^{-k}\Big]+O(\rho^{-2}).$$

由于 $\sum_{\nu,\,\mu=1}^{m}\alpha_\nu\alpha_\mu(w_\nu w_\mu)^{-k}=\big(\sum_{\nu=1}^{m}\beta_\nu x_\nu^k\big)^2=1$, 于上式中令 $\rho\to\infty$,便得

$$|c_{2k-1}|\leqslant D_{kk}.$$

引理证毕.

3. 定理的证明

当 $k=8$ 时,由引理得

$$|c_{15}| \leqslant D_{88} \leqslant |c_6|^2 + 4(|c_1c_4c_6| + |c_2c_3c_6|) + 6|c_1^2c_2c_6|$$
$$+ 4(|c_1c_4| + |c_2c_3|)^2 + 12|c_1^2c_2(c_1c_4 + c_2c_3)| + 9|c_1^2c_2|^2$$
$$+ 4|c_5|^2 + 12|(c_2^2 + 2c_1c_3)c_5| + 9|c_2^2 + 2c_1c_3|^2$$
$$+ 16|c_1^3c_5| + 24|c_1^3(c_2^2 + 2c_1c_3)| + 16|c_1|^6 + 9|c_4|^2$$
$$+ 72|c_1c_2c_4| + 144|c_1c_2|^2 + 16|c_3|^2 + 80|c_1^2c_3|$$
$$+ 100|c_1|^4 + 25|c_2|^2 + 36|c_1|^2 + 1. \tag{6}$$

由于 $|c_1| = |b_1| \leqslant 1$，$|c_2| = |b_2|$，$|c_3| \leqslant 1$，$|c_5| \leqslant 2$，于是，由(3)式便得

$$|c_{15}| \leqslant 81 + 184|b_1|^2 + 148|b_1|^4 + 16|b_1|^6 + 89|b_2|^2 + 189|b_1b_2|^2$$
$$+ |c_6|^2 + 4|c_4c_6| + 10|c_2c_6| + 13|c_4|^2 + 92|c_2c_4| + 9|b_2|^4. \tag{7}$$

但是，根据面积定理，有

$$|b_1|^2 + 2|b_2|^2 + 4|b_4|^2 + 6|b_6|^2 \leqslant 1,$$

因而，$|b_1|^4 \leqslant (1 - 2|b_2|^2)|b_1|^2$，$|b_1|^6 \leqslant |b_1|^2(1 - 4|b_2|^2 + 4|b_2|^4)$，于是，由不等式(4)，有

$$|c_{15}| \leqslant 81 + 348|b_1|^2 - 171|b_1b_2|^2 + 89|b_2|^2 + |c_6|^2 + 4|c_4c_6|$$
$$+ 10|c_2c_6| + 13|c_4|^2 + 92|c_2c_4| + 73|b_2|^4. \tag{8}$$

但是，我们知道，$c_2 = -b_2$，$c_3 = -(b_3 + b_1^2)$，$c_4 = -(b_4 + 3b_1b_2)$，

$$c_6 = -(b_6 + 5b_1b_4 + 10b_1^2b_2 + 5b_2b_3),$$

$$|c_2c_4| \leqslant |b_2b_4| + 3|b_2|^2,$$
$$|c_4|^2 \leqslant |b_4|^2 + 6|b_2b_4| + 9|b_1b_2|^2,$$
$$|c_2c_6| \leqslant |b_2b_6| + 5|b_2b_4| + 5|b_1b_2|^2 + 5|b_2|^2,$$
$$|c_4c_6| \leqslant |b_4b_6| + 5|b_4|^2 + 25|b_2b_4| + 3|b_2b_6| + 15|b_1b_2|^2 + 15|b_2|^2,$$
$$|c_6|^2 \leqslant |b_6|^2 + 25|b_4|^2 + 10|b_4b_6| + 20|b_2b_6| + 100|b_2b_4| + 75|b_1b_2|^2$$
$$+ 25|b_2|^2,$$

将这些不等式代入(5)式，便得

$$|c_{15}| \leqslant 81 + 348|b_1|^2 + 631|b_2|^2 + 420|b_2b_4| + 42|b_2b_6|$$
$$+ 14|b_4b_6| + 58|b_4|^2 + |b_6|^2. \tag{9}$$

利用不等式 $2|xy| \leqslant \eta|x|^2 + \frac{1}{\eta}|y|^2$，$\eta > 0$，$x$，$y$ 是实数，有

$$420|b_2 b_4| \leqslant 35|b_2|^2 + 1\,260|b_4|^2,\quad 42|b_2 b_6| \leqslant |b_2|^2 + 441|b_6|^2,$$
$$14|b_4 b_6| \leqslant |b_4|^2 + 49|b_6|^2,$$

于是，由不等式(6)，便得

$$|c_{15}| \leqslant 81 + 348|b_1|^2 + 668|b_2|^2 + 1\,319|b_4|^2 + 490|b_6|^2$$
$$\leqslant 415 + 14|b_1|^2 \leqslant 429.$$

等号当且仅当 $|b_1|=1$ 时发生，因而，等号当且仅当 $g(\zeta)=\zeta+\frac{\eta}{\zeta}$，$|\eta|=1$ 时成立. 定理证完.

◇参考文献◇

[1] Springer, G., *Trans. Amer. Math. Soc.*, **70**(1951), 421-450.

[2] Kodai, Y., *Math. Sem. Rep.*, **28**(1977), 253-261.

[3] Schober, G., *Proc. Amer. Math. Soc.*, **67**(1977), 111-116.

[4] 任福尧，复旦学报，**4**(1978)，93-96.

[5] 夏道行，*Science Record*, **4**(1951), 351-362.

[6] Fitzgerald, Ch., *Arch. Rat. Mech. Anal.*, **46**(1972), 356-368.

The Distortion Theorems for Bieberbach Class and Grunsky Class of Univalent Functions

REN FUYAO

1. Introduction and Notations

Let B_1 denote the class of all of the functions $f(\zeta)$, analytic and univalent in the unit disk $|\zeta|<1$, such that $f(0)=0$ and $f(\zeta_1)f(\zeta_2)\neq 1$ for any ζ_1 and ζ_2 in $|\zeta|<1$. These are the functions of the so-called Bieberbach or Bieberbach-Eilenberg class. Let G_1 denote the class of all of the functions $f(\zeta)$, analytic and univalent in the unit disk $|\zeta|<1$, such that $f(0)=0$ and $f(\zeta_1)\overline{f(\zeta_2)}\neq -1$ for any ζ_1 and ζ_2 in $|\zeta|<1$, these are the functions of the so-called Grunsky class.

In this paper, by applying the area principle of Tao-Shing Shah[1], we obtained another necessary and sufficient condition, and proved some distortion theorems for these classes, particularly exponentiated Golusin inequalities and Fitzgerald inequalities.

For the convenience of assertion, we introduce some notations. Let $f(\zeta)$ be the analytic and univalent function in $|\zeta|<1$ and $f(0)=0$. We define two functions of ζ and z in the unit disk by

$$f(\zeta, z)=f'(0)(f(\zeta)-f(z))\zeta z/(\zeta-z)f(\zeta)f(z)$$

and by

$$F(\zeta, z)=\ln f(\zeta, z),$$

* Originally published in *Chin. Ann. of Math.* Vol. 3, No. 6, (1982), 691 - 703.

taking the single-valued branch of the function which vanishes at $\zeta=0$. Since $f(\zeta)$ is univalent in the disk, these functions are regular in the unit disk, and hence the following expansion is valid in the unit disk

$$F(\zeta, z)=\sum_{m, n=1}^{\infty} d_{mn}\zeta^m z^n.$$

Similarly, if $f\in B_1$, we define

$$\varphi(\zeta, z)=1-f(\zeta)f(z)$$

and

$$\Phi(\zeta, z)=\ln\varphi(\zeta, z)=\sum_{m, n=1}^{\infty} C_{mn}\zeta^m z^n,$$

taking the single-valued branch of the function which vanishes at $\zeta=0$. Let $f\in G_1$, we define

$$\psi(\zeta, z)=1+f(\zeta)\overline{f(\bar{z})}$$

and

$$\Psi(\zeta, z)=\ln\psi(\zeta, z)=\sum_{m, n=1}^{\infty} B_{mn}\zeta^m z^n$$

which vanishes at $\zeta=0$.

Moreover, we define also

$$D_n(\zeta)=\sum_{m=1}^{\infty} d_{mn}\zeta^m,\quad C_n(\zeta)=\sum_{m=1}^{\infty} C_{mn}\zeta^m,\quad E_n(\zeta)=\sum_{m=1}^{\infty} B_{mn}\zeta^m,$$

$$D_n^{(p)}(\zeta)=\frac{\mathrm{d}^p}{\mathrm{d}\zeta^p}D_n(\zeta),\ C_n^{(p)}(\zeta)=\frac{\mathrm{d}^p}{\mathrm{d}\zeta^p}C_n(\zeta),\ E_n^{(p)}(\zeta)=\frac{\mathrm{d}^p}{\mathrm{d}\zeta^p}E_n(\zeta),$$

and

$$K(\zeta, \bar{z})=-\ln(1-\zeta\bar{z}),\quad K^{(p, q)}(\zeta, z)=\frac{\partial^{p+q}}{\partial\zeta^p\partial\bar{z}^q}K(\zeta, \bar{z}),$$

where p and q are arbitrary nonnegative integers.

2. On the Bieberbach Function

Theorem 1 *Let $f \in B_1$, $\{x_n\}$, $\{x'_n\}$, $\{y_n\}$ and $\{y'_n\}$ be arbitrary sequences of complex numbers such that*

$$\begin{aligned} X &= \sum_{n=1}^{\infty} |x_n|^2/n < \infty, \quad X' = \sum_{n=1}^{\infty} |x'_n|^2/n < \infty, \\ Y &= \sum_{n=1}^{\infty} |y_n|^2/n < \infty, \quad Y' = \sum_{n=1}^{\infty} |y'_n|^2/n < \infty. \end{aligned} \tag{2.1}$$

Then we have

$$\begin{aligned} &\left| \sum_{m,n=1}^{\infty} (C_{mn}x_m + d_{mn}y_m)x'_n \right|^2 + \left| \sum_{m,n=1}^{\infty} (d_{mn}x_m + C_{mn}y_m)x'_n \right|^2 \\ &+ \left| \sum_{m,n=1}^{\infty} (C_{mn}x_m + d_{mn}y_m)y'_n \right|^2 + \left| \sum_{m,n=1}^{\infty} (d_{mn}x_m + C_{mn}y_m)y'_n \right|^2 \\ &\leqslant (X+Y)(X'+Y'). \end{aligned} \tag{2.2}$$

Here the equality holds if and only if for any positive integer n,

$$\begin{aligned} \sum_{m=1}^{\infty} (C_{mn}x_m + d_{mn}y_m) &= \lambda_{11}\overline{x'_n}/n, \\ \sum_{m=1}^{\infty} (d_{mn}x_m + C_{mn}y_m) &= \lambda_{12}\overline{x'_n}/n, \\ \sum_{m=1}^{\infty} (C_{mn}x_m + d_{mn}y_m) &= \lambda_{21}\overline{y'_n}/n, \\ \sum_{m=1}^{\infty} (d_{mn}x_m + C_{mn}y_m) &= \lambda_{22}\overline{y'_n}/n, \end{aligned} \tag{2.3}$$

where λ_{ij} $(i, j = 1, 2)$ are constants such that

$$(|\lambda_{11}|^2 + |\lambda_{12}|^2)X^2 + (|\lambda_{21}|^2 + |\lambda_{22}|^2)Y^2 = (Y+X)(Y'+X'). \tag{2.4}$$

Proof By [1], since $f \in B_1$, for any $\{x_n\}$ and $\{y_n\}$ with

$$X=\sum_{n=1}^{\infty}|x_n|^2/n<\infty,\ Y=\sum_{n=1}^{\infty}|y_n|^2/n<\infty,$$

we have

$$\sum_{n=1}^{\infty}n\left\{\left|\sum_{m=1}^{\infty}(C_{mn}x_m+d_{mn}y_m)\right|^2+\left|\sum_{m=1}^{\infty}(d_{mn}x_m+C_{mn}y_m)\right|^2\right\}\leqslant X+Y. \tag{2.5}$$

Here the equality holds if and only if the area of the complement of the union of the image $f(|\zeta|<1)$ and the image $1/f(|\zeta|<1)$ vanishes.

By Cauchy inequality and using (2.5), we get

the left-hand side of (2.2)

$$\leqslant(X'+Y')\sum_{n=1}^{\infty}n\left(\left|\sum_{m=1}^{\infty}(C_{mn}x_m+d_{mn}y_m)\right|^2+\left|\sum_{m=1}^{\infty}(d_{mn}x_m+C_{mn}y_m)\right|^2\right)$$
$$\leqslant(X'+Y')(X+Y). \tag{2.6}$$

By the necessary and sufficient condition for which the equality holds of Cauchy inequality, it follows that the equality holds for the first inequality of (2.6) if and only if (2.3) is valid. In this case, the left-hand side of (2.2) equals to

$$(|\lambda_{11}|^2+|\lambda_{12}|^2)(X')^2+(|\lambda_{21}|^2+|\lambda_{22}|^2)(Y')^2.$$

Since the necessary and sufficient condition for which the equality holds for second inequality of (2.6), is that the area complementary to the union of image $f(|\zeta|<1)$ and image $1/f(|\zeta|<1)$ vanishes, it follows that (2.4) is valid. This completes the proof.

Corollary 1 *Let $f\in B_1$, sequences $\{x_n\}$ and $\{x'_n\}$ satisfy the condition* (2.1). *We have*

$$\left|\sum_{m,n=1}^{\infty}d_{mn}x_mx'_n\right|^2+\left|\sum_{m,n=1}^{\infty}C_{mn}x_mx'_n\right|^2\leqslant X\cdot X', \tag{2.7}$$

where $X=\sum_{n=1}^{\infty}|x_n|^2/n$, $X'=\sum_{n=1}^{\infty}|x'_n|^2/n$. The equality holds if and only if for all $n\geqslant 1$,

$$\begin{cases} C_{mn} x_m = \lambda_1 \overline{x}'_n / n, \\ d_{mn} x_m = \lambda_2 \overline{x}'_n / n, \end{cases} \tag{2.8}$$

where λ_1 *and* λ_2 *are constants such that*

$$|\lambda_1|^2 + |\lambda_2|^2 = X/X'. \tag{2.9}$$

Theorem 2 *Let* $f \in B_1$, p, q *be any nonnegative integers, and let* $\zeta_1, \cdots, \zeta_N; \zeta'_1, \cdots, \zeta'_{N'}$ *be two systems of distinguished points in* $|\zeta| < 1$. *Then for any nonzero complex numbers* $\{\eta_\mu\}$ *and* $\{\eta'_\nu\}$ ($\mu = 1, \cdots, N$; $\nu = 1, \cdots, N'$), *we have*

1)

$$\sum_{n=1}^{\infty} n \left\{ \left| \sum_{\mu=1}^{N} \eta_\mu C_n^{(p)}(\zeta_\mu) + \sum_{\nu=1}^{N'} \eta'_\nu D_n^{(q)}(\zeta'_\nu) \right|^2 + \left| \sum_{\mu=1}^{N} \eta_\mu D^{(p)}(\zeta_\mu) + \sum_{\nu=1}^{N'} \eta'_\nu C_n^{(q)}(\zeta'_\nu) \right|^2 \right\}$$
$$\leqslant \sum_{\mu,\nu=1}^{N} \eta_\mu \overline{\eta}_\nu K^{(p,p)}(\zeta_\mu, \overline{\zeta}_\nu) + \sum_{\mu,\nu=1}^{N'} \eta'_\mu \overline{\eta}'_\nu K^{(q,q)}(\zeta'_\mu, \overline{\zeta}'_\nu). \tag{2.10}$$

The equality holds if and only if the area of the complement of the union of image $f(|\zeta| < 1)$ *and image* $1/f(|\zeta| < 1)$ *to be vanished.*

2)

$$\left| \sum_{\mu=1}^{N} \sum_{\nu=1}^{N'} \eta_\mu \eta'_\nu F^{(p,q)}(\zeta_\mu, \zeta'_\nu) \right|^2 + \left| \sum_{\mu=1}^{N} \sum_{\nu=1}^{N'} \eta_\mu \eta'_\nu \Phi^{(p,q)}(\zeta_\mu, \zeta'_\nu) \right|^2$$
$$\leqslant \sum_{\mu,\nu=1}^{N} \eta_\mu \overline{\eta}_\nu K^{(p,p)}(\zeta_\mu, \overline{\zeta}_\nu) + \sum_{\mu,\nu=1}^{N'} \eta'_\mu \overline{\eta}'_\nu K^{(q,q)}(\zeta'_\mu, \overline{\zeta}'_\nu). \tag{2.11}$$

The equality holds only if

$$\begin{cases} \sum_{\mu=1}^{N} \sum_{\nu=1}^{N'} \eta_\mu \eta'_\nu \Phi^{(p,q)}(\zeta_\mu, \zeta'_\nu) = \lambda_1 \sum_{\mu,\nu=1}^{N'} \eta'_\mu \overline{\eta}'_\nu K^{(q,p)}(\zeta'_\mu, \overline{\zeta}'_\nu), \\ \sum_{\mu=1}^{N} \sum_{\nu=1}^{N'} \eta_\mu \eta'_\nu F^{(p,q)}(\zeta_\mu, \zeta'_\nu) = \lambda_2 \sum_{\mu,\nu=1}^{N'} \eta'_\mu \overline{\eta}'_\nu K^{(q,p)}(\zeta'_\mu, \overline{\zeta}'_\nu), \end{cases} \tag{2.12}$$

where λ_1 *and* λ_2 *are constants such that*

$$|\lambda_1|^2 + |\lambda_2|^2 = \sum_{\mu,\nu=1}^{N} \eta_\mu \overline{\eta}_\nu K^{(p,p)}(\zeta_\mu, \overline{\zeta}_\nu) / \sum_{\mu,\nu=1}^{N'} \eta'_\mu \overline{\eta}'_\nu K^{(q,q)}(\zeta'_\mu, \overline{\zeta}'_\nu). \tag{2.13}$$

To prove these, we set

$$x_n = \sum_{\mu=1}^{N} \eta_\mu \frac{\partial^p}{\partial \zeta_\mu^p} \zeta_\mu^n = n(n-1)\cdots(n-p+1) \sum_{\mu=1}^{N} \eta_\mu \zeta_\mu^{n-p},\ n \geqslant p,$$

$$x'_n = y_n = \sum_{\nu=1}^{N'} \eta'_\nu \frac{\partial^q}{\partial \zeta_\nu^q} \zeta'^n_\nu = n(n-1)\cdots(n-q+1) \sum_{\nu=1}^{N'} \eta'_\nu \zeta'^{n-q}_\nu,\ q \geqslant n.$$

It is obvious that

$$X = \sum_{n=1}^{\infty} |x_n|^2/n = \sum_{\mu,\nu=1}^{N} \eta_\mu \bar{\eta}_\nu K^{(p,p)}(\zeta_\mu, \bar{\zeta}_\nu) < \infty,$$

$$X' = Y = \sum_{n=1}^{\infty} |x'_n|^2/n = \sum_{\mu,\nu=1}^{N'} \eta'_\mu \bar{\eta}'_\nu K^{(q,q)}(\zeta'_\mu, \bar{\zeta}'_\nu) < \infty.$$

Therefore, (2.10) follows from inequality (2.5) and the equality holds if and only if the area of the complement of the union of image $f(|\zeta|<1)$ and image $1/f(|\zeta|<1)$ to be vanished. From (2.7), (2.8) and (2.9), we get the second part of the theorem at once. This completes the proof.

This theorem has several corollaries.

Corollary 1 *Let* $f \in B_1$, $\{\zeta_\mu\}$ *and* $\{\zeta'_\nu\}$ *be the two systems of distinguished points in the unit disk* $|\zeta|<1$, $\mu=1, \cdots, N$; $\nu=1, \cdots, N'$, *then we have*

$$\left| \sum_{\mu=1}^{N} \sum_{\nu=1}^{N'} \eta_\mu \eta'_\nu \left(\frac{f'(\zeta_\mu) f'(\zeta'_\nu)}{(f(\zeta_\mu) - f(\zeta'_\nu))^2} - \frac{1}{(\zeta_\mu - \zeta'_\nu)^2} \right) \right|^2 + \left| \sum_{\mu=1}^{N} \sum_{\nu=1}^{N'} \eta_\mu \eta'_\nu \frac{f'(\zeta_\mu) f'(\zeta'_\nu)}{(1 - f(\zeta_\mu) f(\zeta'_\nu))^2} \right|^2$$

$$\leqslant \sum_{\mu,\nu=1}^{N} \eta_\mu \bar{\eta}_\nu (1 - \zeta_\mu \bar{\zeta}_\nu)^{-2} \cdot \sum_{\mu,\nu=1}^{N'} \eta'_\mu \bar{\eta}'_\nu (1 - \zeta'_\mu \bar{\zeta}'_\nu)^{-2}. \tag{2.14}$$

Here the equality holds only if

$$\sum_{\mu=1}^{N} \sum_{\nu=1}^{N'} \eta_\mu \eta'_\nu \left(\frac{f'(\zeta_\mu) f'(\zeta'_\nu)}{(f(\zeta_\mu) - f(\zeta'_\nu))^2} - \frac{1}{(\zeta_\mu - \zeta'_\nu)^2} \right) = \lambda_1 \sum_{\mu,\nu=1}^{N} \eta'_\mu \bar{\eta}'_\nu (1 - \zeta'_\mu \bar{\zeta}'_\nu)^{-2},$$

$$\sum_{\mu=1}^{N} \sum_{\nu=1}^{N'} \eta_\mu \eta'_\nu \left(\frac{f'(\zeta_\mu) f'(\zeta'_\nu)}{(1 - f(\zeta_\mu) f(\zeta'_\nu))^2} \right) = \lambda_2 \sum_{\mu,\nu=1}^{N'} \eta'_\mu \bar{\eta}'_\nu (1 - \zeta'_\mu \bar{\zeta}'_\nu)^{-2},$$

where λ_1, λ_2 *are constants such that*

$$|\lambda_1|^2+|\lambda_2|^2=\sum_{\mu,\nu=1}^{N}\eta_\mu\bar{\eta}_\nu(1-\zeta_\mu\bar{\zeta}_\nu)^{-2}/\sum_{\mu,\nu=1}^{N'}\eta'_\mu\bar{\eta}'_\nu(1-\zeta'_\mu\bar{\zeta}'_\nu)^{-2}.$$

Corollary 2 *Suppose $f\in B_1$, ζ is any point in the unit disk, then we have*

$$|\{f,\zeta\}|^2+36|f'(\zeta)/(1-f(\zeta)^2)|^4\leqslant 36(1-|\zeta|^2)^{-4}, \tag{2.15}$$

where $\{f,\zeta\}=(f''(\zeta)/f'(\zeta))'-\frac{1}{2}(f''(\zeta)/f'(\zeta))^2$ denotes the Schwarz derivative of the function f at ζ. The equality holds only if f satisfies the following equation

$$\{f,\zeta\}=6\lambda_1(f'(\zeta))^2/\lambda_2(1-f(\zeta)^2),$$

where λ_1, λ_2 are constants such that

$$|\lambda_1|^2+|\lambda_2|^2=1.$$

We know from [8] that

$$\{f,\zeta\}=6\frac{\partial^2}{\partial\zeta\partial z}\left(\ln\frac{f(\zeta)-f(z)}{\zeta-z}\right)\Bigg|_{\zeta=z}=\left(\frac{f''(\zeta)}{f'(\zeta)}\right)'-\frac{1}{2}\left(\frac{f''(\zeta)}{f'(\zeta)}\right)^2.$$

Set $N'=N=1$, $\eta'_1=\eta_1=\sqrt{6}$, $\zeta_1=\zeta$, $\zeta'_1=z$ in above corollary and let $z\to\zeta$, we obtain this corollary at once.

Theorem 3 *Suppose f is regular in the unit disk, $f(0)=0$, $f'(0)\neq 0$. Then the necessary and sufficient condition for $f\in B_1$ is that*

$$\frac{1}{\pi}\iint_{|z|<1}\left|\frac{f'(\zeta)f'(z)}{(f(\zeta)-f(z))^2}-\frac{1}{(\zeta-z)^2}\right|^2 d\sigma_z+\frac{1}{\pi}\iint_{|z|<1}\left|\frac{f'(\zeta)f'(z)}{(1-f(\zeta)f(z))^2}\right|^2 d\sigma_z \leqslant(1-|\zeta|^2)^{-2} \tag{2.16}$$

holds for any ζ in the unit disk.

Moreover, if $f\in B_1$, the equality of (2.16) holds if and only if the area of the complement of the union of the image $f(|\zeta|<1)$ and the image $1/f(|\zeta|<1)$ vanishes.

Now we prove the necessity. Assume $f\in B_1$. Since

$$\ln f(\zeta,z)=\sum_{m,n=1}^{\infty}d_{mn}\zeta^m z^n=\sum_{n=1}^{\infty}D_n(\zeta)z^n,$$

$$\ln \varphi(\zeta, z) = \sum_{m, n=1}^{\infty} C_{mn} \zeta^m z^n = \sum_{n=1}^{\infty} C_n(\zeta) z^n,$$

taking the differentiation on both sides of these last expressions for z, and ζ, then we get

$$\frac{f'(\zeta)f'(z)}{(f(\zeta)-f(z))^2} - \frac{1}{(\zeta-z)^2} = \sum_{n=1}^{\infty} nD'_n(\zeta) z^{n-1},$$

$$\frac{f'(\zeta)f'(z)}{(1-f(\zeta)f(z))^2} = \sum_{n=1}^{\infty} nC'_n(\zeta) z^n.$$

Therefore, by virtue of

$$\frac{1}{\pi} \iint_{|z|<1} z^n \bar{z}^{n'} \mathrm{d}\sigma_z = \begin{cases} 0, & n' \neq n, \\ 1/(n+1), & n' = n, \end{cases} \tag{2.17}$$

for n, $n'=1, 2, \cdots$, we obtain that the left-hand side of (2.16) equals to

$$\sum_{n=1}^{\infty} n(|D'_n(\zeta)|^2 + |C'_n(\zeta)|^2).$$

Applying Theorem 2, we can conclude that the inequality is true.

Now we show the sufficiency of the condition (2.16). Assuming condition (2.16) to be satisfied, we first show that $f'(\zeta)$ doesn't vanishing in $|\zeta|<1$. Otherwise, there exists at least a point ζ_0 in $|\zeta|<1$ such that $f'(\zeta_0) \neq 0$. It follows from (2.16) that

$$\frac{1}{\pi} \iint_{|z|<1} |\zeta_0 - z|^{-4} \mathrm{d}\sigma_z \leqslant (1-|\zeta_0|^{-2})^2.$$

But this is impossible, because for any sufficiently small positive number we have

$$\frac{1}{\pi} \iint_{|z|<1} |z-\zeta_0|^{-4} \mathrm{d}\sigma_z \geqslant \frac{1}{\pi} \iint_{|z-\zeta_0| \leqslant \eta} |\zeta_0 - z|^{-4} \mathrm{d}\sigma_z > \frac{1}{\eta^2}.$$

Letting $\eta \to 0$ in the last inequality, we get

$$\frac{1}{\pi} \iint_{|z|<1} |\zeta_0 - z|^{-4} \mathrm{d}\sigma_z \geqslant \infty.$$

So $f^{j}(\zeta) \neq 0$ for any ζ in $|\zeta|<1$.

Secondly we show the univalence of $f(\zeta)$ in $|\zeta|<1$. If it isn't true, then there exist at least ζ_0 and z_0 in the unit disk such that

$$f(\zeta_0)=f(z_0),\ f'(\zeta_0)\neq 0,\ f'(z_0)\neq 0.$$

Therefore, in some neighborhood at z_0, we have

$$f(z)=f(\zeta_0)+f'(z_0)(z-z_0)+O(|z-z_0|^2).$$

In view of this last expression, for any sufficiently small positive number η, we conclude that

$$\frac{1}{\pi}\iint_{|z|<1}\left|\frac{f'(\zeta_0)f'(z)}{(f(\zeta)-f(\zeta_0))^2}-\frac{1}{(\zeta-z)^2}\right|^2 d\sigma_z$$

$$>\frac{1}{\pi}\iint_{|z-z_0|<\eta}\left|\frac{f'(\zeta_0)(f'(z_0)+O(|z-z_0|^2))}{(z-z_0)^2(f'(z_0)+O(|z-z_0|))^2}-\frac{1}{(\zeta_0-z)^2}\right|^2 d\sigma_z>c/\eta^2,$$

where c is a nonzero constant. Letting $\eta\rightarrow+0$, from the last inequality and (2.16), we get $(1-|\zeta_0|^2)^{-2}\geqslant+\infty$. This is also impossible. Hence $f(\zeta)$ is univalent in

$$|\zeta|<1.$$

Finally, we show $f(\zeta_1)f(\zeta_2)\neq 1$ for any ζ_1 and ζ_2 in $|\zeta|<1$. To do this, we assume that there are ζ_0 and z_0 in the unit disk such that $f(\zeta_0)f(z_0)=1$. Then it follows from (2.16) that

$$\frac{1}{\pi}\iint_{|z|<1}\left|\frac{f'(\zeta_0)f'(z)}{(1-f(z)f(\zeta_0))^2}\right|^2 d\sigma_z\leqslant(1-|\zeta_0|^2)^{-2}.$$

But this is impossible, as proved above. Thus we complete the proof of the sufficiency. Therefore, the assertion of the theorem is true.

Remark For Bieberbach-Eilenberg class B_1, we know the inequalities (2.5) are also the sufficient conditions, but that sufficient conditions is too difficult to check. The condition (2.16) is easier to check than (2.5). On the other hand, Theorem 2 really is the generalization of Bazilevic theorem too.

Theorem 4 *Let $f\in B_1$, both $\{\zeta_\mu\}$ $(\mu=1,\cdots,N)$ and $\{\zeta'_\nu\}$ $(\nu=1,\cdots,N')$ be the distinguished points in $|\zeta|<1$, and l be any nonnegative integer. Then we have*

$$\left|\sum_{\mu=1}^{N}\sum_{\nu=1}^{N'}\eta_\mu\eta'_\nu(F(\zeta_\mu,\zeta'_\nu))^l\right|^2+\left|\sum_{\mu=1}^{N}\sum_{\nu=1}^{N'}\eta_\mu\eta'_\nu(\Phi(\zeta_\mu,\zeta'_\nu))^l\right|^2$$
$$\leqslant\sum_{\mu,\nu=1}^{N}\eta_\mu\bar{\eta}_\nu(K(\zeta_\mu,\bar{\zeta}_\nu))^l\cdot\sum_{\mu,\nu=1}^{N'}\eta'_\mu\bar{\eta}'_\nu(K(\zeta'_\mu,\bar{\zeta}'_\nu))^l \tag{2.18}$$

for any nonzero complex numbers $\{\eta_\mu\}$ *and* $\{\eta'_\nu\}$.

To prove this, we will proceed, as in [2, 3]. We now set

$$x_n(\mu)=\eta_\mu^{\frac{1}{l}}\zeta_\mu^n,\ x'_n(\nu)=\eta_\nu'^{\frac{1}{l}}\zeta_\nu'^n$$

for all $n\geqslant 1$, then the left-hand side of (2.18) equals to

$$I_1=\left|\sum_{\mu=1}^{N}\sum_{\nu=1}^{N'}\Big(\sum_{m,n=1}^{\infty}d_{mn}x'_m(\nu)x_n(\mu)\Big)^l\right|^2+\left|\sum_{\mu=1}^{N}\sum_{\nu=1}^{N'}\Big(\sum_{m,n=1}^{\infty}C_{mn}x'_m(\nu)x_n(\mu)\Big)^l\right|^2.$$

By Cauchy inequality once more, we get

$$I_1=\left|\sum_{n_1}\cdots\sum_{n_l}\Big(\sum_{\mu=1}^{N}\prod_{j=1}^{l}x_{n_j}(\mu)\Big)\cdot\sum_{m_1}\cdots\sum_{m_l}\Big(\sum_{\nu=1}^{N'}\prod_{j=1}^{l}d_{m_jn_j}x'_{m_j}(\nu)\Big)\right|^2$$
$$+\left|\sum_{n_1}\cdots\sum_{n_l}\Big(\sum_{\mu=1}^{N}\prod_{j=1}^{l}x_{n_j}(\mu)\Big)\cdot\sum_{m_1}\cdots\sum_{m_l}\Big(\sum_{\nu=1}^{N'}\prod_{j=1}^{l}C_{m_jn_j}x'_{m_j}(\nu)\Big)\right|^2$$
$$\leqslant\Big(\sum_{n_1}\cdots\sum_{n_l}\prod_{j=1}^{l}\frac{1}{n_j}\Big|\sum_{\mu=1}\prod_{j=1}^{l}x_{n_j}(\mu)\Big|^2\Big)\cdot I_2,$$

where

$$I_2=\sum_{n_1}\cdots\sum_{n_l}\prod_{j=1}^{l}n_j\Big|\sum_{m_1}\cdots\sum_{m_l}\sum_{\nu=1}^{N'}\prod_{j=1}^{l}d_{m_jn_j}x'_{m_j}(\nu)\Big|^2$$
$$+\sum_{n_1}\cdots\sum_{n_l}\prod_{j=1}^{l}n_j\Big|\sum_{m_1}\cdots\sum_{m_l}\sum_{\nu=1}^{N'}\prod_{j=1}^{l}C_{m_jn_j}x_{m_j}(\nu)\Big|^2.$$

Using the following Shah inequalities once more

$$\sum_{n=1}^{\infty}n\Big(\Big|\sum_{m=1}^{\infty}d_{mn}y_m\Big|^2+\Big|\sum_{m=1}^{\infty}C_{mn}y_m\Big|^2\Big)\leqslant\sum_{n=1}^{\infty}|y_n|^2/n,$$

it follows that

$$I_2=\prod_{j=1}^{l-1}\sum_{n_j}n_j\sum_{n_l}n_l\Big(\Big|\sum_{m_l}d_{m_ln_l}\Big(\sum_{m_l}\cdots\sum_{m_{l-1}}\sum_{\nu=1}^{N'}\prod_{j=1}^{l-1}d_{m_jn_j}x'_{m_j}(\nu)\Big)x'_{m_l}(\nu)\Big|^2$$
$$+\Big(\Big|\sum_{m_l}C_{m_ln_l}\Big(\sum_{m_l}\cdots\sum_{m_{l-1}}\sum_{\nu=1}^{N'}\prod_{j=1}^{l-1}C_{m_jn_j}x'_{m_j}(\nu)\Big)x'_{m_l}(\nu)\Big|^2\Big)$$

$$\leqslant \prod_{j=1}^{l-1} \sum_{n_j} n_j \sum_{n_l} \frac{1}{n_l} \Big(\Big| \sum_{m_l} \cdots \sum_{m_{l-1}} \sum_{\nu=1}^{N'} \prod_{j=1}^{l-1} d_{m_j n_j} x'_{m_j}(\nu) x'_{n_l}(\nu) \Big|^2$$

$$+ \Big| \sum_{m_1} \cdots \sum_{m_{l-1}} \sum_{\nu=1}^{N'} \prod_{j=1}^{l-1} C_{m_j n_j} x'_{m_j}(\nu) x'_{n_l}(\nu) \Big|^2 \Big)$$

$$\leqslant \prod_{j=2}^{l} \sum_{n_j} \frac{1}{n_j} \sum_{n_1} n_1 \Big(\Big| \sum_{m_1} d_{m_1 n_1} \Big(\sum_{\nu=1}^{N'} \prod_{j=1}^{l} x'_{n_j}(\nu) \Big) \Big|^2 + \Big| \sum_{m_1} C_{m_1 n_1} \Big(\sum_{\nu=1}^{N'} \prod_{j=1}^{l} x'_{n_j}(\nu) \Big) \Big|^2 \Big)$$

$$\leqslant \prod_{j=1}^{l} \sum_{n_j} \frac{1}{n_j} \Big| \sum_{\nu=1}^{N'} x'_{n_j}(\nu) \Big|^2.$$

Therefore, we obtain the left-hand side of (2.18)

$$I_1 \leqslant \Big(\sum_{n_1} \cdots \sum_{n_l} \prod_{j=1}^{l} \frac{1}{n_j} \Big| \sum_{\mu=1}^{N} \prod_{j=1}^{l} x_{n_j}(\mu) \Big|^2 \Big) \cdot \Big(\sum_{n_1} \cdots \sum_{n_l} \prod_{j=1}^{l} \frac{1}{n_j} \Big| \sum_{\nu=1}^{N'} \prod_{j=1}^{l} x'_{n_j}(\nu) \Big|^2 \Big).$$

Since

$$\sum_{n_1} \cdots \sum_{n_l} \prod_{j=1}^{l} \frac{1}{n_j} \Big| \sum_{\mu=1}^{N} \prod_{j=1}^{l} x_{m_j}(\mu) \Big|^2 = \sum_{\mu,\nu=1}^{N} \Big(\sum_{n=1}^{\infty} \frac{1}{n} x_n(\mu) \overline{x_n(\nu)} \Big)^l$$

$$= \sum_{\mu,\nu=1}^{N} \eta_\mu \bar{\eta}_\nu (K(\zeta_\mu, \bar{\zeta}_\nu))^l,$$

and similarly

$$\sum_{n_1} \cdots \sum_{n_l} \prod_{j=1}^{l} \frac{1}{n_j} \Big| \sum_{\nu=1}^{N'} \prod_{j=1}^{l} x_{n_j}(\nu) \Big|^2 = \sum_{\mu,\nu=1}^{N'} \eta_\mu \bar{\eta}_\nu (K(\zeta'_\mu, \bar{\zeta}'_\nu))^l,$$

we get (2.18) at once, when substituting these two relations into the last inequality. This completes the proof.

From this theorem, we obtain immediately the following:

Corollary 1 *Let* $f \in B_1$, *l be any nonnegative real number, and* $\{\zeta_\mu\}$ ($\mu=1, \cdots, N$) *be an arbitrary system of distinguished points in* $|\zeta|<1$. *We have*

$$\Big| \sum_{\mu,\nu=1}^{N} \eta_\mu \eta_\nu (F(\zeta_\mu, \zeta_\nu))^l \Big|^2 + \Big| \sum_{\mu,\nu=1}^{N} \eta_\mu \eta_\nu (\Phi(\zeta_\mu, \zeta_\nu))^l \Big|^2$$

$$\leqslant \Big(\sum_{\mu,\nu=1}^{N} \eta_\mu \bar{\eta}_\nu (K(\zeta_\mu, \zeta_\nu))^l \Big)^2 \qquad (2.19)$$

for any complex constants $\{\eta_\mu\}$.

In particular, when $l=1$, inequality (2.19) *reduces to the inequality* (3.5) *of* [1].

Corollary 2 *Let $f\in B_1$, p be any complex number, and $\{\zeta_\mu\}$ ($\mu=1,\cdots,N$) be distinguished points. Then for any complex numbers $\{\eta_\mu\}$, we have*

$$\left|\sum_{\mu,\nu=1}^{N}\eta_\mu\eta_\nu(f(\zeta_\mu,\zeta_\nu))^p\right|\leqslant\sum_{\mu,\nu=1}^{N}\eta_\mu\bar{\eta}_\nu(1-\zeta_\mu\bar{\zeta}_\nu)^{-|p|},\qquad(2.20)$$

$$\left|\sum_{\mu,\nu=1}^{N}\eta_\mu\eta_\nu(\varphi(\zeta_\mu,\zeta_\nu))^p\right|\leqslant\sum_{\mu,\nu=1}^{N}\eta_\mu\bar{\eta}_\nu(1-\zeta_\mu\bar{\zeta}_\nu)^{-|p|}.\qquad(2.21)$$

Proof By virtue of Taylor expansion of exponential function and using (2.19), it follows that

$$\left|\sum_{\mu,\nu=1}^{N}\eta_\mu\eta_\nu(f(\zeta_\mu,\zeta_\nu))^p\right|\leqslant\sum_{l=0}^{\infty}\frac{1}{l!}|p|^l\left|\sum_{\mu,\nu=1}^{N}\eta_\mu\eta_\nu(F(\zeta_\mu,\zeta_\nu))^l\right|$$

$$\leqslant\sum_{l=0}^{\infty}\frac{1}{l!}|p|^l\sum_{\mu,\nu=1}^{N}\eta_\mu\bar{\eta}_\nu(K(\zeta_\mu,\zeta_\nu))^l=\sum_{\mu,\nu=1}^{N}\eta_\mu\bar{\eta}_\nu(1-\zeta_\mu\zeta_\nu)^{-|p|}.$$

Similarly, we get (2.21).

Theorem 5 *Let $f\in B_1$, l be nonnegative integer, and both $\{\zeta_\mu\}$ ($\mu=1,2,\cdots,N$) and $\{\zeta'_\nu\}$ ($\nu=1,2,\cdots,N'$) be arbitrary systems of distinguished points in $|\zeta|<1$. Then*

$$\left|\sum_{\mu=1}^{N}\sum_{\nu=1}^{N'}\eta_\mu\eta'_\nu(\ln f(\zeta_\mu,\zeta'_\nu)(\varphi(\zeta_\mu,\zeta'_\nu))^{\varepsilon})^l\right|$$

$$\leqslant\sum_{\mu,\nu=1}^{N}\eta_\mu\bar{\eta}_\nu(K(\zeta_\mu,\bar{\zeta}_\nu))^l\cdot\sum_{\mu,\nu=1}^{N'}\eta'_\mu\bar{\eta}'_\nu(K(\zeta'_\mu,\bar{\zeta}'_\nu))^l\qquad(2.22)$$

holds for any nonzero constants $\{\eta_\mu\}$ and $\{\eta'_\nu\}$, $\varepsilon=1,-1$.

In particular, while $l=1$, this theorem reduces to the corresponding inequality of [1].

Proof We first see that the left-hand side of (2.22) equals to

$$I=\left|\sum_{\mu,\nu=1}^{N}\eta_\mu\eta'_\nu\left(\sum_{m,n=1}^{\infty}(d_{mn}+\varepsilon C_{mn})\zeta'^{m}_{\nu}\zeta^n_\mu\right)^l\right|.$$

By means of the same method as in above Theorem 4, using Cauchy inequality again and again and applying the following Shah inequality

$$\sum_{n=1}^{\infty} n \left| \sum_{m=1}^{\infty} (d_{mn} + \varepsilon C_{mn}) x'_m \right|^2 \leqslant \sum_{n=1}^{\infty} |x'_n|^2 / n,$$

we can obtain

$$I \leqslant \sum_{\mu,\nu=1}^{N} \eta_\mu \bar{\eta}_\nu (K(\zeta_\mu, \bar{\zeta}_\nu))^l \cdot \sum_{\mu,\nu=1}^{N'} \eta'_\mu \bar{\eta}'_\nu (K(\zeta'_\mu, \bar{\zeta}'_\nu))^l.$$

This completes the proof.

Corollary 1 *Suppose $f \in B_1$, $\varepsilon = 1, -1$, and p be any complex number, then*

$$\left| \sum_{\mu,\nu=1}^{N} \eta_\mu \eta_\nu (f(\zeta_\mu, \zeta_\nu)(\varphi(\zeta_\mu, \zeta_\nu))^\varepsilon)^p \right| \leqslant \sum_{\mu,\nu=1}^{N} \eta_\mu \bar{\eta}_\nu (1 - \zeta_\mu \bar{\zeta}_\nu)^{-|p|} \quad (2.23)$$

for any nonzero complex constants $\{\eta_\mu\}$.

Theorem 6 *Let $f \in B_1$, l be any nonnegative integer and $\{\zeta_\mu)$ $(\mu = 1, \cdots, N)$ be a system of distinguished points in $|\zeta| < 1$. Let $P_n(t)$ denote the n-th Faber polynomials generated by the function f, $\varepsilon = 1, -1$, and*

$$g_n^{(\varepsilon)}(\zeta) = P_n(1/f(\zeta)) - (\zeta^{-n} + \varepsilon \bar{\zeta}^n),$$

$$h_n(\zeta) = P_n(f(\zeta)) + 2n\gamma_n, \ \ln(f(\zeta)/\zeta) = -2\sum_{n=1}^{\infty} \gamma_n \zeta^n.$$

Then for any nonzero complex constants $\{\eta_\mu\}$ we have

$$\frac{l}{2} \sum_{n=1}^{\infty} \left(\left| \sum_{\mu=1}^{N} \eta_\mu g_n^{(\varepsilon)}(\zeta_\mu) \right|^2 + \left| \sum_{\mu=1}^{N} \eta_\mu h_n(\zeta_\mu) \right|^2 \right) / n$$

$$\leqslant \sum_{\mu,\nu=1}^{N} \eta_\mu \bar{\eta}_\nu \ln(|f(\zeta_\mu, \zeta_\nu)|^{\varepsilon l} / |1 - \zeta_\mu \bar{\zeta}_\nu|^l); \quad (2.24)$$

$$\frac{l}{2} \sum_{n=1}^{\infty} \left| \sum_{\mu=1}^{N} (g_n^{(\varepsilon)}(\zeta_\mu) + \varepsilon' h_n(\zeta_\mu)) \right|^2 / n$$

$$\leqslant \sum_{\mu,\nu=1}^{N} \eta_\mu \bar{\eta}_\nu \ln(|f(\zeta_\mu, \zeta_\nu)(\varphi(\zeta_\mu, \zeta_\nu))^{\varepsilon'}|^{\varepsilon l} / |1 - \zeta_\mu \bar{\zeta}_\nu|^l). \quad (2.25)$$

Here the equalities hold if and only if the area of the complement of the union of image $f(|\zeta| < 1)$ and $1/f(|\zeta| < 1)$ vanishes. $\varepsilon' = 1, -1$.

Proof Using the properties of the Faber polynomials, we get

$$-g_n^{(\varepsilon)}(\zeta)=n\sum_{m=1}^{\infty}d_{mn}\zeta^m+\varepsilon\bar{\zeta}^n,$$

$$-h_n(\zeta)=n\sum_{m=1}^{\infty}C_{mn}\zeta^m.$$

Therefore, applying (2. 10) of Theorem 2 (here, we choose $p=q=0$), we obtain

$$\begin{aligned}
&\sum_{\mu,\nu=1}^{N}\eta_\mu\bar{\eta}_\nu K(\zeta_\mu,\bar{\zeta}_\nu)+\sum_{\mu,\nu=1}^{N'}\eta'_\mu\bar{\eta}'_\nu K(\zeta'_\mu,\bar{\zeta}'_\nu)\\
\geqslant&\sum_{n=1}^{\infty}n\Big(\Big|\sum_{\mu=1}^{N}\eta_\mu\Big(\sum_{m=1}^{\infty}d_{mn}\zeta_\mu^m\Big)+\sum_{\nu=1}^{N'}\eta'_\nu\Big(\sum_{m=1}^{\infty}C_{mn}\zeta'^m_\nu\Big)\Big|^2\\
&+\Big|\sum_{\mu=1}^{N}\eta_\mu\Big(\sum_{m=1}^{\infty}C_{mn}\zeta_\mu^m\Big)+\sum_{\nu=1}^{N'}\eta'_\nu\Big(\sum_{m=1}^{\infty}d_{mn}\zeta'^m_\nu\Big)\Big|^2\Big)\\
=&\sum_{n=1}^{\infty}\Big(\Big|\sum_{\mu=1}^{N}\eta_\mu g_n^{(\varepsilon)}(\zeta_\mu)+\sum_{\nu=1}^{N'}\eta'_\nu h_n(\zeta'_\nu)\Big|^2+\Big|\sum_{\mu=1}^{N}\eta_\mu h_n(\zeta_\mu)+\sum_{\nu=1}^{N'}\eta'_\nu g_n^{(\varepsilon)}(\zeta'_\nu)\Big|^2\Big)/n\\
&-2\varepsilon\mathrm{Re}\Big\{\sum_{\mu,\nu=1}^{N}\eta_\mu\bar{\eta}_\nu F(\zeta_\mu,\zeta_\nu)+\sum_{\mu=1}^{N}\sum_{\nu=1}^{N'}\bar{\eta}_\mu\eta'_\nu\Phi(\zeta_\mu,\zeta'_\nu)\\
&+\sum_{\mu,\nu=1}^{N'}\eta'_\mu\bar{\eta}'_\nu F(\zeta'_\mu,\zeta'_\nu)+\sum_{\mu=1}^{N}\sum_{\nu=1}^{N'}\eta_\mu\bar{\eta}'_\nu\Phi(\zeta_\mu,\zeta'_\nu)\Big\}\\
&-\sum_{\mu,\nu=1}^{N}\eta_\mu\bar{\eta}_\nu K(\bar{\zeta}_\mu,\zeta_\nu)-\sum_{\mu,\nu=1}^{N'}\eta'_\mu\bar{\eta}'_\nu K(\bar{\zeta}'_\mu,\zeta'_\nu),
\end{aligned}$$

where $\{\zeta_\mu\}$ and $\{\zeta'_\nu\}$ both are the systems of the distinguished points in $|\zeta|<1$, $\{\eta_\mu\}$ and $\{\eta_\nu\}$ both are any complex constants.

We know that if $A(\zeta,z)=A(z,\zeta)$, then[2]

$$\mathrm{Re}\Big\{\sum_{\mu,\nu=1}^{N}\eta_\mu\bar{\eta}_\nu A(\zeta_\mu,z_\nu)\Big\}=\sum_{\mu,\nu=1}^{N}\eta_\mu\bar{\eta}_\nu\mathrm{Re}\{A(\zeta_\mu,z_\nu)\}$$

and $K(\zeta,\bar{z})=\overline{K(\bar{\zeta},z)}$. Therefore, it follows from the above inequality that

$$\begin{aligned}
&\frac{l}{2}\sum_{n=1}^{N}\Big\{\Big|\sum_{\mu=1}^{N}\eta_\mu g_n^{(\varepsilon)}(\zeta_\mu)+\sum_{\nu=1}^{N'}\eta'_\nu h_n(\zeta'_\nu)\Big|^2+\Big|\sum_{\mu=1}^{N}\eta_\mu h_n(\zeta_\mu)+\sum_{\nu=1}^{N'}\eta'_\nu g_n^{(\varepsilon)}(\zeta'_\nu)\Big|^2\Big\}/n\\
\leqslant&\sum_{\mu,\nu=1}^{N}\eta_\mu\bar{\eta}_\nu\ln(|f(\zeta_\mu,\zeta_\nu)|^{\varepsilon}/|1-\zeta_\mu\bar{\zeta}_\nu|)^l
\end{aligned}$$

$$+\sum_{\mu,\nu=1}^{N'}\eta'_\mu\bar{\eta}'_\nu\ln(|f(\zeta'_\mu,\zeta'_\nu)|^{\varepsilon}/|1-\zeta'_\mu\bar{\zeta}'_\nu|)^{l}+\sum_{\mu=1}^{N}\sum_{\nu=1}^{N'}\eta_\mu\bar{\eta}'_\nu\ln|\varphi(\zeta_\mu,\zeta'_\nu)|^{\varepsilon l}$$

$$+\sum_{\mu=1}^{N}\sum_{\nu=1}^{N'}\bar{\eta}_\mu\eta'_\nu|\varphi(\zeta_\mu,\zeta'_\nu)|^{\varepsilon l}. \tag{2.26}$$

Particularly, if we set $\eta'_\mu=0$ ($\mu=1,\cdots,N'$), it follows (2.24) from (2.26); if we set $N'=N$, $\eta'_\mu=\varepsilon'\eta_\mu$, $\zeta'_\mu=\zeta_\mu$ for $\mu=1,\cdots,N$, it follows from (2.26) that (2.25) is valid. This completes the proof.

Theorem 7 *Let $f\in B_1$, l be nonnegative integer and $\{\zeta_\mu\}$ ($\mu=1,\cdots,N$) be the distinguished points in $|\zeta|<1$. Let $\sum_{\mu,\nu=1}^{N}a_{\mu\nu}\eta_\mu\bar{\eta}_\nu\geqslant 0$ for any complex numbers $\{\zeta_\mu\}$ ($\mu=1,\cdots,N$) and $\varepsilon=1,-1$. Then we have*

$$\sum_{\mu,\nu=1}^{N}\eta_\mu\bar{\eta}_\nu a_{\mu\nu}\left|\frac{f(\zeta_\mu)f(\zeta_\nu)}{\zeta_\mu\zeta_\nu}\right|^{\varepsilon l}\cdot\exp\left\{\frac{l}{2}\sum_{n=1}^{\infty}\frac{1}{n}(g_n^{(\varepsilon)}(\zeta_\mu)\overline{g_n^{(\varepsilon)}(\zeta_\nu)}+h_n(\zeta_\mu)\overline{h_n(\zeta_\nu)})\right\}$$

$$\leqslant\sum_{\mu,\nu=1}^{N}a_{\mu\nu}\eta_\mu\bar{\eta}_\nu\left|\frac{f'(0)(f(\zeta_\mu)-f(\zeta_\nu))}{\zeta_\mu-\zeta_\nu}\right|^{\varepsilon l}/|1-\zeta_\mu\bar{\zeta}_\nu|^{l}, \tag{2.27}$$

$$\sum_{\mu,\nu=1}^{N}a_{\mu\nu}\eta_\mu\bar{\eta}_\nu\left|\frac{f(\zeta_\mu)f(\zeta_\nu)}{\zeta_\mu\zeta_\nu}\right|^{\varepsilon l}\cdot\exp\left\{\frac{l}{2}\sum_{n=1}^{\infty}\frac{1}{n}(g_n^{(\varepsilon)}(\zeta_\mu)\right.$$

$$\left.+\varepsilon' h_n(\zeta_\mu))\overline{(g_n^{(\varepsilon)}(\zeta_\nu)+\varepsilon' h_n(\zeta_\nu))}\right\}$$

$$\leqslant\sum_{\mu,\nu=1}^{N}a_{\mu\nu}\eta_\mu\bar{\eta}_\nu\left|\frac{f'(0)(f(\zeta_\mu)-f(\zeta_\nu))}{\zeta_\mu-\zeta_\nu}\right|^{\varepsilon l}/|1-\zeta_\mu\bar{\zeta}_\nu|^{l}, \tag{2.28}$$

where $\{\eta_\mu)$ ($\mu=1,\cdots,N$) are any nonzero constants.

Proof Set

$$a_{\mu,\nu}^{(1)}=\ln(|f(\zeta_\mu,\zeta_\nu)|^{\varepsilon l}/|1-\zeta_\mu\bar{\zeta}_\nu|^{l}),$$

$$a_{\mu,\nu}^{(2)}=\frac{l}{2}\sum_{n=1}^{\infty}(g_n^{(\varepsilon)}(\zeta_\mu)\overline{g_n^{(\varepsilon)}(\zeta_\nu)}+h_n(\zeta_\mu)\overline{h_n(\zeta_\nu)})/n$$

and $a_{\mu,\nu}^{(3)}=a_{\mu,\nu}^{(1)}-a_{\mu,\nu}^{(2)}$ for $\mu,\nu=1,\cdots,N$.

Evidently the matrix $(a_{\mu,\nu}^{(2)})$ is positive semi-definite. Applying Theorem 6, it follows that the matrices $(a_{\mu,\nu}^{(1)})$ and $(a_{\mu,\nu}^{(3)})$ for $\mu,\nu=1,\cdots,N$ are both positive semi-definite. So from [4, p. 314, Lemma 1] we have

$$\sum_{\mu,\nu=1}^{N} a_{\mu\nu} \exp(a_{\mu,\nu}^{(2)})(\exp(a_{\mu,\nu}^{(3)})-1)\eta_\mu \bar{\eta}_\nu \geqslant 0$$

for any nonzero simultaneously complex numbers $\{\eta_\mu\}$. Namely

$$\sum_{\mu,\nu=1}^{N} a_{\mu\nu}\eta_\mu\bar{\eta}_\nu |f(\zeta_\mu, \zeta_\nu)|^{\varepsilon l}/|1-\zeta_\mu\bar{\zeta}_\nu|^l \geqslant \sum_{\mu,\nu=1}^{N} \eta_\mu\bar{\eta}_\nu a_{\mu\nu} \exp\{a_{\mu,\nu}^{(2)}\}.$$

Substituting $\eta_\mu |f(\zeta_\mu)/\zeta_\mu|^{\varepsilon l}$ for η_μ in the last inequalities, we obtain at once the inequalities (2.27). Similarly, just as we have proved above, we can verify the inequalities (2.28). Thus we complete the proof.

3. On the Grunsky Functions

Similarly, by virtue of the basic inequality for $f \in G_1$ from [1], we can prove the following results:

Theorem 1 *Suppose $f \in G_1$, p, q are any nonnegative integers and $\{\zeta_\mu\}$ $(\mu = 1, \cdots, N)$, $\{\zeta'_\nu\}$ $(\nu = 1, \cdots, N')$ are both the systems of the distinguished points in $|\zeta| < 1$. Then for any nonzero simultaneously complex constants $\{\eta_\mu\}$ and $\{\eta'_\nu\}$, we have*

1)

$$\sum_{n=1}^{\infty} n\Big(\Big|\sum_{\mu=1}^{N}\overline{\eta_\mu E_n^{(p)}(\zeta_\mu)} + \sum_{\nu=1}^{N'}\eta'_\nu D_n^{(q)}(\zeta'_\nu)\Big|^2 + \Big|\sum_{\mu=1}^{N}\overline{\eta_\mu D_n^{(p)}(\zeta_\mu)} + \sum_{\nu=1}^{N'}\eta'_\nu E_n^{(q)}(\zeta'_\nu)\Big|^2\Big)$$
$$\leqslant \sum_{\mu,\nu=1}^{N}\eta_\mu\bar{\eta}_\nu K^{(p,p)}(\zeta_\mu, \bar{\zeta}_\nu) + \sum_{\mu,\nu=1}^{N'}\eta'_\mu\bar{\eta}'_\nu K^{(q,q)}(\zeta'_\mu, \bar{\zeta}'_\nu), \qquad (3.1)$$

where the equality holds if and only if the area of the complement of the union of $f(|\zeta|<1)$ and $1/f(|\zeta|<1)$ vanishes.

2)

$$\Big|\sum_{\mu=1}^{N}\sum_{\nu=1}^{N'}\eta_\mu\eta'_\nu F^{(p,q)}(\zeta_\mu, \zeta'_\nu)\Big| + \Big|\sum_{\mu=1}^{N}\sum_{\nu=1}^{N'}\eta_\mu\bar{\eta}'_\nu \Psi^{(p,q)}(\zeta_\mu, \bar{\zeta}'_\nu)\Big|$$
$$\leqslant \Big(\sum_{\mu,\nu=1}^{N}\eta_\mu\bar{\eta}_\nu K^{(p,p)}(\zeta_\mu, \bar{\zeta}_\nu)\cdot\sum_{\mu,\nu=1}^{N'}\eta'_\mu\bar{\eta}'_\nu K^{(q,q)}(\zeta'_\mu, \bar{\zeta}'_\nu)\Big)^{\frac{1}{2}} \qquad (3.2)$$

of which equality holds only if for $n=1, 2, \cdots$.

$$\sum_{m=1}^{\infty}\left(d_{mn}\sum_{\mu=1}^{N'}\eta'_{\mu}\frac{\partial^{q}}{\partial\zeta'^{q}_{\mu}}\zeta'^{m}_{\mu}+\overline{e_{mn}\sum_{\mu=1}^{N'}\eta'_{\mu}\frac{\partial^{q}}{\partial\zeta'^{q}_{\mu}}\zeta'^{m}_{\mu}}\right)=\frac{\lambda}{n}\sum_{\mu=1}^{N}\eta_{\mu}\frac{\partial^{p}}{\partial\zeta^{p}_{\mu}}\zeta^{n}_{\mu}, \tag{3.3}$$

λ being constant such that

$$|\lambda|^{2}=\sum_{\mu,\nu=1}^{N'}\eta'_{\nu}\bar{\eta}'_{\nu}K^{(q,q)}(\zeta'_{\mu},\bar{\zeta}'_{\nu})/\sum_{\mu,\nu=1}^{N}\eta_{\mu}\bar{\eta}_{\nu}K^{(p,p)}(\zeta_{\mu},\bar{\zeta}_{\nu}).$$

Corollary 1 *Let $f\in G_1$, $\{\zeta_{\mu}\}$ ($\mu=1,\cdots,N$) and $\{\zeta'_{\nu}\}$ ($\nu=1,\cdots,N'$) be both the systems of the distinguished points in $|\zeta|<1$, then*

$$\left|\sum_{\mu=1}^{N}\sum_{\nu=1}^{N'}\eta_{\mu}\eta'_{\nu}F(\zeta_{\mu},\zeta'_{\nu})\right|+\left|\sum_{\mu=1}^{N}\sum_{\nu=1}^{N'}\eta_{\mu}\bar{\eta}'_{\nu}\Psi(\zeta_{\mu},\bar{\zeta}'_{\nu})\right|$$
$$\leqslant\left(\sum_{\mu,\nu=1}^{N}\eta_{\mu}\bar{\eta}_{\nu}K(\zeta_{\mu},\bar{\zeta}_{\nu})\cdot\sum_{\mu,\nu=1}^{N'}\eta'_{\mu}\bar{\eta}'_{\nu}K(\zeta'_{\mu},\bar{\zeta}'_{\nu})\right)^{1/2}. \tag{3.4}$$

In particular, while $N'=N$, $\zeta'_{\mu}=\zeta$, $\eta'_{\mu}=\eta_{\mu}$, it reduces to the inequalities (4.1) of [1].

Corollary 2 *Under the assumption of above Corollary 1, we have*

$$\left|\sum_{\mu=1}^{N}\sum_{\nu=1}^{N'}\eta_{\mu}\eta'_{\nu}\left(\frac{f'(\zeta_{\mu})f'(\zeta'_{\nu})}{(f(\zeta_{\mu})-f(\zeta'_{\nu}))^{2}}-\frac{1}{(\zeta_{\mu}-\zeta'_{\nu})^{2}}\right)\right|$$
$$+\left|\sum_{\mu=1}^{N}\sum_{\nu=1}^{N'}\eta_{\mu}\eta'_{\nu}\frac{f'(\zeta_{\mu})\overline{f'(\zeta'_{\nu})}}{(1+f(\zeta_{\mu})\overline{f(\zeta'_{\nu})})^{2}}\right|$$
$$\leqslant\left(\sum_{\mu,\nu=1}^{N}\eta_{\mu}\bar{\eta}_{\nu}(1-\zeta_{\mu}\bar{\zeta}_{\nu})^{-2}\cdot\sum_{\mu,\nu=1}^{N'}\eta'_{\mu}\bar{\eta}'_{\nu}(1-\zeta'_{\mu}\bar{\zeta}'_{\nu})^{-2}\right)^{1/2}. \tag{3.5}$$

This strenthens the corresponding results of [4].

Corollary 3 *Let $f\in G_1$, then for any point ζ in the unit disk we have*

$$|\{f,\zeta\}|+6|f'(\zeta)|^{2}/(1+|f(\zeta)|^{2})^{2}\leqslant 6(1-|\zeta|^{2})^{-2}. \tag{3.6}$$

This was proved by Beresniewiez-Rajca, Olga[4].

Theorem 2 *Suppose $f(\zeta)$ is regular in the unit disk, $f(0)=0$ and $f'(0)\neq 0$, then the necessary and sufficient condition for $f\in G_1$ is that*

$$\frac{1}{\pi}\iint_{|z|<1}\left(\left|\frac{f'(\zeta)f'(z)}{(f(\zeta)-f(z))^{2}}-\frac{1}{(\zeta-z)^{2}}\right|^{2}+\left|\frac{f'(\zeta)\overline{f'(z)}}{(1+f(\zeta)\overline{f(z)})^{2}}\right|^{2}\right)d\sigma_{z}$$

$$\leqslant (1-|\zeta|^2)^{-2} \tag{3.7}$$

for any point ζ *in* $|\zeta|<1$. *Moreover, if* $f\in G_1$, *then the equality of* (3.7) *holds if and only if the area of the complement of the union of* $f(|\zeta|<1)$ *and* $1/f(|\zeta|<1)$ *vanishes.*

Theorem 3 *Let* $f\in G_1$, l *be any nonnegative integer, and* $\{\zeta_\mu\}$ ($\mu=1,\cdots,N$), $\{\zeta'_\nu\}$ ($\nu=1,\cdots,N'$) *be any distinguished points in* $|\zeta|<1$, *then we have*

$$\left|\sum_{\mu=1}^{N}\sum_{\nu=1}^{N'}\eta_\mu\eta'_\nu(F(\zeta_\mu,\zeta'_\nu))^l\right|^2+\left|\sum_{\mu=1}^{N}\sum_{\nu=1}^{N'}\eta_\mu\eta'_\nu(\Psi(\zeta_\mu,\bar{\zeta}'_\nu))^l\right|^2$$

$$\leqslant\sum_{\mu,\nu=1}^{N}\eta_\mu\bar{\eta}_\nu(K(\zeta_\mu,\bar{\zeta}_\nu))^l\cdot\sum_{\mu,\nu=1}^{N'}\eta'_\mu\bar{\eta}'_\nu(K(\zeta'_\mu,\bar{\zeta}'_\nu))^l \tag{3.8}$$

for any complex numbers $\{\eta_\mu\}$ *and* $\{\eta'_\nu\}$.

Corollary *Let* $f\in G_1$, p *be any complex number and let* $\{\zeta_\mu\}$ ($\mu=1,\cdots,N$) *be any distinguished points in* $|\zeta|<1$, *then we have*

$$\left|\sum_{\mu,\nu=1}^{N}\eta_\mu\eta_\nu(f(\zeta_\mu,\zeta_\nu))^p\right|\leqslant\sum_{\mu,\nu=1}^{N}\eta_\mu\bar{\eta}_\nu(1-\zeta_\mu\bar{\zeta}_\nu))^{-|p|}, \tag{3.9}$$

$$\left|\sum_{\mu,\nu=1}^{N}\eta_\mu\eta_\nu(\psi(\zeta_\mu,\zeta_\nu))^p\right|\leqslant\sum_{\mu,\nu=1}^{N}\eta_\mu\bar{\eta}_\nu(1-\zeta_\mu\bar{\zeta}_\nu))^{-|p|}, \tag{3.10}$$

for any nonzero complex numbers $\{\eta_\mu\}$.

Theorem 4 *Let* $f\in G_1$, $\varepsilon=1,-1$, p *be any complex number and* l *be any nonnegative integer. Let* $\{\zeta_\mu\}$ ($\mu=1,\cdots,N$) *and* $\{\zeta'_\nu\}$ ($\nu=1,\cdots,N'$) *be any distinguished points in* $|\zeta|<1$, *then we have*

$$\left|\sum_{\mu=1}^{N}\sum_{\nu=1}^{N'}\eta_\mu\eta'_\nu[\ln(f(\zeta_\mu,\zeta'_\nu)(\psi(\zeta_\mu,\zeta'_\nu))^\varepsilon)]^l\right|^2$$

$$\leqslant\sum_{\mu,\nu=1}^{N}\eta_\mu\bar{\eta}_\nu(K(\zeta_\mu,\bar{\zeta}_\nu))^l\cdot\sum_{\mu,\nu=1}^{N'}\eta'_\mu\bar{\eta}'_\nu(K(\zeta'_\mu,\bar{\zeta}'_\nu))^l, \tag{3.11}$$

$$\left|\sum_{\mu,\nu=1}^{N}\eta_\mu\eta_\nu(f(\zeta_\mu,\zeta_\nu)(\psi(\zeta_\mu,\zeta_\nu))^\varepsilon)^p\right|\leqslant\sum_{\mu,\nu=1}^{N}\eta_\mu\bar{\eta}_\nu(1-\zeta_\mu\bar{\zeta}_\nu)^{-|p|} \tag{3.12}$$

for any complex numbers $\{\eta_\mu\}$ *and* $\{\eta'_\nu\}$.

Theorem 5 *Let* $f\in G_1$, $\varepsilon=1,-1$, $l>0$ *and* $(a_{\mu,\nu})$ *for* $\mu,\nu=1,\cdots,$

N be positive semi-definite Hermite matrix. Let $P_n(t)$ be the n-th Faber polynomials generated by function f, and

$$g_n^{(\varepsilon)}(\zeta)=P_n(1/f(\zeta))-(\zeta^{-n}+\varepsilon\bar{\zeta}^n),$$

$$h_n^*(\zeta)=P_n(-f(\zeta))+2\gamma_n,\quad \ln(f(\zeta)/\zeta)=2\sum_{n=0}^{\infty}\gamma_n\zeta^n$$

for $n\geqslant 1$, then we have

$$\frac{l}{2}\sum_{n=1}^{\infty}\left(\left|\sum_{\mu=1}^{N}\eta_\mu g_n^{(\varepsilon)}(\zeta_\mu)\right|^2+\left|\sum_{\mu=1}^{N}\eta_\mu h_n^*(\zeta_\mu)\right|^2\right)/n$$
$$\leqslant\sum_{\mu,\nu=1}^{N}\eta_\mu\bar{\eta}_\nu\ln(|f(\zeta_\mu,\zeta_\nu)|^{\varepsilon l}/|1-\zeta_\mu\bar{\zeta}_\nu|^l),\tag{3.13}$$

$$\sum_{\mu,\nu=1}^{N}\eta_\mu\bar{\eta}_\nu a_{\mu\nu}\left|\frac{f(\zeta_\mu)f(\zeta_\nu)}{\zeta_\mu\zeta_\nu}\right|^{\varepsilon l}\cdot\exp\left\{\frac{l}{2}\sum_{n=1}^{\infty}\frac{1}{n}(g_n^{(\varepsilon)}(\zeta_\mu)\overline{g_n^{(\varepsilon)}(\zeta_\nu)}+h_n^*(\zeta_\mu)\overline{h_n^*(\zeta_\nu)})\right\}$$
$$\leqslant\sum_{\mu,\nu=1}^{N}\eta_\mu\bar{\eta}_\nu a_{\mu\nu}|f'(0)(f(\zeta_\mu)-f(\zeta_\nu))|^{\varepsilon l}/|\zeta_\mu-\zeta_\nu|^l|1-\zeta_\mu\bar{\zeta}_\nu|^l,\tag{3.14}$$

where $\{\zeta_\mu)$ $(\mu=1,\cdots,N)$ are any distinguished points in $|\zeta|<1$, $\{\eta_\mu\}$ $(\mu=1,\cdots,N)$ are any complex numbers.

◇ References ◇

[1] Xia Dioxing, *Fudan Journal*, **2**(1956), 133 - 145.

[2] Hu Ke, *Jiangxi Normal College J.*, **1**(1979), 5 - 14.

[3] Ren Fuyao, *Fudan Journal*, **3**(1979), 69 - 75.

[4] Beresniewiez-Rajca, Olga, *Demonstratio Math.*, **9**: 3(1976), 307 - 319.

[5] Лебедев, Н. А., Принцип площадеий в теории одиолистиых функций, 1975.

[6] Bieberbach, L., *Math. Ann.*, **77**(1916), 153 - 172.

[7] Grunsky, H. *Math. Ann.*, **108**(1933), 190 - 196.

[8] Schober, G., Univalent Functions-Selected Topics, 1975.

[9] Еазилевич, И. Е., *Матем сб.*, **74**(1967), 133 - 146.

Bieberbach 函数族和 Grunsky 函数族的另一种充要条件及偏差定理

任福尧

摘 要

本文主要利用夏道行[1]对于 Bieberbach 函数族和 Grunsky 函数族的面积定理证明这两种函数族的另一充要条件，以及指数化的 Golusin 不等式和 Fitzgerald 不等式等深刻的偏差定理.

关于具有拟共形扩张的比伯霸赫函数族*

任福尧

1. 引 言

近10多年来,人们对具有拟共形扩张的种种单叶函数之度量的和几何的性质之研究表现了很大的兴趣,如 С. Л. Крушкаль[13], В. Я. Гутлянский[1-2], R. Kühnau[6-9], O. Lehto[11], J. O. Mcleavey[10], M. Schiffer 和 G. Schober[12],等等.

本文的目的在于用具有拟共形扩张的面积原理方法,研究两类具有拟共形扩张的比伯霸赫(L. Bieberbach)函数,给出了这种函数族的 Golusin 不等式、Grunsky 不等式,指数化的 Golusin 偏差定理和 Fitzgerald 不等式,以及 Schwarz 导数的估计等一系列结果. 当 $k\to 1$ 时,它们就退化成关于比伯霸赫函数族的相应的结果[3],[5].

设 $f(z)$, $f(0)=0$ 是单位圆 $|z|<1$ 上的正则的单叶函数,对 $|z_i|<1$, $i=1, 2$, $f(z_1)\cdot f(z_2)\neq 1$. 记这种函数的全体为 B,称为 Bieberbach 函数族[16]. 若 $f\in B$, $g(z)=f(z)^{-1}$,且 $f(|z|<1)\cup g(|z|<1)$ 无外点,则称 $f\in\widetilde{B}$.

设 $f(z)$是闭复平面上的同胚映照,在 $|z|<1$ 内 $f(z)\in\widetilde{B}$,在 $|z|>1$ 上是 K-拟共形映照. 记这种函数的全体为 $S_B(K)$. 设 $f(\zeta)$是闭平面上的同胚映照,在 $|\zeta|>1$ 上, $g(z)=f(\zeta)^{-1}$ $(\zeta=z^{-1})\in\widetilde{B}$;在 $|\zeta|<1$ 上, $f(\zeta)$是 K-拟共形映照,记这种函数的全体为 $\Sigma_B(K)$.

---

* 原载《数学学报》,1982年,第25卷,第4期,441-455.

2. Golusin 不等式和 Grunsky 不等式

引理 设 $f\in\Sigma_B(K)$，$Q(w)$是 $f(|\zeta|\leqslant 1)$ 上的解析函数，且存在 ρ，$1<\rho$，使得在圆环 $1<|\zeta|<\rho$ 上，

$$q(\zeta)=Q\circ f(\zeta)=\sum_{n=-\infty}^{\infty}c_n\zeta^n,$$

$$q_2(\zeta)=Q\circ\frac{1}{f(\zeta)}=\sum_{n=-\infty}^{\infty}d_n\zeta^n,$$

则

$$\frac{1}{2}k^2\Big(\sum_{n=-\infty}^{\infty}n|c_n|^2+\sum_{n=-\infty}^{\infty}n|d_n|^2\Big)\geqslant(1-k^2)\sum_{n=1}^{\infty}n|c_{-n}|^2, \tag{2.1}$$

等号当且仅当

$$q(\zeta)-\sum_{n=0}^{\infty}c_n\zeta^n=\begin{cases}k\varepsilon\sum\limits_{n=1}^{\infty}\bar{c}_n\zeta^{-n}, & |\zeta|>1,\\ k\varepsilon\sum\limits_{n=1}^{\infty}\bar{c}_n(\bar{\zeta})^n, & |\zeta|\leqslant 1\end{cases} \tag{2.2}$$

时成立，其中 $k=(K-1)/(K+1)$，$|\varepsilon|=1$.

证 证明的方法取自文献[1]. 据假设，曲线 $f(|\zeta|=1)$ 与曲线 $1/f(|\zeta|=1)$ 相重，但是方向相反，因而，利用格林公式和积分号下的变数变换，有

$$\begin{aligned}\sigma(q(|\zeta|<1))&=\lim_{\rho\to1+0}\sigma(q(|\zeta|\leqslant\rho))=\lim_{\rho\to1+0}\sigma(q_2(|\zeta|\geqslant\rho))\\&=\frac{1}{2\mathrm{i}}\lim_{\rho\to1+0}\oint_{|\zeta|=\rho}\overline{q(\zeta)}\mathrm{d}q(\zeta)=\frac{-1}{2\mathrm{i}}\lim_{\rho\to1+0}\oint_{|\zeta|=\rho}\overline{q_2(\zeta)}\mathrm{d}q_2(\zeta)\\&=\frac{1}{4\mathrm{i}}\lim_{\rho\to1+0}\Big[\oint_{|\zeta|=\rho}\overline{q(\zeta)}\mathrm{d}q(\zeta)+\oint_{|\zeta|=\rho}\overline{q_2(\zeta)}\mathrm{d}q_2(\zeta)\Big]\\&=\frac{1}{2}\sum_{n=-\infty}^{\infty}n(|c_n|^2+|d_n|^2).\end{aligned} \tag{2.3}$$

另一方面，在 $q(\zeta)$的映照下，$|\zeta|<1$ 的像之黎曼曲面的面积为

$$\sigma(q(|\zeta|<1))=\iint_{|\zeta|<1}(|q_\zeta|^2-|q_{\bar{\zeta}}|^2)\mathrm{d}\sigma_\zeta$$

$$=\iint_{|\zeta|<1}\frac{1-|\mu(\zeta)|^2}{|u(\zeta)|^2}|q_{\bar{\zeta}}|^2\mathrm{d}\sigma_\zeta \geqslant \frac{1-k^2}{k^2}\iint_{|\zeta|<1}|q_{\bar{\zeta}}|^2\mathrm{d}\sigma_\zeta. \tag{2.4}$$

我们定义函数

$$g(\zeta)=q(\zeta)-\sum_{n=0}^{\infty}c_n\zeta^n \quad (|\zeta|<\infty)$$

$$=\begin{cases} q(\zeta)-\sum_{n=0}^{\infty}c_n\zeta^n, & |\zeta|\leqslant\rho, \\ \sum_{n=1}^{\infty}c_{-n}\zeta^{-n}, & |\zeta|\geqslant\rho. \end{cases}$$

易见，$g(\infty)=0$；$g_{\bar{\zeta}}=q_{\bar{\zeta}}$，$|\zeta|<\rho$；$g_{\bar{\zeta}}=0$，$|\zeta|>\rho$. 由文献[17]，易知

$$g(\zeta)=T_E q_{\bar{\zeta}},\ g_\zeta=\Pi_E q_{\bar{\zeta}},$$

其中，$E=\{\zeta \mid |\zeta|<1\}$，

$$T_E h(\zeta)=\frac{1}{\pi}\iint_B\frac{h(z)}{\zeta-z}\mathrm{d}\sigma_z,\ \Pi_E h(\zeta)=-\frac{1}{\pi}\iint_B\frac{h(z)}{(z-\zeta)^2}\mathrm{d}\sigma_z.$$

于是，由文献[15, p. 87]的引理，有

$$\begin{aligned}\iint_{|\zeta|<\infty}|q_{\bar{\zeta}}|^2\mathrm{d}\sigma_\zeta &= \iint_{|\zeta|<\infty}|\Pi q_{\bar{\zeta}}|^2\mathrm{d}\sigma_\zeta \\ &= \iint_{|\zeta|<\infty}|g_\zeta|^2\mathrm{d}\sigma_\zeta \geqslant \iint_{|\zeta|>1}|g_\zeta|^2\mathrm{d}\sigma_\zeta \\ &= \sum_{n=1}^{\infty}n|c_{-n}|^2. \end{aligned}\tag{2.5}$$

由关系式(2.3)，(2.4)和(2.5)即得(2.1). 其等号当且仅当 $|\mu(\zeta)|=k$，$g_\zeta(\zeta)=0$ 在单位圆 $|\zeta|<1$ 上几乎处处成立时成立. 由文献[1]可知，此时，当且仅当(2.2)满足时成立. 引理证完.

推论 设 $f\in\Sigma_B(K)$，且在 $|\zeta|>1$，

$$f(\zeta)=b_{-1}\zeta+\sum_{n=0}^{\infty}b_n\zeta^{-n},\ \frac{1}{f(\zeta)}=\sum_{n=1}^{\infty}a_n\zeta^{-n},$$

则

$$k^2\sum_{n=1}^{\infty}n|a_n|^2+(2-k^2)\sum_{n=1}^{\infty}n|b_n|^2\leqslant k^2|b_{-1}|^2. \tag{2.6}$$

定理 1 设 $f\in\Sigma_B(K)$,且对 $|\zeta|$, $|\zeta'|>1$,

$$\ln\frac{f(\zeta)-f(\zeta')}{\zeta-\zeta'}=\ln f'(\infty)+\sum_{m,\,n=1}^{\infty}c_{mn}\zeta^{-m}\zeta'^{-n},$$

$$\ln\frac{f(\zeta)-f(\zeta')^{-1}}{\zeta}=\sum_{m=0,\,n=1}^{\infty}b_{mn}\zeta'^{-m}\zeta^{-n}+\ln f'(\infty).$$

又复数列 $\{x_n\}$, $\{x_n'\}$ 若满足条件

$$\sum_{n=1}^{\infty}\frac{|x_n|^2}{n}<\infty,\quad\sum_{n=1}^{\infty}\frac{|x'_n|^2}{n}<\infty,\tag{2.7}$$

则

$$k^2\sum_{n=1}^{\infty}n\left|\sum_{m=1}^{\infty}b_{mn}x_m\right|^2+(2-k^2)\sum_{n=1}^{\infty}n\left|\sum_{m=1}^{\infty}c_{mn}x_m\right|^2\leqslant k^2\sum_{n=1}^{\infty}\frac{|x_n|^2}{n},\tag{2.8}$$

$$\begin{aligned}k^2\left|\sum_{m,\,n=1}^{\infty}b_{nm}x'_nx_m\right|^2&+(2-k^2)\left|\sum_{m,\,n=1}^{\infty}c_{mn}x_mx'_n\right|^2\\&\leqslant k^2\sum_{n=1}^{\infty}\frac{|x_n|^2}{n}\cdot\sum_{n=1}^{\infty}\frac{|x'_n|^2}{n}.\end{aligned}\tag{2.9}$$

当 $k\to1$ 时,(2.8)就退化成[3]的结果.

证 于引理中令 $Q(w)=\sum_{m=1}^{N}x_mF_m(w)/m$, $F_m(t)$ 是 $f(\zeta)(|\zeta|>1)$ 的第 n 次 Faber 多项式,于是,由 Faber 多项式的性质[18],我们有

$$q(\zeta)=\sum_{m=1}^{N}\frac{x_m}{m}F_m(f(\zeta))=\sum_{m=1}^{N}\frac{x_m}{m}\zeta^m-\sum_{n=1}^{\infty}\Big(\sum_{m=1}^{N}c_{mn}x_m\Big)\zeta^{-n},$$

$$q_2(\zeta)=\sum_{m=1}^{N}\frac{x_m}{m}F_m\Big(\frac{1}{f(\zeta)}\Big)=-\sum_{n=0}^{\infty}\Big(\sum_{m=1}^{N}b_{nm}x_m\Big)\zeta^{-n},$$

将它们的系数代入(2.1)式,便得

$$\begin{aligned}k^2\Big(\sum_{n=1}^{N}\frac{|x_n|^2}{n}-\sum_{n=1}^{\infty}n\left|\sum_{m=1}^{N}c_{mn}x_m\right|^2-\sum_{n=1}^{\infty}n\left|\sum_{m=1}^{N}b_{nm}x_m\right|^2\Big)&\\\geqslant2(1-k^2)\sum_{n=1}^{\infty}n\left|\sum_{m=1}^{N}c_{mn}x_m\right|^2,&\end{aligned}$$

整理之,便得

$$k^2\sum_{n=1}^{\infty}n\left|\sum_{m=1}^{N}b_{nm}x_m\right|^2+(2-k^2)\sum_{n=1}^{\infty}n\left|\sum_{m=1}^{N}c_{mn}x_m\right|^2$$
$$\leqslant k^2\sum_{n=1}^{N}\frac{|x_n|^2}{n}.$$

在条件(2.7)下,容易证明上式对 $N\to+\infty$ 也是成立的. 故(2.8)式是真实的.

利用(2.8)和柯西不等式便得

$$k^2\left|\sum_{m,n=1}^{\infty}b_{nm}x'_nx_m\right|^2+(2-k^2)\left|\sum_{m,n=1}^{\infty}c_{mn}x_mx'_n\right|^2$$
$$\leqslant\sum_{n=1}^{\infty}\frac{|x'_n|^2}{n}\left(k^2\sum_{n=1}^{\infty}n\left|\sum_{m=1}^{\infty}b_{nm}x_m\right|^2+(2-k^2)\sum_{n=1}^{\infty}n\left|\sum_{m=1}^{\infty}c_{mn}x_m\right|^2\right)$$
$$\leqslant k^2\sum_{n=1}^{\infty}\frac{|x'_n|^2}{n}\cdot\sum_{n=1}^{\infty}\frac{|x_n|^2}{n}.$$

故定理是成立的.

定理 2 设 $f\in S_B(K)$,且对 $|z|$, $|z'|<1$,

$$\ln\frac{f(z)-f(z')}{z-z'}=\sum_{m,n=0}^{\infty}d_{mn}z^mz^n,$$

$$\ln(1-f(z')f(z))=\sum_{m,n=1}^{\infty}a_{mn}z'^mz^n,$$

又复数 $\{x_n\}$, $\{x'_n\}$ 满足条件(2.7),则

$$k^2\sum_{n=1}^{\infty}n\left|\sum_{m=1}^{\infty}a_{mn}x_m\right|^2+(2-k^2)\sum_{n=1}^{\infty}n\left|\sum_{m=1}^{\infty}d_{mn}x_m\right|^2$$
$$\leqslant k^2\sum_{n=1}^{\infty}\frac{|x_n|^2}{n},\tag{2.10}$$

$$k^2\left|\sum_{m,n=1}^{\infty}a_{mn}x_mx'_n\right|^2+(2-k^2)\left|\sum_{m,n=1}^{\infty}d_{mn}x_mx'_n\right|^2$$
$$\leqslant k^2\sum_{n=1}^{\infty}\frac{|x'_n|^2}{n}\cdot\sum_{n=1}^{\infty}\frac{|x_n|^2}{n}.\tag{2.11}$$

当 $k\to1$ 时, (2.10)就退化成[5]的相应结果.

证 令 $g(\zeta)=f(z)^{-1}$, $z=\zeta^{-1}$,由于 $f\in S_B(K)$,因而, $g(\zeta)\in\Sigma_B(K)$. 事实上,只要证明 $g(\zeta)$在 $|\zeta|<1$ 内是 K -拟共形映照就行了. 由于 g 和 f 的

复伸张分别为

$$\mu_g(\zeta)=g_{\bar{\zeta}}/g_{\zeta}, \quad u_f(z)=f_{\bar{z}}/f_z,$$

且

$$g_{\bar{\zeta}}(\zeta)=f_{\bar{z}}(z)\left(\frac{\bar{z}}{f(z)}\right)^2, \quad g_{\zeta}(\zeta)=f_z(z)\left(\frac{z}{f(z)}\right)^2,$$

所以，$|\mu_g(\zeta)|=|\mu_f(z)|\leqslant k<1$. 故 $g(\zeta)\in\Sigma_B(K)$.

另一方面，由于 $\zeta=z^{-1}$, $\zeta'=z'^{-1}$,

$$\ln g'(\infty)+\sum_{m,\,n=1}^{\infty}c_{mn}\zeta'^{-m}\zeta^{-n}=\ln\frac{g(\zeta)-g(\zeta')}{\zeta-\zeta'}$$

$$=\ln\frac{f(z)-f(z')}{z-z'}-\ln\frac{f(z)}{z}-\ln\frac{f(z')}{z'}=\sum_{m,\,n=1}^{\infty}d_{mn}z'^m z^n,$$

$$\sum_{m=0,\,n=1}^{\infty}b_{mn}\zeta'^{-m}\zeta^{-n}=-\ln g'(\infty)+\ln(g(\zeta)-f(z'))\zeta^{-1}$$

$$=\ln f'(0)-\ln\frac{f(z)}{z}+\ln(1-f(z)f(z'))$$

$$=\ln f'(0)-\ln\frac{f(z)}{z}+\sum_{m,\,n=1}^{\infty}a_{mn}z'^m z^n.$$

比较上两关系式中 $z'^m z^n=\zeta'^{-m}\zeta^{-n}$ 项的系数，便得

$$d_{mn}=c_{mn}, \quad a_{mn}=b_{mn} \quad (m, n\geqslant 1),$$

$$d_{mn}=d_{nm}, \quad c_{mn}=c_{nm}, \quad a_{mn}=a_{nm}, \quad b_{mn}=b_{nm} \quad (m, n\geqslant 1).$$

由是，由不等式(2.8)和(2.9)即得不等式(2.10)和(2.11). 由此得证定理.

3. 加强的偏差定理

定理 3 设 $f(\zeta)\in\Sigma_B(K)$, $\{\zeta_\mu\}$, $\{\zeta'_\nu\}$($\mu=1, 2, \cdots, N$; $\nu=1, 2, \cdots, N'$) 是 $|\zeta|>1$ 中的点列，$\{\eta_\mu\}$, $\{\eta'_\nu\}$是不同时为零的任何复数，

$$C_n(\zeta)=\sum_{m=1}^{\infty}c_{mn}\zeta^{-m}, \quad B_n(\zeta)=\sum_{m=1}^{\infty}b_{mn}\zeta^{-m},$$

p, q 是非负整数，则

1) $$k^2\sum_{n=1}^{\infty}n\left|\sum_{\mu=1}^{N}\eta_\mu B_n^{(p)}(\zeta_\mu)\right|^2+(2-k^2)\sum_{n=1}^{\infty}n\left|\sum_{\mu=1}^{N}\eta_\mu C_n^{(p)}(\zeta_\mu)\right|^2$$
$$\leqslant k^2\sum_{\mu,\nu=1}^{N}\eta_\mu\bar{\eta}_\nu\frac{\partial^{p+p}}{\partial\zeta_\mu^p\partial\bar{\zeta}_\nu^p}\ln\frac{\zeta_\mu\bar{\zeta}_\nu}{\zeta_\mu\bar{\zeta}_\nu-1}. \tag{3.1}$$

2) $$k^2\left|\sum_{\mu=1}^{N}\sum_{\nu=1}^{N'}\eta_\mu\eta'_\nu\frac{\partial^{p+q}}{\partial\zeta_\mu^p\partial\zeta'^q_\nu}\ln\left(1-\frac{1}{f(\zeta_\mu)f(\zeta'_\nu)}\right)\right|^2$$
$$+(2-k^2)\left|\sum_{\mu=1}^{N}\sum_{\nu=1}^{N'}\eta_\mu\eta'_\nu\frac{\partial^{p+q}}{\partial\zeta_\mu^p\partial\zeta'^q_\nu}\ln\frac{f(\zeta_\mu)-f(\zeta'_\nu)}{f'(\infty)(\zeta_\mu-\zeta'_\nu)}\right|^2$$
$$\leqslant k^2\left(\sum_{\mu,\nu=1}^{N}\eta_\mu\bar{\eta}_\nu\frac{\partial^{p+p}}{\partial\zeta_\mu^\nu\partial\bar{\zeta}_\nu^p}\ln\frac{\zeta_\mu\bar{\zeta}_\nu}{\zeta_\mu\bar{\zeta}_\nu-1}\right)$$
$$\cdot\left(\sum_{\mu,\nu=1}^{N'}\eta'_\mu\bar{\eta}'_\nu\frac{\partial^{q+q}}{\partial\zeta'_\mu\partial\bar{\zeta}'^q_\nu}\ln\frac{\zeta'_\mu\bar{\zeta}'_\nu}{\zeta'_\mu\bar{\zeta}'_\nu-1}\right). \tag{3.2}$$

其中

$$B_n^{(p)}(\zeta)=\frac{\mathrm{d}^p}{\mathrm{d}\zeta^p}B_n(\zeta),\quad C_n^{(p)}(\zeta)=\frac{\mathrm{d}^p}{\mathrm{d}\zeta^p}C_n(\zeta).$$

证 于(2.8)和(2.9)中,令

$$x_m=\sum_{\mu=1}^{N}\eta_\mu\frac{\mathrm{d}^p}{\mathrm{d}\zeta_\mu^p}\zeta_\mu^{-m},\quad x'_n=\sum_{\nu=1}^{N'}\eta'_\nu\frac{\mathrm{d}^q}{\mathrm{d}\zeta'^q_\nu}(\zeta'_\nu)^{-n}.$$

由文献[14, p. 125],易知$\{x_m\}$, $\{x'_n\}$满足条件(2.7),将它们代入(2.8)和(2.9)并交换求和与微分的次序,便得到(3.1)和(3.2).

推论 3.1 在定理 3 的条件下,则有

$$k^2\left|\sum_{\mu=1}^{N}\sum_{\nu=1}^{N'}\eta_\mu\eta'_\nu\frac{f'(\zeta_\mu)f'(\zeta'_\nu)}{(f(\zeta_\mu)f(\zeta'_\nu)-1)^2}\right|^2+(2-k^2)\left|\sum_{\mu=1}^{N}\sum_{\nu=1}^{N'}\eta_\mu\eta'_\nu\left[\frac{f'(\zeta_\mu)f'(\zeta'_\nu)}{(f(\zeta_\mu)-f(\zeta'_\nu))^2}-\frac{1}{(\zeta_\mu-\zeta'_\nu)^2}\right]\right|^2\leqslant k^2\sum_{\mu,\nu=1}^{N}\eta_\mu\bar{\eta}_\nu\frac{1}{(\zeta_\mu\bar{\zeta}_\nu-1)^2}\cdot\sum_{\mu,\nu=1}^{N'}\eta'_\mu\bar{\eta}'_\nu\frac{1}{(\zeta'_\mu\bar{\zeta}'_\nu-1)^2}. \tag{3.3}$$

推论 3.2 设 $f\in\Sigma_B(K)$,则对任何 ζ_0, $|\zeta_0|>1$, 有

$$36k^2\left|\frac{f'(\zeta_0)}{(f(\zeta_0)^2-1)}\right|^4+(2-k^2)|\{f,\zeta_0\}|^2\leqslant\frac{36k^2}{(|\zeta_0|^2-1)^4}. \tag{3.4}$$

其证明只要注意到 Schwarz 导数$\{f, \zeta_0\}$有下述关系式：

$$\{f, \zeta_0\} = 6\lim_{\zeta\to\zeta_0}\frac{\partial^2}{\partial\zeta\partial\zeta_0}\ln\frac{f(\zeta)-f(\zeta_0)}{\zeta-\zeta_0}.$$

推论 3.3 若 $f\in\Sigma_B(K)$，则对任何 ζ，$|\zeta|>1$，有

$$\frac{k^2}{\pi}\iint_{|z|>1}\left|\frac{f'(\zeta)f'(z)}{(f(\zeta)f(z)-1)^2}\right|^2 d\sigma_z + \frac{(2-k^2)}{\pi}\iint_{|z|>1}\left|\frac{f'(\zeta)f'(z)}{(f(\zeta)-f(z))^2}-\frac{1}{(\zeta-z)^2}\right|^2 d\sigma_z \leqslant \frac{k^2}{(|\zeta|^2-1)^2}. \tag{3.5}$$

证 对于 $|\zeta|$，$|z|>1$，由于

$$\ln\frac{f(\zeta)-f(z)}{\zeta-z}-\ln f'(\infty)=\sum_{m,n=1}^{\infty}c_{mn}\zeta^{-m}z^{-n}=\sum_{n=1}^{\infty}C_n(\zeta)z^{-n},$$

$$\ln\frac{f(\zeta)-f(z)^{-1}}{\zeta}-\ln f'(\infty)=\sum_{n=0}^{\infty}\left(\sum_{m=1}^{\infty}b_{nm}\zeta^{-m}\right)z^{-n}=\sum_{n=0}^{\infty}B_n(\zeta)z^{-n},$$

所以，

$$\frac{f'(\zeta)f'(z)}{(f(\zeta)-f(z))^2}-\frac{1}{(\zeta-z)^2}=-\sum_{n=1}^{\infty}nC'_n(\zeta)z^{-(n+1)},$$

$$\frac{f'(\zeta)f'(z)}{(f(\zeta)f(z)-1)^2}=\sum_{n=1}^{\infty}nB'_n(\zeta)z^{-(n+1)}.$$

注意到

$$\frac{1}{\pi}\iint_{|z|>1}z^{-(n+1)}(\bar{z})^{-(n'+1)}d\sigma_z=\frac{1}{\pi}\iint_{|z|<1}z^{n-1}(\bar{z})^{n'-1}d\sigma = \begin{cases}0, & n'\neq n,\\ 1/n, & n'=n,\end{cases} \tag{3.6}$$

由上述二关系式，积分之，便得

$$\frac{1}{\pi}\iint_{|z|>1}\left|\frac{f'(\zeta)f'(z)}{(f(\zeta)-f(z))^2}-\frac{1}{(\zeta-z)^2}\right|^2 d\sigma_z=\sum_{n=1}^{\infty}n|C_n(\zeta)|^2,$$

$$\frac{1}{\pi}\iint_{|z|>1}\left|\frac{f'(\zeta)f'(z)}{(f(\zeta)f(z)-1)^2}\right|^2 d\sigma_z=\sum_{n=1}^{\infty}n|B'_n(\zeta)|^2.$$

于(3.1)中取 $p=1$, $N'=N=1$, $\eta_\mu=\eta'_\nu=1$，便得

$$k^2\sum_{n=1}^{\infty}n|B'_n(\zeta)|^2+(2-k^2)\sum_{n=1}^{\infty}n|C'_n(\zeta)|^2\leqslant\frac{1}{(|\zeta|^2-1)^2}.$$

故推论 3.3 是真实的.

推论 3.4 在定理 3 的假设下,则有

$$k^2\left|\sum_{\mu=1}^{N}\sum_{\nu=1}^{N'}\eta_\mu\eta'_\nu\ln\left(1-\frac{1}{f(\zeta_\mu)f(\zeta'_\nu)}\right)\right|^2+(2-k^2)\left|\sum_{\mu=1}^{N}\sum_{\nu=1}^{N'}\eta_\mu\eta'_\nu\ln\frac{f(\zeta_\mu)-f(\zeta'_\nu)}{f'(\infty)(\zeta_\mu-\zeta'_\nu)}\right|^2$$
$$\leqslant k^2\sum_{\mu,\nu=1}^{N}\eta_\mu\bar{\eta}_\nu\ln\frac{\zeta_\mu\bar{\zeta}_\nu}{\zeta_\mu\bar{\zeta}_\nu-1}\cdot\sum_{\mu,\nu=1}^{N'}\eta'_\mu\bar{\eta}'_\nu\ln\frac{\zeta'_\mu\bar{\zeta}'_\nu}{\zeta'_\mu\bar{\zeta}'_\nu-1}. \tag{3.7}$$

定理 4 设 $f\in S_B(K)$, $\{z_\mu\}'$, $\{z'_\nu\}$ ($\mu=1, 2, \cdots, N$; $\nu=1, 2, \cdots, N'$) 是 $|z|<1$ 中的点列，$\{\eta_\mu\}$, $\{\eta'_\nu\}$是不同时为零的任何复数，p, q 是非负整数，则

$$k^2\sum_{n=1}^{\infty}n\left|\sum_{\mu=1}^{N}\eta_\mu A_n^{(p)}(\zeta_\mu)\right|^2+(2-k^2)\sum_{n=1}^{\infty}n\left|\sum_{\mu=1}^{N}\eta_\mu D_n^{(p)}(\zeta_\mu)\right|^2$$
$$\leqslant k^2\sum_{\mu,\nu=1}^{N}\eta_\mu\bar{\eta}_\nu\frac{\partial^{p+p}}{\partial z_\mu^p\partial\bar{z}_\nu^p}\ln\frac{1}{1-z_\mu\bar{z}_\nu}, \tag{3.8}$$

$$k^2\left|\sum_{\mu=1}^{N}\sum_{\nu=1}^{N'}\eta_\mu\eta'_\nu\frac{\partial^{p+q}}{\partial z_\mu^p\partial z'^q_\nu}\ln(1-f(z_\mu)f(z'_\nu))\right|^2$$
$$+(2-k^2)\left|\sum_{\mu=1}^{N}\sum_{\nu=1}^{N'}\eta_\mu\eta'_\nu\ln\frac{(f(z_\mu)-f(z'_\nu))z_\mu z'_\nu}{f'(0)(z_\mu-z'_\nu)f(z_\mu)f(z'_\nu)}\right|^2$$
$$\leqslant k^2\left(\sum_{\mu,\nu=1}^{N}\eta_\mu\bar{\eta}_\nu\frac{\partial^{p+p}}{\partial z_\mu^p\partial\bar{z}_\nu^p}\ln\frac{1}{1-z_\mu\bar{z}_\nu}\right)\left(\sum_{\mu,\nu=1}^{N'}\eta'_\mu\bar{\eta}'_\nu\frac{\partial^{q+q}}{\partial z'^q_\mu\partial\bar{z}'^q_\nu}\ln\frac{1}{1-z'_\mu\bar{z}'_\nu}\right), \tag{3.9}$$

其中 $$A_n(z)=\sum_{m=1}^{\infty}a_{mn}z^m,\ D_n(z)=\sum_{m=1}^{\infty}d_{mn}z^m.$$

由定理 3 的证明可知,这定理的证明只要于定理 2 的(2.10)和(2.11)中令

$$x_m=\sum_{\mu=1}^{N}\eta_\mu\frac{\mathrm{d}^p}{\mathrm{d}z^p}z^m,\quad x'_n=\sum_{\nu=1}^{N'}\eta'_\nu\frac{\mathrm{d}^q}{\mathrm{d}z'^q}z'^n$$

就行了.

类似推论 3.1 至推论 3.4 的证明,便可得到下述推论.

推论 4.1 在定理 4 的假设下,则有

$$k^2\left|\sum_{\mu=1}^{N}\sum_{\nu=1}^{N'}\eta_\mu\eta'_\nu\frac{f'(z_\mu)f'(z'_\nu)}{(1-f(z_\mu)f(z'_\nu))^2}\right|^2$$
$$+(2-k^2)\left|\sum_{\mu=1}^{N}\sum_{\nu=1}^{N'}\eta_\mu\eta'_\nu\left[\frac{f'(z_\mu)f'(z'_\nu)}{(f(z_\mu)-f(z'_\nu))^2}-\frac{1}{(z_\mu-z'_\nu)^2}\right]\right|^2$$
$$\leqslant k^2\sum_{\mu,\nu=1}^{N}\eta_\mu\bar{\eta}_\nu\frac{1}{(1-z_\mu\bar{z}_\nu)^2}\cdot\sum_{\mu,\nu=1}^{N'}\eta'_\mu\bar{\eta}'_\nu\frac{1}{(1-z'_\mu\bar{z}'_\nu)}. \tag{3.10}$$

推论 4.2 设 $f\in S_B(K)$,对 $|z|<1$ 中任何 z,则

$$36k^2\left|\frac{f'(z)^2}{(1-f(z)^2)^2}\right|^2+(2-k^2)|\{f,z\}|^2\leqslant\frac{36k^2}{(1-|z|^2)^4}. \tag{3.11}$$

推论 4.3 设 $f\in S_B(K)$,则对 $|z|<1$ 中的 z,有

$$\frac{k^2}{\pi}\iint_{|\zeta|<1}\left|\frac{f'(z)f'(\zeta)}{(1-f(z)f(\zeta))^2}\right|^2 d\sigma_\zeta$$
$$+\frac{(2-k^2)}{\pi}\iint_{|\zeta|<1}\left|\frac{f'(z)f'(\zeta)}{(f(z)-f(\zeta))^2}-\frac{1}{(z-\zeta)^2}\right|^2 d\sigma_\zeta$$
$$\leqslant\frac{k^2}{(1-|z|^2)^2}. \tag{3.12}$$

4. 指数化的 Golusin 偏差定理

定理 5 设 $f\in\Sigma_B(K)$, $\{\zeta_\mu\}$, $\{\zeta'_\nu\}$ ($\mu=1,2,\cdots,N;\nu=1,2,\cdots,N'$) 是 $|\zeta|>1$ 中的点列, $\{\eta_\mu\}$, $\{\eta'_\nu\}$ 是不同时为零的任何复数, l 是自然数, p 是任何复数,则有

$$\left|\sum_{\mu=1}^{N}\sum_{\nu=1}^{N'}\eta_\mu\eta'_\nu\left(\ln\left(1-\frac{1}{f(\zeta_\mu)f(\zeta'_\nu)}\right)^{kp}\right)^l\right|^2$$
$$+\left|\sum_{\mu=1}^{N}\sum_{\nu=1}^{N'}\eta_\mu\eta'_\nu\left(\ln\left(\frac{f(\zeta_\mu)-f(\zeta'_\nu)}{f'(\infty)(\zeta_\mu-\zeta'_\nu)}\right)^{p\sqrt{2-k^2}}\right)^l\right|^2$$
$$\leqslant\sum_{\mu,\nu=1}^{N}\eta_\mu\bar{\eta}_\nu\left(\ln\left(1-\frac{1}{\zeta_\mu\bar{\zeta}_\nu}\right)^{-k|p|}\right)^l$$

$$\cdot\sum_{\mu,\nu=1}^{N'}\eta'_\mu\bar{\eta}'_\nu\left(\ln\left(1-\frac{1}{\zeta'_\mu\bar{\zeta}'_\nu}\right)^{-k|p|}\right)^l. \tag{4.1}$$

证 令 $X_m(\mu)=\eta_\mu^{1/l}\zeta_\mu^{-m}$，$X_n(\nu)=\eta'^{1/l}_\nu\zeta'^{-n}_\nu$，利用柯西不等式，我们有

$$\left|\sum_{\mu=1}^{N}\sum_{\nu=1}^{N'}\eta_\mu\eta'_\nu\left(kp\ln\left(1-\frac{1}{f(\zeta_\mu)f(\zeta'_\nu)}\right)\right)^l\right|^2$$

$$=(|p|k)^{2l}\left|\sum_{\mu=1}^{N}\sum_{\nu=1}^{N'}\eta_\mu\eta'_\nu\left(\sum_{m,n=1}^{\infty}b_{mn}\zeta_\mu^{-m}\zeta'^{-n}_\nu\right)^l\right|^2$$

$$\leqslant|p|^{2l}\sum_{m_1}\cdots\sum_{m_l}\prod_{j=1}^{l}\frac{1}{m_j}\left|\sum_{\mu=1}^{N}\prod_{j=1}^{l}X_{m_j}(\mu)\right|^2$$

$$\cdot\sum_{m_1}\cdots\sum_{m_l}\prod_{j=1}^{l}k^2m_j\left|\sum_{n_1}\cdots\sum_{n_l}\sum_{\nu=1}^{N'}\prod_{j=1}^{l}b_{m_jn_j}X_{n_j}(\nu)\right|^2,$$

$$\left|\sum_{\mu=1}^{N}\sum_{\nu=1}^{N'}\eta_\mu\eta'_\nu\left(p\sqrt{2-k^2}\ln\frac{f(\zeta_\mu)-f(\zeta'_\nu)}{f'(\infty)(\zeta_\mu-\zeta'_\nu)}\right)^l\right|^2$$

$$=(|p|\sqrt{2-k^2})^{2l}\left|\sum_{\mu=1}^{N}\sum_{\nu=1}^{N'}\eta_\mu\eta'_\nu\left(\sum_{m,n=1}^{\infty}c_{mn}\zeta_\mu^{-m}\zeta'^{-n}_\nu\right)^l\right|^2$$

$$\leqslant|p|^2\sum_{m_1}\cdots\sum_{m_l}\prod_{j=1}^{l}\frac{1}{m_j}\left|\sum_{\mu=1}^{N}\prod_{j=1}^{l}X_{m_j}(\mu)\right|^2$$

$$\cdot\sum_{m_1}\cdots\sum_{m_l}\prod_{j=1}^{l}(2-k^2)m_j\left|\sum_{n_1}\cdots\sum_{n_l}\sum_{\nu=1}^{N'}\prod_{j=1}^{l}c_{m_jn_j}X_{n_j}(\nu)\right|^2.$$

由是，(4.1)式的左端

$$\leqslant|p|^{2l}\sum_{m_1}\cdots\sum_{m_l}\prod_{j=1}^{l}\frac{1}{m_j}\left|\sum_{\mu=1}^{N}\prod_{j=1}^{l}X_{m_j}(\mu)\right|^2$$

$$\cdot\left\{\sum_{m_1}\cdots\sum_{m_l}\prod_{j=1}^{l}k^2m_j\left|\sum_{n_1}\cdots\sum_{n_l}\sum_{\nu=1}^{N'}\prod_{j=1}^{l}b_{m_jn_j}X_{n_j}(\nu)\right|^2\right. \tag{4.2}$$

$$\left.+\sum_{m_1}\cdots\sum_{m_l}\prod_{j=1}^{l}(2-k^2)m_j\left|\sum_{n_1}\cdots\sum_{n_l}\sum_{\nu=1}^{N'}\prod_{j=1}^{l}c_{m_jn_j}X_{n_j}(\nu)\right|^2\right\}.$$

但是，利用不等式(2.8)，上式右端花括号中的项

$$\leqslant\sum_{m_1}\cdots\sum_{m_{l-1}}\prod_{j=1}^{l-1}k^2m_j\sum_{m_l=1}^{\infty}\frac{k^2}{m_l}\left|\sum_{n_1}\cdots\sum_{n_{l-1}}\sum_{\nu=1}^{N'}\prod_{j=1}^{l-1}b_{m_jn_j}X_{n_j}(\nu)X_{m_l}(\nu)\right|^2$$

$$+\sum_{m_1}\cdots\sum_{m_{l-1}}\prod_{j=1}^{l-1}(2-k^2)m_j\sum_{m_l=1}^{\infty}\frac{k^2}{m_l}\Big|\sum_{n_1}\cdots\sum_{n_{l-1}}\sum_{\nu=1}^{N'}\prod_{j=1}^{l-1}c_{m_jn_j}X_{n_j}(\nu)X_{m_l}(\nu)\Big|^2$$

$$\leqslant\sum_{m_l}\cdots\sum_{m_2}\prod_{j=2}^{l}\frac{k^2}{m_j}\Big[\sum_{m_1=1}^{\infty}k^2m_1\Big|\sum_{n_1=1}^{\infty}b_{m_1n_1}\Big(\sum_{\nu=1}^{N'}X_{n_1}(\nu)\prod_{j=2}^{l}X_{m_j}(\nu)\Big)\Big|^2$$

$$+\sum_{m_1=1}^{\infty}(2-k^2)m_1\Big|\sum_{n_1=1}^{\infty}c_{m_1n_1}\Big(\sum_{\nu=1}^{N'}X_{n_l}(\nu)\prod_{j=2}^{l}X_{m_j}(\nu)\Big)\Big|^2\Big]$$

$$\leqslant\sum_{m_l}\cdots\sum_{m_1}\prod_{j=1}^{l}\frac{k^2}{m_j}\Big|\sum_{\nu=1}^{N'}\prod_{j=1}^{l}X_{m_j}(\nu)\Big|^2$$

$$=\sum_{\mu,\nu=1}^{N'}\prod_{j=1}^{l}\sum_{m_j=1}^{\infty}\frac{k^2}{m_j}X_{m_j}(\mu)\overline{X_{m_j}(\nu)}$$

$$=\sum_{\mu,\nu=1}^{N'}\left(\sum_{m=1}^{\infty}\frac{k^2}{m}X_m(\mu)\overline{X_m(\nu)}\right)^l$$

$$=\sum_{\mu,\nu=1}^{N'}\eta'_\mu\bar{\eta}'_\nu\left(-k^2\ln\left(1-\frac{1}{\zeta_\mu\bar{\zeta}'_\nu}\right)\right)^l. \tag{4.3}$$

因此,由(4.2)和(4.3)便得(4.1)的左端

$$\leqslant\sum_{\mu,\nu=1}^{N}\eta_\mu\bar{\eta}_\nu\left(\ln\left(1-\frac{1}{\zeta_\mu\bar{\zeta}_\nu}\right)^{-k|p|}\right)^l\sum_{\mu,\nu=1}^{N'}\eta'_\mu\bar{\eta}'_\nu\left(\ln\left(1-\frac{1}{\zeta'_\mu\bar{\zeta}'_\nu}\right)^{-k|p|}\right)^l.$$

故定理5是真实的.

推论 5.1 设 $f\in\Sigma_B(K)$, 对任何复数 p 和$\{\eta_\mu\}$,以及 $|\zeta|>1$ 中的点列$\{\zeta_\mu\}$ $(\mu=1,2,\cdots,N)$,则

$$\left|\sum_{\mu,\nu=1}^{N}\eta_\mu\eta_\nu\left(\frac{f(\zeta_\mu)-f(\zeta_\nu)}{f'(\infty)(\zeta_\mu-\zeta_\nu)}\right)^{p\sqrt{2-k^2}}\right|\leqslant\sum_{\mu,\nu=1}^{N}\eta_\mu\bar{\eta}_\nu\left(1-\frac{1}{\zeta_\mu\bar{\zeta}_\nu}\right)^{-k|p|}, \tag{4.4}$$

$$\left|\sum_{\mu,\nu=1}^{N}\eta_\mu\eta_\nu\left(1-\frac{1}{f(\zeta_\mu)f(\zeta_\nu)}\right)^{kp}\right|\leqslant\sum_{\mu,\nu=1}^{N}\eta_\mu\bar{\eta}_\nu\left(1-\frac{1}{\zeta_\mu\bar{\zeta}_\nu}\right)^{-k|p|}. \tag{4.5}$$

证 于定理5中令 $N'=N$, $\eta'_\mu=\eta_\nu$, $\zeta'_\mu=\zeta_\nu$, 此时,(4.1)便成为

$$\left|\sum_{\mu,\nu=1}^{N}\eta_\mu\eta_\nu\left(\ln\left(1-\frac{1}{f(\zeta_\mu)f(\zeta_\nu)}\right)^{kp}\right)^l\right|^2+\left|\sum_{\mu,\nu=1}^{N}\eta_\mu\eta_\nu\left(\ln\left(\frac{f(\zeta_\mu)-f(\zeta_\nu)}{f'(\infty)(\zeta_\mu-\zeta_\nu)}\right)^{p\sqrt{2-k^2}}\right)^l\right|^2$$

$$\leqslant\left(\sum_{\mu,\nu=1}^{N}\eta_\mu\bar{\eta}_\nu\left(\ln\left(1-\frac{1}{\zeta_\mu\bar{\zeta}_\nu}\right)^{-k|p|}\right)^l\right)^2. \tag{4.6}$$

由是,我们有

$$\begin{aligned}&\left|\sum_{\mu,\nu=1}^{N}\eta_\mu\eta_\nu\left(\frac{f(\zeta_\mu)-f(\zeta_\nu)}{f'(\infty)(\zeta_\mu-\zeta_\nu)}\right)^{p\sqrt{2-k^2}}\right|\\&\leqslant\sum_{l=0}^{\infty}\frac{1}{l!}\left|\sum_{\mu,\nu=1}^{N}\eta_\mu\eta_\nu\ln\left(\frac{f(\zeta_\mu)-f(\zeta_\nu)}{f'(\infty)(\zeta_\mu-\zeta_\nu)}\right)^{p\sqrt{2-k^2}}\right|\\&\leqslant\sum_{\mu,\nu=1}^{N}\eta_\mu\bar{\eta}_\nu\sum_{l=0}^{\infty}\frac{1}{l!}\left(\ln\left(1-\frac{1}{\zeta_\mu\bar{\zeta}_\nu}\right)^{-k|p|}\right)^{l}\\&=\sum_{\mu,\nu=1}^{N}\eta_\mu\bar{\eta}_\nu\left(1-\frac{1}{\zeta_\mu\bar{\zeta}_\nu}\right)^{-k|p|}.\end{aligned}$$

同理,我们有

$$\begin{aligned}\left|\sum_{\mu,\nu=1}^{N}\eta_\mu\eta_\nu\left(1-\frac{1}{f(\zeta_\mu)f(\zeta_\nu)}\right)^{kp}\right|&\leqslant\sum_{l=0}^{\infty}\frac{1}{l!}\left|\sum_{\mu,\nu=1}^{N}\eta_\mu\eta_\nu\left(\ln\left(1-\frac{1}{f(\zeta_\mu)f(\zeta_\nu)}\right)^{kp}\right)^{l}\right|\\&\leqslant\sum_{l=0}^{\infty}\frac{1}{l!}\sum_{\mu,\nu=1}^{N}\eta_\mu\bar{\eta}_\nu\left(\ln\left(1-\frac{1}{\zeta_\mu\bar{\zeta}_\nu}\right)^{-k|p|}\right)^{l}\\&=\sum_{\mu,\nu=1}^{N}\eta_\mu\bar{\eta}_\nu\left(1-\frac{1}{\zeta_\mu\bar{\zeta}_\nu}\right)^{-k|p|}.\end{aligned}$$

推论 5.2 设 $f\in\Sigma_B(K)$,则对 $|\zeta|>1$ 中的点 ζ,

$$\left(1-\frac{1}{|\zeta|^2}\right)^{\frac{k}{\sqrt{2-k^2}}}\leqslant\left|\frac{f'(\zeta)}{f'(\infty)}\right|\leqslant\left(1-\frac{1}{|\zeta|^2}\right)^{-\frac{k}{\sqrt{2-k^2}}},\tag{4.7}$$

$$\left(1-\frac{1}{|\zeta|^2}\right)\leqslant\left|\left(\frac{1}{f(\zeta)}\right)^2-1\right|\leqslant\left(1-\frac{1}{|\zeta|^2}\right)^{-1}.\tag{4.8}$$

定理 6 设 $f\in\Sigma_B(K)$,对 $|\zeta|>1$ 中的点列 $\{\zeta_\mu\}$ $(\mu=1,2,\cdots,N)$,以及任何复数 p 和 $\{\eta_\mu\}$,则得

$$\begin{aligned}&\left|\sum_{\mu,\nu=1}^{N}\eta_\mu\eta_\nu\left(1-\frac{1}{f(\zeta_\mu)f(\zeta_\nu)}\right)^{k^2p}\right|+\left|\sum_{\mu,\nu=1}^{N}\eta_\mu\eta_\nu\left(\frac{f(\zeta_\mu)-f(\zeta_\nu)}{f'(\infty)(\zeta_\mu-\zeta_\nu)}\right)^{kp\sqrt{2-k^2}}\right|\\&\qquad\leqslant\frac{3}{2}\sum_{\mu,\nu=1}^{N}\eta_\mu\bar{\eta}_\nu\left(1-\frac{1}{\zeta_\mu\bar{\zeta}_\nu}\right)^{-k^2|p|}.\end{aligned}\tag{4.9}$$

证 由定理 5 的证明,易知,(4.9)的左端

$$\leqslant\sum_{l=0}^{\infty}\frac{|p|^l}{l!}\Big[k^{2l}\Big|\sum_{\mu,\nu=1}^{N}\eta_\mu\eta_\nu\Big(\sum_{m,n=1}^{\infty}b_{mn}\zeta_\mu^{-m}\zeta_\nu^{-n}\Big)^l\Big|$$

$$+k^l(2-k^2)^{l/2}\Big|\sum_{\mu,\nu=1}^{N}\eta_\mu\eta_\nu\Big(\sum_{m,n=1}^{\infty}c_{mn}\zeta_\mu^{-m}\zeta_\nu^{-n}\Big)^l\Big|\Big].$$

令 $X_m(\mu)=\eta_\mu^{1/l}\zeta_\mu^{-m}$，逐次利用柯西不等式，上式右端

$$\leqslant\sum_{l=0}^{\infty}\frac{|p|^l}{l!}k^l\Big(\sum_{m_1}\cdots\sum_{m_l}\prod_{j=1}^{l}\frac{1}{m_j}\Big|\sum_{\mu=1}^{N}\prod_{j=1}^{l}X_{m_j}(\mu)\Big|^2\Big)^{\frac{1}{2}}$$

$$\cdot\Big[\Big(\sum_{m_1}\cdots\sum_{m_l}\prod_{j=1}^{l}k^2m_j\Big|\sum_{n_1}\cdots\sum_{n_l}\sum_{\nu=1}^{N}\prod_{j=1}^{l}b_{m_jn_j}X_{n_j}(\nu)\Big|^2\Big)^{\frac{1}{2}}$$

$$+\Big(\sum_{m_1}\cdots\sum_{m_l}\prod_{j=1}^{l}(2-k^2)m_j\Big|\sum_{n_1}\cdots\sum_{n_l}\sum_{\nu=1}^{N}\prod_{j=1}^{l}c_{m_jn_j}X_{n_j}(\nu)\Big|^2\Big)^{\frac{1}{2}}\Big]$$

$$\leqslant\sum_{l=0}^{\infty}\frac{|p|^l}{l!}k^{2l}\sum_{m_1}\cdots\sum_{m_l}\prod_{j=1}^{l}\frac{1}{m_j}\Big|\sum_{\mu=1}^{N}\prod_{j=1}^{l}X_{m_j}(\mu)\Big|^2$$

$$+\frac{1}{2}\sum_{l=0}^{\infty}\frac{|p|^l}{l!}\Big[\sum_{m_1}\cdots\sum_{m_l}\prod_{j=1}^{l}k^2m_j\Big|\sum_{n_1}\cdots\sum_{n_l}\sum_{\nu=1}^{N}\prod_{j=1}^{l}b_{m_jn_j}X_{n_j}(\nu)\Big|^2$$

$$+\sum_{m_1}\cdots\sum_{m_l}\prod_{j=1}^{l}(2-k^2)m_j\Big|\sum_{n_1}\cdots\sum_{n_l}\sum_{\nu=1}^{N}\prod_{j=1}^{l}c_{m_jn_j}X_{n_j}(\nu)\Big|^2\Big].$$

由(4.3)，上式右端

$$\leqslant\frac{3}{2}\sum_{l=0}^{\infty}\frac{|p|^l}{l!}\sum_{\mu,\nu=1}^{N}\eta_\mu\bar{\eta}_\nu\Big(\sum_{m=1}^{\infty}\frac{k^2}{m}(\zeta_\mu\bar{\zeta}_\nu)^{-m}\Big)^l$$

$$=\frac{3}{2}\sum_{\mu,\nu=1}^{N}\eta_\mu\bar{\eta}_\nu\exp\Big\{k^2|p|\sum_{m=1}^{\infty}\frac{1}{m}(\zeta_\mu\bar{\zeta}_\nu)^{-m}\Big\}$$

$$=\frac{3}{2}\sum_{\mu,\nu=1}^{N}\eta_\mu\bar{\eta}_\nu(1-1/\zeta_\mu\bar{\zeta}_\nu)^{-k^2|p|}.$$

类似地，对函数族 $S_B(K)$ 有下述定理.

定理 7 设 $f\in S_B(K)$，$\{z_\mu\}$，$\{z'_\nu\}$ $(\mu=1,2,\cdots,N;\nu=1,2,\cdots,N')$ 是 $|z|<1$ 中的点列，l 是任何非负整数，p 是任何复数，则对任何复数 $\{\eta_\mu\}$，$\{\eta'_\nu\}$ 成立着

$$\Big|\sum_{\mu=1}^{N}\sum_{\nu=1}^{N'}\eta_\mu\eta'_\nu(kp\ln(1-f(z_\mu)f(z'_\nu)))^l\Big|^2$$

$$+\left|\sum_{\mu=1}^{N}\sum_{\nu=1}^{N'}\eta_\mu\eta'_\nu\left(p\sqrt{2-k^2}\ln\frac{f'(0)(f(z_\mu)-f(z'_\nu))z_\mu z'_\nu}{(z_\mu-z'_\nu)f(z_\mu)f(z'_\nu)}\right)^l\right|^2$$
$$\leqslant\sum_{\mu,\nu=1}^{N}\eta_\mu\bar{\eta}_\nu(-k|p|\ln(1-z_\mu\bar{z}_\nu))^l\cdot\sum_{\mu,\nu=1}^{N'}\eta'_\mu\bar{\eta}'_\nu(-k|p|\ln(1-z'_\mu\bar{z}'_\nu))^l.$$
(4.10)

推论 7.1 在上述定理的假设下,则

$$\left|\sum_{\mu,\nu=1}^{N}\eta_\mu\eta_\nu\left(\frac{f'(0)(f(z_\mu)-f(z_\nu))z_\mu z_\nu}{(z_\mu-z_\nu)f(z_\mu)f(z_\nu)}\right)^{p\sqrt{2-k^2}}\right|\leqslant\sum_{\mu,\nu=1}^{N}\eta_\mu\bar{\eta}_\nu(1-z_\mu\bar{z}_\nu)^{-k|p|},$$
(4.11)

$$\left|\sum_{\mu,\nu=1}^{N}\eta_\mu\eta_\nu(1-f(z_\mu)f(z'_\nu))^{kp}\right|\leqslant\sum_{\mu,\nu=1}^{N}\eta_\mu\bar{\eta}_\nu(1-z_\mu\bar{z}_\nu)^{-k|p|}. \quad (4.12)$$

特别地,对 $f\in S_B(K)$, $|z|<1$, 则

$$(1-|z|^2)^{\frac{k}{\sqrt{2-k^2}}}\leqslant\left|f'(0)f'(z)\left(\frac{z}{f(z)}\right)^2\right|\leqslant(1-|z|^2)^{-\frac{k}{\sqrt{2-k^2}}},$$
$$(1-|z|^2)\leqslant|1-f(z)^2|\leqslant(1-|z|^2)^{-1}.$$

定理 8 设 $f\in S_B(K)$, $\{z_\mu\}$ $(\mu=1,2,\cdots,N)$ 是 $|z|<1$ 中的任何点列, $\{\eta_\mu\}$是任何复数,则

$$\left|\sum_{\mu,\nu=1}^{N}\eta_\mu\eta_\nu(1-f(z_\mu)f(z_\nu))^{k^2p}\right|+\left|\sum_{\mu,\nu=1}^{N}\eta_\mu\eta_\nu\left(\frac{f'(0)(f(z_\mu)-f(z_\nu))z_\mu z_\nu}{(z_\mu-z_\nu)f(z_\mu)f(z_\nu)}\right)^{pk\sqrt{2-k^2}}\right|$$
$$\leqslant\frac{3}{2}\sum_{\mu,\nu=1}^{N}\eta_\mu\bar{\eta}_\nu(1-z_\mu\bar{z}_\nu)^{-k^2|p|}. \quad (4.13)$$

5. Fitzgerald 不等式

定理 9 设 $f\in\Sigma_B(K)$, $\{\zeta_\mu\}$ $(\mu=1,2,\cdots,N)$ 是 $|\zeta|>1$ 中的点列, $\sum_{\mu,\nu=1}^{N}a_{\mu\nu}\eta_\mu\bar{\eta}_\nu\geqslant0$ 对任何复数$\{\eta_\mu\}$成立,又设

$$g_n^{(\varepsilon)}(\zeta)=F_n(f(\zeta))-\left(\zeta^n+\frac{\varepsilon k}{\sqrt{2-k^2}}(\bar{\zeta})^{-n}\right),\ \varepsilon=1,-1,$$

$$h_n(\zeta)=F_n\left(\frac{1}{f(\zeta)}\right)+2n\gamma_n,\quad \ln\frac{f(\zeta)}{\zeta}=2\sum_{n=0}^{\infty}\gamma_n\zeta^{-n},$$

则

$$\begin{aligned}&\frac{1}{2}\sum_{n=1}^{\infty}\frac{1}{n}\left[k^2\left|\sum_{\mu=1}^{N}\eta_\mu h_n(\zeta_\mu)\right|^2+(2-k^2)\left|\sum_{\mu=1}^{N}\eta_\mu g_n^{(\varepsilon)}(\zeta_\mu)\right|^2\right]\\&\leqslant\sum_{\mu,\nu=1}^{N}\eta_\mu\bar{\eta}_\nu\ln\left|\frac{f(\zeta_\mu)-f(\zeta_\nu)}{f'(\infty)(\zeta_\mu-\zeta_\nu)}\right|^{\varepsilon k\sqrt{2-k^2}}\cdot\frac{1}{|1-(\zeta_\mu\bar{\zeta}_\nu)^{-1}|^{k^2}},\end{aligned}\tag{5.1}$$

$$\begin{aligned}&\sum_{\mu,\nu=1}^{N}\eta_\mu\bar{\eta}_\nu a_{\mu\nu}\exp\left\{\frac{l}{2}\sum_{n=1}^{\infty}\frac{1}{n}\left[k^2h_n(\zeta_\mu)\overline{h_n(\zeta_\nu)}+(2-k^2)g_n^{(\varepsilon)}(\zeta_\mu)\overline{g_n^{(\varepsilon)}(\zeta_\nu)}\right]\right\}\\&\leqslant\sum_{\mu,\nu=1}^{N}a_{\mu\nu}\eta_\mu\bar{\eta}_\nu\left|\frac{f(\zeta_\mu)-f(\zeta_\nu)}{f'(\infty)(\zeta_\mu-\zeta_\nu)}\right|^{\varepsilon lk\sqrt{2-k^2}}\cdot\frac{1}{|1-(\zeta_\mu\bar{\zeta}_\nu)^{-1}|^2}.\end{aligned}\tag{5.2}$$

证 于定理 3 中令 $p=0$，则有

$$\begin{aligned}-k^2\sum_{\mu,\nu=1}^{N}\eta_\mu\bar{\eta}_\nu\ln(1-(\zeta_\mu\bar{\zeta}_\nu)^{-1})&\geqslant\sum_{n=1}^{\infty}n\Big[k^2\Big|\sum_{\mu=1}^{N}\eta_\mu\Big(\sum_{m=1}^{\infty}b_{nm}\zeta_\mu^{-m}\Big)\Big|^2\\&\quad+(2-k^2)\Big|\sum_{\mu=1}^{N}\eta_\mu\Big(\sum_{m=1}^{\infty}c_{mn}\zeta_\mu^{-m}\Big)\Big|^2\Big].\end{aligned}\tag{5.3}$$

但是，

$$\begin{aligned}&(2-k^2)\sum_{n=1}^{\infty}n\left|\sum_{\mu=1}^{N}\eta_\mu\Big(\sum_{m=1}^{\infty}c_{mn}\zeta_\mu^{-m}\Big)\right|^2\\&=(2-k^2)\sum_{n=1}^{\infty}n\left|\sum_{\mu=1}^{N}\eta_\mu\left(\sum_{m=1}^{\infty}c_{mn}\zeta_\mu^{-m}+\frac{\varepsilon k}{n\sqrt{2-k^2}}(\bar{\zeta}_\mu)^{-n}\right)\right|^2\\&\quad-2\varepsilon k\sqrt{2-k^2}\,\mathrm{Re}\left\{\sum_{\mu,\nu=1}^{N}\eta_\mu\bar{\eta}_\nu\ln\frac{f(\zeta_\mu)-f(\zeta_\nu)}{f'(\infty)(\zeta_\mu-\zeta_\nu)}\right\}\\&\quad+k^2\sum_{\mu,\nu=1}^{N}\eta_\mu\bar{\eta}_\nu\ln(1-(\bar{\zeta}_\mu\zeta_\nu)^{-1}),\end{aligned}$$

$$\mathrm{Re}\left\{\sum_{\mu,\nu=1}^{N}\eta_\mu\bar{\eta}_\nu\ln\frac{f(\zeta_\mu)-f(\zeta_\nu)}{f'(\infty)(\zeta_\mu-\zeta_\nu)}\right\}=\sum_{\mu,\nu=1}^{N}\eta_\mu\bar{\eta}_\nu\ln\left|\frac{f(\zeta_\mu)-f(\zeta_\nu)}{f'(\infty)(\zeta_\mu-\zeta_\nu)}\right|,$$

$$\sum_{m=1}^{\infty}c_{mn}\zeta^{-m}=\frac{1}{n}(\zeta^n-F_n(f(\zeta))),$$

$$\frac{1}{n}F_n\left(\frac{1}{f(\zeta)}\right)=-\Big(\sum_{m=1}^{\infty}b_{mn}\zeta^{-m}+2\gamma_n\Big).$$

因此,我们有

$$\sum_{n=1}^{\infty}\frac{1}{n}\left[k^2\left|\sum_{\mu=1}^{N}\eta_\mu h_n(\zeta_\mu)\right|^2+(2-k^2)\left|\sum_{\mu=1}^{N}\eta_\mu g_n^{(\varepsilon)}(\zeta_\mu)\right|^2\right]$$
$$\leqslant 2\varepsilon k\sqrt{2-k^2}\sum_{\mu,\nu=1}^{N}\eta_\mu\bar{\eta}_\nu\ln\left|\frac{f(\zeta_\mu)-f(\zeta_\nu)}{f'(\infty)(\zeta_\mu-\zeta_\nu)}\right|$$
$$+2k^2\sum_{\mu,\nu=1}^{N}\eta_\mu\bar{\eta}_\nu\ln\frac{1}{|1-(\zeta_\mu\bar{\zeta}_\nu)^{-1}|}.$$

这就是(5.1)式.

下面来证明(5.2),将(5.1)式改写成下述二次型的形式:

$$\sum_{\mu,\nu=1}^{N}\eta_\mu\bar{\eta}_\nu\ln\left|\frac{f(\zeta_\mu)-f(\zeta_\nu)}{f'(\infty)(\zeta_\mu-\zeta_\nu)}\right|^{\varepsilon lk\sqrt{2-k^2}}\cdot\frac{1}{|1-(\zeta_\mu\bar{\zeta}_\nu)^{-1}|^{lk^2}}$$
$$\geqslant\sum_{\mu,\nu=1}^{N}\eta_\mu\bar{\eta}_\nu\left\{\frac{l}{2}\sum_{n=1}^{\infty}\frac{1}{n}\left[k^2h_n(\zeta_\mu)\overline{h_n(\zeta_\nu)}+(2-k^2)g_n^{(\varepsilon)}(\zeta_\mu)\overline{g_n^{(\varepsilon)}(\zeta_\nu)}\right]\right\},\tag{5.4}$$

由是,若令

$$a_{\mu\nu}^{(1)}=\ln\left|\frac{f(\zeta_\mu)-f(\zeta_\nu)}{f'(\infty)(\zeta_\mu-\zeta_\nu)}\right|^{\varepsilon kl\sqrt{2-k^2}}\cdot\frac{1}{|1-(\zeta_\mu\bar{\zeta}_\nu)^{-1}|^{lk^2}},$$
$$a_{\mu\nu}^{(2)}=\frac{l}{2}\sum_{n=1}^{\infty}\frac{1}{n}\left[k^2h_n(\zeta_\mu)\overline{h_n(\zeta_\nu)}+(2-k^2)g_n^{(\varepsilon)}(\zeta_\mu)\overline{g_n^{(\varepsilon)}(\zeta_\nu)}\right],$$
$$a_{\mu\nu}^{(3)}=a_{\mu\nu}^{(1)}-a_{\mu\nu}^{(2)}\quad(\mu,\nu=1,2,\cdots,N),$$

则这 3 个矩阵都是爱尔米特半正定型,即

$$(a_{\mu\nu}^{(i)})\geqslant 0\quad(i=1,2,3).$$

于是,由文献[14, p. 314],便得

$$\sum_{\mu,\nu=1}^{N}\eta_\mu\bar{\eta}_\nu a_{\mu\nu}e^{a_{\mu\nu}^{(2)}}(e^{a_{\mu\nu}^{(3)}}-1)\geqslant 0,$$

也即

$$\sum_{\mu,\nu=1}^{N}\eta_\mu\bar{\eta}_\nu a_{\mu\nu}e^{a_{\mu\nu}^{(1)}}\geqslant\sum_{\mu,\nu=1}^{N}\eta_\mu\bar{\eta}_\nu a_{\mu\nu}e^{a_{\mu\nu}^{(2)}}.$$

这就是我们所要求的(5.2)式，即 Fitzgerald 不等式.

对于函数族 $S_B(K)$，我们有下述定理.

定理 10 设 $f\in S_B(K)$，$\{z_\mu\}$ $(\mu=1, 2, \cdots, N)$ 是 $|z|<1$ 中的点列，$\sum_{\mu,\nu=1}^{N} a_{\mu\nu}\eta_\mu\bar{\eta}_\nu\geqslant 0$ 对任何复数 $\{\eta_\mu\}$ 成立，又设

$$G_n^{(\varepsilon)}(z)=F_n\left(\frac{1}{f(z)}\right)-\left(z^{-n}+\frac{\varepsilon k}{\sqrt{2-k^2}}(\bar{z})^n\right),\ \varepsilon=1,-1,$$

$$H_n(z)=F_n(f(z))+2n\gamma_n,\quad \ln\frac{f(z)}{z}=-\sum_{n=0}^{\infty}2\gamma_n z^n,$$

则对任何正数 l 成立着

$$\frac{l}{2}\sum_{n=1}^{\infty}\frac{1}{n}\left[k^2\left|\sum_{\mu=1}^{N}\eta_\mu H_n(z_\mu)\right|^2+(2-k^2)\left|\sum_{\mu=1}^{N}\eta_\mu G_n^{(\varepsilon)}(\zeta_\mu)\right|^2\right]$$
$$\leqslant\sum_{\mu,\nu=1}^{N}\eta_\mu\bar{\eta}_\nu\ln\left|\frac{f'(0)(f(z_\mu)-f(z_\nu))z_\mu z_\nu}{(z_\mu-z_\nu)f(z_\mu)f(z_\nu)}\right|^{\varepsilon kl\sqrt{2-k^2}}\frac{1}{|1-z_\mu\bar{z}_\nu|^{lk^2}},\tag{5.5}$$

$$\sum_{\mu,\nu=1}^{N}\eta_\mu\bar{\eta}_\nu a_{\mu\nu}\exp\left\{\frac{l}{2}\sum_{n=1}^{\infty}\frac{1}{n}[k^2H_n(z_\mu)\overline{H_n(z_\nu)}+(2-k^2)G_n^{(\varepsilon)}(z_\mu)\overline{G_n^{(\varepsilon)}(z_\nu)}]\right\}$$
$$\leqslant\sum_{\mu,\nu=1}^{N}\eta_\mu\bar{\eta}_\nu a_{\mu\nu}\left|\frac{f'(0)(f(z_\mu)-f(z_\nu))z_\mu z_\nu}{(z_\mu-z_\nu)f(z_\mu)f(z_\nu)}\right|^{\varepsilon kl\sqrt{2-k^2}}\cdot\frac{1}{|1-z_\mu\bar{z}_\nu|^{lk^2}}.\tag{5.6}$$

结束语 本文的方法对 Grunsky 函数族也同样是适用的，即对单位圆 $|z|<1$ 中的正则的单叶函数 $f(z)$，且对 $|z_i|<1$ $(i=1, 2)$，$f(z_1)\overline{f(z_2)}\neq -1$ 的函数也是适用的.

何成奇同志曾对引理的证明提出过宝贵的意见，作者在此表示深切的谢意.

◇参考文献◇

[1] Гутлянский В. Я., О принципе площадей для одногокласса квазиконформных отображений *ДАН*, 212(1973), 540 - 543.

[2] Гутлянский В. Я., Щепетев В. А., Обобщенная георема-площадей для одногокласса квазнкон-формных отображений. *ДАН*, 218 (1974), 509 - 512.

[3] 任福尧,Bieberbach 函数族和 Grunsky 函数族的另一充要条件和偏差定理,科学通报,9(1981),516-519.

[4] 胡克,解析函数的若干性质,江西师院学报(自然科学版),1(1979),5-14.

[5] 夏道行,单叶函数论中的面积原理 II,复旦大学学报,2(1956),133-145.

[6] Kühnau R. Wertannahmeprobleme bei quassikonformen abbil dungen mit ortsabhängiger Dilatationsbeschrankung, *Diese Nachr.*, 40(1969), 1-11.

[7] ——, Verzerrungssatze und Koeffizientenbedingungen vom Grunsyschen Typ fur quasikonforme Abbildungen, *Math. Nachr.*, 48(1971), 77-105.

[8] ——, Zur analytischen Darstellung gewisser Extremal funktionen der quasikonformen Abbildung, *Math. Nachr.*, 60(1974), 53-62.

[9] ——, Schlichte konforme Abbildungen auf nichtüberlapppende Gebiete mit gemeinsamer quasikonformer Fortsetzung, *Math. Nachr.*, 86(1978), 175-180.

[10] Mcleavey J. O., Extremal problems in classes of analytic univalent functions with quasiconformal extensions, *Trans. Amer. Math. Soc.*, 195(1974), 327-343.

[11] Lehto O. Schlicht functions with a quasiconformal extension, *Ann. Acad. Sci. Fenn. Ser.* A 150(1971), 1-10.

[12] Schiffer M., Schober G., Coefficient problems and generalized Grunsky inequalities for schlicht functions with quasiconformal extensions, *Arch. Rational Mech. Anal.*, 60 (1975/76), 205-228.

[13] Крушкаль С. Л., Некоторые локальные теоремы для квазикон-конформных отображений рима новых поверхностей, *ДАН*, 199(1971), 269-272.

[14] Лебедев Н. А., Принцип площадей в теории однолистных функций. Издательство наука, Москва, 1975.

[15] Ahlfors L., Quasiconformal Mapping, Princeton, Van Nostrand, 1966.

[16] Bieberbach L. Uber einige Extremal probleme in Gebeite der konformen Abbildung, *Math. Ann.*, 77(1915-1916), 153-172.

[17] Векуа И. Н., Обобщенные аналитические функций, Москва, 1959.

[18] Schiffer M., Faber polynomials in the theory of univalent Functions. *Bull Amer. Math. Soc.*, 54(1948), 503-517.

Stronger Distortion Theorems of Univalent Functions and Its Applications*

REN FUYAO

Abstract: The purpose of this paper is to improve and to generalize the famous Fitzgerald inequalities, the Bazilevic inequalities and the Hayman regular theorem, by means of the representation theorem of continuous linear functionals on the space of continuous functions.

1. Introduction

The research on the Bieberbach conjecture due to Fitzgerald in 1971 is very important, for it not only improved the estimate for the coefficients of univalent functions, but enabled us to see more deeply the importance of applying the "exponentaited" Grunsky inequalities. Concretely speaking, he obtained the following "exponentaited" distortion theorems about the difference quotients and their reciprocals:

Let $f(z)=z+a_2z^2+a_3z^3+\cdots$ be in S, namely $f(z)$ is regular and univalent in the unit disk $\Delta=\{z: |z|<1\}$, and, let β_μ ($\mu=1, 2, 3, \cdots, N$) be arbitrary complex numbers which are not all zeros, and z_μ ($\mu=1, 2, 3, \cdots, N$) be arbitrary points in the unit disk, then

$$\left|\sum_{\mu,\nu=1}^{N}\beta_\mu\beta_\nu\left(\frac{f(z_\mu)-f(z_\nu)}{z_\mu-z_\nu}\frac{z_\mu z_\nu}{f(z_\mu)f(z_\nu)}\right)^{\varepsilon}\right|\leqslant\sum_{\mu,\nu=1}^{N}\beta_\mu\bar{\beta}_\nu(1-z_\mu\bar{z}_\nu)^{-1}\quad(\varepsilon=1,-1),\tag{1.1}$$

* Originally published in *Chin. Ann. of Math. B*, Vol. 4, No. 4, (1983), 425 - 441.

$$\sum_{\mu,\nu=1}^{N}\beta_\mu\bar{\beta}_\nu\left|\frac{f(z_\mu)-f(z_\nu)}{z_\mu-z_\nu}\frac{1}{1-z_\mu\bar{z}_\nu}\right|^l\geqslant\left|\sum_{\mu=1}^{N}\beta_\mu\left|\frac{f(z_\mu)}{z_\mu}\right|^l\right|^2 \tag{1.2}$$

for $l=1, 2$, where $(f(z_\mu)-f(z_\nu))/(z_\mu-z_\nu)$ is interpreted as $f'(z)$ when $z_\mu=z_\nu$.

In fact, Xia Daoshing already established a result stronger than (1.2) early in 1951. From (1.2) we have the following Fitzgerald's coefficient inequalities:

Let $f(z)=z+a_2z^2+a_3z^3+\cdots$ be in S,

$$a_{p,q}=\sum_{k=1}^{p+q-1}\beta_k(p, q)|a_k|^2-|a_pa_q|^2, \tag{1.3}$$

where

$$\beta_k(p, q)=\begin{cases}p-|k-q|, & |k-q|<p\ (p\leqslant q),\\ 0, & \text{otherwise}\end{cases} \tag{1.4}$$

then

$$(a_{p,q})_{2\leqslant p,q\leqslant N}\geqslant 0. \tag{1.5}$$

It means that the matrices $(a_{p,q})$ are positive semidefinite. We well know that from (1.5) a series of important results can be obtained. In 1979, Kung Seng first improved the Fitzgerald's distortion theorems (1.2) and coefficient inequalities (1.5). In the same year, Hu Ke improved Kung Seng's results.

The aim of this paper is to make further improvements on the above mentioned Fitzgerald inequalities, together with the Bazilevic inequality and Hayman's regularity theorem by means of the representation theorem of continuous linear functionals on the space of continuous functions.

Suppose K is a bounded closed set in the complex plane. Denote by $c(K)$ the set of all bounded continuous functions on K. Endowed with the norm

$$\|f\|=\max_{z\in K}|f(z)|,\ f(z)\in c(K),$$

$c(K)$ is a Banach space.

Denote by $c^*(K)$ the conjugate space of $c(K)$. Let L_1, L_2 and $L\in c^*(K)$, $h(z, \zeta)$ be bounded continuous functions defined on $K\times K$.

We define

$$L_1L_2(h(z,\zeta))=L_2(L_1(h(z,\zeta))),$$
$$|L|^2(h(z,\zeta))=L(\overline{L(h(z,\zeta))}),$$

where z is operated by the linear functionals ahead of ζ. Here and later we suppose that $\Phi(w)=\sum_{n=0}^{\infty}c_n w^n$ is an integal function and define $\Phi^{+}(w)=\sum_{n=0}^{\infty}|c_n|w^n$.

2. Strengthened Distortion Theorems

Is this section, we shall strengthen the Grunsky distortion theorem and the improved Fitzgerald distortion theorem due to Kung Seng, Hu Ke.

Lemma *Suppose* $\sum_{\mu,\nu=1}^{N}a_{\mu\nu}x_\mu\bar{x}_\nu\geqslant 0$ $(\alpha_{\nu\mu}=\bar{\alpha}_{\mu\nu})$ *is non-negative Hermitian, then*

$$\left|\sum_{\mu,\nu=1}^{N}\alpha_{\mu\nu}x_\mu\bar{y}_\nu\right|^2\leqslant\sum_{\mu,\nu=1}^{N}\alpha_{\mu\nu}x_\mu\bar{x}_\nu\cdot\sum_{\mu,\nu=1}^{N}\alpha_{\mu\nu}y_\mu\bar{y}_\nu \tag{2.1}$$

holds for any complex numbers $\{x_\mu\}$ *and* $\{y_\mu\}$ $(\mu=1,2,\cdots,N)$.

Proof Since $\sum\alpha_{\mu\nu}\lambda_\mu\bar{\lambda}_\nu$ is non-negative Hermitian,

$$0\leqslant\sum_{\mu,\nu=1}^{N}\alpha_{\mu\nu}(e^{i\beta}x_\mu-\lambda y_\mu)\overline{(e^{i\beta}x_\nu-\lambda y_\nu)}$$
$$=\sum_{\mu,\nu=1}^{N}\alpha_{\mu\nu}x_\mu\bar{x}_\nu-2\lambda\mathrm{Re}\left(e^{i\beta}\sum_{\mu,\nu=1}^{N}\alpha_{\mu\nu}x_\mu\bar{y}_\nu\right)+\lambda^2\sum_{\mu,\nu=1}^{N}\alpha_{\mu\nu}y_\mu\bar{y}_\nu$$

holds for any real numbers β and λ. This is a non-negative quadratic form of λ, so

$$\left[\mathrm{Re}\left(e^{i\beta}\sum_{\mu,\nu=1}^{N}\alpha_{\mu\nu}x_\mu\bar{y}_\nu\right)\right]^2\leqslant\sum_{\mu,\nu=1}^{N}\alpha_{\mu\nu}x_\mu\bar{x}_\nu\cdot\sum_{\mu,\nu=1}^{N}\alpha_{\mu\nu}y_\mu\bar{y}_\nu.$$

Take $\beta=-\arg\left(\sum_{\mu,\nu=1}^{N}\alpha_{\mu\nu}x_\mu\bar{y}_\nu\right)$, then (2.1) follows.

Theorem 1 *Suppose* $f\in S$, $\sum_{\mu,\nu=1}^{N}\alpha_{\mu\nu}x_{\mu}\bar{x}_{\nu}\geqslant 0$ $(\alpha_{\mu\nu}=\bar{\alpha}_{\nu\mu},\ \mu,\nu=1,2,\cdots,N)$, $l>0$, *m is a natural number. Denote by* $F_n(t)$ *the n-th Faber polynomial of* $f(z)$, $\varepsilon=1,-1$,

$$g_n^{(\varepsilon)}(z)=F_n(1/f(z))-(z^{-n}+\bar{z}\varepsilon^{n}),\tag{2.2}$$

$$F(z,\zeta)=\frac{f(z)-f(\zeta)}{z-\zeta}\cdot\frac{z\zeta}{f(z)f(\zeta)},\ f(z,\zeta)=\frac{f(z)-f(\zeta)}{z-\zeta},\tag{2.3}$$

then

$$\left|\sum_{\mu,\nu=1}^{N}\alpha_{\mu\nu}L_1^{(\mu)}\overline{L_2^{(\nu)}}\left\{\Phi\left[\left(\frac{l}{2}\sum_{n=1}^{\infty}\frac{1}{n}g_n^{(\varepsilon)}(z)\overline{g_n^{(\varepsilon)}(\zeta)}\right)^{m}\right]\right\}\right|^{2}$$
$$\leqslant\sum_{\mu,\nu=1}^{N}\alpha_{\mu\nu}L_1^{(\mu)}\overline{L_1^{(\nu)}}\left\{\Phi^{+}\left[\left(\ln\frac{|F(z,\zeta)|^{l\varepsilon}}{|1-z\bar{\zeta}|^{l}}\right)^{m}\right]\right\}$$
$$\cdot\sum_{\mu,\nu=1}^{N}\alpha_{\mu\nu}L_2^{(\mu)}\overline{L_2^{(\nu)}}\left\{\Phi^{+}\left[\left(\ln\frac{|F(z,\zeta)|^{l\varepsilon}}{|1-z\bar{\zeta}|^{l}}\right)^{m}\right]\right\}\tag{2.4}$$

holds for any integral function $\Phi(w)$ *and continuous linear functionals* $L_1^{(\mu)}$, $L_2^{(\mu)}$ *and* $L^{(\mu)}\in c^{*}(|z|\leqslant\rho<1)$ $(\mu=1,2,\cdots,N)$. *Particularly, we have*

$$\sum_{\mu,\nu=1}^{N}\alpha_{\mu\nu}L^{(\mu)}\overline{L^{(\nu)}}\left(\left|\frac{f(z)f(\zeta)}{z\zeta}\right|^{l\varepsilon}\exp\left\{\frac{1}{2}\sum_{n=1}^{N}\frac{1}{n}g_n^{(\varepsilon)}(z)\overline{g_n^{(\varepsilon)}(\zeta)}\right\}\right)$$
$$\leqslant\sum_{\mu,\nu=1}^{N}\alpha_{\mu\nu}L^{(\mu)}\overline{L^{(\nu)}}\left(\frac{|f(z,\zeta)|^{l\varepsilon}}{|1-z\bar{\zeta}|^{l}}\right).\tag{2.5}$$

It is obvious that (2.5) contains the result (1.1) in [5].

Proof We first consider the Löwner-Golusin function[12, 13]

$$f(z)=\lim_{t\to+\infty}e^{t}f(z,t),$$

where $f(z,t)=e^{-t}(z+a_2(t)z^2+a_3(t)z^3+\cdots)$ satisfies the differential equation and initial condition

$$\frac{\partial}{\partial t}f(z,t)=-f(z,t)\frac{1+k(t)f(z,t)}{1-k(t)f(z,t)},\ f(z,0)=z.\tag{2.6}$$

It is well known that thus obtained function set S^{*} is a dense subset of S.

For $f(z) \in S^*$ and $F(\zeta') = 1/f(\zeta'^{-1})$, we have[12]

$$\begin{cases} \ln \dfrac{F(z') - F(\zeta')}{z' - \zeta'} = \ln F(z, \zeta) = -2\displaystyle\int_0^\infty h(z, t) h(\zeta, t) \mathrm{d}t, \\ \ln(1 - 1/z'\bar{\zeta}') = \ln(1 - z\bar{\zeta}) = -2\displaystyle\int_0^\infty h(z, t) \overline{h(\zeta, t)} \mathrm{d}t, \end{cases} \tag{2.7}$$

where $z' = z^{-1}$, $\zeta' = \zeta^{-1}$, $|z| < 1$, $|\zeta| < 1$, $\ln 1 = 0$ and

$$h(z, t) = k(t) f(z, t) / (1 - k(t) f(z, t)).$$

It follows that

$$\ln \frac{|F(z, \zeta)|^\varepsilon}{|1 - z\bar{\zeta}|} = \begin{cases} 4\displaystyle\int_0^\infty \operatorname{Im} h(z, t) \operatorname{Im} h(\zeta, t) \mathrm{d}t, & \varepsilon = 1, \\ 4\displaystyle\int_0^\infty \operatorname{Re} h(z, t) \operatorname{Re} h(\zeta, t) \mathrm{d}t, & \varepsilon = -1. \end{cases} \tag{2.8}$$

Suppose

$$\ln F(z, \zeta) = \sum_{m, n=1}^\infty \gamma_{mn} z^m \zeta^n. \tag{2.9}$$

By the properties of Faber polynomials[15], we have

$$g_n^{(\varepsilon)}(z) = -n \sum_{p=1}^\infty \gamma_{pn} z^p - \varepsilon \bar{z}^n. \tag{2.10}$$

Hence

$$\sum_{n=1}^\infty \frac{1}{n} g_n^{(\varepsilon)}(z) \overline{g_n^{(\varepsilon)}(\zeta)} = \sum_{p, q=1}^\infty \left(\sum_{n=1}^\infty n \gamma_{pn} \bar{\gamma}_{qn} \right) z^p \bar{\zeta}^q + \sum_{n=1}^\infty \frac{1}{n} (\bar{z}\zeta)^n$$
$$+ \varepsilon \left(\sum_{p, n=1}^\infty \gamma_{pn} z^p \zeta^n + \overline{\sum_{p, n=1}^\infty \gamma_{pn} \zeta^p z^n} \right).$$

From $\gamma_{np} = \gamma_{pn}$ and Milin's lemma[10]

$$\sum_{n=1}^\infty n \gamma_{pn} \bar{\gamma}_{qn} = \begin{cases} 0, & q \neq p, \\ p^{-1}, & q = p, \end{cases}$$

it follows that

$$\sum_{n=1}^\infty \frac{1}{n} g_n^{(\varepsilon)}(z) \overline{g_n^{(\varepsilon)}(\zeta)} = 2\operatorname{Re}\left\{ \sum_{p=1}^\infty \frac{1}{p} (z\bar{\zeta})^p + \varepsilon \sum_{p, n=1}^\infty \gamma_{pn} z^p \zeta^n \right\}$$
$$= 2\ln(|F(z, \zeta)|^\varepsilon / |1 - z\bar{\zeta}|).$$

comparing this expression with (2.8), we obtain

$$\frac{1}{2}\sum_{n=1}^{\infty}\frac{1}{n}g_n^{(\varepsilon)}(z)\overline{g_n^{(\varepsilon)}(\zeta)}=4\int_0^{\infty}g(z,t)g(\zeta,t)\mathrm{d}t, \tag{2.11}$$

where $g(z, t)=\mathrm{Im}h(z, t)$ when $\varepsilon=1$ and $g(z, t)=\mathrm{Re}h(z, t)$ when $\varepsilon=-1$.

According to Riesz representation theorem, for any $L\in c^*(|z|\leqslant\rho<1)$ there exists a complex measure $\mu(z)$ on $|z|\leqslant\rho$, which satisfies the condition $|\mu|=\|L\|$, so that[17]

$$L(\varphi)=\int_{|z|\leqslant\rho}\varphi(z)\mathrm{d}\mu(z),\ \varphi\in C(|z|\leqslant\rho<1). \tag{2.12}$$

Then from (2.11), Fubini theorem, Cauchy inequality and (2.1), we have

$$\begin{aligned}
&\left|\sum_{\mu,\nu=1}^{N}\alpha_{\mu\nu}L_1^{(\mu)}\overline{L_2^{(\nu)}}\left\{\left[\frac{l}{2}\sum_{n=1}^{\infty}\frac{1}{n}g_n^{(\varepsilon)}(\varepsilon)\overline{g_n^{(\varepsilon)}(\zeta)}\right]^m\right\}\right|^2\\
&=\left|\sum_{\mu,\nu=1}^{N}\alpha_{\mu\nu}L_1^{(\mu)}\overline{L_2^{(\nu)}}\left(4l\int_0^{\infty}g(z,t)g(\zeta,t)\mathrm{d}t\right)^m\right|^2\\
&\leqslant\left\{\int_0^{\infty}\cdots\int_0^{\infty}(4l)^m\left[\sum_{\mu,\nu=1}^{N}\alpha_{\mu\nu}L_1^{(\mu)}\left(\prod_{j=1}^{m}g(z,t_j)\right)\overline{L_1^{(\nu)}\left(\prod_{j=1}^{m}g(z,t_j)\right)}\right]^{1/2}\right.\\
&\quad\left.\cdot\left[\sum_{\mu,\nu=1}^{N}\alpha_{\mu\nu}L_2^{(\mu)}\left(\prod_{j=1}^{m}g(\zeta,t_j)\right)\overline{L_2^{(\nu)}\left(\prod_{j=1}^{m}g(\zeta,t_j)\right)}\right]^{1/2}\prod_{j=1}^{m}\mathrm{d}t_j\right\}^2\\
&\leqslant\sum_{\mu,\nu=1}^{N}\alpha_{\mu\nu}L_1^{(\mu)}\overline{L_1^{(\nu)}}\left(4l\int_0^{\infty}g(z,t)g(\zeta,t)\mathrm{d}t\right)^m\\
&\quad\cdot\sum_{\mu,\nu=1}^{N}\alpha_{\mu\nu}L_2^{(\mu)}\overline{L_2^{(\nu)}}\left(4l\int_0^{\infty}g(z,t)g(\zeta,t)\mathrm{d}t\right)^m.
\end{aligned}$$

Using (2.8), we deduce

$$\begin{aligned}
&\left|\sum_{\mu,\nu=1}^{N}\alpha_{\mu\nu}L_1^{(\mu)}\overline{L_2^{(\nu)}}\left(\frac{l}{2}\sum_{n=1}^{\infty}\frac{1}{n}g_n^{(\varepsilon)}(z)\overline{g_n^{(\varepsilon)}(\zeta)}\right)^m\right|^2\\
&\leqslant\sum_{\mu,\nu=1}^{N}\alpha_{\mu\nu}L_1^{(\mu)}\overline{L_1^{(\nu)}}\left(\ln\frac{|F(z,\zeta)|^{l\varepsilon}}{|1-z\bar{\zeta}|^l}\right)^m\cdot\sum_{\mu,\nu=1}^{N}\alpha_{\mu\nu}L_2^{(\mu)}\overline{L_2^{(\nu)}}\left(\ln\frac{|F(z,\zeta)|^{l\varepsilon}}{|1-z\bar{\zeta}|^l}\right)^m.
\end{aligned} \tag{2.13}$$

Then for any integral function $\Phi(w)$, by Cauchy inequality,

$$\left|\sum_{\mu,\nu=1}^{N}\alpha_{\mu\nu}L_1^{(\mu)}\overline{L_2^{(\nu)}}\left\{\Phi\left[\left(\frac{l}{2}\sum_{n=1}^{\infty}\frac{1}{n}g_n^{(\varepsilon)}(z)\overline{g_n^{(\varepsilon)}(\zeta)}\right)^m\right]\right\}\right|^2$$

$$\leqslant \left\{ \sum_{p=1}^{\infty} |C_p| \left[\sum_{\mu,\nu=1}^{N} \alpha_{\mu\nu} L_1^{(\mu)} \overline{L_1^{(\nu)}} \left(\ln \frac{|F(z,\zeta)|^{l\varepsilon}}{|1-z\bar{\zeta}|^{l}} \right)^{mp} \right. \right.$$

$$\left. \left. \cdot \sum_{\mu,\nu=1}^{N} \alpha_{\mu\nu} L_2^{(\mu)} \overline{L_2^{(\nu)}} \left(\ln \frac{|F(z,\zeta)|^{l\varepsilon}}{|1-z\bar{\zeta}|^{l}} \right)^{mp} \right]^{1/2} \right\}^2$$

$$\leqslant \sum_{\mu,\nu=1}^{N} \alpha_{\mu\nu} L_1^{(\mu)} \overline{L_1^{(\nu)}} \{ \Phi^+ [(\ln|F(z,\zeta)|^{l\varepsilon}/|1-z\bar{\zeta}|^{l})^m] \}$$

$$\cdot \sum_{\mu,\nu=1}^{N} \alpha_{\mu\nu} L_2^{(\mu)} \overline{L_2^{(\nu)}} \{ \Phi^+ [(\ln|F(z,\zeta)|^{l\varepsilon}/|1-z\bar{\zeta}|^{l})^m] \}$$

holds. So (2.4) is true for any $f \in S^*$.

Especially, we choose $\Phi(w) = \exp\{w\}$, $m=1$, $L_1^{(\mu)} = L_2^{(\mu)} = L'^{(\mu)}$, then

$$\left| \sum_{\mu,\nu=1}^{N} \alpha_{\mu\nu} L'^{(\mu)} \overline{L'^{(\nu)}} \left(\exp\left\{ \frac{l}{2} \sum_{n=1}^{\infty} \frac{1}{n} g_n^{(\varepsilon)}(z) \overline{g_n^{(\varepsilon)}(\zeta)} \right\} \right) \right|$$

$$\leqslant \sum_{\mu,\nu=1}^{N} \alpha_{\mu\nu} L'^{(\mu)} \overline{L'^{(\nu)}} (|F(z,\zeta)|^{l\varepsilon}/|1-z\bar{\zeta}|^{l}). \tag{2.14}$$

The terms in the absolute value sign on the left hand side are

$$\sum_{p=1}^{\infty} \frac{1}{p!} \left(\frac{l}{2}\right)^p \sum_{n_1,\cdots,n_p=1}^{\infty} \left(\prod_{j=1}^{p} n_j^{-1}\right) \sum_{\mu,\nu=1}^{N} \alpha_{\mu\nu} L'^{(\mu)} \left(\prod_{j=1}^{p} g_{n_j}^{(\varepsilon)}(z)\right) \overline{L'^{(\nu)} \left(\prod_{j=1}^{p} g_{n_j}^{(\varepsilon)}(\zeta)\right)}.$$

It is obvious that they are non-negative. If we choose $L'^{(\mu)}(|z/f(z)|^{l\varepsilon}\varphi) = L^{(\mu)}(\varphi)$, $\varphi \in C(|z| \leqslant \rho < 1)$, then

$$\sum_{\mu,\nu=1}^{N} \alpha_{\mu\nu} L^{(\mu)} \overline{L^{(\nu)}} \left(\left| \frac{f(z)f(\zeta)}{z\zeta} \right|^{l\varepsilon} \exp\left\{ \frac{l}{2} \sum_{n=1}^{\infty} \frac{1}{n} g_n^{(\varepsilon)}(z) \overline{g_n^{(\varepsilon)}(\zeta)} \right\} \right)$$

$$\leqslant \sum_{\mu,\nu=1}^{N} \alpha_{\mu\nu} L^{(\mu)} \overline{L^{(\nu)}} (|f(z,\zeta)|^{l\varepsilon}/|1-z\bar{\zeta}|^{l}).$$

So (2.5) is also true for any $f \in S^*$.

As S^* is dense in S, by the continuity of linear functionals, we know that (2.4) and (2.5) are true for any $f \in S$.

Especially, we choose $L^{(\mu)}(\varphi) = \lambda_\mu \varphi(z_\mu)$. From (2.5), we obtain[5]

$$\sum_{\mu,\nu=1}^{N} \alpha_{\mu\nu} \lambda_\mu \bar{\lambda}_\nu \left| \frac{f(z_\mu) f(z_\nu)}{z_\mu z_\nu} \right|^{l\varepsilon} \exp\left\{ \frac{l}{2} \sum_{n=1}^{\infty} \frac{1}{n} g_n^{(\varepsilon)}(z_\mu) \overline{g_n^{(\varepsilon)}(z_\nu)} \right\}$$

$$\leqslant \sum_{\mu,\nu=1}^{N} \alpha_{\mu\nu}\lambda_{\mu}\bar{\lambda}_{\nu} \frac{|f(z_{\mu}, z_{\nu})|^{l\varepsilon}}{|1-z_{\mu}\bar{z}_{\nu}|^{l}}. \tag{2.15}$$

Analogously, from (2.4) we can obtain

$$\left|\sum_{\mu,\nu=1}^{N} \alpha_{\mu\nu}\lambda_{\mu}\overline{\lambda'_{2}} \left|\frac{f(z_{\mu})f(z'_{\nu})}{z_{\mu}z'_{\nu}}\right|^{l\varepsilon} \exp\left\{\frac{l}{2}\sum_{n=1}^{\infty}\frac{1}{n}g_{n}^{(\varepsilon)}(z_{\mu})\overline{g_{n}^{(\varepsilon)}(z'_{\nu})}\right\}\right|^{2}$$
$$\leqslant \sum_{\mu,\nu=1}^{N} \alpha_{\mu\nu}\lambda_{\mu}\bar{\lambda}_{\nu} \frac{|f(z_{\mu}, z_{\nu})|^{l\varepsilon}}{|1-z_{\mu}\bar{z}_{\nu}|^{l}} \cdot \sum_{\mu,\nu=1}^{N} \alpha_{\mu\nu}\lambda'_{\mu}\overline{\lambda'_{\nu}} \frac{|f(z'_{\mu}, z'_{\nu})|^{l\varepsilon}}{|1-z'_{\mu}\bar{z}'_{\nu}|^{l}}, \tag{2.16}$$

which is stronger than (2.15). Thus Thorem 1 is proved.

Theorem 2 *Under the same hypotheses of Theorem* 1, *we have*

$$\left|\sum_{\mu,\nu=1}^{N} \alpha_{\mu\nu}L_{1}^{(\mu)}\overline{L_{2}^{(\nu)}}\left\{\Phi\left[\left(\frac{l}{n(1+n\varepsilon\operatorname{Re}\gamma_{mn})}\operatorname{Re}g_{n}^{(\varepsilon)}(z)\operatorname{Re}g_{n}^{(\varepsilon)}(\zeta)\right)^{m}\right]\right\}\right|^{2} \leqslant A_{1}\cdot A_{2}, \tag{2.17}$$

$$\left|\sum_{\mu,\nu=1}^{N} \alpha_{\mu\nu}L_{1}^{(\mu)}\overline{L_{2}^{(\nu)}}\left\{\Phi\left[\left(\frac{l}{n(1-n\varepsilon\operatorname{Re}\gamma_{mn})}\operatorname{Im}g_{n}^{(\varepsilon)}(z)\operatorname{Im}g_{n}^{(\varepsilon)}(\zeta)\right)^{m}\right]\right\}\right|^{2} \leqslant A_{1}\cdot A_{2}, \tag{2.18}$$

$$\left|\sum_{\mu,\nu=1}^{N} \alpha_{\mu\nu}L_{1}^{(\mu)}\overline{L_{2}^{(\nu)}}\left\{\Phi\left[\left(\frac{l}{n\sqrt{1-(n\operatorname{Re}\gamma_{mn})^{2}}}\operatorname{Re}g_{n}^{(\varepsilon)}(z)\operatorname{Im}g_{n}^{(\varepsilon)}(\zeta)\right)^{m}\right]\right\}\right|^{2} \leqslant A_{1}\cdot A_{2}, \tag{2.19}$$

here

$$A_{i}=\sum_{\mu,\nu=1}^{N} \alpha_{\mu\nu}L_{i}^{(\mu)}\overline{L_{i}^{(\nu)}}\left\{\Phi^{+}\left[\left(\ln\frac{|F(z,\zeta)|^{l\varepsilon}}{|1-z\bar{\zeta}|^{l}}\right)^{m}\right]\right\} \quad (i=1, 2).$$

Particularly, we have

$$\sum_{\mu,\nu=1}^{N} \alpha_{\mu\nu}L^{(\mu)}\overline{L^{(\nu)}}\left(\left|\frac{f(z)f(\zeta)}{z\zeta}\right|^{l\varepsilon}\exp\left\{\frac{l}{n(1+n\varepsilon\operatorname{Re}\gamma_{mn})}\operatorname{Re}g_{n}^{(\varepsilon)}(z)\operatorname{Re}g_{n}^{(\varepsilon)}(\zeta)\right\}\right)$$
$$\leqslant \sum_{\mu,\nu=1}^{N} \alpha_{\mu\nu}L^{(\mu)}\overline{L^{(\nu)}}\left(\frac{|f(z,\zeta)|^{l\varepsilon}}{|1-z\bar{\zeta}|^{l}}\right)=B, \tag{2.20}$$

$$\sum_{\mu,\nu=1}^{N} \alpha_{\mu\nu}L^{(\mu)}\overline{L^{(\nu)}}\left(\left|\frac{f(z)f(\zeta)}{z\zeta}\right|^{l\varepsilon}\exp\left\{\frac{l}{n(1-n\varepsilon\operatorname{Re}\gamma_{mn})}\operatorname{Im}g_{n}^{(\varepsilon)}(z)\operatorname{Im}g_{n}^{(\varepsilon)}(\zeta)\right\}\right)\leqslant B, \tag{2.21}$$

$$\left|\sum_{\mu,\nu=1}^{N}\alpha_{\mu\nu}L^{(\mu)}\overline{L^{(\nu)}}\left(\left|\frac{f(z)f(\zeta)}{z\zeta}\right|^{l\varepsilon}\exp\left\{\frac{l}{n\sqrt{1-(n\mathrm{Re}\,\gamma_{nn})^2}}\mathrm{Re}\,g_n^{(\varepsilon)}(z)\mathrm{Im}\,g_n^{(\varepsilon)}(\zeta)\right\}\right)\right| \leqslant B. \tag{2.22}$$

It is obvious that (2.20) and (2.21) contain (1.2) and (1.3) in [5].

Proof Suppose

$$h(z,t)=\sum_{p=1}^{\infty}b_p(t)z^p.$$

It was proved by Golusin that[11]

$$\begin{cases}\gamma_{pq}=-2\displaystyle\int_0^{\infty}b_p(t)b_q(t)\mathrm{d}t \quad (p,q=1,2,3,\cdots),\\ \displaystyle\int_0^{\infty}b_p(t)\overline{b_q(t)}\mathrm{d}t=\begin{cases}0, & q\neq p,\\ 1/2p, & q=p \quad (p,q=1,2,\cdots).\end{cases}\end{cases} \tag{2.23}$$

Substituting (2.23) into (2.10), we obtain

$$\begin{aligned}\frac{1}{n}g_n^{(\varepsilon)}(z)&=2\left(\int_0^{\infty}\sum_{p=1}^{\infty}z^pb_p(t)b_n(t)\mathrm{d}t-\varepsilon\bar{z}^n\int_0^{\infty}b_n(t)\overline{b_n(t)}\mathrm{d}t\right)\\&=2\int_0^{\infty}b_n(t)\left(\sum_{p=1}^{\infty}b_p(t)z^p-\varepsilon\sum_{p=1}^{\infty}\overline{b_p(t)z^p}\right)\mathrm{d}t.\end{aligned}$$

Hence

$$\frac{1}{n}\mathrm{Re}\,g_n^{(\varepsilon)}(z)=\begin{cases}4\displaystyle\int_0^{\infty}\mathrm{Im}\,b_n(t)\mathrm{Im}\,h(z,t)\mathrm{d}t & (\varepsilon=1),\\ 4\displaystyle\int_0^{\infty}\mathrm{Re}\,b_n(t)\mathrm{Re}\,h(z,t)\mathrm{d}t & (\varepsilon=-1),\end{cases} \tag{2.24}$$

$$\frac{1}{n}\mathrm{Im}\,g_n^{(\varepsilon)}(z)=\begin{cases}4\displaystyle\int_0^{\infty}\mathrm{Re}\,b_n(t)\mathrm{Im}\,h(z,t)\mathrm{d}t & (\varepsilon=1),\\ 4\displaystyle\int_0^{\infty}\mathrm{Im}\,b_n(t)\mathrm{Re}\,h(z,t)\mathrm{d}t & (\varepsilon=-1).\end{cases} \tag{2.25}$$

Define

$$b_n^{(\varepsilon)}(t)=\begin{cases}\mathrm{Im}\,b_n(t) & (\varepsilon=1),\\ \mathrm{Re}\,b_n(t) & (\varepsilon=-1),\end{cases}\qquad g^{(\varepsilon)}(z,t)=\begin{cases}\mathrm{Im}\,h(z,t) & (\varepsilon=1),\\ \mathrm{Re}\,h(z,t) & (\varepsilon=-1).\end{cases}$$

From (2.24), (2.1) and Cauchy inequality, it follows that

$$
\begin{aligned}
&\left|\sum_{\mu,\nu=1}^{N}\alpha_{\mu\nu}L_1^{(\mu)}\overline{L_2^{(\nu)}}\left(\frac{l}{n(1+n\varepsilon\operatorname{Re}\gamma_{mn})}\operatorname{Re}g_n^{(\varepsilon)}(z)\operatorname{Re}g_n^{(\varepsilon)}(\zeta)\right)^m\right| \\
=&\left|\sum_{\mu,\nu=1}^{N}\alpha_{\mu\nu}L_1^{(\mu)}\overline{L_2^{(\nu)}}\left(\frac{16ln}{1+n\varepsilon\operatorname{Re}\gamma_{mn}}\int_0^\infty\int_0^\infty b_n^{(\varepsilon)}(t)b_n^{(\varepsilon)}(t')g^{(\varepsilon)}(z,t)g^{(\varepsilon)}(\zeta,t')\mathrm{d}t\mathrm{d}t'\right)^m\right| \\
\leqslant&\left(\frac{16nl}{1+n\varepsilon\operatorname{Re}\gamma_{mn}}\right)^m\int_0^\infty\cdots\int_0^\infty\left|\prod_{j=1}^m b_n^{(\varepsilon)}(t_j)b_n^{(\varepsilon)}(t_j')\right| \\
&\cdot\left(\sum_{\mu,\nu=1}^{N}\alpha_{\mu\nu}L_1^{(\mu)}\left(\prod_{j=1}^m g^{(\varepsilon)}(z,t_j)\right)\overline{L_1^{(\nu)}\left(\prod_{j=1}^m g^{(\varepsilon)}(\zeta,t_j)\right)}\right)^{1/2} \\
&\cdot\left(\sum_{\mu,\nu=1}^{N}\alpha_{\mu\nu}L_2^{(\mu)}\left(\prod_{j=1}^m g^{(\varepsilon)}(z,t_j')\right)\overline{L_2^{(\nu)}\left(\prod_{j=1}^m g^{(\varepsilon)}(\zeta,t_j')\right)}\right)^{1/2} \\
&\cdot\left(\prod_{j=1}^m \mathrm{d}t,\ \mathrm{d}t_j'\right)\leqslant\left(\frac{16nl}{1+n\varepsilon\operatorname{Re}\gamma_{mn}}\right)^m \\
&\cdot\left(\int_0^\infty\cdots\int_0^\infty\left|\prod_{j=1}^m b_n^{(s)}(t_j')\right|^2\cdot\sum_{\mu,\nu=1}^{N}\alpha_{\mu\nu}L_1^{(\mu)}\left(\prod_{j=1}^m g^{(\varepsilon)}(z,t_j)\right)\overline{L_1^{(\nu)}\left(\prod_{j=1}^m g^{(\varepsilon)}(\zeta,t_j)\right)}\prod_{j=1}^m\mathrm{d}t_j\mathrm{d}t_j'\right)^{1/2} \\
&\cdot\left(\int_0^\infty\cdots\int_0^\infty\left|\prod_{j=1}^m b_n^{(s)}(t_j)\right|^2\cdot\sum_{\mu,\nu=1}^{N}\alpha_{\mu\nu}L_2^{(\mu)}\left(\prod_{j=1}^m g^{(\varepsilon)}(z,t_j')\right)\overline{L_2^{(\nu)}\left(\prod_{j=1}^m g^{(\varepsilon)}(\zeta,t_j')\right)}\prod_{j=1}^m\mathrm{d}t_j\mathrm{d}t_j'\right)^{1/2} \\
=&\left(\frac{16nl}{1+n\varepsilon\operatorname{Re}\gamma_{mn}}\right)^m\left(\int_0^\infty|b_n^{(\varepsilon)}(t)|^2\mathrm{d}t\right)^m \\
&\cdot\sum_{\mu,\nu=1}^{N}\alpha_{\mu\nu}L_1^{(\mu)}\overline{L_1^{(\nu)}}\left(\int_0^\infty g^{(\varepsilon)}(z,t)g^{(\varepsilon)}(\zeta,t)\mathrm{d}t\right)^m \\
&\cdot\sum_{\mu,\nu=1}^{N}\alpha_{\mu\nu}L_2^{(\mu)}\overline{L_2^{(\nu)}}\left(\int_0^\infty g^{(\varepsilon)}(z,t)g^{(\varepsilon)}(\zeta,t)\mathrm{d}t\right)^m.
\end{aligned}
\tag{2.26}
$$

As we have

$$
\begin{cases}
\displaystyle\int_0^\infty|\operatorname{Im}b_n(t)|^2\mathrm{d}t=\frac{1}{2}\int_0^\infty(|b_n(t)|^2-\operatorname{Re}b_n^2(t))\mathrm{d}t=\frac{1}{4n}(1+n\varepsilon\operatorname{Re}\gamma_{mn}),\\
\displaystyle\int_0^\infty|\operatorname{Re}b_n(t)|^2\mathrm{d}t=\frac{1}{2}\int_0^\infty(|b_n(t)|^2+\operatorname{Re}b_n^2(t))\mathrm{d}t=\frac{1}{4n}(1-n\varepsilon\operatorname{Re}\gamma_{mn}),
\end{cases}
\tag{2.27}
$$

we obtain from (2.26), (2.27) and (2.8)

$$\left|\sum_{\mu,\nu=1}^{N}\alpha_{\mu\nu}L_1^{(\mu)}\overline{L_2^{(\nu)}}\left(\frac{l}{n(1+n\varepsilon\operatorname{Re}\gamma_{nn})}\operatorname{Re}g_n^{(\varepsilon)}(z)\operatorname{Re}g_n^{(\varepsilon)}(\zeta)\right)^m\right|^2\leqslant D_1\cdot D_2,\tag{2.28}$$

here

$$D_4=\sum_{\mu,\nu=1}^{N}\alpha_{\mu\nu}L_i^{(\mu)}L_i^{(\nu)}\left(\ln\frac{|F(z,\zeta)|^{l\varepsilon}}{|1-z\bar{\zeta}|^l}\right)^m\quad(i=1,2).$$

Analogously, we can obtain from (2.24), (2.25) and (2.27).

$$\left|\sum_{\mu,\nu=1}^{N}\alpha_{\mu\nu}L_1^{(\mu)}\overline{L_2^{(\nu)}}\left(\frac{l}{n(1-n\varepsilon\operatorname{Re}\gamma_{nn})}\operatorname{Im}g_n^{(\varepsilon)}(z)\operatorname{Im}g_n^{(\varepsilon)}(\zeta)\right)^m\right|^2\leqslant D_1D_2,\tag{2.29}$$

$$\left|\sum_{\mu,\nu=1}^{N}\alpha_{\mu\nu}L_1^{(\mu)}\overline{L_2^{(\nu)}}\left(\frac{l}{n\sqrt{1-(n\operatorname{Re}\gamma_{nn})^2}}\operatorname{Re}g_n^{(\varepsilon)}(z)\operatorname{Im}g_n^{(\varepsilon)}(\zeta)\right)^m\right|^2\leqslant D_1D_2.\tag{2.30}$$

Then, using the Cauchy inequality, (2.17), (2.18) and (2.19) follow from (2.28), (2.29) and (2.30). Moreover, we know that (2.20), (2.21) and (2.22) also hold.

Particularly, we choose $L^{(\mu)}(\varphi)=\lambda_\mu\varphi(z_\mu)$, then[5]

$$\sum_{\mu,\nu=1}^{N}\alpha_{\mu\nu}\lambda_\mu\bar{\lambda}_\nu\left|\frac{f(z_\mu)f(z_\nu)}{z_\mu z_\nu}\right|^{l\varepsilon}\exp\left\{\frac{l\operatorname{Re}g_n^{(\varepsilon)}(z_\mu)\operatorname{Re}g_n^{(\varepsilon)}(z_\nu)}{n(1+n\varepsilon\operatorname{Re}\gamma_{nn})}\right\}$$
$$\leqslant\sum_{\mu,\nu=1}^{N}\alpha_{\mu\nu}\lambda_\mu\bar{\lambda}_\nu|f(z_\mu,z_\nu)|^{l\varepsilon}/|1-z_\mu\bar{z}_\nu|^l,\tag{2.31}$$

$$\sum_{\mu,\nu=1}^{N}\alpha_{\mu\nu}\lambda_\mu\bar{\lambda}_\nu\left|\frac{f(z_\mu)f(z_\nu)}{z_\mu z_\nu}\right|^{l\varepsilon}\exp\left\{\frac{l\operatorname{Im}g_n^{(\varepsilon)}(z_\mu)\operatorname{Im}g_n^{(\varepsilon)}(z_\nu)}{n(1-n\varepsilon\operatorname{Re}\gamma_{nn})}\right\}$$
$$\leqslant\sum_{\mu,\nu=1}^{N}\alpha_{\mu\nu}\lambda_\mu\bar{\lambda}_\nu|f(z_\mu,z_\nu)|^{\varepsilon}/|1-z_\mu\bar{z}_v|^l.\tag{2.32}$$

Analogously, from (2.17), (2.18) and (2.22), we have

$$\left|\sum_{\mu,\nu=1}^{N}\alpha_{\mu\nu}\lambda_\mu\bar{\lambda}'_\nu\left|\frac{f(z_\mu)f(z'_\nu)}{z_\mu z'_\nu}\right|^{l\varepsilon}\exp\left\{\frac{l\operatorname{Re}g_n^{(\varepsilon)}(z_\mu)\operatorname{Re}g_n^{(\varepsilon)}(z'_\nu)}{n(1+n\varepsilon\operatorname{Re}\gamma_{nn})}\right\}\right|^2\leqslant BB',\tag{2.33}$$

$$\left|\sum_{\mu,\nu=1}^{N}\alpha_{\mu\nu}\lambda_{\mu}\bar{\lambda}'_{\nu}\left|\frac{f(z_{\mu})f(z'_{\nu})}{z_{\mu}z'_{\nu}}\right|^{l\varepsilon}\exp\left\{\frac{l\operatorname{Im}g_n^{(\varepsilon)}(z_{\mu})\operatorname{Im}g_n^{(\varepsilon)}(z'_{\nu})}{n(1-n\varepsilon\operatorname{Re}\gamma_{nn})}\right\}\right|^2\leqslant BB', \tag{2.34}$$

$$\left|\sum_{\mu,\nu=1}^{N}\alpha_{\mu\nu}\lambda_{\mu}\bar{\lambda}_{\nu}\left|\frac{f(z_{\mu})f(z'_{\nu})}{z_{\mu}z'_{\nu}}\right|^{l\varepsilon}\exp\left\{\frac{l\operatorname{Re}g_n^{(\varepsilon)}(z_{\mu})\operatorname{Im}g_n^{(\varepsilon)}(z'_{\nu})}{n\sqrt{1-(n\operatorname{Re}\gamma_{nn})^2}}\right\}\right|^2\leqslant BB', \tag{2.35}$$

here

$$B=\sum_{\mu,\nu=1}^{N}\alpha_{\mu\nu}\lambda_{\mu}\bar{\lambda}_{\nu}\frac{|f(z_{\mu},z_{\nu})|^{l\varepsilon}}{|1-z_{\mu}\bar{z}_{\nu}|^{l}},\quad B'=\sum_{\mu,\nu=1}^{N}\alpha_{\mu\nu}\lambda'_{\mu}\overline{\lambda'_{\nu}}\frac{|f(z'_{\mu},z'_{\nu})|^{l\varepsilon}}{|1-z'_{\mu}\bar{z}'_{\nu}|^{l}}.$$

Apparently, (2.33), (2.34) are stronger than (2.31) and (2.32).

From (2.7) and (2.1), imitating the proof of Theorem 1, it is not difficult to prove the following theorem.

Theorem 3 *Suppose* $f\in S$, $\sum_{\mu,\nu=1}^{N}\alpha_{\mu\nu}x_{\mu}\bar{x}_{\nu}\geqslant 0$ $(\bar{\alpha}_{\nu\mu}=\alpha_{\mu\nu},\ \mu,\nu=1,2,\cdots,N)$, *$m$ is a natural number, p is a complex number, then for any integral function Φ and linear functionals* $L_1^{(\mu)}$, $L_2^{(\mu)}$ *and* $L^{(\mu)}\in c^*(D)$, *where D is an arbitrary closed subset of* $|z|<1$ *and* $\mu=1,2,\cdots,N$, *we have*

$$\begin{aligned}&\left|\sum_{\mu,\nu=1}^{N}\alpha_{\mu\nu}L_1^{(\mu)}L_2^{(\nu)}\{\Phi[(p\ln F(z,z'))^m]\}\right|^2\\ \leqslant&\sum_{\mu,\nu=1}^{N}\alpha_{\mu\nu}L_1^{(\mu)}\overline{L_1^{(\nu)}}\{\Phi^+[(|p|\ln(1-z\bar{z}')^{-1})^m]\}\\ &\cdot\sum_{\mu,\nu=1}^{N}\alpha_{\mu\nu}L_2^{(\mu)}\overline{L_2^{(\nu)}}\{\Phi^+[(|p|\ln(1-z\bar{z}')^{-1})^m]\}.\end{aligned}\tag{2.36}$$

Particularly, for $F(\zeta)\in\Sigma$, $|\zeta_{\mu}|>1$, $|\zeta'_{\mu}|>1$, *we have*

$$\begin{aligned}&\left|\sum_{\mu,\nu=1}^{N}\alpha_{\mu\nu}\lambda_{\mu}\lambda'_{\nu}\Phi\{[p\ln((F(\zeta_{\mu})-F(\zeta'_{\nu}))/(\zeta_{\mu}-\zeta'_{\nu}))]^m\}\right|^2\\ \leqslant&\sum_{\mu,\nu=1}^{N}\alpha_{\mu\nu}\lambda_{\mu}\bar{\lambda}_{\nu}\Phi^+\{[\ln(1-1/\zeta_{\mu}\bar{\zeta}_{\nu})^{-|p|}]^m\}\\ &\cdot\sum_{\mu,\nu=1}^{N}\alpha_{\mu\nu}\lambda'_{\mu}\bar{\lambda}'_{\nu}\Phi^+\{[\ln(1-1/\zeta'_{\mu}\bar{\zeta}'_{\nu})^{-|p|}]^m\}.\end{aligned}\tag{2.37}$$

Apparently, (2.37) contains (1.1) and some results corresponding to (1.1) in [2, 8].

3. The Improvements of Bazilevic Inequality and Hayman's Regularity Theorem

Theorem 1 *Suppose* $f \in S$, $\{\gamma_{kn}\}$ *are the Grunsky coefficients of*

$$g(\zeta)=1/f(z)=\zeta+\sum_{n=0}^{\infty} b_n\zeta^{-n},$$

$z=1/\zeta$, $\theta_0=\theta_0(f)$ *and* α_f *are the Hayman direction and Hayman constant of* f *respectively*, $\ln(f(z)/z)=2\sum_{n=1}^{\infty}\gamma_n z^n$. *If* $\sum_{k=1}^{\infty}|\eta_k|^2/k<\infty$, *then we have the sharp inequality*

$$\begin{aligned}&\left(\frac{1}{2}\ln\frac{1}{\alpha_f}-\sum_{n=1}^{\infty} n\left|\gamma_n-\frac{1}{n}e^{-in\theta_0}\right|^2\right)\left(\sum_{k=1}^{\infty}\frac{1}{k}|\eta_k|^2-\sum_{n=1}^{\infty} n\left|\sum_{k=1}^{\infty}\eta_k\gamma_{kn}\right|^2\right)\\ &\geqslant\left|\sum_{k=1}^{\infty}\eta_k\left[\left(\frac{1}{k}e^{-ik\theta_0}-\gamma_k\right)+\sum_{n=1}^{\infty} n\overline{\left(\gamma_n-\frac{1}{n}e^{-in\theta_0}\right)}\gamma_{kn}\right]\right|^2.\end{aligned}\tag{3.1}$$

This improves the famous Bazilevic inequality[14] and Golusin inequality. Particularly, it contains[6]

$$\begin{aligned}&\left(\frac{1}{2}\ln\frac{1}{\alpha_f}-\sum_{n=1}^{\infty} n\left|\gamma_n-\frac{1}{n}e^{-in\theta_0}\right|^2\right)\left(1-\sum_{n=1}^{\infty} n|b_n|^2\right)\\ &\geqslant\left|(\gamma_1-e^{-i\theta_0}+\sum_{n=1}^{\infty} n\overline{\left(\gamma_n-\frac{1}{n}e^{-in\theta_0}\right)}b_n\right|^2.\end{aligned}\tag{3.2}$$

Proof From (2.4), it follows that for any $L\in c^*(|z|\leqslant\rho<1)$,

$$|L|^2\left(\ln\frac{|F(z,\zeta)|^{l\varepsilon}}{|1-z\zeta|^l}-\frac{l}{2}\sum_{n=1}^{\infty}\frac{1}{n}g_n^{(\varepsilon)}(z)\overline{g_n^{(\varepsilon)}(\zeta)}\right)\geqslant 0.\tag{3.3}$$

Let $L(\varphi)=L'(\varphi)+x\varphi(z_0)$, where $\varphi\in C(|z|\leqslant\rho<1)$, $L'\in c^*(|z|\leqslant\rho<1)$, x is a real number, z_0 is an arbitrary fixed point in the unit disk. Assume $\varphi(z,\zeta)$ is a continuous function on $(|z|<1)\times(|\zeta|<1)$, which satisfies the condition

$$\overline{L'(\varphi(z, z_0))} = L'(\overline{\varphi(z_0, z)}).$$

Since $|L|^2\varphi(z, \zeta) \geqslant 0$,

$$|L'|^2\varphi(z, \zeta) + 2x\mathrm{Re}[L'\overline{(\varphi(z_0, z))}] + x^2\overline{\varphi(z_0, z_0)} \geqslant 0. \tag{3.4}$$

Take $L=1$ (i. e. $L(\varphi)=\varphi$). It follows from $|L|^2\varphi(z, \zeta) \geqslant 0$ that $\varphi(z, \zeta) \geqslant 0$. So from (3.4), we obtain

$$\overline{\varphi(z_0, z)}\,|L'|^2(\varphi(z, \zeta)) \geqslant \{\mathrm{Re}[L'\overline{(\varphi(z_0, z_0))}]\}^2. \tag{3.5}$$

Substituting $\varphi(z, \zeta) = \ln(|F(z, \zeta)|^{l\varepsilon}/|1-z\zeta|^l) - \frac{l}{2}\sum_{n=1}^{\infty}\frac{1}{n}g_n^{(\varepsilon)}(z)\overline{g_n^{(\varepsilon)}(\zeta)}$ into (3.5), we have

$$\begin{aligned}&\left(\ln\frac{|F(z_0, z_0)|^{l\varepsilon}}{(1-|z_0|^2)^l} - \frac{l}{2}\sum_{n=1}^{\infty}\frac{1}{n}|g_n^{(\varepsilon)}(z_0)|^2\right)\\ &\cdot|L'|^2\left(\ln\frac{|F(z, \zeta)|^{l\varepsilon}}{|1-z\zeta)^l} - \frac{l}{2}\sum_{n=1}^{\infty}\frac{1}{n}g_n^{(\varepsilon)}(z)\overline{g_n^{(\varepsilon)}(\zeta)}\right)\\ &\geqslant\left\{\mathrm{Re}\left[L'\left(\ln\frac{|F(z_0, z)|^{l\varepsilon}}{|1-z_0\bar{z}|^l} - \frac{l}{2}\sum_{n=1}^{\infty}\frac{1}{n}\overline{g_n^{(\varepsilon)}(z_0)}g_n^{(\varepsilon)}(z)\right)\right]\right\}^2.\end{aligned} \tag{3.6}$$

Apparently, (3.6) is stronger than (2.7). In particular, taking $\varepsilon=1$, $l=2$, we have

$$\begin{aligned}&\left(\ln\frac{|F(z_0, z_0)|^2}{(1-|z_0|^2)^2} - \sum_{n=1}^{\infty}\frac{1}{n}|g_n^{(1)}(z_0)|^2\right)\cdot|L'|^2\left(\ln\left|\frac{F(z, \zeta)}{1-z\bar{\zeta}}\right|^2 - \sum_{n=1}^{\infty}\frac{1}{n}g_n^{(1)}(z)\overline{g_n^{(1)}(\zeta)}\right)\\ &\geqslant\left\{\mathrm{Re}\left[L'\left(\ln\left|\frac{F(z_0, z)}{1-z_0\bar{z}}\right|^2 - \sum_{n=1}^{\infty}\frac{1}{n}\overline{g_n^{(1)}(z_0)}g_n^{(1)}(z)\right)\right]\right\}^2.\end{aligned} \tag{3.7}$$

Assume $z_0 = re^{i\theta_0}$. As

$$0 < \alpha_f = \lim_{r\to1}(1-r)^2|f(z_0)|/r,\quad \lim_{r\to1} 1/f(z_0) = 0,$$

$$\lim_{r\to1} g_n^{(1)}(z_0) = 2n\left(\gamma_n - \frac{1}{n}e^{-in\theta_0}\right)^{[15]},\quad \left|\frac{zf'(z)}{f(z)}\right| \leqslant \frac{1+|z|}{1-|z|},$$

we deduce

$$\lim_{r\to1}\left(\ln\frac{|F(z_0, z_0|^2}{(1-r^2)^2} - \sum_{n=1}^{\infty}\frac{1}{n}|g_n^{(1)}(z_0)|^2\right)$$

$$\leqslant 4\left(\frac{1}{2}\ln\frac{1}{\alpha_f}-\sum_{n=1}^{\infty}n\left|\gamma_n-\frac{1}{n}e^{-in\theta_0}\right|^2\right), \tag{3.8}$$

$$\lim_{r\to 1}\left(\ln\left|\frac{F(z_0, z)}{1-z_0\bar{z}}\right|^2\right)=\ln\left|\frac{z}{f(z)(1-\bar{z}_0 z)^2}\right|^2. \tag{3.9}$$

By the arbitrariness of z_0 and the continuity of L', it follows from (3.7), (3.8) and (3.9) that

$$\left(\frac{1}{2}\ln\frac{1}{\alpha_f}-\sum_{n=1}^{\infty}n\left|\gamma_n-\frac{1}{n}e^{-in\theta_0}\right|^2\right)|L'|^2\left(\ln\left|\frac{F(z, \zeta)}{1-z\bar{\zeta}}\right|^2-\sum_{n=1}^{\infty}\frac{1}{n}g_n^{(\varepsilon)}(z)\overline{g_n^{(\varepsilon)}(\zeta)}\right)$$
$$\geqslant\left\{\mathrm{Re}\left[L'\left(\ln\left|\frac{z/f(z)}{(1-z_0\bar{z})^2}\right|-\sum_{n=1}^{\infty}\overline{\left(\gamma_n-\frac{1}{n}e^{-in\theta_0}\right)g_n^{(1)}(z)}\right)\right]\right\}^2. \tag{3.10}$$

Assume

$$L'(z^k)=\begin{cases}\eta_k e^{i\alpha}, & 1\leqslant k\leqslant N,\ 0\leqslant\alpha<+\infty,\\ 0, & k>N.\end{cases}$$

This can be done if we take $d\mu'(z)=e^{i\alpha}\sum_{p=1}^{N}\frac{p+1}{2\pi\rho^{p+1}}\eta_p e^{-ip\theta}dr d\theta$, $z=re^{i\theta}$ in (2.12). Then

$$|L'|^2\left(\ln\left|\frac{F(z, \zeta)}{1-z\bar{\zeta}}\right|^2\right)=|L'|^2(\ln F(z, \zeta)+\overline{\ln F(z, \zeta)}-\ln(1-z\bar{\zeta})(1-\bar{z}\zeta))$$
$$=\sum_{p, q=1}^{\infty}\bar{\gamma}_{pq}\overline{L'(z^p)L'(\bar{\zeta}^q)}+\sum_{p, q=1}^{\infty}\gamma_{pq}\overline{L'(\bar{z}^p)}L'(\zeta^q)+\sum_{p=1}^{\infty}\frac{1}{p}\overline{L'(z^p)}L'(\zeta^p)$$
$$+\sum_{p=1}^{\infty}\frac{1}{p}\overline{L'(\bar{z}^p)L'(\bar{\zeta}^p)}=\sum_{k=1}^{N}\frac{1}{k}|\eta_k|^2, \tag{3.11}$$

$$|L'|^2\left(\ln\left(\left|\frac{z}{f(z)}\right|\cdot|1-\bar{z}_0 z|^{-2}\right)\right)$$
$$=\sum_{p=1}^{\infty}L'\left(\frac{1}{p}((\bar{z}_0 z)^p+(z_0\bar{z})^p)-(\gamma_p z^p+\overline{\gamma_p z^p})\right)$$
$$=e^{i\alpha}\sum_{k=1}^{N}\left(\frac{1}{k}e^{-ik\theta_0}-\gamma_k\right)\eta_k. \tag{3.12}$$

$$L'(g_n^{(1)}(z))=-L'\left(n\sum_{p=1}^{\infty}\gamma_{pn}z^p+\bar{z}^n\right)=-ne^{id}\sum_{k=1}^{N}\eta_k\gamma_{kn},$$

$$|L'|^2\Big(\sum_{n=1}^{\infty}\frac{1}{n}g_n^{(1)}(z)g_n^{(1)}(\zeta)\Big)=\sum_{n=1}^{\infty}\frac{1}{n}|L'(g_n^{(1)}(z))|^2=\sum_{n=1}^{\infty}n\Big|\sum_{k=1}^{N}\eta_k\gamma_{kn}\Big|^2.$$

From (3.10) and the arbitrariness of α, we obtain

$$\Big(\frac{1}{2}\ln\frac{1}{\alpha_f}-\sum_{n=1}^{\infty}n\Big|\gamma_n-\frac{1}{n}e^{-in\theta_0}\Big|^2\Big)\Big(\sum_{k=1}^{N}\frac{1}{k}|\eta_k|^2-\sum_{n=1}^{\infty}n\Big|\sum_{k=1}^{N}\eta_k\gamma_k\Big|^2\Big)$$
$$\geqslant\sum_{k=1}^{N}\Big|\eta_k\Big[\Big(\frac{1}{k}e^{-ik\theta_0}-\gamma_k\Big)+\sum_{n=1}^{\infty}n\Big(\gamma_n-\frac{1}{n}e^{-in\theta_0}\Big)\gamma_{kn}\Big]\Big|^2. \tag{3.13}$$

Therefore (3.1) is true. Furthermore, when $\alpha_f=1$, it follows from

$$\sum_{n=1}^{\infty}n\Big|\gamma_n-\frac{1}{n}e^{-in\theta_0}\Big|^2\leqslant\frac{1}{2}\ln\frac{1}{\alpha_f}$$

that

$$\gamma_n=\frac{1}{n}e^{-in\theta_0}\quad(n=1,2,3,\cdots).$$

So the equality of (3.13) holds.

Finally, note that $\gamma_{1n}=-b_n$, so it is obvious that (3.1) contain (3.2). This completes the proof.

Theorem 2 *Suppose $f\in S$, $\theta_0(f)$ and α_f are the Hayman direction and Hayman constant of f respectively, $\{\gamma_{kn}\}$ are the Grunsky coefficients of f, $\ln(f(z)/z)=2\sum_{n=1}^{\infty}\gamma_n z^n$, then*

$$\left\{\frac{1}{2}\ln\frac{1}{\alpha_f}-\frac{2n\Big[\mathrm{Re}\Big(\gamma_n-\frac{1}{n}e^{-in\theta_0}\Big)\Big]^2}{1+n\mathrm{Re}\,\gamma_{nn}}\right\}\left(\sum_{k=1}^{N}\frac{1}{k}|\eta_k|^2-\frac{n\Big|\sum_{k=1}^{N}\eta_k\gamma_{kn}+\frac{1}{n}\eta_n\Big|^2}{4(1+n\mathrm{Re}\,\gamma_{nn})}\right)$$
$$\geqslant\left|\sum_{k=1}^{N}\eta_k\Big(\frac{1}{k}e^{-ik\theta_0}-\gamma_k\Big)+\frac{n\Big(\sum_{k=1}^{N}\eta_k\gamma_{kn}+\frac{1}{n}\eta_n\Big)\cdot\mathrm{Re}\Big(\gamma_n-\frac{1}{n}e^{-in\theta_0}\Big)}{1+n\mathrm{Re}\,\gamma_{nn}}\right|^2, \tag{3.14}$$

$$\left\{\frac{1}{2}\ln\frac{1}{\alpha_f}-\frac{2n\Big[\mathrm{Im}\Big(\gamma_n-\frac{1}{n}e^{-in\theta_0}\Big)\Big]^2}{1-n\mathrm{Re}\,\gamma_{nn}}\right\}\left(\sum_{k=1}^{N}\frac{1}{k}|\eta_k|^2-\frac{n\Big|\sum_{k=1}^{N}\eta_k\gamma_{kn}-\frac{1}{n}\eta_n\Big|^2}{4(1-n\mathrm{Re}\,\gamma_{nn})}\right)$$

$$\geqslant \left| \sum_{k=1}^{N} \eta_k \left(\frac{1}{k} e^{-ik\theta_0} - \gamma_k \right) - i \frac{n \left(\sum_{k=1}^{N} \eta_k \gamma_{kn} - \frac{1}{n} \eta_n \right) \operatorname{Im} \left(\gamma_n - \frac{1}{n} e^{-in\theta_0} \right)}{1 - n \operatorname{Re} \gamma_{nn}} \right|^2, \tag{3.15}$$

$$1 - \alpha_f \exp \left\{ \left| \frac{4n \operatorname{Re} \left(\gamma_n - \frac{1}{n} e^{-in\theta_0} \right) \operatorname{Im} \left(\gamma_n - \frac{1}{n} e^{-in\theta_0} \right)}{\sqrt{1 - (n \operatorname{Re} \gamma_{nn})^2}} \right| \right\} \geqslant 0, \tag{3.16}$$

where $\eta_n = 0$ *when* $n > N$. *Especially, when* $n = 1$, (3.14), (3.15) *improve the corresponding results of* [9].

Proof Assume $\overline{L_2} = L_1 = L$, $m = 1$ in (2.17) and (2.18), then

$$|L|^2 \left(\ln \frac{|F(z, \zeta)|^{l\varepsilon}}{|1 - z\bar{\zeta}|^l} - \frac{l}{n(1 + n\varepsilon \operatorname{Re} \gamma_{nn})} \operatorname{Re} g_n^{(\varepsilon)}(z) \operatorname{Re} g_n^{(\varepsilon)}(\zeta) \right) \geqslant 0, \tag{3.17}$$

$$|L|^2 \left(\ln \frac{|F(z, \zeta)|^{l\varepsilon}}{|1 - z\bar{\zeta}|^l} - \frac{l}{n(1 - n\varepsilon \operatorname{Re} \gamma_{nn})} \operatorname{Im} g_n^{(\varepsilon)}(z) \operatorname{Im} g_n^{(\varepsilon)}(\zeta) \right) \geqslant 0. \tag{3.18}$$

Imitating the proof of Theorem 1, from (3.17) and (3.18), it is not difficult to prove

$$\left\{ \frac{1}{2} \ln \frac{1}{\alpha_f} - \frac{2n \left[\operatorname{Re} \left(\gamma_n - \frac{1}{n} e^{-in\theta_0} \right) \right]^2}{1 + n \operatorname{Re} \gamma_{nn}} \right\} |L'|^2 \left(\ln \frac{|F(z, \zeta)|^2}{|1 - z\bar{\zeta}|^2} - \frac{2 \operatorname{Re} g_n^{(1)}(z) \operatorname{Re} g_n^{(1)}(\zeta)}{n(1 + n \operatorname{Re} \gamma_{nn})} \right)$$
$$\geqslant \left\{ \operatorname{Re} \left[L' \left(\ln \frac{|z/f(z)|}{|1 - \bar{z}_0 z|^2} - \frac{\operatorname{Re} g_n^{(1)}(z_0) \operatorname{Re} g_n^{(1)}(z)}{n(1 + n \operatorname{Re} \gamma_{nn})} \right) \right] \right\}^2, \tag{3.19}$$

$$\left\{ \frac{1}{2} \ln \frac{1}{\alpha_f} - \frac{2n \left[\operatorname{Im} \left(\gamma_n - \frac{1}{n} e^{-in\theta_0} \right) \right]^2}{1 - n \operatorname{Re} \gamma_{nn}} \right\} |L'|^2 \left(\ln \left| \frac{F(z, \zeta)}{1 - z\bar{\zeta}} \right|^2 - \frac{2 \operatorname{Im} g_n^{(1)}(z) \operatorname{Im} g_n^{(1)}(\zeta)}{n(1 - n \operatorname{Re} \gamma_{nn})} \right)$$
$$\geqslant \left\{ \operatorname{Re} \left[L' \left(\ln \frac{|z/f(z)|}{|1 - z\bar{z}_0|^2} - \frac{\operatorname{Im} g_n^{(1)}(z_0) \operatorname{Im} g_n^{(1)}(z)}{n(1 - n \operatorname{Re} \gamma_{nn})} \right) \right] \right\}^2, \tag{3.20}$$

where $z_0 = e^{i\theta_0}$. Assume

$$L'(z^k)=\begin{cases}\eta_k e^{i\alpha}, & 1\leqslant k\leqslant N,\ 0\leqslant\alpha<+\infty,\\ 0, & k>N,\end{cases}$$

it is easy to obtain

$$L'(\operatorname{Re} g_n^{(1)}(z))=\frac{1}{2}L'(g_n^{(1)}(z)+\overline{g_n^{(1)}(z)})=-\frac{n}{2}e^{i\alpha}\left(\sum_{k=1}^{N}\eta_k\gamma_{kn}+\frac{1}{n}\eta_n\right),$$

$$L'(\operatorname{Im} g_n^{(1)}(z))=\frac{1}{2i}L'(g_n^{(1)}(z)-\overline{g_n^{(1)}(z)})=-\frac{n}{2i}e^{i\alpha}\left(\sum_{k=1}^{N}\eta_k\gamma_{kn}-\frac{1}{n}\eta_n\right),$$

where $\eta_n=0$ when $n>N$. Then (3.14) and (3.15) can be easily deduced by substituting (3.11), (3.12) and the preceding two expressions into (3.19) and (3.20).

Finally, assume $L_2(\varphi)=L(\varphi)=\varphi(z_0)\,|z_0|<1$, from (2.19), we have

$$\left|\frac{\operatorname{Re} g_n^{(1)}(z_0)\cdot\operatorname{Im} g_n^{(1)}(z_0)}{n\sqrt{1-(n\operatorname{Re}\gamma_{nn})^2}}\right|\leqslant\ln\left|\frac{z_0^2 f'(z_0)}{(1-|z_0|^2)f(z_0)^2}\right|.$$

Set $z_0=re^{i\theta_0}$, $\theta_0=\theta_0(f)$ and let $r\to 1$, then (3.16) follows. Thus, Theorem 2 is proved.

4. The Improvements of Fitzgerald's Coefficients

Theorem 1 *Suppose* $f(z)=z+a_2z^2+a_3z_3+\cdots\in S$, $\{\gamma_{pq}\}$ $(p, q=1, 2, \cdots)$ *are the Grunsky coefficients of* $g(1/z)=f^{-1}(z)$, $\ln(f(z)/z)=2\sum_{n=1}^{\infty}\gamma_n z^n$, *then*

$$\sum_{p,q=2}^{N}a_{pq}\bar{x}_p x_q\geqslant\sum_{t=1}^{\infty}\frac{1}{t!}\left(\frac{2}{n(1+n\operatorname{Re}\gamma_{nn})}\right)^t\left|\sum_{p=2}^{N}D_{p,n}^{(t)}x_p\right|^2$$
$$+P_n\left|\sum_{p=2}^{N}\left[p^2-|a_p|^2-\sum_{i=1}^{\infty}\frac{1}{t!}\left(\frac{4\operatorname{Re}\left(\gamma_n-\frac{1}{n}e^{-in\theta_0}\right)}{1+n\operatorname{Re}\gamma_{nn}}\right)^t D_{p,n}^{(t)}\right]x_p\right|^2, \tag{4.1}$$

$$\sum_{p,q=2}^{N}a_{pq}\bar{x}_p x_q\geqslant\sum_{t=1}^{\infty}\frac{1}{t!}\left(\frac{2}{n(1-n\operatorname{Re}\gamma_{nn})}\right)^t\left|\sum_{p=2}^{N}E_{p,n}^{(t)}x_p\right|^2$$

$$+Q_n\left|\sum_{p=2}^{N}\left[p^2-|a_p|^2-\sum_{i=1}^{\infty}\frac{1}{t!}\left(\frac{4\mathrm{Im}\left(\gamma_n-\frac{1}{n}\mathrm{e}^{-in\theta_0}\right)}{1-n\mathrm{Re}\,\gamma_{nn}}\right)^t E_{p,n}^{(t)}\right]x_p\right|^2, \tag{4.2}$$

where a_{pq}, $\beta_k(p, q)$ *are defined as in* (1.3), (1.4), *and besides, we define*

$$G_n(z)=F_n\left(\frac{1}{f(z)}\right)-\frac{1}{z^n}-z^n,\ [G_n(z)]^t=\sum_{p=t}^{\infty}G_{p,n}^{(t)}z^p,$$

$$H_n(z)=F_n\left(\frac{1}{f(z)}\right)-\frac{1}{z^n}+z^n,\ [H_n(z)]^t=\sum_{p=t}^{\infty}H_{p,n}^{(t)}z^p,$$

$$D_{p,n}^{(t)}=2^{-t}\sum_{r=0}^{t}\binom{t}{r}\sum_{k=1}^{N-t+r}a_kG_{p-k,n}^{(t-r)}\cdot\overline{\sum_{k=1}^{N-r}a_{k'}G_{p-k',n}^{(r)}},$$

$$E_{p,n}^{(t)}=(2\mathrm{i})^{-t}\sum_{r=0}^{t}\binom{t}{r}\sum_{k=1}^{N-t+r}a_kH_{p-k,n}^{(t-r)}\overline{\sum_{k'=1}^{N-r}a_{k'}H_{p-k',n}^{(r)}},$$

$$P_n=\alpha_f^2\left(1-\alpha_f^2\exp\left\{\frac{8n\left[\mathrm{Re}\left(\gamma_n-\frac{1}{n}\mathrm{e}^{-in\theta_0}\right)\right]^2}{1+n\mathrm{Re}\,\gamma_{nn}}\right\}\right)^{-1},$$

$$Q_n=\alpha_f^2\left(1-\alpha_f^2\exp\left\{\frac{8n\left[\mathrm{Im}\left(\gamma_n-\frac{1}{n}\mathrm{e}^{-in\theta_0}\right)\right]^2}{1-n\mathrm{Re}\,\gamma_{nn}}\right\}\right)^{-1}.$$

Particularly, we have[7]

$$\sum_{p,q=2}^{N}a_{pq}\bar{x}_px_q\geqslant\frac{\alpha_f^2}{1-\alpha_f^2}\left|\sum_{p=1}^{N}(p^2-|a_p|^2)x_p\right|^2. \tag{4.3}$$

(4.1) and (4.2) extend and improve the corresponding results of [4, 9].

Proof From (2.20) and (2.21), we have

$$|L|^2\left(\left|\frac{f(z)-f(\zeta)}{z-\zeta}\right|^2\cdot|1-z\bar{\zeta}|^{-2}-\left|\frac{f(z)f(\zeta)}{z\zeta}\right|^2\exp\left\{\frac{2\mathrm{Re}\,g_n^{(1)}(z)\mathrm{Re}\,g_n^{(1)}(\zeta)}{n(1-n\mathrm{Re}\,\gamma_{nn})}\right\}\right)\geqslant 0, \tag{4.4}$$

$$|L|^2\left(\left|\frac{f(z)-f(\zeta)}{z-\zeta}\right|^2\cdot|1-z\bar{\zeta}|^{-2}-\left|\frac{f(z)f(\zeta)}{z\zeta}\right|^2\exp\left\{\frac{2\mathrm{Im}\,g_n^{(1)}(z)\mathrm{Im}\,g_n^{(1)}(\zeta)}{n(1-n\mathrm{Re}\,\gamma_{nn})}\right\}\right)\geqslant 0. \tag{4.5}$$

Imitating the proof of Theorem 1 in section 3 we can deduce

$$|L'|^2\left(\left|\frac{f(z)-f(\zeta)}{z-\zeta}\right|^2/|1-z\bar{\zeta}|^2-\left|\frac{f(z)f(\zeta)}{z\zeta}\right|^2\exp\left\{\frac{2\mathrm{Re}\,g_n^{(1)}(z)\mathrm{Re}\,g_n^{(1)}(\zeta)}{n(1+n\mathrm{Re}\,\gamma_{nn})}\right\}\right)$$
$$\geqslant P_n\left\{\mathrm{Re}\left[L'\left(|1-\bar{z}_0z|^{-4}-\left|\frac{f(z)}{2}\right|^2\exp\left\{\frac{2\mathrm{Re}\,g_n^{(1)}(z_0)\mathrm{Re}\,g_n^{(1)}(z)}{n(1+n\mathrm{Re}\,\gamma_{nn})}\right\}\right)\right]\right\}^2, \tag{4.6}$$

$$|L'|^2\left(\left|\frac{f(z)-f(\zeta)}{z-\zeta}\right|^2|1-z\bar{\zeta}|^{-2}-\left|\frac{f(z)f(\zeta)}{z\zeta}\right|^2\exp\left\{\frac{2\mathrm{Im}\,g_n^{(1)}(z)\mathrm{Im}\,g_n^{(1)}(\zeta)}{n(1-n\mathrm{Re}\,\gamma_{nn})}\right\}\right)$$
$$\geqslant Q_n\left\{\mathrm{Re}\left[L'\left(|1-\bar{z}_0z|^{-4}-\left|\frac{f(z)}{2}\right|^2\exp\left\{\frac{2\mathrm{Im}\,g_n^{(1)}(z_0)\mathrm{Im}\,g_n^{(1)}(z)}{n(1-n\mathrm{Re}\,\gamma_{nn})}\right\}\right)\right]\right\}^2, \tag{4.7}$$

here $z_0=\mathrm{e}^{i\theta_0}$, $g_n^{(1)}(z_0)=2n\left(\gamma_n-\frac{1}{n}\mathrm{e}^{-in\theta_0}\right)$.

It is not difficult to know that for any complex number sequence x_1, x_2, $\cdots$, x_N and real number α, there exists

$$L'(z^{k-1}\bar{z}^{k'-1})=\begin{cases}x_k\mathrm{e}^{i\alpha}, & 1\leqslant k'=k\leqslant N,\\ 0, & k'\neq k,\ k'=k>N.\end{cases}$$

So after some calculation, we have

$$L'(|1-\bar{z}_0z|^{-4})=\mathrm{e}^{i\alpha}\sum_{k=1}^{N}k^2x_k, \tag{4.8}$$

$$L'\left(\left|\frac{f(z)}{z}\right|^2\right)=\mathrm{e}^{i\alpha}\sum_{k=1}^{N}|a_k|^2x_k, \tag{4.9}$$

$$|L'|^2\left(\left|\frac{f(z)-f(\zeta)}{(z-\zeta)(1-z\zeta)}\right|^2\right)$$
$$=\sum_{k,k'=1}^{\infty}\sum_{p,p'=0}^{\infty}\bar{a}_ka_{k'}\sum_{t=1}^{k}\sum_{t'=1}^{k'}\overline{L'(z^{k-t+p}\bar{z}^{k'-t'+p'})}L'(\zeta^{t'+p-1}\bar{\zeta}^{t+p'-1}).$$

The terms on the right hand side of the above expression don't vanish and $L'(z^{m-1}\bar{z}^{m-1})=x_m\mathrm{e}^{i\alpha}$, $L'(\zeta^{n-1}\bar{\zeta}^{n-1})=x_n\mathrm{e}^{i\alpha}$ if and only if

$$m=k-t+p+1=k'-t'+p'+1,\ n=t+p'=t'+p. \tag{$*$}$$

We can rewrite (*) as

$$k=k',\ p'=n-t,\ p'=m-k+t-1,\ t'=(n-m)+k-t+1,$$

then it follows that

$$|L'|^2\left(\left|\frac{f(z)-f(\zeta)}{(z-\zeta)(1-z\bar{\zeta})}\right|^2\right)=\sum_{m,\,n=1}^{N}\left(\sum_{k=1}^{n+m-1}K(k)|a_k|^2\right)\bar{x}_m x_n.$$

When $1\leqslant k\leqslant m-n$, as $1\leqslant t'\leqslant -t+1\leqslant 0$, this is impossible, so $K(k)=0$; when $m-n<k\leqslant m$, as $1\leqslant t'\leqslant n-m+k$, $K(k)=k+n-m$; when $m<k\leqslant n-m$, as $n\geqslant t'\geqslant 1+k-m$, $K(k)=n+m-k$. Therefore, when $n\leqslant m$

$$K(k)=\beta_k(n,\,m)=\begin{cases}n-|k-m|, & |k-m|<n,\\ 0, & \text{otherwise}.\end{cases}$$

Hence

$$|L'|^2\left(\left|\frac{f(z)-f(\zeta)}{(z-\zeta)(1-\bar{\zeta}z)}\right|^2\right)=\sum_{m,\,n=1}^{N}\left(\sum_{k=1}^{n+m-1}\beta_k(n,\,m)|a_k|^2\right)\bar{x}_m x_n, \tag{4.10}$$

$$|L'|^2\left(\left|\frac{f(z)f(\zeta)}{z\zeta}\right|^2\right)=\left|L'\left(\left|\frac{f(z)}{z}\right|^2\right)\right|^2=\left|\sum_{k=1}^{N}|a_k|^2x_k\right|^2. \tag{4.11}$$

On the other hand, as

$$L'\left(\left|\frac{f(z)}{z}\right|^2[\operatorname{Re}g_n^{(1)}(z)]^t\right)=L'\left(\left|\frac{f(z)}{z}\right|^2[\operatorname{Re}G_n(z)]^t\right)$$
$$=\sum_{r=0}^{t}2^{-t}\binom{t}{r}\sum_{k,\,k'=1}^{\infty}\sum_{p=t-r}^{\infty}\sum_{p'=r}^{\infty}a_k\overline{a_{k'}}G_{p',\,n}^{(t-r)}\overline{G_{p,\,n}^{(r)}}L'(z^{k-1+p}\bar{z}^{k'-1+p'})=\sum_{m=2}^{N}D_{m,\,n}^{(t)}x_m e^{i\alpha},$$

we have

$$|L'|^2\left(\left|\frac{f(z)f(\zeta)}{z\zeta}\right|^2\exp\left\{\frac{2\operatorname{Re}g_n^{(1)}(z)\operatorname{Re}g_n^{(1)}(\zeta)}{n(1+n\operatorname{Re}\gamma_{nn})}\right\}\right)$$
$$=\sum_{t=0}^{\infty}\frac{1}{t!}\left(\frac{2}{n(1+n\operatorname{Re}\gamma_{nn})}\right)^t\left|L'\left(\left|\frac{f(z)}{z}\right|^2[\operatorname{Re}g_n^{(1)}(z)]^t\right)\right|^2$$
$$=\left|\sum_{k=1}^{N}|a_k|^2x_k\right|^2+\sum_{t=1}^{\infty}\frac{1}{t!}\left(\frac{2}{n(1+n\operatorname{Re}\gamma_{nn})}\right)^t\left|\sum_{k=2}^{N}D_{k,\,n}^{(t)}x_k\right|^2, \tag{4.12}$$

$$L'\left(\left|\frac{f(z)}{z}\right|^2 \exp\left\{\frac{2\mathrm{Re}\, g_n^{(1)}(z_0)\mathrm{Re}\, g_n^{(1)}(z)}{n(1+n\mathrm{Re}\,\gamma_{nn})}\right\}\right)$$

$$=\sum_{t=0}^{\infty}\frac{1}{t!}\left(\frac{2\mathrm{Re}\, g_n^{(1)}(z_0)}{n(1+n\mathrm{Re}\,\gamma_{nn})}\right)^t L'\left(\left|\frac{f(z)}{z}\right|^2[\mathrm{Re}\, g_n^{(1)}(z)]^t\right) \tag{4.13}$$

$$=\mathrm{e}^{\mathrm{i}\alpha}\left(\sum_{k=1}^{N}|a_k|^2 x_k+\sum_{t=0}^{\infty}\frac{1}{t!}\left(\frac{2\mathrm{Re}\, g_n^{(1)}(z_0)}{n(1+n\mathrm{Re}\,\gamma_{nn})}\right)^t\sum_{k=2}^{N}D_{k,n}^{(t)}x_k\right).$$

Then (4. 1) follows by substituting (4. 8), (4. 10), (4. 12) and (4. 13) into (4. 6).

The existence of L' still remains to be proved. For this purpose, we prove that for any k $(1\leqslant k\leqslant N)$, there exists $L_k\in c^*(|z|\leqslant\rho<1)$, such that

$$L_k(z^{p-1}\bar{z}^{q-1})=\begin{cases}1, & \text{if } p=k,\ q=k,\\ 0, & \text{otherwise.}\end{cases} \tag{4.14}$$

Assume that G_k is the linear subspace generated by $z^{p-1}\bar{z}^{q-1}(|p-k|+|q-k|=0)$, and $x_0=|z|^{2(k-1)}$, then

$$d(x_0,G_k)>0.$$

If it is not true, then there exists

$$\sum_{p,q=1}^{\infty}C_{pq}z^{p-1}\bar{z}^{q-1}\quad(C_{kk}=0),$$

which converges to x_0 uniformly in $|z|\leqslant\rho\leqslant1$, i. e.

$$x_0=\sum_{p,q=1}^{\infty}C_{pq}z^{p-1}\bar{z}^{q-1}.$$

Set $z=r\mathrm{e}^{\mathrm{i}\theta}(0\leqslant\theta<2\pi)$, and act $L(\varphi)=\frac{1}{2\pi}\int_0^{2\pi}\varphi(\theta)\mathrm{d}\theta$ on both sides of the above expression, then it follows that

$$r^{2(k-1)}=\sum_{p=1}^{\infty}C_{pq}r^{2(p-1)}\quad(0\leqslant r\leqslant\rho).$$

Therefore we deduce $1=0$, this is impossible. So $d(x_0,G_k)>0$.

By the continuation theorem of functionals[16], there must exists linear functional L_k which satisfies condition (4. 14). Then

$$L' = e^{i\alpha}\sum_{k=1}^{N} x_k L_k \in c^*(|z|\leqslant\rho<1)$$

is the linear functional we seek for. Therefore, (4.1) is true.

Imitating the proof of (4.12) and (4.13), we have

$$|L'|^2\left(\left|\frac{f(z)f(\zeta)}{z\zeta}\right|^2\exp\left\{\frac{2\operatorname{Im}g_n^{(1)}(z)\operatorname{Im}g_n^{(1)}(\zeta)}{n(1-n\operatorname{Re}\gamma_{nn})}\right\}\right)$$

$$=\left|\sum_{k=1}^{N}|a_k|^2x_k\right|^2+\sum_{t=1}^{\infty}\frac{1}{t!}\left(\frac{2}{n(1-n\operatorname{Re}\gamma_{nn})}\right)^t\left|\sum_{k=2}^{N}E_{k,n}^{(t)}x_k\right|^2, \quad (4.15)$$

$$L'\left(\left|\frac{f(z)}{z}\right|^2\exp\left\{\frac{2\operatorname{Im}g_n^{(1)}(z)\operatorname{Im}g_n^{(1)}(\zeta)}{n(1-\operatorname{Re}\gamma_{nn})}\right\}\right)$$

$$=e^{i\alpha}\left(\sum_{k=1}^{N}|a_k|^2x_k+\sum_{t=1}^{\infty}\frac{1}{t!}\left(\frac{2\operatorname{Im}g_n^{(1)}(z_0)}{n(1-n\operatorname{Re}\gamma_{nn})}\right)^t\sum_{k=2}^{N}E_{k,n}^{(t)}x_k\right). \quad (4.16)$$

Substitute (4.8), (4.10), (4.15) and (4.16) into (4.7), and note that α is arbitrary, then (4.2) can be easily obtained.

Finally, from (4.4) it follows that

$$|L|^2\left(\left|\frac{f(z)-f(\zeta)}{(z-\zeta)(1-z\bar{\zeta})}\right|^2-\left|\frac{f(z)f(\zeta)}{z\zeta}\right|^2\right)\geqslant 0,$$

hence

$$|L'|^2\left(\left|\frac{f(z)-f(\zeta)}{(z-\zeta)(1-z\bar{\zeta})}\right|^2-\left|\frac{f(z)f(\zeta)}{z\zeta}\right|^2\right)$$

$$\geqslant\frac{\alpha_f^2}{1-\alpha_f^2}\left\{\operatorname{Re}\left[L'\left(|1-\bar{z}_0z|^{-4}-\left|\frac{f(z)}{z}\right|^2\right)\right]\right\}^2. \quad (4.17)$$

Substitute (4.8), (4.9), (4.10) and (4.11) into (4.17), then (4.3) follows by the arbitrariness of α. Thus Theorem 1 is proved.

5. The Grunsky Inequalities of the Inverses of Meromorphic Univalent Functions

Theorem 1 *Suppose* $F(\zeta)=\zeta+b_1\zeta^{-1}+b_2\zeta^{-2}+\cdots\in\Sigma'$, $G(w)=F^{-1}(w)=w+B_1w^{-1}+B_2w^{-2}+\cdots$,

$$\ln\frac{G(w)-G(\xi)}{w-\xi}=\sum_{p,q=1}^{\infty}\lambda_{pq}w^{-p}\xi^{-q}. \tag{5.1}$$

Then we have

$$\left|\sum_{m,n=1}^{\infty}\lambda_{mn}x_m x_n\right|\leqslant\sum_{n=1}^{\infty}\frac{1}{n}\left|x_n+\sum_{m=n+2}^{\infty}B_m^{(-n)}x_m\right|^2, \tag{5.2}$$

where $B_n^{(-k)}$ *is defined as follows*:

$$[G(w)]^{-k}=w^{-k}+\sum_{n=k+2}^{\infty}B_n^{(-k)}w^{-n}. \tag{5.3}$$

Proof By Theorem 3 of section 2, for $F\in\Sigma$, $L_\mu\in c^*(1<r\leqslant|\zeta|\leqslant R<\infty)$, we have

$$\left|\sum_{\mu,\nu=1}^{N}L_\mu L_\nu\ln\left(\frac{z-\zeta}{F(z)-F(\zeta)}\right)\right|\leqslant\sum_{\mu,\nu=1}^{N}L_\mu\bar{L}_\nu(-\ln(1-1/z\bar{\zeta})).$$

For $L'_\mu\in c^*(F(r\leqslant|\zeta|\leqslant R))$, therefore we have

$$\left|\sum_{\mu,\nu=1}^{N}L'_\mu L'_\nu\ln\left(\frac{G(w)-G(\xi)}{w-\xi}\right)\right|\leqslant B, \tag{5.4}$$

here

$$B=\sum_{k=1}^{\infty}\frac{1}{k}\sum_{\mu,\nu=1}^{N}L'_\mu\overline{L'_\nu}([G(w)\overline{G(\xi)}]^{-k}. \tag{5.5}$$

Suppose that G_μ is the linear subspace generated by $\{w^{-n}\}$ $(n\neq\mu,\ n=1,2,3,\cdots)$, then $d=d(w^{-\mu},G_\mu)>0$. If it is not true, then there exists a power series $\sum_{n=1}^{\infty}C'_n w^{-n}$ $(C'_\mu=0)$ which converges to w^{-n} uniformly in $F(r\leqslant|\zeta|\leqslant R)$, i. e.

$$w^{-\mu}=\sum_{n=1}^{\infty}C'_n w^{-n}\quad(C'_\mu=0). \tag{5.6}$$

Assume $C_\rho=F(|\zeta|=\rho)(r<\rho<R)$, $L(\varphi)=\frac{1}{2\pi}\oint_{c_\rho}w^\mu\varphi(w)\mathrm{d}w$. Acting on both sides of (5.6) with operator $L(\varphi)$, and using Cauchy formula, we deduce $1=0$, this is impossible, so $d>0$.

Then, by the continuation theorem of functionals, we know that there exists $L'_\mu \in c^*(r \leqslant |\zeta| \leqslant R)$, such that

$$L'_\mu(w^{-n}) = \begin{cases} x_\mu, & n = \mu, \\ 0, & n \neq \mu. \end{cases} \tag{5.7}$$

Substituting (5.7), (5.1) and (5.3) into (5.4), (5.5), after some simple calculation, we can easily obtain (5.2). This completes the proof of Theorem 1.

Corollary *Suppose* $F(\zeta) = \zeta + b_1\zeta^{-1} + b_2\zeta^{-2} + \cdots \in \Sigma'$, $\zeta = F^{-1}(w) = w + B_1 w^{-1} + B_2 w^{-2} + \cdots$, $\{\gamma_{pq}\}(p, q = 1, 2, 3, \cdots)$ *are the Grunsky coefficients of* $G(w)$, *then*

$$\left| a^2\left(B_3 + \frac{1}{2}B_1^2\right) + 2abB_2 + b^2 B_1 \right| \leqslant |b|^2 + \frac{1}{2}|a|^2, \tag{5.8}$$

$$\begin{aligned} &\left| a^2\left(B_5 + B_1 B_2 + B_2^2 + \frac{1}{3}B_1^3\right) + 2ab(B_4 + B_1 B_2) + b^2\left(B_8 + \frac{1}{2}B_1^2\right) \right. \\ &\left. + c^2 B_1 + 2(acB_3 + bcB_2) \right| \leqslant |c + aB_3^{(-1)}|^2 + \frac{1}{2}|b|^2 + \frac{1}{3}|a|^2. \end{aligned} \tag{5.9}$$

We are convinced that these Grunsky inequalities will be of significance to the proof of Springer conjecture.

The author would like to express his thanks to Professor Xia Daoshing and Professor Yan Shaozong for their valuable suggestions.

◇ References ◇

[1] Fitzgerald, C. H. *Arch. Rat. Mech. Anal.*, 46(1972), 356 - 368.

[2] Xia Dao-Shing, *Science Record*, 4(1951), 351 - 362.

[3] 龚升,中国科学,3(1979), 237 - 246.

[4] 龚升,科学通报,15(1980), 673 - 675.

[5] 胡克,数学年刊,1: 3 - 4(1980), 421 - 428.

[6] 胡克,数学年刊,3: 3(1982), 293 - 302.

[7] 胡克,江西师院学报(自然科学版),1(1980), 1 - 6.

[8] 任福尧,复旦学报,3(1979), 69 - 75.

[9] 任福光,复旦学报,1(1980), 1-14.
[10] Milin, M. Dokl. Akad. Nauk, SSSR, 154(1964), 264-267.
[11] Golusin, G. M., *Mat. Sb. N. S.*, 29(1951), 197.
[12] Golusin, G. M., 复变函数几何理论,科学出版社,1956.
[13] Loewner, K., *Math. Ann.*, 89(1923) 103-121.
[14] Bazilevich, I. E., *Mat. Sb. N. S.*, 74(1967), 133-146.
[15] Schober, G., Univalent Functions-Selected Topics, 1975.
[16] 夏道行等,实变函数论与泛函分析(下册),人民教育出版社,1979.
[17] Schwarz, D., Linear Operations, Port I: General Theory, 1958.

Conformal Mapping of Non-overlapping Domain with Quasiconformal Extension*

Ren Fuyao

Let $\Sigma_r(K)$ $(0<r<1)$ denote a class of functions $w=f(z)$ which are homeomorphic mappings such that they map the sphere of z onto the sphere of w and that they are regular and univalent both in $|z|<r$ and $|z|>1/r$, while they are K-quasiconformal mappings in $r<|z|<1/r$ and have the form $f(z)=z+o(|z|)$ in the neighborhood at $z=\infty$. It was investigated first by R. Kühnau[1]. In the present paper we get series of distortion theorems for this class of functions and imporve Kühnau's results. We prove the following theorems.

Theorem 1 *Let* $f\in\Sigma_r(K)$, $\alpha_n=\dfrac{1-r^{4n}}{r^{2n}(1-k^2r^{4n})}$, $\beta=\dfrac{1}{1-k^2r^{4n}}$, $k=(K-1)/(K+1)$, λ *and* h *be arbitrary complex numbers, and let the sequences of complex-numbers* $\{x_n\}$, $\{x'_n\}$ $(n=0,1,\cdots,\infty)$ *and* $\{X_n\}$, $\{X'_n\}$ $(n=1,\cdots,\infty)$ *satisfy the following condition*:

$$\sum_{n=1}^{\infty}\frac{\alpha_n}{n}|\xi_n|^2<\infty. \tag{1}$$

Then we have

1)

$$2\Big|h\Big[\sum_{m,n=0}^{\infty}a_{mn}x'_m x_n+\sum_{m=0,n=1}^{\infty}b_{mn}x'_m X_n+(1-k^2)\sum_{n=1}^{\infty}\frac{\beta_n}{n}x'_n X_n$$
$$-\frac{4k^2}{1-k^2}x_0x'_0\ln r\Big]+\lambda\Big[\sum_{m,n=1}^{\infty}c_{mn}X_mX'_n+\sum_{m=0,n=1}^{\infty}b_{mn}x_mX'_n$$

---

* Originally published in *Kexue Tongbao*, Special Issue, (1983), 100 - 105.

$$+(1-k^2)\sum_{n=1}^{\infty}\frac{\beta_n}{n}x_nX_n\Big]\Big| \leqslant k\Big[\Big(\sum_{n=1}^{\infty}\frac{\alpha_n}{n}|x_n|^2 \tag{2}$$

$$-\frac{4}{1-k^2}|x_0|^2\ln r\Big)+|h^2|\Big(\sum_{n=1}^{\infty}\frac{\alpha_n}{n}|x'_n|^2-\frac{4}{1-k^2}|x'_0|^2\ln r\Big)$$

$$+\sum_{n=1}^{\infty}\frac{\alpha_n}{n}(|X_n|^2+|\lambda|^2|X'_n|^2)\Big].$$

2)

$$(2|h|-|h|^2)\Big|\sum_{m,\,n=0}^{\infty}\tilde{a}_{mn}x'_mx_n+\sum_{m=0,\,n=1}^{\infty}\tilde{b}_{mn}x'_mX_n\Big|^2A'^{-1}$$

$$+(2|\lambda|-|\lambda|^2)\Big|\sum_{m,\,n=1}^{\infty}c_{mn}X_mX'_n+\sum_{m=0,\,n=1}^{\infty}\tilde{b}_{mn}x_mX'_n\Big|^2B'^{-1} \tag{3}$$

$$\leqslant k^2(A+B),$$

for $0<|h|,\ |\lambda|\leqslant 2$; *in particular*,

$$\Big|\sum_{m,\,n=1}^{\infty}c_{mn}X_mX'_n\Big|^2\leqslant k^2BB',$$

$$\Big|\sum_{m,\,n=0}^{\infty}\tilde{a}_{mn}x_mx'_n\Big|^2\leqslant k^2AA',$$

$$\Big|\sum_{m=0,\,n=1}^{\infty}\tilde{b}_{mn}x'_mX_n\Big|^2\leqslant k^2A'B.$$

3)

$$\sum_{n=1}^{\infty}\frac{n}{\alpha_n}\Big|\sum_{m=1}^{\infty}c_{mn}X_m\Big|^2\leqslant k^2B,$$

$$\sum_{n=1}^{\infty}\frac{n}{\alpha_n}\Big|\sum_{m=0}^{\infty}\tilde{a}_{mn}x_m\Big|^2\leqslant k^2A, \tag{4}$$

$$\sum_{n=1}^{\infty}\frac{n}{\alpha_n}\Big|\sum_{m=1}^{\infty}\tilde{b}_{mn}x'_m\Big|^2\leqslant k^2A',$$

where

$$F(z,\zeta)=\ln\frac{f(z)-f(\zeta)}{z-\zeta}=\sum_{m,\,n=0}^{\infty}a_{mn}z^m\zeta^n\quad(|z|,\ |\zeta|<r),$$

$$G(z,\zeta)=\ln\frac{f(z)-f(\zeta)}{z-\zeta}=\sum_{m,\,n=1}^{\infty}c_{mn}z^{-m}\zeta^{-n}\quad(|z|,\ |\zeta|>1/r),$$

$$H(z,\zeta)=\ln\frac{f(\zeta)-f(z)}{\zeta}=\sum_{m=0,\,n=1}^{\infty}b_{mn}z^m\zeta^{-n}\quad(|z|<r,\ |\zeta|>1/r),$$

$$\widetilde{a}_{mn}=\begin{cases}a_{mn}, & m+n\neq 0,\\ a_{00}-\dfrac{4k^2}{1-k^2}\ln r, & m=n=0,\end{cases}$$

$$\widetilde{b}_{mn}=\begin{cases}b_{mn}, & m\neq n,\\ b_{mn}+(1-k^2)\beta_n/n, & m=n,\end{cases}$$

$$A=\sum_{n=1}^{\infty}\frac{\alpha_n}{n}|x_n|^2-\frac{4k^2}{1-k^2}\ln r,\quad B=\sum_{n=1}^{\infty}\frac{\alpha_n}{n}|X_n|^2,$$

$$A'=\sum_{n=1}^{\infty}\frac{\alpha_n}{n}|x'_n|^2-\frac{4k^2}{1-k^2}\ln r,\quad B'=\sum_{n=1}^{\infty}\frac{\alpha_n}{n}|X'_n|^2.$$

Particularly, for $h=\lambda=1$, $x'_n=x_n$ *and* $X'_n=X_n\,(n=1,2,\cdots)$, *follow immediately Kühnau's results from the above inequalities.*

Corollary *Let* $f\in\Sigma_r(K)$, $f(\zeta)=\zeta+\sum_{n=0}^{\infty}b_n\zeta^{-n}(|\zeta|>1/r)$, $f(z)=\sum_{n=1}^{\infty}a_nz^n(|z|<r)$, *and* $\ln f(z)/z=2\sum_{n=0}^{\infty}\gamma_nz^n$. *Then we have*

$$\sum_{n=1}^{\infty}\frac{n}{\alpha_n}|b_n|^2\leqslant k^2\alpha_1,\quad \sum_{n=1}^{\infty}\frac{n}{\alpha_n}|\gamma_n|^2\leqslant\frac{k^2}{1-k^2}\ln 1/r. \tag{5}$$

In particular, we get

$$|b_1|\leqslant k\alpha_1,\quad |a_2/a_1|\leqslant 2k\alpha_1\sqrt{-\ln r/(1-k^2)}. \tag{6}$$

Theorem 2 *Let* $f\in\Sigma_r(K)$, $\{z_\mu\}$ *and* $\{z'_\nu\}$ $(\mu=1,\cdots,p;\ \nu=1,\cdots,p')$ *be some distinguished points in* $|z|<r$, *and* $\{\zeta_\mu\}$, $\{\zeta'_\nu\}$ *be some distinguished points in* $1/r<|\zeta|<\infty$. *Then the following inequalities hold:*

$$\left|\sum_{\mu=1}^{p}\sum_{\nu=1}^{p'}\eta_\mu\eta'_\nu\left(\ln\frac{f(z_\mu)-f(z'_\nu)}{z_\mu-z'_\nu}-\frac{4k^2}{1-k^2}\ln r\right)\right|^2\leqslant k^2AA', \tag{7}$$

$$\left|\sum_{\mu=1}^{p}\sum_{\nu=1}^{q}\eta_\mu\lambda_\nu\left(\ln\frac{f(\zeta_\nu)-f(z_\mu)}{\zeta_\nu-z_\mu}-k^2\sum_{n=1}^{\infty}\frac{\alpha_n}{n}\left(\frac{r^2z_\mu}{\zeta_\nu}\right)^n\right)\right|^2\leqslant k^2AB, \tag{8}$$

$$\left|\sum_{\mu=1}^{q}\sum_{\nu=1}^{q'}\lambda_\mu\lambda'_\nu\ln\frac{f(\zeta_\mu)-f(\zeta'_\nu)}{\zeta_\mu-\zeta'_\nu}\right|^2\leqslant k^2BB', \tag{9}$$

where $\{\eta_\mu\}_1^\infty$, $\{\eta'_\nu\}_1^\infty$, $\{\lambda_\mu\}_1^\infty$ *and* $\{\lambda'_\nu\}_1^\infty$ *are arbitrary complex numbers and*

$$A=\sum_{\mu,\nu=1}^{p}\eta_\mu\bar{\eta}_\nu\left(\sum_{n=1}^{\infty}\frac{\alpha_n}{n}(z_\mu\bar{z}_\nu)^n-\frac{4}{1-k^2}\ln r\right),$$

$$B=\sum_{\mu,\nu=1}^{q}\lambda_\mu\bar{\lambda}_\nu\left(\sum_{n=1}^{\infty}\frac{\alpha_n}{n}(\zeta_\mu\bar{\zeta}_\nu)^{-n}\right),$$

$$A'=\sum_{\mu,\nu=1}^{p'}\eta'_\mu\bar{\eta}'_\nu\left(\sum_{n=1}^{\infty}\frac{\alpha_n}{n}(z'_\mu\bar{z}'_\nu)^n-\frac{4}{1-k^2}\ln r\right),$$

$$B'=\sum_{\mu,\nu=1}^{q'}\lambda'_\mu\bar{\lambda}'_\nu\left(\sum_{n=1}^{\infty}\frac{\alpha_n}{n}(\zeta'_\mu\bar{\zeta}'_\nu)^{-n}\right).$$

Especially, for $|z|<r$, *we have*

$$\left|\ln\frac{f(z)}{z}-\frac{4k^2}{1-k^2}\ln r\right|^2\leqslant\frac{4k^2}{1-k^2}\left(\sum_{n=1}^{\infty}\frac{\alpha_n}{n}|z|^{2n}-\frac{4}{1-k^2}\ln r\right)\ln\frac{1}{r};\qquad(10)$$

for $|\zeta|>1/r$, *we have*

$$\left|\ln\frac{f(\zeta)}{\zeta}\right|^2\leqslant\frac{-4k^2}{1-k^2}\left(\sum_{n=1}^{\infty}\frac{\alpha_n}{n}|\zeta|^{-2n}\right)\ln r,\qquad(11)$$

$$|\ln f'(\zeta)|\leqslant k\sum_{n=1}^{\infty}\frac{\alpha_n}{n}|\zeta|^{-2n}.\qquad(12)$$

Theorem 3 *On the assumption in Theorem 2, the following inequalities hold*:

$$\left|\sum_{\mu=1}^{p}\sum_{\nu=1}^{p'}\eta_\mu\eta'_\nu\left[\frac{f'(z_\mu)f'(z'_\nu)}{(f(z_\mu)-f(z'_\nu))^2}-\frac{1}{(z_\mu-z'_\nu)^2}\right]\right|^2\leqslant k^2DD',\qquad(13)$$

$$\left|\sum_{\mu=1}^{q}\sum_{\nu=1}^{q'}\lambda_\mu\lambda'_\nu\left[\frac{f'(\zeta_\mu)f'(\zeta'_\nu)}{(f(\zeta_\mu)-f(\zeta'_\nu))^2}-\frac{1}{(\zeta_\mu-\zeta'_\nu)^2}\right]\right|^2\leqslant k^2EE',\qquad(14)$$

$$\left|\sum_{\mu=1}^{p}\sum_{\nu=1}^{q'}\eta_\mu\lambda'_\nu\left[\frac{f'(z_\mu)f'(\zeta'_\nu)}{(f(z_\mu)-f(\zeta'_\nu))^2}-(1-k^2)\sum_{n=1}^{\infty}n\beta_n z_\mu^{n-1}\zeta'^{-(n+1)}_\nu\right]\right|^2\leqslant k^2DE',\qquad(15)$$

where

$$D=\sum_{\mu,\nu=1}^{p}\eta_\mu\bar{\eta}_\nu\left(\sum_{n=1}^{\infty}n\alpha_n(z_\mu\bar{z}_\nu)^{n-1}\right),$$

$$D'=\sum_{\mu,\nu=1}^{p'}\eta'_{\mu}\bar{\eta}'_{\nu}\left(\sum_{n=1}^{\infty}n\alpha_n(z'_{\mu}\bar{z}'_{\nu})^{n-1}\right),$$

$$E=\sum_{\mu,\nu=1}^{q}\lambda_{\mu}\bar{\lambda}_{\nu}\left(\sum_{n=1}^{\infty}n\alpha_n(\zeta_{\mu}\bar{\zeta}_{\nu})^{-n-1}\right),$$

$$E'=\sum_{\mu,\nu=1}^{q'}\lambda'_{\mu}\bar{\lambda}'_{\nu}\left(\sum_{n=1}^{\infty}n\alpha_n(\zeta'_{\mu}\bar{\zeta}'_{\nu})^{-n-1}\right).$$

Theorem 4 *Suppose $f\in\Sigma_r(K)$. Then we have*

$$|\{f,z\}|\leqslant 6k\left[\frac{r^2}{(r^2-|z|^2)^2}-(1-k^2)\sum_{n=1}^{\infty}\frac{n(r|z|)^{2n}}{|z|^2(1-k^2r^{4n})}\right],\quad(16)$$

$$|\{f,\zeta\}|\leqslant 6k\left[\frac{r^2}{(r^2|\zeta|^2-1)^2}-(1-k^2)\sum_{n=1}^{\infty}\frac{n(r/|\zeta|)^{2n}}{|\zeta|^2(1-k^2r^{4n})}\right],\quad(17)$$

for $|z|<r$ and $|\zeta|>1/r$, where

$$\{f,z\}=\frac{f'''(z)}{f'(z)}-\frac{3}{2}\left(\frac{f''(z)}{f'(z)}\right)^2$$

is the Schwarz derivative of f at point z.

Theorem 5 *Let $f\in\Sigma_r(K)$. Then the following inequalities hold:*

$$\frac{1}{\pi}\iint_{|z|>1/r}\left|\frac{f'(\zeta)f'(z)}{(f(\zeta)-f(z))^2}-\frac{1}{(\zeta-z)^2}\right|^2\mathrm{d}\sigma_z$$
$$\leqslant k^2\left[\frac{r^2}{(r^2|\zeta|^2-1)^2}-(1-k^2)\sum_{n=1}^{\infty}\frac{n}{|\zeta|^2(1-k^2r^{4n})}\left(\frac{r}{|\zeta|}\right)^{2n}\right]$$
$$-(1-k^2)\sum_{n=1}^{\infty}\frac{r^{6n}}{n(1-r^{4n})}|n\zeta^{n-1}-F'_n(f(\zeta))f'(\zeta)|^2\quad(|\zeta|>1/r),\quad(18)$$

$$\frac{1}{\pi}\iint_{|\zeta|<r}\left|\frac{f'(\zeta)f'(z)}{(f(\zeta)-f(z))^2}-\frac{1}{(\zeta-z)^2}\right|^2\mathrm{d}\sigma_\zeta$$
$$\leqslant k^2\left[\frac{r^2}{(r^2-|z|^2)^2}-(1-k^2)\sum_{n=1}^{\infty}\frac{n(r|z|)^{2n}}{|z|^2(1-k^2r^{4n})}\right]$$
$$-(1-k^2)\sum_{n=1}^{\infty}\frac{r^{6n}}{n(1-r^{4n})}\left|nz^{-(n+1)}-F'_n\left(\frac{f'(0)}{f(z)}\right)\left(\frac{f'(0)}{f(z)}\right)'\right|^2\quad(|z|<r),\quad(19)$$

$$\frac{1}{\pi}\iint_{|\zeta|>1/r}\left|\frac{f'(\zeta)f'(z)}{(f(\zeta)-f(z))^2}-(1-k^2)\sum_{n=1}^{\infty}\frac{nz^{n-1}\zeta^{-(n+1)}}{1-k^2r^{4n}}\right|^2 d\sigma_\zeta$$
$$\leqslant k^2\left[\frac{r^2}{(r^2-|z|^2)^2}-(1-k^2)\sum_{n=1}^{\infty}\frac{n(r|z|)^{2n}}{|z|^2(1-k^2r^{4n})}\right]$$
$$-(1-k^2)\sum_{n=1}^{\infty}\frac{r^{6n}}{n(1-r^{4n})}\left|F_n'(f(z))f'(z)-(1-k^2)\frac{nz^{n-1}}{1-k^2r^{4n}}\right|^2$$
$$(|z|<r), \tag{20}$$

where $F_n(t)$ is the n-th Faber polynomial of f.

Theorem 6 *Let $f\in\Sigma_r(K)$, $|z_\mu|<r$ and $|\zeta_\mu|>1/r$ ($\mu=1,\cdots,N$). Then for arbitrary complex numbers p and $\{\eta_\mu\}$ ($\mu=1,\cdots,N$), we have*

$$\left|\sum_{\mu,\nu=1}^{N}\eta_\mu\eta_\nu\left[\frac{f(\zeta_\mu)-f(\zeta_\nu)}{\zeta_\mu-\zeta_\nu}\right]^p\right|$$
$$\leqslant\sum_{\mu,\nu=1}^{N}\eta_\mu\bar{\eta}_\nu\left(1-\frac{1}{r^2\zeta_\mu\bar{\zeta}_\nu}\right)^{-|p|k}\prod_{n=1}^{\infty}\exp\left\{-\frac{k|p|(1-k^2)}{n(1-k^2r^{4n})}\left(\frac{r^2}{\zeta_\mu\bar{\zeta}_\nu}\right)^n\right\}, \tag{21}$$

$$\left|\sum_{\mu,\nu=1}^{N}\eta_\mu\eta_\nu\left(\frac{f(z_\mu)-f(z_\nu)}{z_\mu-z_\nu}\cdot\frac{z_\mu z_\nu f'(0)}{f(z_\mu)f(z_\nu)}\right)^p\right| \tag{22}$$
$$\leqslant\sum_{\mu,\nu=1}^{N}\eta_\mu\bar{\eta}_\nu\left(1-\frac{z_\mu\bar{z}_\nu}{r^2}\right)^{-|p|k}\prod_{n=1}^{\infty}\exp\left\{-\frac{k|p|(1-k^2)}{n(1-k^2r^{4n})}(r^2z_\mu\bar{z}_\nu)^n\right\},$$

$$\left|\sum_{\mu,\nu=1}^{N}\eta_\mu\eta_\nu\left[\left(1-\frac{f(z_\mu)}{f(\zeta_\nu)}\right)\exp\left\{(1-k^2)\sum_{n=1}^{\infty}\frac{z_\mu^n\zeta_\nu^{-n}}{n(1-k^2r^{4n})}\right\}\right]^p\right| \tag{23}$$
$$\leqslant\sum_{\mu,\nu=1}^{N}\eta_\mu\bar{\eta}_\nu\left(1-\frac{z_\mu}{r^2\bar{\zeta}_\nu}\right)^{-|p|k}\prod_{n=1}^{\infty}\exp\left\{-\frac{k|p|(1-k^2)}{n(1-k^2r^{4n})}\left(\frac{r^2z_\mu}{\bar{\zeta}_\nu}\right)^n\right\}.$$

Theorem 7 *Let $f\in\Sigma_r(K)$, $|z_\mu|<r$ and $|\zeta_\mu|>1/r$ ($\mu=1,\cdots,N$), $\varepsilon=1,-1$, and let*

$$A_n^{(\varepsilon)}(z)=F_n(1/f(z))-(z^{-n}-\varepsilon k\alpha_n\bar{z}^n),\quad |z|<r,$$
$$C_n^{(\varepsilon)}(\zeta)=F_n(f(\zeta))-(\zeta^{-n}-\varepsilon k\alpha_n\bar{\zeta}^{-n}),\quad |\zeta|>1/r,$$

α_n be defined as in Theorem 1. If $\sum_{\mu,\nu=1}^{N}\alpha_{\mu\nu}\eta_\mu\bar{\eta}_\nu\geqslant 0$ ($\bar{a}_{\nu\mu}=a_{\mu\nu}$, $\mu,\nu=1,\cdots,N$) for any complex numbers $\{\eta_\mu\}$ ($\mu=1,\cdots,N$) and l be a positive number,

we have

$$\sum_{\mu,\ \nu=1}^{N} \alpha_{\mu\nu}\eta_{\mu}\bar{\eta}_{\nu}\exp\left\{\frac{l}{2}\sum_{n=1}^{\infty}\frac{1}{n\alpha_{n}}C_{n}^{(\varepsilon)}(\zeta_{\mu})\overline{C_{n}^{(\varepsilon)}(\zeta_{\nu})}\right\}$$

$$\leqslant \sum_{\mu,\ \nu=1}^{N} \alpha_{\mu\nu}\eta_{\mu}\bar{\eta}_{\nu}\left|\frac{f(\zeta_{\mu})-f(\zeta_{\nu})}{\zeta_{\mu}-\zeta_{\nu}}\right|^{-\varepsilon lk}\exp\left\{lk^{2}\operatorname{Re}\left(\sum_{n=1}^{\infty}\frac{\alpha_{n}}{n}(\zeta_{\mu}\bar{\zeta}_{\nu})^{-n}\right)\right\}, \quad (24)$$

$$\sum_{\mu,\ \nu=1}^{N} \alpha_{\mu\nu}\eta_{\mu}\bar{\eta}_{\nu}\exp\left\{\frac{l}{2}\sum_{n=1}^{\infty}\frac{1}{n\alpha_{n}}A_{n}^{(\varepsilon)}(z_{\mu})\overline{A_{n}^{(\varepsilon)}(z_{\nu})}\right\}$$

$$\leqslant \sum_{\mu,\ \nu=1}^{N} \alpha_{\mu\nu}\eta_{\mu}\bar{\eta}_{\nu}\left|\frac{f'(0)(f(z_{\mu})-f(z_{\nu}))z_{\mu}z_{\nu}}{(z_{\mu}-z_{\nu})f(z_{\mu})f(z_{\nu})}\right|^{-\varepsilon lk}\exp\left\{lk^{2}\operatorname{Re}\left(\sum_{n=1}^{\infty}\frac{\alpha_{n}}{n}(z_{\mu}\bar{z}_{\nu})^{n}\right)\right\}. \quad (25)$$

◇ References ◇

[1] Kühnau, R., *Math. Nachr.*, 86(1978), 175 - 200.

关于串联式管系共振脉冲射流的瞬态特性*

任福尧

On the Transic Specific Properties of the Resonant Pulse Jet for the Pipeline System Connected in Series

Ren Fu-yao

Abstract: In this paper we deal with the transic specific properties of the resonant pulse jet for the pipeline system connected in series of theoretical fluid. We give the transitive functions of the continuous jet connected in parallel and the export specific properties of the pulse jet system and the resonant pulse jet system.

一、前　言

文献[1]在对理想流体管系给出其固有频率的基础上，设计了串联式和分流式共振脉冲射流的适用装置，经测试瞬态压力表明，其共振射流具有强烈的脉动特性，在喷嘴处共振射流的峰值压力比压力平均值提高两倍左右，试验证明共振射流的打击力与打击效果都比普通连续射流提高两倍左右. 文献[5]用流体阻抗法和特征线法对串接式管系研究了压力比与频率的关系. 他们都没有用解析方法对共振射流进行过瞬态分析. 在[2]中我们曾对终端为盲端的均匀单管的共振射流之瞬态特性进行了分析. 本文将用流体管路的传输理论和 Z -变换方法，对理想流体的串联式和分流式管系之共振射流的瞬态进行理论分析，以便为进一步提高共振射流的压力比和打击力提供理论依据.

---

* 本文为中国科学院科学基金资助的课题. 原载《应用数学学报》，1987 年，第 10 卷，第 1 期，81 - 90.

二、连续分流式射流的传递函数

我们讨论由三均匀管段所组成的分叉管系，设这些均匀管段的截面积为 A_i、长度为 l_i，如图 1 所示. 设它们各自的流阻、流感、流容和渗漏分别为 R_i，L_i，C_i 和 G_i，传播因子为 $\gamma_i(s)$、特性阻抗为 $Z_{c_i}(s)$、始端阻抗为 Z_{i1}、终端阻抗为 Z_{i2}.

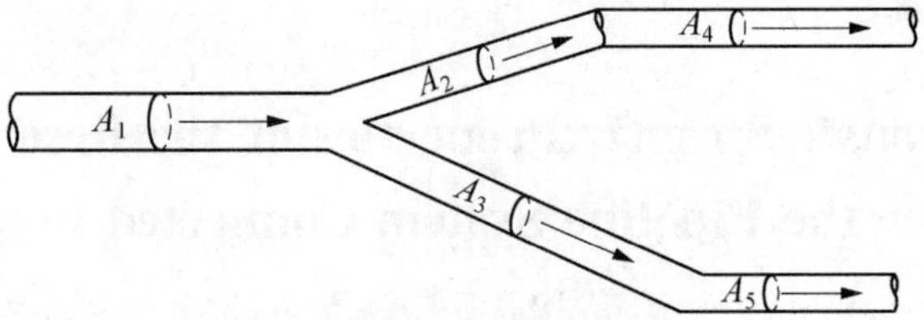

图 1 分流(叉)管系示意图

我们熟知，对每个管段成立着下述关系式：

$$\begin{cases} P_{i2}(s)=P_{i1}(s)\cosh\gamma_i(s)l_i-Z_{c_i}(s)Q_{i1}(s)\sinh\gamma_i(s)l_i, \\ Q_{i2}(s)=-(P_{i1}(s)/Z_{c_i}(s))\sinh\gamma_i(s)l_i+Q_{i1}(s)\cosh\gamma_i(s)l_i, \end{cases} \tag{1}$$

$$Z_{i2}(s)=\frac{P_{i1}(s)\cosh\gamma_i(s)l_i-Z_{c_i}(s)Q_{i1}(s)\sinh\gamma_i(s)l_i}{-P_{i1}(s)\sinh\gamma_i(s)l_i+Z_{c_i}(s)Q_{i1}(s)\cosh\gamma_i(s)l_i}\cdot Z_{c_i}(s), \tag{2}$$

$$P_{i2}(s)=P_{i1}(s)\cdot Z_{i2}(s)/(Z_{i2}(s)\cosh\gamma_i(s)l_i+Z_{c_i}(s)\sinh\gamma_i(s)l_i), \tag{3}$$

其中 $i=1,\ 2,\ \cdots,\ 5$，$P_{i1}(s)=P_i(0,\ s)$，$Q_{i1}(s)=Q_i(0,\ s)$，$P_{i2}(s)=P_i(l_i,\ s)$，$Q_{i2}(s)=Q_i(l_i,\ s)$ 分别表示在始端与终端的压力与流量，

$$\gamma_i(s)=\sqrt{(L_is+R_i)(C_is+G_i)}, \tag{4}$$

$$Z_{c_i}(s)=\sqrt{(L_is+R_i)/(C_is+G_i)}. \tag{5}$$

设分叉处的边界条件为

$$P_{12}(s)=P_{21}(s)=P_{31}(s),\ Q_{12}(s)=Q_{21}(s)+Q_{31}(s), \tag{6}$$

则

$$Z_{12}(s)=Z_{21}(s)\cdot Z_{31}(s)/(Z_{21}(s)+Z_{31}(s)). \tag{7}$$

设在串联处的边界条件为

$$P_{22}(s)=P_{41}(s),\ Q_{22}(s)=Q_{41}(s),\\ P_{32}(s)=P_{51}(s),\ Q_{32}(s)=Q_{51}(s). \tag{8}$$

则

$$Z_{22}(s)=Z_{41}(s),\ Z_{32}(s)=Z_{51}(s). \tag{9}$$

设

$$G_i(s)=Z_{i2}(s)/(Z_{i2}(s)\cosh\gamma_i(s)l_i+Z_{c_i}\cdot\sinh\gamma_i(s)l_i), \tag{10}$$

$$\begin{cases}G_{421}(s)=G_4(s)G_2(s)G_1(s),\\ G_{531}(s)=G_5(s)G_3(s)G_1(s)\end{cases} \tag{11}$$

分别称为各管段（$i=1, 2, \cdots, 5$）和串联管 4.2.1 及 5.3.1 的传递函数，由(3)，(7)，(9)，(10)和(11)，对分叉管系不难证明：

$$P_{42}(s)=G_{421}(s)P_{11}(s),\ P_{52}(s)=G_{531}(s)\cdot P_{11}(s), \tag{12}$$

其中 $Z_{12}=Z_{21}\cdot Z_{31}/(Z_{21}+Z_{31})$. 特别当 5 管和 3 管退化成没有时，则得串联管系 1-2-4，此时，

$$P_{42}(s)=G_{421}(s)P_{11}(s). \tag{13}$$

三、脉冲射流系统的输出特性

在上述管系中，若输入以 T_1 为周期的冲击脉冲信号序列 $p_{11}(0)$，$p_{11}(T_1)$，$p_{11}(2T_1)$，…，$p_{11}(nT_1)$，…，即输入压力为

$$p_{11}^*(t)=\sum_{n=0}^{\infty}p_{11}(nT_1)\delta(t-nT_1), \tag{14}$$

其中 $\delta(t)$为单位脉冲函数，其 Laplace 变换，即 $p_{11}(t)$的 Z-变换为

$$P_{11}^*(z_1)=Z[p_{11}(t)]=\sum_{n=0}^{\infty}p_{11}(nT_1)z_1^{-n}\quad (z_{T_1}=\mathrm{e}^{T_1 s}). \tag{15}$$

由于对应于输入为 $p_{11}(nT_1)\delta(t-nT_1)$ 的输出为

$$G_{421}(s)p_{11}(nT_1)\mathrm{e}^{-nT_1 s},$$

所以由这管系传输的线性可加性,则对应于 $p_{11}^{*}(t)$ 的输出 $P_{42}(s)$, $P_{52}(s)$ 分别为

$$\begin{cases} P_{42}(s)=\sum_{n=0}^{\infty} G_{421}(s) p_{11}(nT_1) \mathrm{e}^{-nT_1 s}=G_{421}(s) \cdot P_{11}^{*}(z_1), \\ P_{52}(s)=\sum_{n=0}^{\infty} G_{531}(s) p_{11}(nT_1) \mathrm{e}^{-nT_1 s}=G_{531}(s) \cdot P_{11}^{*}(z_1), \end{cases} \tag{16}$$

这里 $P_{11}^{*}(z_1)$ 由(15)式所定义.

为了便于研究输出压力 $p_{42}(t)$ 和 $p_{52}(t)$ 的瞬态特性,我们应用 Z-变换这一方法,考虑下述分流式具有依次连接的脉冲系统,如图 2 所示.

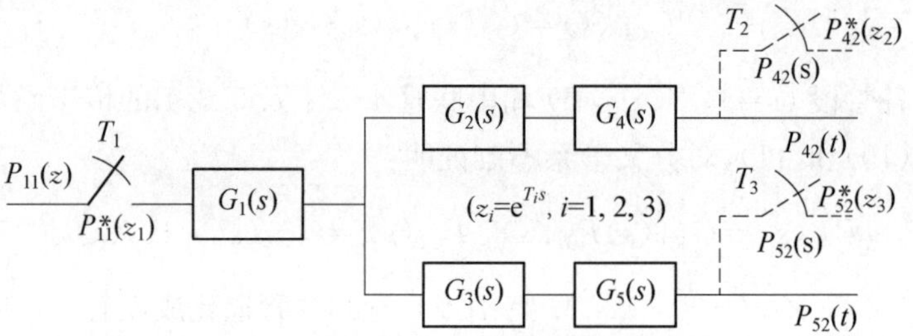

图 2 分流-串联式具有依次连接的脉冲系统

设 $p_{42}(t)$, $p_{52}(t)$, $g_{421}(t)$ 和 $g_{531}(t)$ 分别是 $P_{42}(s)$, $P_{52}(s)$, $G_{421}(s)$ 和 $G_{531}(s)$ 的 Laplace 逆变换,又设 $P_{42}^{*}(z_2)$ 和 $G_{421}^{*}(z_2)$ 是以 T_2 为周期的 $p_{42}(t)$ 与 $g_{421}(t)$ 的 Z-变换,即

$$\begin{aligned} P_{42}^{*}(z_2) &= Z[p_{42}(t)]=\sum_{n=0}^{\infty} p_{42}(nT_2) z_2^{-n} \quad (z_2=\mathrm{e}^{T_2 s}), \\ G_{421}^{*}(z_2) &= Z[g_{421}(t)]=\sum_{n=0}^{\infty} g_{421}(nT_2) z_2^{-n}; \end{aligned} \tag{17}$$

同理,设 $P_{52}^{*}(z_3)$ 与 $G_{531}^{*}(z_3)$ 是以 T_3 为周期的 $p_{52}(t)$ 与 $g_{531}(t)$ 的 Z-变换(这里 $z_3=\mathrm{e}^{T_3 s}$),也可得出类似(17)的式子.

我们特别讨论等幅值的冲击脉冲信号系统,即

$$p_{11}(t)=P_{0\eta}(t),\ \eta(t)=\begin{cases}1, & t>0, \\ 0, & t<0,\end{cases} \tag{18}$$

则 $p_{11}^*(t)=p_0\sum_{n=0}^{\infty}\delta(t-nT_1)$,

$$P_{11}^*(z_1)=P_0/(1-z_1^{-1})\quad (z_1=\mathrm{e}^{T_1 s}). \tag{19}$$

若

$$T_1/T_2=b/q,\ b,\ q=1,\ 2,\ \cdots, \tag{20}$$

由[3]可知

$$P_{11}^*(z_1)=P_0\left(1+\sum_{n=1}^{q-1}z_2^{-n(b/q)}\right)/(1-z_2^{-b}), \tag{21}$$

其中 $z_2=\mathrm{e}^{T_2 s}$. 特别当 $T_1/T_2=k$ (k 为整数),即 $T_2=(1/k)T_1$ 时,(21)退化成

$$P_{11}^*(z_1)=P_0/(1-z_2^{-k}).$$

此时,则

$$P_{42}(s)=P_0G_{421}(s)/(1-z_2^{-k}). \tag{22}$$

由于[3]

$$P_{42}^*(z_2)=\frac{1}{T_2}\sum_{n=-\infty}^{\infty}P_{42}(s+n\mathrm{j}\omega_{T_2}),\ \omega_{T_2}=2\pi/T_2,$$

$$G_{421}^*(z_2)=\frac{1}{T_2}\sum_{n=-\infty}^{\infty}G_{421}(s+n\mathrm{j}\omega_{T_2}),$$

则得

$$P_{42}^*(z_2)=P_0G_{421}^*(z_2)/(1-z_2^{-k}), \tag{23}$$

式中 $\mathrm{j}=\sqrt{-1}$. 同理,特别当 $T_1/T_3=k'$ 时,则有

$$P_{52}^*(z_3)=P_0G_{531}^*(z_3)/(1-z_3^{-k'})\quad (z_3=\mathrm{e}^{T_3 s}). \tag{24}$$

利用 Z -变换的计算公式[3],我们有

$$G_{421}^*(z_2)=\sum_{s_n}\mathrm{Res}[G_{421}(p)/(1-\mathrm{e}^{-T_2(s-p)}),\ p=s_n], \tag{25}$$

这里 s_n 表示 $G_{421}(p)$ 的极点, $\mathrm{Res}[F(p),\ p=s_n]$ 表示函数 $F(p)$ 在 $p=s_n$ 点的留数. 由是根据 Z -变换的反演公式,便得

$$p_{42}(mT_2)=\lim_{\rho\to+\infty}\frac{1}{2\pi \mathrm{j}}\int_{\Gamma_\rho}P_{42}^*(z)z^{m-1}\mathrm{d}z, \tag{26}$$

若 $p_{42}(0)=g_{421}(0)p_{11}(0)=0$，则[2]还有

$$\begin{aligned}\Delta p_{42}(mT_2)&=p_{42}((m+1)T_2)-p_{42}(mT_2)\\&=\lim_{\rho\to+\infty}\frac{1}{2\pi \mathrm{j}}\int_{\Gamma_\rho}P_{42}^*(z)(z-1)z^{m-1}\mathrm{d}z,\end{aligned} \tag{27}$$

其中 $\Gamma_\rho=\{z: |z|=\rho\}$, $m=1, 2, 3, \cdots$. 由(26)和(27)，我们就能获得管 4 的输出压力序列 $\{p_{42}(mT_2)\}$ 及其相邻两者之改变量. 同理，对 $p_{52}(mT_3)$ 及其变差 $\Delta p_{52}(mT_3)$ 也有与(26)和(27)类似的表达式.

下面仅讨论串联管系 1 - 2 - 4 的情况. 由文献[1]可知，此串联管系的端点阻抗为

$$Z_{11}=a^2/sV_1,\ Z_{42}=(L_n/A_n)s, \tag{28}$$

这里 L_n 与 A_n 分别表示管 4 终端喷嘴的长度与截面积，a 表示声速，V_1 表示管 1 之始端容腔的体积.

由(10)和(11)，我们有

$$\begin{aligned}G_{421}(s)=\Big[&\Big(\cosh\gamma_4 l_4+\frac{Z_{c_4}}{Z_{42}}\cdot\sinh\gamma_4 l_4\Big)\Big(\cosh\gamma_2 l_2+\frac{Z_{c_2}}{Z_{22}}\sinh\gamma_2 l_2\Big)\\&\cdot\Big(\cosh\gamma_1 l_1+\frac{Z_{c_1}}{Z_{12}}\sinh\gamma_1 l_1\Big)\Big]^{-1}.\end{aligned} \tag{29}$$

由关系式(2)和(9)可知，Z_{12}, Z_{22} 和 Z_{42} 之间满足下述关系式：

$$\begin{cases}Z_{12}=Z_{c_1}\cdot(Z_{11}+\mathrm{j}Z_{c_1}\cdot\tan\mathrm{j}\gamma_1 l_1)/(\mathrm{j}Z_{11}\cdot\tan\mathrm{j}\gamma_1 l_1+Z_{c_1}),\\Z_{22}=Z_{c_2}\cdot(Z_{12}+\mathrm{j}Z_{c_2}\cdot\tan\mathrm{j}\gamma_2 l_2)/(\mathrm{j}Z_{12}\cdot\tan\mathrm{j}\gamma_2 l_2+Z_{c_2}),\\Z_{42}=Z_{c_4}\cdot(Z_{22}+\mathrm{j}Z_{c_4}\cdot\tan\mathrm{j}\gamma_4 l_4)/(\mathrm{j}Z_{22}\cdot\tan\mathrm{j}\gamma_4 l_4+Z_{c_4}).\end{cases} \tag{30}$$

由于流体是理想的，因而，

$$\gamma_i(s)=s/a,\ Z_{c_i}(s)=a/A_i\quad(i=1, 2, 4). \tag{31}$$

为了求得 $G_{421}^*(z)$，首先得讨论 $G_{421}(s)$ 的所有极点，即 $[G_{421}(s)]^{-1}$ 的所有零点. 注意到 $\cosh\zeta=\cos\mathrm{j}\zeta$, $\sinh\zeta=-\mathrm{j}\sin\mathrm{j}\zeta$ 及(28)和(31)，易知 $\cosh\gamma_4 l_4+(Z_{c_4}/Z_{42})\cdot\sinh\gamma_4 l_4=0$ 等价于

$$\tan(\mathrm{j}l_4 s/a) = -(A_4 L_n / A_n l_4) \cdot (\mathrm{j}l_4 s/a). \tag{32}$$

设 $x = \mathrm{j}l_4 s/a$, $\alpha = A_4 L_n / A_n l_4$，则(32)就成为 $\tan x = -\alpha x$. 易知它有无穷多个解，如图 3 所示：

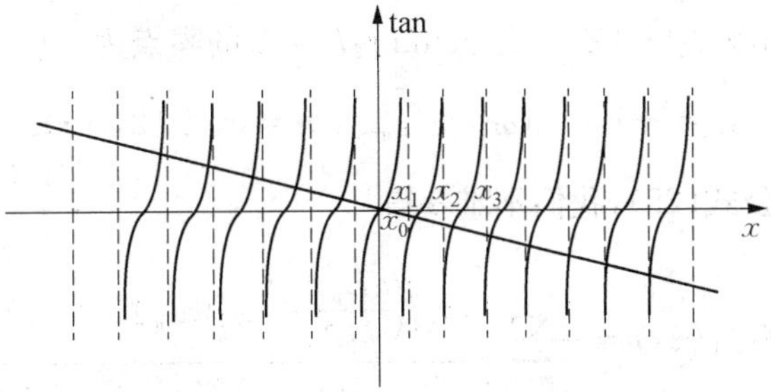

图 3　方程 $\tan x = -\alpha x$ 的图解

设其解为 $x_n \geqslant 0$, $n = 0, 1, 2, \cdots$. 由是，则得 $s_n = \mp \mathrm{j}(a x_n / l_4)$ $(n = 0, 1, 2, \cdots)$ 恰是 $\cosh\gamma_4 l_4 + (Z_{c_4}/Z_{42})\sinh\gamma_4 l_4 = 0$ 的一级零点.

关于 $\cosh\gamma_1 l_1 + (Z_{c_1}/Z_{12})\sinh\gamma_1 l_1$ 的零点，它等价于 $\tan \mathrm{j}\gamma_1 l_1 = -\mathrm{j}(Z_{12}/Z_{c_1})$. 利用(30)中的第一个方程式，并注意到 $Z_{11} = a^2/sV_1 \neq 0$，则得 $(\tan(\mathrm{j}l_1 s/a))^2 = -1$. 令 $s = \mathrm{j}\omega$, ω 是实数，则得 $[\tan(\omega l_1/a)]^2 = -1$. 但这是不可能的. 故它无零点.

最后，讨论 $\cosh\gamma_2 l_2 + (Z_{c_2}/Z_n)\sinh\gamma_2 l_2 = 0$ 的解，它等价于 $\tan \mathrm{j}\gamma_2 l_2 = -\mathrm{j}(Z_{22}/Z_{c_2})$. 由(30)的第三个关系式中解出 Z_{22} 并代入前述方程式，则得

$$\tan \mathrm{j}\gamma_2 l_2 = -\mathrm{j}(Z_{c_4}/Z_{c_2}) \cdot (Z_{42} - \mathrm{j}Z_{c_4} \cdot \tan \mathrm{j}\gamma_4 l_4)/(-\mathrm{j}Z_{42} \cdot \tan \mathrm{j}\gamma_4 l_4 + Z_{c_4}).$$

注意到(28)，(31)，令 $s = \mathrm{j}\omega$, $x = \omega l_2/a$, $y = \omega l_4/a$，则得

$$\tan x = \left(\frac{A_2}{A_4}\right)\left(\frac{y + \beta \cdot \tan y}{y \cdot \tan y - \beta}\right) \quad \left(\beta = \frac{l_4 A_{11}}{L_n A_4}\right). \tag{33}$$

不难证明它有可列无穷多解. 事实上，若令

$$\varphi(y) = (y + \beta \cdot \tan y)/(y \cdot \tan y - \beta),$$

则易知

$$(y\tan y - \beta)^2 \varphi'(y) = -(\beta + \beta^2 \sec^2 y + \beta \tan^2 y + y^2 \sec^2 y) < 0.$$

这说明 $\varphi(y)$ 在 $-\infty < y < +\infty$ 上关于 y 是严格单调下降的，因而(33)式的右端项关于 $\omega \in (-\infty, +\infty)$ 是严格单调下降的. 但是，$\tan x$ 在 $\omega \in (-\infty, +\infty)$ 是无限振荡的，因此，方程(33)关于 ω 有可列无穷多解. 而且，若 ω 是(33)的解，则 $-\omega$ 也是它的解. 因此，设(33)的解为 $\omega = 0, \pm\omega_1, \pm\omega_2, \cdots$，其中 $0 < \omega_1 < \omega_2 < \cdots$，则 $\cosh\gamma_2 l_2 + (Z_{c_1}/Z_{22})\sinh\gamma_2 l_2 = 0$ 的零点为

$$s_n = \pm \mathrm{j}\omega_n \quad (\omega_n < \omega_{n+1},\ n = 0,\ 1,\ 2,\ \cdots).$$

经计算，由公式(25)，我们不难获得

$$G_{421}^{*}(z_2) = -\sum_{m=1}^{\infty}\left[\frac{\pm\mathrm{j}\left(\dfrac{aL_n x_m}{l_4 A_n}\right) g_2(x_m) g_1(x_m)}{g_4(x_m)(1 - \mathrm{e}^{\pm\mathrm{j}(a x_m/l_4)T_2} \cdot z_2^{-1})}\right.$$
$$\left. \pm\mathrm{j}\frac{\left(\dfrac{L_n\omega_n}{A_n} + \dfrac{a}{A_4}\cdot\tan\dfrac{l_4\omega_m}{a}\right)\dfrac{a}{A_4}\cdot h_4(\omega_m) h_1(\omega_m)}{h_2(\omega_m)(1 - \mathrm{e}^{\pm\mathrm{j}T_2\omega_m}\cdot z_2^{-1})}\right], \tag{34}$$

其中，

$$-g_4(x_m) = \left(\frac{L_n}{A_n} + \frac{l_4}{A_4}\right)\cos x_m - \frac{L_n}{A_n}x_m \cdot \sin x_m,$$

$$\left[\left(\frac{aL_n x_m}{l_4 A_n} - \frac{a}{A_4}\tan x_m\right)\frac{a}{A_4}\right][g_2(x_m)]^{-1}$$
$$= \left(\frac{aL_n x_m}{l_4 A_n} - \frac{a}{A_4}\tan x_m\right)\frac{a}{A_4}\cdot\cos x_m + \frac{a}{A_2}\left(\frac{a}{A_4} - \frac{aL_n x_m}{l_4 A_n}\tan x_m\right)\sin x_m,$$

$$\left(\frac{a^2}{V_1} + \frac{a^2 x_m}{l_4 A_1}\tan\frac{l_1 x_m}{l_4}\right)[g_1(x_m)]^{-1}$$
$$= \left(\frac{a^2}{V_1} + \frac{a^2 x_m}{A_1 l_4}\tan\frac{l_1 x_m}{l_4}\right)\cos\frac{l_1 x_m}{l_4} + \left(\frac{a^2}{V_1}\tan\frac{l_1 x_m}{l_4} - \frac{a^2 x_m}{l_4 A_1}\right)\sin\frac{l_1 x_m}{l_4},$$

$$h_4(\omega_m) = \cos\frac{l_4\omega_m}{a} + \frac{aA_m}{A_4 L_n\omega_m}\sin\frac{l_4\omega_m}{a},$$

$$\left(\frac{a^2}{V_1} + \frac{a\omega_m}{A_1}\tan\frac{l_1\omega_m}{a}\right)[h_1(\omega_m)]^{-1}$$
$$= \left(\frac{a^2}{V_1} + \frac{a\omega_m}{A_1}\tan\frac{l_1\omega_m}{a}\right)\cos\frac{l_1\omega_m}{a} + \left(\frac{a^2}{V_1}\tan\frac{l_1\omega_m}{a} - \frac{a\omega_m}{A_1}\right)\sin\frac{l_1\omega_m}{a},$$

$$h_2(\omega_m) = \left(\frac{L_n}{A_n} + \frac{l_4}{A_4}\sec^2\frac{l_4\omega_m}{a}\right)\frac{a}{A_4}\cos\frac{l_2\omega_m}{a}$$

$$-\frac{l_2}{A_4}\left(\frac{L_n\omega_m}{A_n}+\frac{a}{A_4}\tan\frac{l_4\omega_m}{a}\right)\sin\frac{l_2\omega_m}{a}-\frac{a}{A_2}\left(\frac{L_n}{A_n}\tan\frac{l_4\omega_m}{a}\right.$$

$$\left.+\frac{L_n\omega_m l_4}{aA_n}\sec^2\frac{l_4\omega_m}{a}\right)\sin\frac{l_2\omega_m}{a}+\frac{l_2}{A_2}\left(\frac{a}{A_4}-\frac{L_n\omega_m}{A_n}\tan\frac{l_4\omega_m}{a}\right)\cos\frac{l_2\omega_m}{a}.$$

由是，将(34)代入(23)，便得

$$P_{42}^*(z_2)=\frac{2P_0}{1-z_2^{-k}}\sum_{k'=1}^{\infty}\left[\sum_{m'=1}^{\infty}\left(\alpha_{m'}\sin\frac{k'ax_mT_2}{l_4}+\beta_{m'}\sin k'T_2\omega_{m'}\right)\right]z_2^{-k'},\quad(35)$$

其中

$$\alpha_m=\frac{aL_nx_mg_2(x_m)g_1(x_m)}{l_4A_ng_4(x_m)},$$

$$\beta_m=\frac{a}{A_4}\left(\frac{L_n\omega_m}{A_n}+\frac{a}{A_4}\tan\frac{l_4\omega_m}{a}\right)h_4(\omega_m)h_1(\omega_m)/h_2(\omega_m).$$

将(35)代入公式(26)，(27)，对自然数 q 便得

$$p_{42}(qT_2)=2P_0\sum_{m=1}^{\infty}\left[\sum_{k''=0}^{[(q-1)/k]}\left(\alpha_m\cdot\sin\frac{aT_2x_m}{l_4}(q-kk'')+\beta_m\cdot\sin(q-kk'')T_2\omega_m\right)\right],\quad(36)$$

其中 $[(q-1)/k]$ 表示 $(q-1)/k$ 的整数部分，

$$\begin{aligned}\Delta p_{42}(qT_2)&=p_{42}((q+1)T_2)-p_{42}(qT_2)\\&=2P_0\sum_{m=1}^{\infty}\left\{\sum_{k''=0}^{[q/k]}\left[\alpha_m\sin(q+1-kk'')\frac{aT_2x_m}{l_4}+\beta_m\sin(q+1-kk'')T_2\omega_m\right]\right.\\&\quad\left.-\sum_{k''=0}^{[(q-1)/k]}\left[\alpha_m\sin(q-kk'')\frac{aT_2x_m}{l_4}+\beta_m\cdot\sin(q-kk'')T_2\omega_m\right]\right\}.\end{aligned}\quad(37)$$

特别地，当 $T_1=T_2$ 即 $k=1$ 时，我们有

$$\begin{aligned}\Delta p_{42}(qT_1)=2P_0\sum_{m=1}^{\infty}&\left\{\left(\alpha_m\sin\frac{aT_2x_m}{l_4}+\beta_m\sin T_2\omega_m\right)\right.\\&+2\sum_{k''=0}^{q-1}\left[\alpha_m\sin\frac{aT_2x_m}{2l_4}\cdot\cos\left(q-k''+\frac{1}{2}\right)\frac{aT_2x_m}{l_4}\right.\\&\left.\left.+\beta_m\sin\frac{T_2\omega_m}{2}\cdot\cos\left(q-k''+\frac{1}{2}\right)T_2\omega_m\right]\right\};\end{aligned}\quad(38)$$

当 $T_1/T_2=k=2$ 时，对 $q'=0, 1, 2, \cdots$，我们有

$$\begin{aligned}\Delta p_{42}(qT_2)&=\Delta p_{42}[(2q'+1)T_2]\\&=2P_0\sum_{m=1}^{\infty}\left\{\sum_{k''=0}^{q'}\left[2\alpha_m\sin\frac{aT_2x_m}{2l_4}\cdot\cos\left(2q'-2k''+\frac{3}{2}\right)\frac{aT_2x_m}{l_4}\right.\right.\\&\quad\left.\left.+2\beta_m\sin\frac{T_2\omega_m}{2}\cos\left(2q'-2k''+\frac{3}{2}\right)T_2\omega_m\right]\right\}.\end{aligned}\tag{39}$$

根据公式(27)，需要证明 $p_{42}(0)=0$. 事实上 $g_{421}(0)=0$，这是因为据 Laplace 变换的第二展开定理，

$$\begin{aligned}g_{421}(t)&=\sum_{n=0}^{\infty}\mathrm{Res}\left[G_{421}(s)\mathrm{e}^{st},\ s_n=\pm\mathrm{j}\frac{ax_m}{l_4},\ \pm\mathrm{j}\omega_m\right]\\&=2\sum_{m=1}^{\infty}\left(\alpha_m\sin\frac{ax_mt}{l_4}+\beta_m\sin\omega_mt\right),\end{aligned}$$

故 $p_{42}(0)=g_{421}(0)p_{11}(0)=0$.

最后，我们计算 $p_{42}(t)$，由 Laplace 变换的第二展开定理[4]，

$$p_{42}(t)=\sum_{s_m}\mathrm{Res}\left[\frac{P_0G_{421}(p)\mathrm{e}^{pt}}{1-\mathrm{e}^{-kT_2p}},\ s_m\right],$$

其中$\{s_m\}$是 $G_{421}(s)/(1-\mathrm{e}^{-kT_2s})$ 的所有极点，实际上它由 $\left\{\pm\frac{ax_m}{l_4}\mathrm{j}\right\}$，$\{\pm\mathrm{j}\omega_m\}$ 和$\{\pm 2m\pi\mathrm{j}/kT_2\}$ $(m=0, 1, 2, \cdots)$ 所组成. 经计算，我们获得

$$\begin{aligned}p_{42}(t)=2P_0\Bigg\{&-\sum_{m=1}^{\infty}\left[\alpha_m\left(\frac{\sin\frac{T_1}{2l_4}\cdot\cos(t+T_1)\frac{ax_m}{l_4}}{1-\cos(aT_1x_m/l_4)}\right)\right.\\&\left.+\beta_m\left(\frac{\sin\frac{T_1}{2}\cdot\cos\left(t+\frac{T_1}{2}\right)\omega_m}{1-\cos T_1\omega_m}\right)\right]\\&+\frac{1}{T_1}\sum_{m=0}^{\infty}(g_{4,m}g_{2,m}g_{1,m})\cos\frac{2m\pi t}{T_1}\Bigg\},\end{aligned}\tag{40}$$

其中

$$g_{4,m}=m\gamma_1/(m\gamma_1\cdot\cos my_4+\gamma_2\sin my_4)$$

$$g_{2,m}=\frac{(m\gamma_1+\gamma_2\tan my_4)\gamma_2}{(m\gamma_1+\gamma_2\tan my_4)\gamma_2\cos my_2+\gamma_3(\gamma_2-m\gamma_1\tan my_4)\sin my_2}$$

$$g_{1,m}=\frac{\gamma_4+m\gamma_5\cdot\tan my_1}{(\gamma_4+m\gamma_5\tan my_1)\cos my_1+(\gamma_4\tan my_1-m\gamma_5)\sin my_1},$$

$$y_i=\frac{2\pi l_i}{aT_1}\quad(i=1,\,2,\,4),\quad \gamma_1=\frac{2\pi L_n}{A_nT_1},\quad \gamma_2=\frac{a}{A_4},$$

$$\gamma_3=\frac{a}{A_2},\quad \gamma_4=\frac{a^2}{V_1},\quad \gamma_5=\frac{2\pi a}{A_1T_1}.$$

四、共振脉冲射流系统的输出特性

如果输入脉冲的频率恰是串联管系的共振频率，则称此射流为串联共振脉冲射流. 为此，我们得计算串联管系的共振频率.

从(30)中消去 Z_{12}，Z_{22}，并将(31)和端点阻抗(28)及 $s=\mathrm{j}\omega$ 代入所得结果便得到共振频率方程式：

$$\left(\frac{A_2}{A_4}\right)\frac{y+\beta\cdot\tan y}{y\cdot\tan y-\beta}=\frac{A(\omega)}{B(\omega)},\tag{41}$$

其中 β 和 y 如(33)中所定义，而

$$A(\omega)=\frac{a}{V_1}+\frac{\omega}{A_1}\tan\frac{\omega l_1}{a}+\frac{A_1}{A_2}\left(\frac{\omega}{A_1}-\frac{a}{V_1}\tan\frac{\omega l_1}{a}\right)\cdot\tan\frac{\omega l_2}{a},$$

$$B(\omega)=-\left(\frac{a}{V_1}+\frac{\omega}{A_1}\tan\frac{\omega l_1}{a}\right)\tan\frac{\omega l_2}{a}+\frac{A_1}{A_2}\left(\frac{\omega}{A_1}-\frac{a}{V_1}\tan\frac{\omega l_1}{a}\right).$$

我们证明方程式(41)有可列无穷个解. 易见(41)两端的函数都是 ω 的奇函数，因此它的根关于原点是对称的. 另外，我们已经证明(41)的左端关于 $\omega\in(-\infty,+\infty)$ 是严格单调下降的. 关于(41)的右端项，经计算，我们有

$$\frac{\mathrm{d}}{\mathrm{d}\omega}\left(\frac{A(\omega)}{B(\omega)}\right)=\frac{1}{B^2(\omega)}\left\{\frac{l_2}{a}\left[\left(\frac{A_1}{A_2}\right)^2\left(\frac{\omega}{A_1}-\frac{a}{V_1}\tan\frac{\omega l_1}{a}\right)^2+\left(\frac{a}{V_1}+\frac{\omega}{A_1}\tan\frac{\omega l_1}{a}\right)^2\right]\right.$$
$$\left.+\frac{A_1}{A_2}\left[\frac{\omega^2 l_1}{A_1^2a}+\frac{a}{V_1^2A_1}(l_1A_1-V_1)\right]\sec^2\frac{\omega l_1}{a}\right\}\sec^2\frac{\omega l_2}{a}.$$

由是，我们获得：若 $l_1A_1-V_1>0$，即管 1 的体积 l_1A_1 大于其始端容腔的体积 V_1，则 $\left(\dfrac{A(\omega)}{B(\omega)}\right)'_\omega>0$，并且当 $\omega=0$ 或者 $\dfrac{a}{l_1}\left(n\pi+\dfrac{\pi}{2}\right)$，或者 $\dfrac{a}{l_2}\left(n\pi+\dfrac{\pi}{2}\right)$ $(n=0,\pm1,\pm2,\cdots)$ 时，$(A(\omega)/B(\omega))'_\omega=+\infty$. 故(41)的右端项在 $\omega\in(-\infty,+\infty)$

是无穷多次振荡的.因此,设其共振圆频率为

$$0<\eta_1<\eta_2<\cdots,$$

则

$$0<\eta_1<\eta_2<\max\left(\frac{a\pi}{2l_1},\ \frac{a\pi}{2l_2}\right).$$

我们特称 η_1 为基本共振频率,其周期为 $2\pi/\eta_1$. 因此,当输入周期 $T_1=2\pi/\eta_1$ 时,由(36),(38)~(40)便得共振冲击脉冲射流的瞬态特性.

五、均匀单管共振冲击脉冲射流的输出特性

在上述串联管系中,若 $A_1=A_2=A_4=A$,则它退化成管长为 $l=l_1+l_2+l_4$ 的均匀单管系统.此时,$Z_i=Z_c\ (i=1,\ 2,\ 4)$. 于(30)中消去 Z_{12}, Z_{22},便得长为 l 的均匀单管之端点阻抗关系式:

$$Z_{42}=Z_c\left(\frac{Z_{11}+\mathrm{j}Z_c\tan\mathrm{j}\gamma l}{\mathrm{j}Z_{11}\cdot\tan\mathrm{j}\gamma l+Z_c}\right). \tag{42}$$

此外,利用关系式(30),不难证明:

$$G_{421}(s)=1/\left(\cosh\gamma l+\frac{Z_c}{Z_{42}}\sinh\gamma l\right), \tag{43}$$

这就是均匀单管的传递函数.在(28)式端点阻抗条件下,注意到 $Z_c=a/A$, $s=\mathrm{j}\omega$,便得其共振频率方程:

$$\left(\frac{AL_n}{V_1A_n}-\frac{1}{A}\right)\tan\frac{\omega l}{a}=\frac{\omega L_n}{aA_n}+\frac{a}{\omega V_1}. \tag{44}$$

显然,方程(44)的两端都是奇函数,而其右端项关于 ω 的导数为 $-\frac{a}{V_1}\cdot\left(\frac{1}{\omega^2}-\frac{V_1L_n}{a^2A_n}\right)$,且当 $\omega=0$ 时其值为 $-\infty$. 因此,(44)有无穷多个解,其基频 $\eta_1<a\pi/2l$.

若令 $T_1=2\pi/\eta_1$,仿照"三、"中的证明,不难证明:对任何自然数 q,我们有

$$p_2(qT_2)=2P_0\sum_{m=1}^{\infty}d_m\left[\sum_{k''=0}^{[(q-1)/k]}\sin\frac{aT_2x'_m}{l}(q-kk'')\right], \tag{45}$$

其中$\{x_m'\}$是方程 $\tan x'=-(AL_n/lA_n)x'$ 的根，$T_1/T_2=k$，

$$-d_m=\frac{aL_nx'_m/A_nl}{\left(\frac{L_n}{A_n}+\frac{l}{A}\right)\cos x'_m-\frac{L_n}{A_n}x'_m\sin x'_m}.$$

$$\Delta p_2(qT_2)=2P_0\sum_{m=1}^{\infty}\left\{d_m\left[\sum_{k''=0}^{[q/k]}\sin\frac{aT_2x'_m}{l}(q+1-kk'')-\sum_{k''=0}^{[(q-1)/k]}\sin\frac{aT_2x'_m}{l}(q-kk'')\right]\right\}. \quad (46)$$

利用(45)，(46)，我们就能进一步讨论其输出特性.

◇参考文献◇

[1] 柳兆荣，共振射流及其在水力采煤中的应用，上海力学，3(1982)，11－19.

[2] 任福尧，关于均匀单管共振射流的瞬态特性分析，纯粹数学与应用数学(待发表).

[3] Джури，Э.，Импульсные Системы Автоматичского Регулирования，МОСКВА，1963.

[4] M. A. 拉甫伦捷夫，Б. A. 沙巴特著，施祥林等译，复变函数论方法及应用，人民教育出版社. 北京，1961 年.

[5] E. 本杰明・维里，管线动力学与脉冲射流，第一届国际射切割技术论文集，1972，A5.

关于 Teichmüller 极值拟共形映照的三个 Reich 猜想*

任福尧

提　要:在附加条件下,应用 Hamilton 条件、极值拟共形映照唯一和 Hahn-Banach 扩张唯一的充分性条件,证明了关于 Teichmüller 极值拟共形映照的三个 Reich 猜想.

关键词:拟共形映照,Teichmüller 映照,Hahn-Banach 扩张

1. 引　言

设 Ω 是一平面区域, $w=f(z)$ 是 Ω 上一个拟共形映照,其复伸长为 $K(z)=f_{\bar{z}}/f_z$, $Q_f(\Omega)$为 Ω 上具有与 f 相同边界值的一切拟共形映照 $g(z)$的全体. 我们称 $Q_f(\Omega)$中使最大复伸长 $k_g=\sup\limits_{s\in Q}|g_{\bar{z}}/g_z|$ 达到最小值的映照为 $Q_f(\Omega)$的极值映照. 若这种极值映照是唯一的,则称它为 $Q_f(\Omega)$的唯一极值映照. 若 $K(z)=k\overline{\varphi_0(z)}/|\varphi_0(z)|$, $\varphi_0(z)$是 Ω 上全纯函数,则称 f 为 Teichmüller 映照.

设 $z=F(w)$ 是将 Ω 映照到 Ω'上的拟共形映照, $f(z)=F^{-1}(z)$, $K(z)=f_{\bar{z}}/f_z$. 又设 T 是 Ω 内一个给定的紧集, $b(w)$是 T 上给定的一个非负的有界可测函数,满足条件

$$0<b_0=\inf_{w\in T}b(w)\leqslant\sup_{w\in T}b(w)\leqslant b_1<1. \tag{1}$$

设 $Q_F(\Omega, T, b)$为 Ω 到 Ω'上满足下列条件的一切拟共形映照 $G(w)$的全体:

$$G(\zeta)\equiv F(\zeta),\ \zeta\in\partial\Omega,$$

* 国家自然科学基金资助课题. 原载《复旦学报(自然科学版)》,1988 年,第 27 卷,第 4 期,439－445.

$$|G_{\bar{w}}/G_w| \leqslant b(w) \text{ 在 } T \text{ 上几乎处处成立}, \tag{2}$$

并设 $F \in Q_F(\Omega, T, b)$. 我们称 $Q_F(\Omega, T, b)$中使最大复伸长

$$\sup_{w\in Q\setminus T}|G_{\bar{w}}/G_w|$$

达到最小值的映照为 $Q_F(\Omega, T, b)$的极值映照. 若这种极值映照是唯一的,则称它为 $Q_F(\Omega, T, b)$的唯一极值映照.

设 $A_K(z)=Kx+\mathrm{i}y$, $z=x+\mathrm{i}y$ $(K>1)$ 为 Ω 上的仿射伸长, $Q_K(\Omega)$为 Ω 上具有 $A_K(z)$相同边界值的一切拟共形映照的全体.

设 $\mathscr{B}(\Omega)=\{\varphi \in \mathscr{L}^1(\Omega): \varphi(z) \text{ 在 } \Omega \text{ 上解析}\}$, $\|\varphi\|=\iint_\Omega|\varphi(z)|\mathrm{d}x\,\mathrm{d}y$, 则 $\mathscr{B}(\Omega)$是一个 Banach 空间. 对于 $K\in \mathscr{L}^\infty(\Omega)$, 我们考虑线性泛函 $\Lambda_K(\varphi)=\iint_\Omega K(z)\varphi(z)\mathrm{d}x\,\mathrm{d}y$, $\varphi\in\mathscr{B}(\Omega)$. 设

$$\|\Lambda_K\|=\sup_{\varphi\in\mathscr{B}(\Omega),\ \|\varphi\|=1}\left|\iint_\Omega K\varphi\,\mathrm{d}x\,\mathrm{d}y\right|=\|K\|_\infty<1, \tag{3}$$

$$\|K\|_\infty=\operatorname*{ess\,sup}_{z\in\Omega}|K(z)|.$$

在这种情况下,若 Λ_K 从 $\mathscr{B}(\Omega)$到 $\mathscr{L}^1(\Omega)$有唯一的 Hahn-Banach 扩张,则称 $K\in$ HBU. 设 $\Lambda_\Omega[\varphi]=\iint_\Omega\varphi(z)\mathrm{d}x\,\mathrm{d}y$, $\varphi\in\mathscr{B}(\Omega)$, $\|\Lambda_\Omega\|=1$. 若 Λ_Ω 从 $\mathscr{B}(\Omega)$到 $\mathscr{L}^1(\Omega)$有唯一的 Hahn-Banach 扩张,则称 $k=(K-1)/(K+1)\in$ HBU.

1978 年,Reich 猜想[1]:若 $k_F>0$, $Q_F(\Omega, T, b)$ 只有唯一极值映照,当且仅当线性泛函 $\Lambda_\tau[\varphi]=\iint_\Omega\tau\varphi\,\mathrm{d}x\,\mathrm{d}y$ $(\varphi\in\mathscr{B}(\Omega))$ 从 $\mathscr{B}(\Omega)$到 $\mathscr{L}^1(\Omega)$有唯一的 Hahn-Banach 扩张,即 $\tau\in$ HBU. 这里

$$\tau(z)=\begin{cases}K(z)/b\circ f(z),\ z\in F(T),\\ K(z)/k_F,\ z\in\Omega'\setminus F(T),\end{cases} \tag{4}$$

其中, $K(z)=f_{\bar{z}}/f_z$, $f=F^{-1}$.

1979 年,Reich 猜想[2]: $A_K(z)$ 是 $Q_K(\Omega)$ 的唯一极值映照,当且仅当 $\|\Lambda_\Omega\|=1$, Λ_Ω 从 $\mathscr{B}(\Omega)$到 $\mathscr{L}^1(\Omega)$只有唯一的 Hahn-Banach 扩张.

1981 年,Reich 指出[3]:当 $\|\Lambda_k\|=\|K\|_\infty$ 时, Λ_K 从 $\mathscr{B}(\Omega)$到 $\mathscr{L}^1(\Omega)$有唯一 Hahn-Banach 扩张的充要条件,是否以 $K(z)$为其复伸长的拟共形映照 $f(z)$

是 $Q_f(\Omega)$的唯一极值的 Teichmüller 映照,仍是一个没有解决的问题.

1983 年,Reich 又指出[4]:若 $K \in \mathrm{HBU}$, $|K(z)|$在 Ω 上是否几乎处处等于常数 $\|K\|_\infty$,仍是一个没有解决的问题.

本文在附加一定条件下,证明了 Reich 的这几个猜想和问题.

2. 关于 $K \in \mathrm{HBU}$ 的必要条件

定理 1 设 $\|\Lambda_K\| = \|K\|_\infty < 1$, 若对 $K(z)$不存在退化的 Hamilton 序列 $\varphi_n \in \mathscr{B}(\Omega)$, 即不存在收敛于零的函数列 $\varphi_n \in \mathscr{B}(\Omega)$, $\|\varphi_n\| = 1$, 使得

$$\lim_{n\to\infty}\iint_\Omega K\varphi_n \,dx\,dy = \|K\|_\infty, \tag{5}$$

则以 $K(z)$为复伸长的拟共形映照 $f(z)$是 $Q_f(\Omega)$的唯一极值的 Teichmüller 映照,且 $K(z) = k\bar{\varphi}_0/|\varphi_0|$, $\varphi_0 \in \mathscr{B}(\Omega)$.

证 如文献[5]所证,因 $\|\Lambda_K\| = \|K\|_\infty$, 根据 Reich 定理[6],则以 $K(z)$为复伸长的拟共形映照 $f(z)$是 $Q_f(\Omega)$的一个极值映照. 由于 $\|\varphi_n\| = 1$, 由解析函数的平均值定理和凝聚原理,则$\{\varphi_n(z)\}$在 Ω 上内闭均匀有界,因而存在子序列$\{\varphi_{n_j}\}$在 Ω 上内闭均匀收敛于 $\varphi_0(z)$. φ_0 在 Ω 上解析, $\|\varphi_0\| < 1$. $\|\varphi_0\| = 0$ 不可能,否则$\{\varphi_{n_j}\}$是退化的 Hamilton 序列. 若 $0 < \|\varphi_0\| < 1$, 则 $\tilde{\varphi}_{n_j} = (\varphi_{n_j} - \varphi_0)/\|\varphi_{n_j} - \varphi_0\|$ 也将是 $K(z)$的退化的 Hamilton 序列. 事实上,只要证明

$$\lim_{j\to\infty}\iint_\Omega K(z)\varphi_{n_j}(z)\,dx\,dy = \|K\|_\infty \tag{6}$$

即可. 由(5),显然有

$$\begin{aligned} k\|\varphi_{n_j} - \varphi_0\| &\geqslant \left|\iint_\Omega K(\varphi_{n_j} - \varphi_0)\,dx\,dy\right| \\ &\geqslant \mathrm{Re}\iint_\Omega K\varphi_{n_j}\,dx\,dy - k\|\varphi_0\| \to k(1 - \|\varphi_0\|). \end{aligned} \tag{7}$$

显然, $\|\varphi_{n_j} - \varphi_0\| \geqslant 1 - \|\varphi_0\|$, 并且由 $\varphi_0 \in \mathscr{B}(\Omega)$ 和 φ_{n_j} 在 Ω 上内闭匀敛于 φ_0, 对任意给定的 $\varepsilon > 0$, 存在闭子集 $T \subset \Omega$ 和自然数 $N(\varepsilon)$, 使得当 $n_j > N(\varepsilon)$ 时,

$$\iint_{\Omega\setminus T}|\varphi_0|\,dx\,dy < \varepsilon, \quad \|\varphi_0\| < \iint_T |\varphi_0|\,dx\,dy + \varepsilon,$$

$$\left|\iint_T |\varphi_{n_j}|\,\mathrm{d}x\,\mathrm{d}y - \iint_T |\varphi_0|\,\mathrm{d}x\,\mathrm{d}y\right| \leqslant \iint_T |\varphi_{n_j} - \varphi_0|\,\mathrm{d}x\,\mathrm{d}y < \varepsilon,$$

从而,当 $n_j > N(\varepsilon)$ 时,有

$$\|\varphi_{n_j} - \varphi_0\| < 4\delta + 1 - \|\varphi_0\|.$$

由是,当令 $\varepsilon \to 0$ 时,即得

$$\lim_{j\to\infty}\|\varphi_{n_j} - \varphi_0\| = 1 - \|\varphi_0\|. \tag{8}$$

于是,由(8)和(7),便得

$$\lim_{j\to\infty}\iint_\Omega K(\varphi_{n_j} - \varphi_0)\,\mathrm{d}x\,\mathrm{d}y = k(1 - \|\varphi_0\|). \tag{9}$$

由(8)和(9)即得(6). 这说明$\{\widetilde{\varphi}_{n_j}\}$是一退化的 Hamilton 序列,但又不可能,故 $\|\varphi_0\| = 1$. 由(9)和(5)便得

$$\lim_{j\to\infty}\iint_\Omega K\varphi_{n_j}\,\mathrm{d}x\,\mathrm{d}y = \iint_\Omega K\varphi_0\,\mathrm{d}x\,\mathrm{d}y = k = \|K\|_\infty. \tag{10}$$

于是,即得 $K(z) = k\overline{\varphi_0(z)}/|\varphi_0|$,这表明 f 是一 Teichmüller 映照. 因 $\varphi_0 \in \mathscr{B}(\Omega)$,根据 Strebel 的唯一性定理,则 f 是$Q_f(\Omega)$的唯一极值的 Teichmüller 映照. 定理证毕.

3. 极值映照唯一性和 $K \in$ HBU 的判别

Reich 在文献[5, 3]中给出了 Teichmüller 型映照为唯一极值映照和 $K \in$ HBU 的判据,在这里我们开拓这些判据.

定理 2 设 $w = f(z)$ 是 Ω 上的拟共形映照,$f_{\bar z}/f_z = K(z) \in \mathscr{L}^\infty(\Omega)$,$\|K\|_\infty \leqslant k < 1$,$|K(z)|$ 在 Ω 上几乎处处不为零. 若存在函数列 $\varphi_n \in \mathscr{B}(\Omega)$,$n = 1, 2, 3, \cdots$,满足条件:

$\{\varphi_n\}$ 在 Ω 上几乎处处收敛于 $\hat\varphi_0(z)$,并且 $\hat\varphi_0 \neq 0$ 和 $K\hat\varphi_0 > 0$ 在 Ω 上几乎处处成立, (11)

$$\lim_{n\to\infty}\iint_\Omega \delta\{\varphi_n\}\,\mathrm{d}x\,\mathrm{d}y = 0, \tag{12}$$

则 f 是唯一极值的,其中 $\delta\{\varphi\} = |K(z)\varphi(z)| - \mathrm{Re}\{K(z)\varphi(z)\}$.

证 同文献[7]中所证，令 $\hat{K}(w)$ 是 $f^{-1}(w)$ 的复伸长，$g(z)\in Q_f(\Omega)$，$K_1(w)$ 是 $g^{-1}(w)$ 的复伸长，$\mu(z)=K_1(f(z))$，$\tau(z)=\mu(z)/\hat{K}(f(z))$，则对 $\varphi_n\in\mathscr{B}(\Omega)$，$n=1,2,\cdots$，有

$$\iint_\Omega\left[1-\frac{(1-|K|)(1+|\mu|)}{(1+|K|)(1-|\mu|)}\right]|\varphi_n|\,\mathrm{d}x\,\mathrm{d}y\leqslant\frac{2}{(1-k)^2}\iint_\Omega\delta\{\varphi_n\}\,\mathrm{d}x\,\mathrm{d}y,\tag{13}$$

$$\iint_\Omega|\varphi_n|\operatorname{Re}(|K|-\theta_n\tau K)\,\mathrm{d}x\,\mathrm{d}y\leqslant\left(\frac{1+k}{1-k}\right)^2\iint_\Omega\delta\{\varphi_n\}\,\mathrm{d}x\,\mathrm{d}y,\tag{14}$$

这里 $$\theta_n=\frac{\varphi_n}{|\varphi_n|}\left[\frac{1-\overline{K\varphi_n}/|\varphi_n|}{1-K\varphi_n/|\varphi_n|}\right]\to\frac{\hat{\varphi}_0}{|\hat{\varphi}_0|}\left[\frac{1-\overline{K\hat{\varphi}_0}/|\hat{\varphi}_0|}{1-K\hat{\varphi}_0/|\hat{\varphi}_0|}\right]=\frac{\hat{\varphi}_0}{|\hat{\varphi}_0|}.$$

令 $n\to\infty$，由(11)～(14)，即得 $|\tau(z)|=1$，$\iint_\Omega|K\hat{\varphi}_0|(1-\operatorname{Re}\tau)\,\mathrm{d}x\,\mathrm{d}y=0$，故 $\tau=1$. 因此，$\hat{K}(w)\equiv K_1(w)$ 几乎处处成立，故 f 是唯一极值的. 定理证毕.

定理 3 设 $w=f(z)$ 是 Ω 上的拟共形映照，$f_{\bar{z}}/f_z=K(z)\in\mathscr{L}^\infty(\Omega)$，$0<k_0\leqslant\operatorname*{ess\,inf}_{z\in\Omega}|K(z)|\leqslant\operatorname*{ess\,sup}_{z\in\Omega}|K(z)|\leqslant k<1$. 若存在函数列 $\varphi_n\in\mathscr{B}(\Omega)$，$n=1,2,3,\cdots$，满足下列条件：

$\{\varphi_n\}$ 在 Ω 上几乎处处收敛于 $\tilde{\varphi}_0$，$\tilde{\varphi}_0\neq0$ 在 Ω 上几乎处处成立，(15)

$$\lim_{n\to\infty}\iint_\Omega\delta\{\varphi_n\}\,\mathrm{d}x\,\mathrm{d}y=0,\tag{16}$$

或者

$\{\varphi_n\}$ 在 Ω 上几乎处处收敛于 $\tilde{\varphi}_0$，$\tilde{\varphi}_0\neq0$ 和 $K\tilde{\varphi}_0>0$ 在 Ω 上几乎处处成立，$\tilde{\varphi}_0\in\mathscr{L}^1_{\mathrm{loc}}(\Omega)$，(17)

$$\iint_\Omega\delta\{\varphi_n\}\,\mathrm{d}x\,\mathrm{d}y\leqslant M,\ n=1,2,3,\cdots,\tag{18}$$

$$\varlimsup_{A\to\infty}\iint_{\bar{\Omega}(n,A)}\delta\{\varphi_n\}\,\mathrm{d}x\,\mathrm{d}y=0 \text{ 关于 } n \text{ 均匀成立},\tag{19}$$

$$\bar{\Omega}(n,A)=\{z\in\Omega:|\varphi_n(z)|>A|\tilde{\varphi}_0(z)|\},$$

则 f 是唯一极值的.

证 同文献[3]中所证，令 $\alpha(z)=\hat{K}(f(z))$，$\mu(z)=K_1(f(z))$，$\rho=$

$|\alpha(z)-\mu(z)|^2$，则有

$$\iint_{\Omega}\rho|\varphi_n|\mathrm{d}x\,\mathrm{d}y \leqslant C^{\frac{1}{2}}\iint_{\Omega}(\rho|\varphi_n|)^{\frac{1}{2}}(\delta\{\varphi_n\})^{\frac{1}{2}}\mathrm{d}x\,\mathrm{d}y, \tag{20}$$

$$\iint_{\Omega}\rho|\varphi_n|\mathrm{d}x\,\mathrm{d}y \leqslant C\iint_{\Omega}\delta\{\varphi_n\}\mathrm{d}x\,\mathrm{d}y, \tag{21}$$

这里 $C=8k(1+k^2)^2/k_0(1+k)^2(1-k)^6.$

由(15),(16)和(21),即知 f 是唯一极值的.

由(17)～(21),则有

$$\iint_{\Omega}\rho|\widetilde{\varphi}_0|\mathrm{d}x\,\mathrm{d}y \leqslant \varliminf_{n\to\infty}J_{1,n}+\varlimsup_{n\to\infty}J_{2,n}+\varlimsup_{n\to\infty}J_{3,n},$$

$$\varliminf_{n\to\infty}J_{1,n}^2 \leqslant CkM\varliminf_{n\to\infty}\iint_{\bar{\Omega}(n,A)}\delta\{\varphi_n\}\mathrm{d}x\,\mathrm{d}y=0,$$

$$\varlimsup_{n\to\infty}J_{2,n}^2 \leqslant AMC\varlimsup_{n\to\infty}\iint_{\bigcup\limits_{i>m}E_i}\rho|\widetilde{\varphi}_0|\mathrm{d}x\,\mathrm{d}y \to 0\ (m\to\infty),$$

$$\varlimsup_{n\to\infty}J_{3,n}^2 \leqslant CM\varlimsup_{n\to\infty}\iint_{\Omega'''}\delta\{\varphi_n\}\mathrm{d}x\,\mathrm{d}y$$

$$=CM\iint_{\Omega'''}(|K\widetilde{\varphi}_0|-\operatorname{Re}K\widetilde{\varphi}_0)\mathrm{d}x\,\mathrm{d}y=0,$$

其中 $\Omega'''=\{z\in\bigcup\limits_{i\leqslant m}E_i:\ |\varphi_n(z)|<A|\widetilde{\varphi}_0(z)|\}$,

$$\Omega=\bigcup_{i=1}^{\infty}E_i,\ E_i\cap E_j=\varnothing,\ i\neq j,\ mE_i<\infty. \tag{22}$$

于是，$\iint_{\Omega}\rho|\widetilde{\varphi}_0|\mathrm{d}x\,\mathrm{d}y=0$. 因此，$f$ 是唯一极值的. 定理证毕.

推论 文献[3]中的定理 A 和定理 B 成立.

由上述定理的证明,即知下述定理成立.

定理 4 设 $w=f(z)$ 是 Ω 上的拟共形映照,

$$f_{\bar{z}}/f_z=K(z)\in\mathscr{L}^{\infty}(\Omega),\ 0<k_0\leqslant\operatorname*{ess\,inf}_{z\in\Omega}|K(z)|\leqslant\|K\|_{\infty}\leqslant k<1.$$

若存在函数列 $\varphi_n\in\mathscr{B}(\Omega)$, $n=1,2,3,\cdots$, 满足条件(15),(18),(19)和对 Ω 的某一如(3.12)式的分解,当 m 充分大时,

$$\varliminf_{n\to\infty}\iint_{\bigcup\limits_{i\leqslant m}E_i}\delta\{\varphi_n\}\mathrm{d}x\,\mathrm{d}y=0, \tag{23}$$

则 f 是唯一极值的.

定理 5 假定存在均匀有界函数列 $\varphi_n \in \mathscr{B}(\Omega)$, $n=1, 2, 3, \cdots$,使得

$\{\varphi_n\}$ 在 Ω 上几乎处处收敛于 φ_0, $\varphi_0 \neq 0$ 在 Ω 上几乎处处成立, $\varphi_0 \in \mathscr{L}^1_{\mathrm{loc}}(\Omega)$, (24)

$$\iint_\Omega (|\varphi_n| - \operatorname{Re}\varphi_n)\,\mathrm{d}x\,\mathrm{d}y \leqslant M, \; n=1, 2, \cdots, \tag{26}$$

或者存在函数列 $\varphi_n \in \mathscr{B}(\Omega)$, $n=1, 2, 3, \cdots$,使得

$\{\varphi_n\}$ 在 Ω 上几乎处处收敛于 φ_0, φ_0 在 Ω 上几乎处处不为零, (25)

$$\lim_{n\to\infty}\iint_\Omega (|\varphi_n| - \operatorname{Re}\varphi_n)\,\mathrm{d}x\,\mathrm{d}y = 0, \tag{27}$$

则仿射伸长 $A_K(z)=Kx+\mathrm{i}y$, $z=x+\mathrm{i}y$, $K>1$ 是唯一极值的.

由文献[3]的定理 C,关于 $K \in \mathrm{HBU}$ 有下述判据.

定理 6 对 $K(z) \in \mathscr{L}^\infty(\Omega)$ 存在函数列 $\varphi_n \in \mathscr{B}(\Omega)$, $n=1, 2, \cdots$,它满足定理 2～4 中任何一个定理的条件,则线性泛函

$$\Lambda_K[\varphi] = \iint_\Omega K(z)\varphi(z)\,\mathrm{d}x\,\mathrm{d}y, \; \varphi \in \mathscr{B}(\Omega) \tag{28}$$

从 $\mathscr{B}(\Omega)$到 $\mathscr{L}^1(\Omega)$上有唯一的 Hahn-Banach 扩张.

定理 7 若存在满足定理 5 中条件的函数列 $\varphi_n \in \mathscr{B}(\Omega)$, $n=1, 2, \cdots$,则线性泛函

$$\Lambda_\Omega[\varphi] = \iint_\Omega \varphi(z)\,\mathrm{d}x\,\mathrm{d}y, \quad \varphi \in \mathscr{B}(\Omega) \tag{29}$$

从 $\mathscr{B}(\Omega)$到 $\mathscr{L}^1(\Omega)$上有唯一的 Hahn-Banach 扩张.

4. 关于 Reich 猜想

定理 8 设 $A_K(z)=Kx+\mathrm{i}y$, $z=x+\mathrm{i}y$ $(K>1)$ 是面积为有限的区域 Ω 上的仿射伸长, $k=(K-1)/(K+1)$, 则 $A_K(z)$是 $Q_K(\Omega)$的唯一极值映照当且仅当 $\|\Lambda_\Omega\|=1$, Λ_Ω 从 $\mathscr{B}(\Omega)$到 $\mathscr{L}^1(\Omega)$有唯一的 Hahn-Banach 扩张时成立.

证 必要性. 令 $K(z)=(A_K)_{\bar z}/(A_K)_z$,则 $K(z)=k$. 由于 A_K 是 $Q_K(\Omega)$的极值映照,根据 Hamilton 定理[8], $\|\Lambda_\Omega\|=1$, 且存在函数列 $\varphi_n \in \mathscr{B}(\Omega)$, $\|\varphi_n\|=1$, 使得

$$\lim_{n\to\infty}\iint_{\Omega}\varphi_n(z)\,dx\,dy=1. \tag{30}$$

若$\{\varphi_n\}$不存在退化的 Hamilton 序列,根据定理 2,则存在子函数列$\{\varphi_{n_j}\}$在Ω上内闭均匀收敛于$\varphi_0\in\mathscr{B}(\Omega)$, $\|\varphi_0\|=1$,使得$K(z)=k\bar{\varphi}_0/|\varphi_0|=k$. 因而,$\bar{\varphi}_0=|\varphi_0|$, $\varphi_0=C>0$. 由(30),我们有

$$\lim_{j\to\infty}\iint_{\Omega}(|\varphi_{n_j}|-\mathrm{Re}\,\varphi_{n_j})\,dx\,dy=0. \tag{31}$$

根据定理 6, Λ_Ω 从$\mathscr{B}(\Omega)$到$\mathscr{L}^1(\Omega)$有唯一的 Hahn-Banach 扩张. 若$\{\varphi_n\}$是退化的 Hamilton 序列,则$\{\varphi_n\}$在Ω上内闭匀敛于零. 于是,$\tilde{\varphi}_n=\varphi_n+1\in\mathscr{B}(\Omega)$在$\Omega$上内闭匀敛于 1,并且由(30),有

$$\begin{aligned}&\lim_{n\to\infty}\iint_{\Omega}(|\varphi_n|-\mathrm{Re}\,\varphi_n)\,dx\,dy\\ \leqslant&\lim_{n\to\infty}\iint_{\Omega}(|\varphi_n|-\mathrm{Re}\,\varphi_n)\,dx\,dy=0.\end{aligned} \tag{32}$$

根据定理 7, Λ_Ω 从$\mathscr{B}(\Omega)$到$\mathscr{L}^1(\Omega)$有唯一的 Hahn-Banach 扩张.

现在证明其充分性. 由于$\|\Lambda_\Omega\|=1$, A_K 是$Q_K(\Omega)$的极值映照,并且存在 Hamilton 序列$\varphi_n\in\mathscr{B}(\Omega)$, $n=1, 2, \cdots$,使得(30)式成立. 若$\{\varphi_n\}$不存在退化的 Hamilton 序列,则有$\{\varphi_{n_j}\}$在Ω上内闭匀敛于$\varphi_0>0$,并且(31)式成立. 根据定理 5, A_K 是$Q_K(\Omega)$的唯一极值映照. 若$\{\varphi_n\}$在Ω上内闭匀敛于 0,则$\tilde{\varphi}_n=\varphi_n+1$在$\Omega$上内闭匀敛于 1,并且有(32)式. 根据定理 5, A_K 是$Q_K(\Omega)$的唯一极值映照. 定理证毕.

类似地,不难证明下述定理.

定理 9 设$w=f(z)$是Ω上的拟共形映照,其复伸长为$K(z)=f_{\bar z}/f_z\in\mathscr{L}^\infty(\Omega)$, $k=1$,则 f 是关于自己边界值映照族$Q_f(\Omega)$的唯一极值的 Teichmüller 映照$K(z)=k\overline{\varphi_0(z)}/|\varphi_0(z)|$, $\varphi_0\in\mathscr{B}(\Omega)$, $\varphi_0\neq 0$ a. e.,当且仅当$|K(z)|$在Ω除去一零测度集所成的集上处处连续,$\arg\overline{K(z)}=\arg\varphi_0(z)$, $\varphi_0\in\mathscr{B}(\Omega)$, $\|\Lambda_K\|=k$, $K\in$ HBU.

定理 10 设$F\in Q_F(\Omega, T, b)$, $K(z)=f_{\bar z}/f_z$, $f=F^{-1}$,

$$\tau(z)=\begin{cases}K(z)/b(f(z)),\ z\in F(T),\\ K(z)/k_F,\ z\in\Omega'\backslash F(T),\ k_F>0,\end{cases}$$

$|\tau(z)|$在Ω'上除去一零测度集之集上处处连续,则 F 是$Q_F(\Omega, T, b)$的唯一

极值的广义 Teichmüller 映照，且

$$K(z)=\begin{cases} b(f(z))\overline{\varphi_0(z)}/|\varphi_0(z)|,\ z\in F(T),\ \varphi_0\in\mathscr{B}(\Omega'),\ \varphi_0\neq 0 \text{ a. e.},\\ \overline{k_F\varphi_0(z)}/|\varphi_0(z)|,\ z\in\Omega'\backslash F(T),\ \Omega'=F(\Omega), \end{cases} \tag{33}$$

当且仅当（$|\tau(z)|$ 在 Ω' 除去一个零测度集之集上处处连续）$\arg\overline{\tau(z)}=\arg\varphi_0(z)$，$\varphi_0\in\mathscr{B}(\Omega')$，$\varphi_0\neq 0$ 在 Ω' 上几乎处处成立，并且 $\|\Lambda_\tau\|=1$，Λ_τ 从 $\mathscr{B}(\Omega')$ 到 $\mathscr{L}^1(\Omega')$ 有唯一的 Hahn-Banach 扩张.

◇参考文献◇

[1] E. Reich, *Arch. Rational Mech. Anal.*, 68 (1978) 99.

[2] E. Reich, *Math. Z.*, 167 (1979) 81.

[3] E. Reich, *Ann. Acad. Sci. Fenn. Ser. AI Math.*, 6 (1981) 289.

[4] E. Reich, *Proc. Amer. Math. Soc.*, 88 (1983) 305.

[5] K. Strebel, *Comment. Math. Helvetivici*, 53 (1978) 301.

[6] E. Reich, K. Strebel, Contribution to Analysis, Academic Press (1974) 375.

[7] E. Reich, *Indiana Univ. Math, J.*, 30 (1981) 441.

[8] R. S. Hamilton, *Trans. Amer. Math. Soc.*, 138 (1969) 399.

On Three Reich's Conjectures on the Teichmüller Extremal Quasiconformal Mappings

Ren Fuyao

Abstract: In this paper, applying Hamilton's condition and the sufficient conditions of extremal quasiconformal mapping being unique and Hahn-Banach extension being unique, under some additional conditions the author prove three Reich's conjectures on the extremal Teichmüller quasiconformal mappings.

Key words: quasiconformal mapping, Teichmüller mapping, Hahn-Banach extension

Extension of a Theorem of Carleson-Duren*

Ren Fu-Yao and Huang Li-feng
(*Department of Mathematics, Fudan University, Shanghai*)

1. Introduction

A theorem of Carleson[1],[2] as generalized by Duren[3] characterizes those positive measure μ on the unit disc $U = \{z \in \mathbb{C}: |z| < 1\}$ for which the H^p norm dominates the $L^q(\mu)$ norm of elements of H^p. Later on, Hasting[5] proved an analogous results with H^p replaced by A^p, the Bergman space of fuctions f which $\int_0^1\int_0^{2\pi} |f(re^{i\theta})|^p r dr d\theta < \infty$. Actually, his result is more general in that it applies to positive measure and positive n-subharmonic functions on the unit polydisc U^n in $\mathbb{C}^n$, the purpose of this article is to generalize the theorems of Duren and Hastings.

2. Extension of the Theorem of Duren

Theorem I *Let μ be a finite, positive measure on* U, *and suppose that the function $\phi(t)$: $[0, \infty) \to \mathbb{R}$ satisfies the following conditions*:

(i) $\phi(0) = 0$, $\phi(t) > 0$, $t > 0$;

(ii) ϕ *is increasing and* $\lim \dfrac{\phi(t)}{t} = \infty$ *or finite*;

(iii) ϕ' *exists and is increasing in* $(0, \infty)$;

* Originally published in *J. Math. Research and Exposition*, Vol. 9, No. 1, (1989), 31 - 39.

(iv) $\lim_{t\to 0}\phi(ct^{-1})\phi(t^2)=0$;

(v) *there exists a constant* $B>0$ *such that*

$$\sup_{t>0}\frac{t\phi'(t)\phi(ct^{-1})}{\phi(c)}\leqslant B$$

for all $c>0$. *Then in order that there exists a constant* $C>0$ *depending only on* ϕ *such that*

$$\phi^{-1}\left\{\int_U \phi(|f(z)|^p)\right\}\mathrm{d}\mu(z)\leqslant C\|f\|_p^p \tag{1}$$

for all $f\in H^p$, $0<p<\infty$, *it is necessary and sufficient that there is a positive constant* A *depending only on* ϕ *such that*

$$\mu(S_h)\leqslant\phi(Ah) \tag{2}$$

for every set S_h *of the form*

$$S_h=\{z=re^{i\theta}: 1-h\leqslant r<1,\ \theta_0\leqslant\theta\leqslant\theta_0+h\}. \tag{3}$$

We need the following lemma which is obtained by elementary calculus.

Lemma *Suppose that function* $\phi(t)$: $[0,\infty)\to\mathbb{R}$ *satisfies the following conditions*:

(i) $\phi(0)=0$, $\phi(t)>0$, $t>0$;

(ii) ϕ *is increasing and*

(iii) ϕ' *exists and is increasing in* $(0,\infty)$.

Then the following properties are true:

$$\phi(ct)\geqslant c\phi(t),\ \text{for } c\geqslant 1, \tag{4}$$

$$\phi(t)/t \text{ is increasing in } (0,\infty), \tag{5}$$

$$\phi(t_1)+\phi(t_2)\leqslant\phi(t_1+t_2) \tag{6}$$

for all $t_1, t_2\in[0,\infty)$.

Proof of necessity Suppose that (1) holds with p, $0<p<\infty$, it is easy to see that

$$\mu(S_h)\leqslant\phi(c\|f\|_p^p)/\phi(\min_{z\in S_h}|f(z)|^p) \tag{7}$$

for all $f \in H^p$ and for every set S_h of the form (3).

Let $z_0 = \rho e^{ia}$, and let $\rho = 1 - h$, and consider the H^p function $f(z) = [5h^2(1 - z_0 z)^{-2}]^{1/p}$, whose norm is $\| f \|_p^p = 5h^2(1-\rho^2)^{-1} < 5h$.

A geometric argument [4, p. 157] shows that $|f(z)|^p \geqslant 1$ in S_h. Therefore, by (7), $\mu(S_h) \leqslant \phi(5ch)/\phi(1)$ and (2) holds with $A = 5c/\phi(1)$ for $\phi(1) < 1$ and $A = 5c$ for $\phi(1) \geqslant 1$.

Proof of sufficiency Suppose that (2) holds for every set S_h of the form (3), we first prove (1) holds with $p=2$. For $f \in H^2$, it is proved in [3] that

$$|f(z)| \leqslant 16^2(\tilde{\varphi}(z) + \|\varphi\|_1), \tag{8}$$

here $\varphi(t) = f(e^{it})$, and $\tilde{\varphi}(z) = \sup\limits_I \frac{1}{|I|}\int_I |\varphi(t)|\,dt$, where the supremum is taken over all intervals I containing I_z of length $|I| < 1$, and I_z be the boundary arc

$$I_z = \left\{e^{it} : \theta - \frac{1}{2}(1-r) \leqslant t \leqslant \theta + \frac{1}{2}(1-r)\right\}$$

for each point $z = re^{i\theta} \neq 0$ in U.

Therefore, by (8), elementary inequality $(a+b)^2 \leqslant 2(a^2+b^2)$ and the downward convexity of ϕ,

$$\begin{aligned}\int_U \phi(|f(z)|^2)\,d\mu(z) &\leqslant \int_U \phi\{[16^2(\tilde{\varphi}(z) + \|\varphi\|_1)]^2\}\,d\mu \\ &\leqslant \frac{1}{2}\left\{\int_U \phi[c_1\tilde{\varphi}(z)^2]\,d\mu + \int_U \phi(c_1\|\varphi\|_2^2)\,d\mu\right\}\end{aligned} \tag{9}$$

with $c_1 = 512\pi^4$.

It suffices then to show that

$$\int_U \phi(c_1\tilde{\varphi}(z)^2)\,d\mu \leqslant \phi(c_2\|\varphi\|_2^2). \tag{10}$$

To do this, let

$$E_s = \{z \in U: \tilde{\varphi}(z) > s/\sqrt{c_1} > 0\},\ a(s) = \mu(E_s),$$

then

$$\int_{\mathrm{U}}\phi(c_1\widetilde{\varphi}(z)^2)\mathrm{d}\mu=-\int_0^{\infty}\phi(s^2)\mathrm{d}a(s)$$

$$\leqslant 2\int_0^{\infty}s\phi'(s^2)a(s)\mathrm{d}s+a(s)\phi(s^2)\mid_{s=0}. \tag{11}$$

We will show that

$$\lim_{s\to 0}a(s)\phi(s^2)=0. \tag{12}$$

Let $\varphi(t)\in \mathrm{L}^1(\partial\mathrm{U})$, and let for $\varepsilon>0$,

$$A_s^{\varepsilon}=\left\{z\in \mathrm{U}:\int_{I_s}|\varphi(t)|\,\mathrm{d}t>s(\varepsilon+I_z)\right\},$$

$$B_s^{\varepsilon}=\{z\in \mathrm{U}:\text{ exists } w\in A_s^{\varepsilon}\text{ such that } I_w\supset I_z\}.$$

It is proved in [4] that

$$E_s=\lim_{\varepsilon\to 0}B_s^{\varepsilon},\quad \mu(E_s)=\lim_{\varepsilon\to 0}(B_s^{\varepsilon}), \tag{13}$$

and there exists a finite number of points $z_1, z_2, \cdots, z_m$ in U such that the arcs I_{z_n} are disjoint and

$$B_s^{\varepsilon}\subset\bigcup_{n=1}^{m}\{z\in \mathrm{U}: I_z\subset I_{z_n}\},$$

and

$$s\sum_{n=1}^{m}(\varepsilon+|I_{z_n}|)<\sum_{n=1}^{m}\int_{I_{z_n}}|\varphi(t)|\,\mathrm{d}t\leqslant 2\pi\|\varphi\|_1, \tag{14}$$

where J_z is the arc of length $5|I_z|$ whose center coincides with that of I_z.

Therefore, since $\mu(S_h)\leqslant\phi(Ah)$, by (6) and (14),

$$\mu(B_s^{\varepsilon})\leqslant\sum_{n=1}^{m}\mu(\{z\in \mathrm{U}: I_z\subset J_{z_n}\})\leqslant\sum_{n=1}^{m}\mu(S_{j_{z_n}})$$

$$\leqslant\sum_{n=1}^{m}\phi(5A|I_{z_n}|)\leqslant\phi\Big(\sum_{n=1}^{m}5A|I_{z_n}|\Big)\leqslant\phi(10\pi A\|\varphi\|_1/s).$$

Letting $\varepsilon\to 0$, it follows from (13) that

$$a(s)=\mu(E_s)\leqslant\phi(A_1\|\varphi\|_1/s) \tag{15}$$

with $A_1=10\pi A$.

Thus, $a(s)\phi(s^2)\leqslant\phi(A_1\|\varphi\|_1/s)\phi(s^2)$, and (12) holds by condition

(iv) of ϕ.

Then from (11) we have

$$\int_U \phi(c_1\tilde{\varphi}(z)^2)\mathrm{d}\mu(z) \leqslant \int_0^s \phi'(s^2)a(s)\mathrm{d}s^2. \tag{16}$$

For $s > 0$, let

$$\psi_s(t) = \begin{cases} \varphi(t) & \text{wherever } |\varphi(t)| > s/2A_1, \\ 0 & \text{otherwise.} \end{cases}$$

Here we assume $A_1 \geqslant 1$. Let $\tilde{\psi}_s(z) = \sup_I \frac{1}{|I|}\int_I |\psi_s(t)|\mathrm{d}t$ defined as $\tilde{\varphi}(z)$, and let $F_s = \{z \in U: \tilde{\psi}_s(z) > s/2\sqrt{c_1} > 0\}$, then it is proved in [4] that $E_s \subset F_s$. Therefore, from (16) and (15) for ψ_s we obtain

$$\begin{aligned} \int_U \phi(c\tilde{\varphi}(z)^2)\mathrm{d}\mu(z) &\leqslant \int_0^\infty \phi'(s^2)\mu(F_s)\mathrm{d}s^2 \\ &\leqslant \int_0^\infty \phi'(s^2)\phi(A_1 \| \psi_s \|_1/s)\mathrm{d}s^2. \end{aligned} \tag{17}$$

Since $\phi(t)/t$ is increasing in $(0, \infty)$, then

$$\begin{aligned} \phi(A_1 \| \psi_s \|_1/s) = &\left[\phi\left(\frac{A_1}{2\pi s}\int_{2A_1|\varphi(t)|>s} |\varphi(t)|\mathrm{d}t\right) \Big/ \left(\frac{A_1}{2\pi s}\int_{2A_1|\varphi(t)|>s} |\varphi(t)|\mathrm{d}t\right)\right] \\ &\cdot \left(\frac{A_1}{2\pi s}\int_{2A_1|\varphi(t)|>s} |\varphi(t)|\mathrm{d}t\right) \\ \leqslant &\left[\phi\left(\frac{A_1}{2\pi s}\int_{2A_1|\varphi(t)|>s} (2A_1|\varphi(t)|^2/s)\mathrm{d}t\right)\right. \\ &\left. \Big/ \left(\frac{A_1}{2\pi s}\int_{2A_1|\varphi(t)|>s} (2A_1|\varphi(t)|^2/s)\mathrm{d}t\right)\right] \\ &\cdot \left[\frac{A_1}{2\pi s}\int_{2A_1|\varphi(t)|>s} |\varphi(t)|\mathrm{d}t\right] \\ \leqslant &\frac{\phi(2A_1^2 \| \varphi \|_2^2/s^2)}{2A_1^2 \| \varphi \|_2^2/s^2} \cdot \left[\frac{A_1}{2\pi s}\int_{2A_1|\varphi(t)|>s} |\varphi(t)|\mathrm{d}t\right]. \end{aligned}$$

Substituting this inequality into (17) we have

$$\int_U \phi(c_1\tilde{\varphi}(z)^2)\mathrm{d}\mu$$

$$\leqslant \frac{1}{2A_1 \| \varphi \|_2^2} \int_0^\infty \phi'(s^2)\phi(2A_1^2 \| \varphi \|_2^2/s^2) s \left(\frac{1}{2\pi} \int_{2A_1 |\varphi(t)|>s} |\varphi(t)| \mathrm{d}t \right) \mathrm{d}s^2.$$

Exchanging the order of integration, since $\sup\limits_{t>0} t\phi'(t)\phi(ct^{-1}) \leqslant B\phi(c)$, we have $\int_U \phi(c_1 \tilde{\varphi}(z)^2) \mathrm{d}\mu(z) \leqslant 2B\phi(2A_1^2 \| \varphi \|_2^2)$, and (10) holds with $c_2 = 4A_1^2 B$.

Substituting (10) into (9), we have

$$\int_U \phi(|f(z)|^2) \mathrm{d}\mu(z) \leqslant \phi(c_2 \| \varphi \|_2^2) + \phi(c_1 \| \varphi \|_2^2) \cdot \mu(\mathrm{U})$$
$$\leqslant \phi(c \| \varphi \|_2^2) = \phi(c \| f \|_2^2). \tag{18}$$

This proves the sufficiency of (2) for $p=2$.

Finally, for arbitrary p, $0<p<\infty$, if $f \in \mathrm{H}^p$, then $f(z) = B(z)[g(z)]^{2/p}$, where $B(z)$ is the Blaschke product and $g(z) \neq 0$, $g \in \mathrm{H}^2$ and $\| f \|_p^p = \| g \|_2^2$. Therefore,

$$\int_U \phi(|f(z)|^p) \mathrm{d}\mu(z) \leqslant \int_U \phi(|g(z)|^2) \mathrm{d}\mu(z).$$

Since $\mu(S_h) \leqslant \phi(Ah)$ for every set S_h of the form (3), then by (18) we have

$$\int_U \phi(|f(z)|^p) \mathrm{d}\mu(z) \leqslant \phi(c \| g \|_2^2) = \phi(c \| f \|_p^p),$$

and this completes the proof of the theorm 1.

Apply theorem 1 to $\phi(t) = t^{q/p}$, $0<p\leqslant q<\infty$, we obtain the following corollaries immediately.

Corollary 1.1 *Let* μ *be a finite, positive measure on* U, *and suppose* $0<p\leqslant q<\infty$. *Then in order that*

$$\left[\int_U |f(z)|^{qp'/p} \mathrm{d}\mu(z) \right]^{p/q} = c \| f \|_{p'}^{p'} \tag{19}$$

for all $f \in \mathrm{H}^{p'}$, $0<p'<\infty$, *it is necessary and sufficient that* $\mu(S_h) \leqslant (Ah)^{q/p}$ *for every set* S_h *of the form* (3).

As in [3], [4] two inequalities follow immediately from above corollary 1.1.

Corollary 1.2 *If* $0<p<q<\infty$, *then* $f \in \mathrm{H}^p$, $0<p<\infty$, *implies*

$$\left[\int_0^1 (1-r)^{q/p-2} M_a^a(r, f) \mathrm{d}r\right]^{p/q} \leqslant c \| f \|_{p'}^{p'}, \tag{20}$$

where $a = qp'/p$ *and* $M_a^a(r, f) = \frac{1}{2\pi}\int_0^{2\pi} |f(re^{i\theta})|^a \mathrm{d}\theta$. *This generalized a theorem of Hardy-Littlewood* [6]:

$$\left[\int_0^1 (1-r)^{q/p-2} M_q^q(r, f) \mathrm{d}r\right]^{1/q} \leqslant c \| f \|_p.$$

Corollary 1.3 *If* $0 < p \leqslant q < \infty$, *and* $f \in H^{p'}$, $0 < p' < \infty$, *then*

$$\left[\int_{-1}^1 (1-r)^{q/p-1} |f(r)|^{qp'/p} \mathrm{d}r\right]^{p/q} \leqslant c \| f \|_{p'}^{p'}. \tag{21}$$

Particularly, for $p' = p$ all of these corollaries reduce to that of Duren in [3].

3. Extension of the Theorem of Hastings

Let $U^n = \{z=(z_1, \cdots, z_n) \in \mathbb{C}^n: |z_j| < 1, 1 \leqslant j \leqslant n\}$ and let σ_n be $2n$-dimensional Lebesgue measure restricted to U^n, normalized so that U^n has measure one.

Theorem 2 *Let* μ *be a finitie, positive measure on* U^n, *and suppose that function* $\phi(t)$ *satisfies the first conditions* (i) ~ (iii) *of theorem* 1, *and there exists a constant* K *such that*

$$\phi(t_1) \cdot \phi(t_2) \leqslant K\phi(t_1 t_2) \tag{22}$$

for arbitrary t_1, $t_2 > 0$. *Then in order that there exists a constant* $c > 0$ *such that*

$$\phi^{-1}\left\{\int_{U^n} \phi(|f(z)|^p) \mathrm{d}\mu(z)\right\} \leqslant c \| f \|_{A^p}^p = c\int_{U^n} |f(z)|^p \mathrm{d}\sigma_n(z) \tag{23}$$

for all $f \in A^p(U^n)$, $0 < p < \infty$, *it is necessary and sufficient that there is a constant* $A > 0$ *such that*

$$\mu(S_h) \leqslant \phi\left(A \prod_{j=1}^n h_j^2\right) \tag{24}$$

for every set S_h *of the form*

$$S_h=\{z=(r_1e^{i\theta_1},\cdots,r_ne^{i\theta_n}):1-h_j\leqslant r_j<1,$$
$$\theta_j^0\leqslant\theta_j\leqslant\theta_\theta^0+h_j,\ 1\leqslant j\leqslant n\}.\tag{25}$$

Proof If inequality (23) holds for $f\in A^p(U^n)$, then for every set S_h of the form (25) we have

$$\mu(S_h)\leqslant\phi(c\|f\|_{A^p}^p)/\phi(\min_{z\in S_h}|f(z)|^p).\tag{26}$$

We assume $f(z)=\left[c_1\prod_{j=1}^n h_j^4(1-\bar{a}_jz_j)^{-4}\right]^{1/p}$, where $a_j=(1-h_j)\exp\{i(\theta_j^0+h_j)/2\}$, $1\leqslant j\leqslant n$, then

$$\|f\|_{A^p}^p<c_1\prod_{j=1}^n h_j^2,\quad |f(z)|^p\geqslant 1,\ z\in S_h.$$

Therefore

$$\mu(S_h)\leqslant\phi\big(cc_1\prod_{j=1}^n h_j^2\big)/\phi(1)$$

and (24) holds with $A=cc_1/\phi(1)$ for $\phi(1)<1$ and $A=cc_1$ for $\phi(1)\geqslant 1$.

Conversely, suppose that (24) holds for every set S_h of the form (25). For $m=(m_1,\cdots,m_n)\in\mathbb{Z}^n$ and $k=(k_1,\cdots,k_n)\in\mathbb{Z}^n$ with $m_j\geqslant 0$ and $1\leqslant k_j\leqslant 2^{m_j+4}$, $1\leqslant j\leqslant n$, set $T_{mk}=\{z=(r_1e^{i\theta_1},\cdots,r_ne^{i\theta_n}):1-2^{-m_j}\leqslant r_j<1-2^{-m_j}-1,\ 2k_j\pi/2^{m_j+4}\leqslant\theta_j\leqslant 2(k_j+1)\pi/2^{m_j+4},\ 1\leqslant j\leqslant n\}$, and let $z^{mk}=(z_1^{mk},\cdots,z_n^{mk})$, where $z_j^{mk}=(1-2^{-m_j})\cdot\exp\{2(k_j+1/2)\pi i/2^{m_j+4}\}$, $1\leqslant j\leqslant n$, and $U_{mk}=\{z=(z_1,\cdots,z_n)\in\mathbb{C}^n:|z_j-z_j^{mk}|\leqslant(7/8)2^{-m_j},\ 1\leqslant j\leqslant n\}$, it is proved in [5] that

$$|f(z)|^p\leqslant c_2\big(\prod_{j=1}^n 4^{m_j}\big)\int_{U_{mk}}|f(z)|^p d\sigma_n(z),\ z\in T_{mk}\tag{27}$$

and

$$\sum_m\sum_k\int_{U_{mk}}|f(z)|^p d\sigma_n(z)\leqslant N\int_{U^n}|f(z)|^p d\sigma_n(z),\tag{28}$$

where $N=(135)^n$. Therefore by (27)

$$\int_{U^n}\phi(|f(z)|^p)d\mu(z)=\sum_{\substack{m=(m_1,\cdots,m_n)\\ m_j\geqslant 0}}\sum_{\substack{k=(k_1,\cdots,k_n)\\ 1\leqslant k_j\leqslant 2^{m_j+4}}}\int_{T_{mk}}\phi(|f(z)|^p)d\mu(z)$$

$$\leqslant\sum_m\sum_k\mu(T_{mk})\phi\left\{c_2\prod_{j=1}^n 4^{m_j}\int_{U_{mk}}|f(z)|^p d\sigma_n(z)\right\}. \tag{29}$$

Since $T_{mk}\subset S_{mk}$ which is the set S_h of the form (25) with $h_j=2^{-m_j}$, $1\leqslant j\leqslant n$, then

$$\mu(T_{mk})\leqslant\mu(S_{mk})=\phi(A\prod_{j=1}^n 2^{-2m_j}). \tag{30}$$

Substituting (30) into (29), by (22), (6) and (28), we have

$$\int_{U^n}\phi(|f|^p)d\mu\leqslant\sum_m\sum_k\phi(A\prod_{j=1}^n 2^{-2m_j})\phi\left\{c_2\prod_{j=1}^n 4^{m_j}\int_{U_{mk}}|f|^p d\sigma_n\right\}$$

$$\leqslant K\sum_m\sum_k\phi\left(Ac_2\int_{U_{mk}}|f|^p d\sigma_n\right)\leqslant K\phi\left(Ac_2N\int_{U^n}|f|^p d\sigma_n\right)$$

$$\leqslant\phi\left(c\int_{U^n}|f|^p d\sigma_n\right)=\phi(c\,\|f\|_{A^p}^p)$$

where $c=Ac_2NK$ for $K\geqslant 1$ and $c=Ac_2N$ for $K<1$.

Hence (23) holds for all $f\in A_p(U^n)$. This completes the proof of the theorem 2.

It follows the following corollaries immediately.

Corollary 2.1 *Let μ be a finite, positive measure on U^n, and suppose $0<p\leqslant q<\infty$. Then in order that there exists a constant $c>0$ such that*

$$\left\{\int_{U^n}|f(z)|^{qp'/p}d\mu(z)\right\}^{p/q}\leqslant c\,\|f\|_{A^{p'}}^{p'} \tag{31}$$

for all $f\in A^{p'}(U^n)$, $0<p'<\infty$, it is necessary and sufficient that there is a constant $A>0$ such that

$$\mu(S_h)\leqslant(A\prod_{j=1}^n h_j)^{2q/p} \tag{32}$$

for every set S_h of the form (25).

Corollary 2.2 *Suppose $0<p\leqslant q<\infty$, if $f\in A^{p'}(U)$, $0<p'<\infty$, then*

$$\left\{\int_{-1}^{1}|f(r)|^{a}(1-r)^{2q/p-1}\mathrm{d}r\right\}^{p/q}\leqslant c\,\|f\|_{A^{p'}(U^n)}^{p'} \tag{33}$$

and

$$\left\{\int(1-r)^{2q/p-1}M_a^a(r,f)\mathrm{d}r\right\}^{p/q}\leqslant c'\,\|f\|_{A^{p'}(U^n)}^{p}, \tag{34}$$

where $a=qp'/p$, *the constnat* c *and* c' *are independent of* f.

Remark All these corollaries also hold for every positive n-subharmonic functions f in U^n if $1\leqslant p\leqslant q<\infty$.

4. Another Example

Let $\phi(t)=t^a/(1+t)$, $t>0$, it is easy to see that for $a\geqslant 2$,

$$\phi(0)=0,\ \phi'(t)>0,\ \phi''(t)>0;\ \phi(t_1)\cdot\phi(t_2)\leqslant\phi(t_1t_2),\ t_1,\ t_2>0;$$
$$\lim_{t\to 0}\phi(ct^{-1})\phi(t^2)=0.$$

Finally, we show

$$\sup_{t>0}\frac{t\phi'(t)\phi(ct^{-1})}{\phi(c)}\leqslant a. \tag{35}$$

Since

$$\frac{t\phi'(t)\phi(ct^{-1})}{\phi(c)}=\left(\frac{1+c}{t+c}\right)\frac{t[a+(a-1)t]}{(1+t)^2},\ \frac{1+c}{t+c}\leqslant\begin{cases}1/t, & \text{for } 0<t<1,\\ 1, & \text{for } t\geqslant 1.\end{cases}$$

therefore

$$\frac{t\phi'(t)\phi(ct^{-1})}{\phi(c)}=g(t)=\begin{cases}\dfrac{a+(a-1)t}{(1+t)^2}, & \text{for } 0<t<1,\\[2mm] \dfrac{at+(a-1)t^2}{(1+t)^2}, & \text{for } t\geqslant 1.\end{cases}$$

In case of $0<t<1$, since $g'(t)<0$, therefore $g(t)\leqslant a$. In case of $t\geqslant 1$, since $g'(t)>0$, hence $g(t)\leqslant a-1$. So (35) is true.

Thus $\phi(t)=t^a/(1+t)$ for $a\geqslant 2$ satisfies all the conditions of theorem 1 and 2. Therefore we have the following corollaries.

Corollary 1.4 *Let* μ *be a finite, positive measure on* U, *and suppose*

$a \geqslant 2$. *Then in order that*

$$\int_U \frac{|f(z)|^{pa}}{1+|f(z)|^p} d\mu(z) \leqslant c \frac{(\|f\|_p^p)}{1+\|f\|_p^p} \tag{36}$$

for all $f \in H^p$, $0 < p < \infty$, *it is necessary and sufficient that*

$$\mu(S_h) \leqslant \frac{(Ah)^a}{1+Ah} \tag{37}$$

for every set S_h *of the form* (3).

Corollary 2.3 *Let* μ *be a finite, positive measure on* U^n, *and suppose* $a \geqslant 2$. *Then there exists a constant* $c > 0$ *such that*

$$\int_{U^n} \frac{|f(z)|^{pa}}{1+|f(z)|^p} d\mu(z) \leqslant c \frac{(\|f\|_{A^p}^p)^a}{1+\|f\|_{A^p}^p} \tag{38}$$

for all $f \in A^p(U^n)$, $0 < p < \infty$, *if and only if there exists a constant* $A > 0$ *such that*

$$\mu(S_h) \leqslant \frac{\left(A\prod_{j=1}^n h_j^2\right)^a}{1+A\prod_{j=1}^n h_j^2} \tag{39}$$

for every set S_h *of the form* (25).

◇ References ◇

[1] Carleson, L., Amer. J. Math., 80(1958), 921 - 930.
[2] ——, Ann. of Math., (2) 76(1962), 547 - 559.
[3] Duren, P. L., Bull. Amer. Math. Soc., 75(1969), 143 - 146.
[4] ——, Pure and Appl. Math., Vol. 38, Academic Press, New York, 1970.
[5] Hastings, W. W., Proc. Amer. Math. Soc., 52(1975), 237 - 241.
[6] Hardy, G. H. and Littlewood, J. E., II, Math. Z., 34(1932), 403 - 439.

Some Inequalities on Quasi-subordinate Functions*

Fuyao Ren, Shigeyoshi Owa and Seiichi Fukui

The object of the present paper is to derive some interesting coefficient estimates for quasi-subordinate functions. Furthermore, a conjecture for quasi-subordinate functions is shown.

1. Introduction

Let $g(z)$ and $f(z)$ be analytic in the unit disk $U=\{z: |z|<1\}$. A function $g(z)$ is said to be subordinate to $f(z)$ if there exists an analytic function $w(z)$ in the unit disk U with $w(0)=0$ and $|w(z)|<1$ $(z \in U)$, such that $g(z)=f(w(z))$. We denote this subordination by $g(z) \prec f(z)$. The concept of subordination can be traced to Lindelöf [1], but Littlewood [2, 3] and Rogosinski [5, 6] introduced the term and discovered the basic properties.

Further, a function $g(z)$ is said to be quasi-subordinate to $f(z)$ in the unit disk U if there exist the functions $\phi(z)$ and $w(z)$ (with constant coefficient zero) which are analytic and bounded by one in the unit disk U, such that

$$g(z)=\phi(z) f(w(z)). \tag{1.1}$$

Also, we denote this quasi-subordination by $g(z) \prec_q f(z)$. It is clear that the

* Originally published in *Bull. Austral. Math. Soc.*, Vol. 43, (1991), 317 - 324.

This research of the authors was completed at Kinki University while the first author was visiting from Fudan University, Shanghai, People's Republic of China.

quasi-subordination is a generalisation of the subordination. The quasi-subordination was introduced by Robertson [4].

2. Coefficient Estimates

We begin with the statement of the following lemma due to Xia and Chang [7].

Lemma *Let* $w(z) = \sum_{n=1}^{\infty} c_n z^n$ *be analytic in the unit disk* U *and* $|w(z)| < 1$ $(z \in U)$. *Then*

$$|c_1| \leqslant 1, \tag{2.1}$$

$$|c_2| \leqslant 1 - |c_1|^2, \tag{2.2}$$

$$|c_3(1 - |c_1|^2) + \bar{c}_1 c_2^2| \leqslant (1 - |c_1|^2)^2 - |c_2|^2. \tag{2.3}$$

Applying the above lemma, we prove

Theorem 1 *Let* $f(z) = \sum_{n=0}^{\infty} a_n z^n$ *and* $g(z) = \sum_{n=0}^{\infty} b_n z^n$ *be analytic in the unit disk* U *and let* $g(z)$ *be quasi-subordinate to* $f(z)$. *Then*

(i) $|b_0| \leqslant |a_0|$ *with equality only if* $g(z) = e^{i\theta} f(w(z))$ (θ *is real*) *in which* $w(z)$ *is analytic with* $w(0) = 0$ *and* $|w(z)| < 1$ $(z \in U)$.

(ii) $|b_1| \leqslant (5/4)\max(|a_0|, |a_1|)$ *with equality only if* $g(z) = \phi(z) f(w(z))$,

$$\phi(z) = e^{i\beta} \frac{z - \alpha}{1 - \bar{\alpha} z} \quad (|\alpha| = 1/2;\ \beta \text{ is real}).$$

In particular, $g(z) = ((1 + 2z)/(2 + z)) f(z)$ *with* $a_0 = a_1$ *is an extremal function.*

(iii) $|b_2| \leqslant (7/27)(1 + 2\sqrt{7})\max(|a_0|, |a_1|, |a_2|)$ *with equality only if*

$$g(z) = e^{i\beta} \frac{z - \alpha}{1 - \bar{\alpha} z} f(e^{i\theta} z) \quad (|\alpha| = t = (\sqrt{7} - 1)/3),$$

where β *and* θ *are real. In particular,* $g(z)=((z+t)/(1+tz))f(z)$, *with* $-a_0=a_1=a_2$ *is an extremal function.*

Proof Note that there exist the functions $\phi(z)=\sum_{n=0}^{\infty}\alpha_n z^n$ and $w(z)=\sum_{n=1}^{\infty}c_n z^n$ such that $g(z)=\phi(z)f(w(z))$. Comparing the coefficients of both sides, we have

$$b_0=\alpha_0 a_0, \tag{2.4}$$

$$b_1=\alpha_0 c_1 a_1+\alpha_1 a_0, \tag{2.5}$$

$$b_2=\alpha_0(c_1^2 a_2+c_2 a_1)+\alpha_1 c_1 a_1+\alpha_2 a_0. \tag{2.6}$$

It follows from (2.4), (2.5), and (2.6) that

$$|b_0|=|\alpha_0 a_0|, \tag{2.7}$$

$$|b_1|\leqslant(|\alpha_0 c_1|+|\alpha_1|)\max(|a_0|, |a_1|), \tag{2.8}$$

$$|b_2|\leqslant(|\alpha_0 c_1^2|+|\alpha_0 c_2|+|\alpha_1 c_1|+|\alpha_2|)\max(|a_0|, |a_1|, |a_2|), \tag{2.9}$$

respectively. Since $|\phi(z)|<1$ $(z\in U)$, we see that the function $\phi_1(z)$ defined by

$$\phi_1(z)=\frac{\phi(z)-\alpha_0}{1-\bar{\alpha}_0\phi(z)}=\sum_{n=1}^{\infty}d_n z^n \tag{2.10}$$

is analytic and bounded by one in the unit disk U. Note that

$$d_1=\frac{\alpha_1}{1-|\alpha_0|^2} \tag{2.11}$$

and

$$d_2=\frac{1}{1-|\alpha_0|^2}\left(\alpha_2+\frac{\alpha_1^2\bar{\alpha}_0}{1-|\alpha_0|^2}\right). \tag{2.12}$$

Therefore, using the Lemma, we have

$$|\alpha_1|\leqslant 1-|\alpha_0|^2 \tag{2.13}$$

and

$$|d_2| \leqslant 1 - |d_1|^2. \tag{2.14}$$

With the aid of (2.11), (2.12) and (2.14), we obtain

$$|\alpha_2| \leqslant (1 - |\alpha_0|^2) - \frac{|\alpha_1|^2}{1 + |\alpha_0|}. \tag{2.15}$$

Thus, it follows from (2.13) that $|b_0| \leqslant |a_0|$ with equality only if $g(z) = e^{i\theta} f(w(z))$.

Further, using (2.1) and (2.13), we obtain

$$\begin{aligned} |b_1| &\leqslant (1 + |\alpha_0| - |\alpha_0|^2) \max(|a_0|, |a_1|) \\ &\leqslant \frac{5}{4} \max(|a_0|, |a_1|) \end{aligned} \tag{2.16}$$

with equality only if

$$g(z) = e^{i\beta} \frac{z - \alpha}{1 - \bar{\alpha} z} f(e^{i\theta} z) \quad (|\alpha| = t = 1/2).$$

In particular, $g(z) = ((z + t)/(1 + tz)) f(z)$ is an extremal function.

Next, let

$$U = |\alpha_0 c_1^2| + |\alpha_0 c_2| + |\alpha_1 c_1| + |\alpha_2|. \tag{2.17}$$

Then, using (2.2) and (2.15), we have

$$U \leqslant |\alpha_0| + |\alpha_1 c_1| + (1 - |\alpha_0|^2) - \frac{|\alpha_1|^2}{1 + |\alpha_0|}. \tag{2.18}$$

Letting $t = |\alpha_0|$, $x = |c_1|$ and $y = |\alpha_1|$, (2.18) can be written in the form

$$U \leqslant 1 + t - t^2 + xy - y^2/(1 + t) \equiv \psi(y). \tag{2.19}$$

Clearly, $y_0 = x(1+t)/2$ is the maximum point of $\psi(y)$. If $y_0 \geqslant 1 - t^2$, that is, if $1 - x/2 \leqslant t \leqslant 1$, $0 \leqslant x \leqslant 1$, then $y \leqslant 1 - t^2$, so

$$\begin{aligned} U &\leqslant \psi(1 - t^2) \\ &= t + (x + t)(1 - t^2) \\ &\leqslant 1 + 2t - t^2 - t^3 \equiv \Lambda(t). \end{aligned} \tag{2.20}$$

It is easy to see that $t_0=(\sqrt{7}-1)/3$ is the maximum point of $\Lambda(t)$ of which the maximum is

$$\Lambda(t_0)=\frac{7(1+2\sqrt{7})}{27}=1.631\,130\,309. \tag{2.21}$$

Hence, when $y_0 \geqslant 1-t^2$, we obtain

$$U \leqslant \Lambda(t_0)=\frac{7(1+2\sqrt{7})}{27}. \tag{2.22}$$

If $y_0=x(1+t)/2<1-t^2$, that is, if $0 \leqslant t<1-x/2$, then

$$U \leqslant \psi(y)=\left(1+\frac{x^2}{4}\right)(1+t)-t^2 \equiv \Omega(t). \tag{2.23}$$

Since $t_0=(1+x^2/4)/2$ is the maximum point of $\Omega(t)$, if $t_0<1-x/2$, or, if $0 \leqslant x<2(\sqrt{2}-1)$, then

$$U \leqslant \Omega(t_0)=1+\frac{x^2}{4}+\frac{1}{4}\left(1+\frac{x^2}{4}\right)^2<10-6\sqrt{2}, \tag{2.24}$$

and, if $t_0 \geqslant 1-x/2$, or, if $2(\sqrt{2}-1) \leqslant x \leqslant 1$, then

$$U \leqslant \Omega(1-x/2)=\frac{8+4x+2x^2-x^3}{8} \leqslant \frac{13}{8}. \tag{2.25}$$

Therefore, it follows from (2.22), (2.24) and (2.25) that

$$|b_2| \leqslant \frac{7(1+2\sqrt{7})}{27}\max(|a_0|, |a_1|, |a_2|).$$

Finally, from the above process of the proof, we know that the extremal function $g(z)=\phi(z)f(w(z))$ occurs only if

$$|\alpha_0|=t=\frac{\sqrt{7}-1}{3},\ |\alpha_1|=1-t^2=\frac{1+2\sqrt{7}}{9},$$

$$|\alpha_2|=1-t^2-\frac{|\alpha_1|^2}{1+t}=\frac{13-\sqrt{7}}{27},\ |c_1|=1,\ c_2=0.$$

Hence, $w(z)=e^{i\theta}z$, $\phi(z)=e^{i\beta}((z-\alpha)/(1-\bar{\alpha}z))$, $|\alpha|=t$. In particular, $g(z)=((z+t)/(1+tz))f(z)$ with $-a_0=a_1=a_2$ is an extremal function.

This completes the proof of Theorem 1. □

Next we derive

Theorem 2 *Let* $g(z)=\sum_{n=0}^{\infty} b_n z^n$ *be quasi-subordinate to* $f(z)=\sum_{n=0}^{\infty} a_n z^n$, *and let* $G(z)=\sum_{n=1}^{\infty} B_n z^n$ *be quasi-subordinate to* $F(z)=\sum_{n=1}^{\infty} A_n z^n$. *Further, let*

$$U_n=\inf\{u_n: |b_n|\leqslant u_n \max(|a_0|, |a_1|, \cdots, |a_n|) \text{ for all } f(z) \text{ and } g(z) \text{ such that } g(z)\prec_q f(z)\}$$

and let $V_n=\inf\{v_n: |B_n|\leqslant v_n \max(|A_1|, |A_2|, \cdots, |A_n|)$ *for all* $F(z)$ *and* $G(z)$ *such that* $G(z)\prec_q F(z)\}$.

Then

$$|b_n|\leqslant U_n \max(|a_0|, |a_1|, \cdots, |a_n|), \tag{2.26}$$

$$|B_n|\leqslant V_n \max(|A_1|, |A_2|, \cdots, |A_n|), \tag{2.27}$$

and

$$V_n\leqslant U_{n-1}. \tag{2.28}$$

Furthermore, for an arbitrary given natural number n, *if the extremal function is* $g(z)=\phi(z)f(w(z))$ *which makes*

$$|b_{n-1}|=U_{n-1}\max(|a_0|, |a_1|, \cdots, |a_{n-1}|)$$

and $z\phi(z)/w(z)$ *is analytic in* U, *then*

$$V_n=U_{n-1}. \tag{2.29}$$

Proof The inequalities (2.26) and (2.27) are clear from the definitions of U_n and V_n. In order to prove (2.28), we first show that if $G(z)\prec_q F(z)$, then $G(z)/z\prec_q F(z)/z$. Indeed, since $G(z)\prec_q F(z)$, there exist the analytic functions $\Phi(z)$ and $w(z)$ (with constant coefficient zero) bounded by one in the unit disk U such that $G(z)=\Phi(z)F(w(z))$. Thus we have

$$\frac{G(z)}{z}=\phi(z)H(w(z)),$$

where $H(z)=F(z)/z$ and $\phi(z)=\Phi(z)w(z)/z$. This shows that $G(z)/z$ is quasi-subordinate to $F(z)/z$.

If $G(z)=\Phi(z)F(w(z))$ is an extremal function of (2.27), then

$$|B_n|=V_n\max(|A_1|, |A_2|, \cdots, |A_n|). \tag{2.30}$$

Since $G(z)/z$ is quasi-subordinate to $F(z)/z$, by (2.27), we get

$$|B_n|\leqslant U_{n-1}\max(|A_1|, |A_2|, \cdots, |A_n|). \tag{2.31}$$

Consequently, we have $V_n\leqslant U_{n-1}$ for all n.

Finally, in order to show (2.29), let $G(z)=zg(z)$, $F(z)=zf(z)$ and $\Phi(z)=z\phi(z)/w(z)$. Since $g(z)=\phi(z)f(w(z))$, we see that $G(z)=\Phi(z)F(w(z))$. Noting that $\Phi(z)$ is analytic in U and bounded by one in U, we conclude that $G(z)\prec_q F(z)$. Since $B_n=b_{n-1}$, $A_1=a_0$, $\cdots$, $A_n=a_{n-1}$, (2.27) gives

$$|b_{n-1}|\leqslant V_n\max(|a_0|, |a_1|, \cdots, |a_{n-1}|). \tag{2.32}$$

But, by the hypothesis

$$|b_{n-1}|=U_{n-1}\max(|a_0|, |a_1|, \cdots, |a_{n-1}|),$$

we have $U_{n-1}\leqslant V_n$. Thus we complete the assertion of Theorem 2.

Corollary *Let $G(z)=\sum_{n=1}^{\infty}B_nz^n$ be quasi-subordinate to $F(z)=\sum_{n=1}^{\infty}A_nz^n$ in U. Then*

$$|A_1|\leqslant|B_1| \tag{2.33}$$

with equality only if $G(z)=e^{i\beta}F(e^{i\theta}z)$,

$$|B_2|\leqslant\frac{5}{4}\max(|A_1|, |A_2|) \tag{2.34}$$

with equality only if $G(z)=\Phi(z)F(w(z))$ in which $\Phi(z)=e^{i\beta}((z-\alpha)/(1-\bar{\alpha}z))$ $(|\alpha|=1/2)$, $w(z)=e^{i\theta}z$. In particular, $G(z)=((1+2z)/(2+z))F(z)$ with $A_1=A_2$ is an extremal function, and

$$|B_3|\leqslant\frac{7(1+2\sqrt{7})}{27}\max(|A_1|, |A_2|, |A_3|), \tag{2.35}$$

with equality only if

$$G(z) = e^{i\beta} \frac{z-\alpha}{1-\bar{\alpha}z} F(e^{i\theta}z) \quad (|\alpha| = t = (\sqrt{7}-1)/3).$$

In particular, $G(z) = ((z+t)/(1+tz))F(z)$ *with* $-A_1 = A_2 = A_3$ *is an extremal function.*

Remark The inequalities (2. 33) and (2. 34) in the corollary were verified by Robertson [4].

3. Conjecture

In view of the last assertion of Theorem 2, it is enough to study the extremal problem (2. 26) if we can prove that the extremal function of the extremal problem (2. 26) for n has the property of Theorem 2, because we can immediately obtain the solution of the extremal problem (2. 27) for $n+1$.

What can we say on the problem (2. 26)? For this problem, it is reasonable to make the following conjecture.

Conjecture If $g(z) = \sum_{n=0}^{\infty} b_n z^n$ is quasi-subordinate to $f(z) = \sum_{n=0}^{\infty} a_n z^n$, then

$$|b_n| \leqslant U_n \max(|a_0|, |a_1|, \cdots, |a_n|), \tag{3. 1}$$

where

$$|U_n| = \max U_n(t) = \max_{0 \leqslant t < 1} (1 + 2t - t^n - t^{n-1}). \tag{3. 2}$$

The equality in (3. 1) holds only if $g(z) = \phi(z) f(w(z))$ in which $\phi(z) = e^{i\beta}((z-\alpha)/(1-\bar{\alpha}z))$, $w(z) = e^{i\theta}z$, and $t = |\alpha|$ is a maximum point of $U_n(t)$ on [0, 1].

We know that it is true for $n = 1, 2$, but, for $n \geqslant 3$, it is also an open problem.

◇ References ◇

[1] E. Lindelöf, Mémoire sur certaines inégalités dans la théorie des fonctions monogénes et

sur quelques propriétiés nouvelles de ces fonctions dans le voisinage d'un point singulier essentiel, *Acta Soc. Sci. Fenn.* 35 (1909), 1 - 35.

[2] J. E. Littlewood, On inequalities in the theory of functions, *Proc. London Math. Soc.* 23 (1925), 481 - 519.

[3] J. E. Littlewood, *Lectures on the Theory of Functions* (Oxford University Press, 1944).

[4] M. S. Robertson, Quasi-subordination and coefficient conjecture, *Bull. Amer. Math. Soc.* 76 (1970), 1 - 9.

[5] W. Rogosinski, On subordinate functions, *Proc. Cambridge Philos. Soc.* 35 (1939), 1 - 26.

[6] W. Rogosinski, On the coefficients of subordinate functions, *Proc. London Math. Soc.* 48 (1943), 48 - 82.

[7] D. Xia and K. Chang, Some inequalities on subordinate functions, *Acta. Math.* 8 (1958), 408 - 412.

A Dynamical System Formed by a Set of Rational Functions*

Wei-Min Zhou and Fu-Yao Ren

In this paper, we introduce a dynamical system formed by a finite set of rational functions. We discuss the dynamical properties of the Julia sets of these systems.

1. Introduction

The theory of the dynamics of rational functions was created by Fatou ([12], [13]) and Julia [17] independently at the beginning of this century. Under the iterating of a *single* rational function R, the nth iterates of R are defined inductively by $R_{n+1}(z) = R(R_n(z))$, $n \geqslant 1$ and $R_1(z) = R(z)$, the complex plane decomposes into two distinct subsets, one is now called *the stable set or the Fatou set* on which the dynamics are well understood, and the other one is now known as *the Julia set* on which the dynamics are quite chaotic and complicated. Fatou and Julia studied extensively the structures of these sets. During the next fifty years, Borlin [6], Guckenheimer [14], and Jakobson [16] reopened this subject, but the rebirth of this field in 1980's was mainly due to the possibility of computer experiments and the remarkable work of Douady and Hubbard ([7], [8], [9]) and Sullivan ([19] to [22]).

Barnsley et al. ([2], [3], [4]) introduced the Iterated Function Systems (IFS) which are formed by a set of contraction mappings on a complete metric

* Originally published in *Current Topics in Analytic Function Theory*, Ed. H. M. Srivastava & S. Owa, World Scientific, Singapore, 1992, 437 - 449.

space (X, d). They used different IFS to produce computer graphic images of attractors of IFS to model things in the real world, such as clouds, smoke, chimneys, leaves, and ferns.

In this paper, we introduce the dynamical system formed by a *set* of rational functions. It may be considered as a generalization of both the work on classical dynamics of analytic functions and the work on IFS. We discuss the dynamical properties of the Julia sets of such systems which are more complicated. The Julia sets of these systems will not have some properties that the classical Julia sets have; for example, they are not completely invariant in general, and the Julia set of the system formed by two or more rational functions may contain interior points, but does not equal to the whole complex plane. However, as we show, they share many intrinsic properties of the classical Julia sets, and also have properties that the attractors of IFS have. Especially, we show that the Julia set of the systems equal to the closure of the set of repelling periodic points.

2. Definitions and Some Classical Results

Let $R=\{R_1, R_2, \cdots, R_M\}$ be a set of rational functions, and put

$$\Sigma_M=\{(j_1, j_2, \cdots, j_n, \cdots) \mid j_i \in \{1, 2, \cdots, M\}, i=1, 2, \cdots, n, \cdots\}.$$

We define the iteration sequence $\{W_\sigma^n\}$ of R for $\sigma=(j_1, j_2, \cdots, j_n, \cdots) \in \Sigma_M$ in the following way:

$$W_\sigma^1(z)=R_{j_1}(z),$$
$$W_\sigma^2(z)=R_{j_2}\circ R_{j_1}(z),$$
$$\cdots\cdots$$
$$W_\sigma^{n+1}(z)=R_{j_{n+1}}\circ W_\sigma^n(z)=R_{j_{n+1}}\circ R_{j_n}\circ\cdots\circ R_{j_1}(z),$$
$$\cdots\cdots$$

and define the inverse sequence $\{W_\sigma^{-n}\}$ as follows:

$$W_\sigma^{-n}(z)=(W_\sigma^n)^{-1}(z)=R_{j_1}^{-1}\circ R_{j_2}^{-1}\circ\cdots\circ R_{j_n}^{-1}(z)$$

for $n \geqslant 1$.

Under the iteration of $\{W_\sigma^n\}$ for all $\sigma \in \Sigma_M$ defined as above, the Riemann sphere $\overline{\mathbb{C}}$ will be decomposed into two sets; as in the classical dynamics, one of these sets will be called the Julia set and the other set will be called the Fatou set.

Definition 1 A point $z_0 \in \overline{\mathbb{C}}$ is a *stable point*, if there exists a neighborhood U of z_0, such that for all $\sigma \in \Sigma_M$, $W_\sigma^n(z)$ is well defined on U for all $n \geqslant 1$, and the iteration sequence $\{W_\sigma^n(z)|U\}$ forms a normal family in the sense of Montel. The set of stable points is called *the Fatou set*, and its complement, the set of unstable points, is called *the Julia set*, denoted by $F(R)$ and $J(R)$ respectively. The Fatou set and the Julia set of $R_i \in R$ will be called *the classical Fatou set* and *the classical Julia set* and denoted by $F(R_i)$ and $J(R_i)$, respectively.

There are several good surveys on the classical Julia set and Fatou set; for example, see the surveys of Blanchard [5], Brolin [6], Douady [10], Eremenko and Lyubich [11], and Milnor [18].

Definition 2 Given $z \in \overline{\mathbb{C}}$, z is called a *periodic point* of system $R = \{R_1, R_2, \cdots, R_M\}$, if there exist $\sigma \in \Sigma_M$ and integer n such that $W_\sigma^n(z) = z$, and $\lambda = (W_\sigma^n)'(z)$ is called the *multiplier* of z. A periodic point z is called *repelling* if its multiplier satisfies $|\lambda| > 1$.

The following theorem due to Montel will play a very important role as it does in the classical dynamics.

Montel's Theorem *Let $\mathscr{F}$ be a family of meromorphic functions defined on a domain U. Suppose there exist points a, b, c in $\overline{\mathbb{C}}$ such that $[\bigcup_{f\in\mathscr{F}} f(U)] \cap \{a, b, c\} = \varnothing$. Then $\mathscr{F}$ is a normal family on U.*

The following result follows directly from Montel's theorem.

Corollary 1 *Let $R = \{R_1, R_2, \cdots, R_M\}$ be a system of rational functions, and let $J(R)$ be the Julia set of the system. If $z \in J(R)$ and U is a neighborhood of z, then the set*

$$E_U = \overline{\mathbb{C}} - \bigcup_{\sigma\in\Sigma_M} \bigcup_{n>0} W_\sigma^n(U)$$

contains at most two points.

Definition 3 The set

$$E_z = \bigcup_U E_U$$

is called *the exceptional point set* of z, where the union is taken over all neighborhoods U of the point z.

It is obvious that E_U is independent of U, provided that U is sufficiently small, and by Montel's theorem it contains at most two points; therefore, E_z contains at most two points, E_z is often empty.

Example 1 If $R = \{p_1, p_2, \cdots, p_M\}$ is a system of polynomials, then the point at ∞ is the exceptional point.

Marty's Theorem ([1], [15]) *A class $\mathscr{F}$ of functions $f(z)$ meromorphic in a domain D of the complex plane is normal in D, if and only if $|f'(z)|/(1+|f(z)|^2)$ is uniformly bounded on any compact subset of D for $f \in \mathscr{F}$.*

3. Main Results

The classical Julia set and the classical Fatou set are completely invariant [6], but the Julia set $J(R)$ and the Fatou set $F(R)$ are not completely invariant in general as we will discuss later on.

Let $R = \{R_1, R_2, \cdots, R_M\}$ be a system of rational functions. Then we have the following proposition.

Proposition 1 *The Fatou set $F(R)$ is a forward invariant open set; that is, if z belongs to $F(R)$, then $R_i(z)$ is contained in $F(R)$ for every $i \in \{1, 2, \cdots, M\}$. The Julia set $J(R)$ is a backward invariant closed set; that is, if z belongs to $J(R)$, then $R_i^{-1}(z)$ is contained in $J(R)$ for every $i \in \{1, 2, \cdots, M\}$.*

Proof The openness of $F(R)$ follows directly from the definition. To show that $F(R)$ is forward invariant, it suffices to show that, if $\xi \in F(R)$, then $R_i(\xi) \in F(R)$ for $i = 1, 2, \cdots, M$. Let $\xi \in F(R)$; there is a neighborhood U of ξ such that for any $\sigma = (j_1, j_2, \cdots, j_n, \cdots) \in \Sigma_M$, $\{W_\sigma^n(z) \mid U\}$ is normal. Therefore, $\{W_\sigma^n | U\}$ is normal for $\tau = (i, \sigma) = (i, j_1, j_2, \cdots, j_n, \cdots) \in \Sigma_M$, where σ goes over Σ_M. Hence $\{W_\sigma^n | R_i(U)\}$ is normal for all σ

$\in \Sigma_M$. Since R_i is a nonconstant holomorphic function, $R_i(U)$ is a neighborhood of $R_i(\xi)$. Therefore, $R_i(\xi) \in F(R)$.

Since $J(R)$ is the complement of $F(R)$, it is closed. Let $\xi \in J(R)$; then $\xi \notin F(R)$. Therefore, for every $R_i \in R$, $i=1, 2, \cdots, M$, $R_i^{-1}(\xi) \notin F(R)$, otherwise $\xi = R_i(R_i^{-1}(\xi)) \in F(R)$, which is a contradiction. It follows that $J(R)$ is backward invariant.

Theorem 1 *Let E_z be the set of exceptional points for $z \in J(R)$.*

(a) *If E_z contains only one point, then each $R_i(z)$ is conjugate to a polynomial by the same Möbius transformation, where $i \in \{1, 2, \cdots, M\}$.*

(b) *If E_z contains two points, then each $R_i(z)$ is conjugate to the map $z \mapsto K_i z^{\pm d_i}$ by the same Möbius transformation, where $d_i = \deg(R_i)$, $i \in \{1, 2, \cdots, M\}$, and K_i are constants. The sign is $+$ if E_z contains fixed points of R_i and $-$ if E_z consists of a periodic orbit of period two of R_i. In both cases, E_z does not depend on the choice of $z \in J(R)$.*

Moreover, all exceptional points are contained in the Fatou set.

Proof First we prove that $R_i^{-1}(E_z) = E_z$ for every $i \in \{1, 2, \cdots, M\}$. If $\xi \in R_i^{-1}(E_z)$, then there is some point $\eta \in E_z$ with $R_i(\xi) = \eta$. If ξ does not belong to E_z, then there exists a neighborhood U of z such that $\xi \in \bigcup_{\sigma \in \Sigma_M} \bigcup_{n \geq 0} W_\sigma^n(U)$. Hence there are some n_0 and $\tau \in \Sigma_M$ such that $\xi \in W_\tau^{n_0}(U)$. It follows that $\eta = R_i(\xi) \in R_i \circ W_\tau^{n_0}(U)$. This contradicts the fact that $\eta \in E_z$. It follows that $R_i^{-1}(E_z) \subset E_z$ for every R_i, $i \in \{1, 2, \cdots, M\}$. Since E_z contains at most two points and R_i^{-1} could not map two distinct points to one point, it follows that $E_z = R_i^{-1}(E_z)$ for every R_i, $i \in \{1, 2, \cdots, M\}$.

To prove (a), we use a Möbius transformation M which maps the point in E_z to the point at infinity. Then $p_i(z) = M \circ R_i \circ M^{-1}(z)$ has no poles in $\mathbb{C}$. Therefore, each p_i is a polynomial for $i \in \{1, 2, \cdots, M\}$.

To prove (b), we use a Möbius transformation M which moves one point of E_z to ∞ and the other point to 0. In this case the conjugate map $p_i(z) = M \circ R_i \circ M^{-1}(z)$ can be one of two forms, where $i \in \{1, 2, \cdots, M\}$. Suppose that both 0 and ∞ are fixed by $p_i(z)$. Then the above arguments show that $p_i(z)$ is a polynomial. Moreover, 0 will be a zero of multiplicity d_i because no other points can be mapped to 0 under $p_i(z)$. Hence $p_i(z)=K_i z^{+d_i}$, where K_i are constants and $d_i=\deg(R_i)$, $i=1, 2, \cdots, M$. If 0 and ∞ form an orbit of period two for $p_i(z)$, similar arguments show that $p_i(z)=K_i z^{-d_i}$, where K_i are constants and $d_i=\deg(R_i)$, $i=1, 2, \cdots, M$.

In case (a), the point in E_z is a common superattractive fixed point for every $R_i(z)$, it is easy to find a common neighborhood U of E_z, such that $R_i(U)\subset U$, for every R_i. Therefore, for every $\sigma\in\Sigma_M$, $\{W_\sigma^n(U)\}$ omits at least three points. By Montel's theorem, $E_z\subset F(R)$.

In case (b), E_z consists of either two superattractive fixed points or a superattractive orbit of period two of each R_i. Hence we can also find some neighborhoods U_1 and U_2 of two points of E_z, such that for each R_i, either $R_i(U_1)\subset U_1$ and $R_i(U_2)\subset U_2$ or $R_i(U_1)\subset U_2$ and $R_i(U_2)\subset U_1$. Therefore, for every $\sigma\in\Sigma_M$, both $\{W_\sigma^n(U_1)\}$ and $\{W_\sigma^n(U_2)\}$ omit at least three points. Thus, $E_z\subset F(R)$.

Finally, if $E_z\neq\varnothing$ for any z, we know that R_i is conjugate to either (a) or (b). In case (a) we take $U=\overline{\mathbb{C}}-\{0\}$. In case (b) we take $U=\overline{\mathbb{C}}-\{0,\infty\}$. Then $E_U=E_z$ for all $z\in J(R)$. Consequently, E_z is independent of z.

Since E_z is independent of z, we will denote the set of exceptional points by $E(R)$.

Theorem 2 *If $z\in\overline{\mathbb{C}}-E(R)$, then the Julia set is contained in the set of accumulation points of the full backward orbits of z for every $\sigma\in\Sigma_M$. That is,*

$$J(R)\subset\{\text{accumulation points of }[\bigcup_{\sigma\in\Sigma_M}\bigcup_{n\geqslant 0}W_\sigma^{-n}(z)]\}.$$

Proof Since $z\in\overline{\mathbb{C}}-E(R)$, we have, for every $\xi\in J(R)$ and every neighborhood U of ξ, $\bigcup_{\sigma\in\Sigma_M}\bigcup_{n\geqslant 0}W_\sigma^n(U)\supset\{z\}$. Thus there are some $\tau\in\Sigma_M$ and $n_0>0$ such that $W_\tau^{n_0}(U)\supset\{z\}$. It follows that there exists a point $\eta\in U$ with $W_\tau^{n_0}(\eta)=z$. Hence $\eta\in\{W_\tau^{-n_0}(z)\}$. Since U may be arbitrarily small, ξ is an accumulation point of $\{\bigcup_{\sigma\in\Sigma_M}\bigcup_{n\geqslant 0}W_\sigma^{-n}(z)\}$.

Proposition 2 *If $z\in J(R)$, then*

$$J(R)=\text{closure}\left\{\bigcup_{\sigma\in\Sigma_M}\bigcup_{n\geq 0}W_\sigma^{-n}(z)\right\}.$$

Proof Proposition 2 follows from Theorem 2 and the backward invariant property of the Julia set $J(R)$.

There are some connections between $J(R)$ and the classical Julia sets $J(R_i)$ of R_i.

Proposition 3 *Let $J(R_i)$ be the Julia set of R_i, where $i=1, 2, \cdots, M$. Then*

$$J(R)=\text{closure}\left\{\bigcup_{\sigma\in\Sigma_M}\bigcup_{n\geq 0}W_\sigma^{-n}\left(\bigcup_{i=1}^{M}J(R_i)\right)\right\}.$$

Therefore

$$J(R)\supset\bigcup_{i=1}^{M}J(R_i).$$

Proof It is obvious that

$$J(R)\supset\bigcup_{i=1}^{M}J(R_i)$$

by the definition of the Julia set. Since $J(R)$ is backward invariant by Proposition 1, we have

$$J(R)\supset\left\{\bigcup_{\sigma\in\Sigma_M}\bigcup_{n\geq 0}W_\sigma^{-n}\left(\bigcup_{i=1}^{M}J(R_i)\right)\right\}.$$

Since $J(R)$ is closed, we have

$$J(R)\supset\text{closure}\left\{\bigcup_{\sigma\in\Sigma_M}\bigcup_{n\geq 0}W_\sigma^{-n}\left(\bigcup_{i=1}^{M}J(R_i)\right)\right\}.$$

The opposite inclusion follows from Theorem 2.

By Proposition 2 and Proposition 3, we can see that the repelling periodic points of R_i, which belong to the classical Julia set $J(R_i)$ according to the classical dynamics theory, also belong to the Julia set $J(R)$. So one can generate the picture of the Julia set $J(R)$ by finding a repelling periodic point of some R_i and then calculating its inverse orbit. Proposition 3 also says that the classical Julia sets $J(R_i)$ can be regarded as generators of the Julia set $J(R)$.

Theorem 3 *The Julia set $J(R)$ is a nonempty perfect set.*

Proof Since $J(R) \supset J(R_i)$, by Proposition 3, $J(R)$ is nonempty. Now suppose on the contrary that the Julia set $J(R)$ is not perfect; then there exist a point $\xi \in J(R)$ and a neighborhood U of ξ such that $U \cap J(R) = \{\xi\}$. Since $\xi \in J(R)$, there exists some $\sigma_0 \in \Sigma_M$ such that $W^n_{\sigma_0}(z)$ is not normal on U. Therefore, by Montel's theorem, $\bigcup_{n\geqslant 1} W^n_{\sigma_0}(U) \supset \overline{\mathbb{C}} \setminus \{\text{at most two points}\}$.

Consequently, we have $\bigcup_{n\geqslant 1} W^n_{\sigma_0}(U) \supset J(R) \setminus \{\text{at most two points}\}$. Since the Fatou set is forward invariant, therefore the set $\bigcup_{n\geqslant 1} W^n_{\sigma_0}(U)$ contains only countably many points by the assumption that $U \cap J(R) = \{\xi\}$. Consequently $J(R)$ is also a countable set. This contradicts the fact that $J(R) \supset J(R_i)$ and $J(R_i)$ is a perfect set.

Theorem 4 *Let $R = \{R_1, R_2, \cdots, R_M\}$. Then*

$$J(R) = \bigcup_{i=1}^{M} R_i^{-1}(J(R)).$$

Proof By Propositon 1,

$$J(R) \supset \bigcup_{i=1}^{M} R_i^{-1}(J(R)).$$

Suppose that $\xi \in J(R)$ and $\xi \notin \bigcup_{i=1}^{M} R_i^{-1}(J(R))$; then we have $\xi \in \bigcap_{i=1}^{M} R_i^{-1}(F(R))$. It is easy to verify that $\bigcap_{i=1}^{M} R_i^{-1}(F(R)) \subset F(R)$.

Theorem 5 *All repelling periodic points belong to the Julia set.*

Proof Let ξ be a repelling periodic point. Then there are some $\sigma = (j_1, j_2, \cdots, j_n, \cdots) \in \Sigma_M$ and a $k > 0$ such that $W^k_\sigma(\xi) = \xi$. Let $\varphi(x) = R_{j_k} \circ R_{j_{k-1}} \circ \cdots \circ R_{j_1}(x)$; then ξ is a repelling fixed point of $\varphi(x)$. Therefore, ξ belongs to the classical Julia set $J(\varphi)$ of φ. If $\xi \notin J(R)$, then there is a neighborhood U of ξ such that for any $\sigma \in \Sigma_M$, $\{W^n_\sigma \mid U\}$ is normal. Take $\tau = \overline{(j_1, j_2, \cdots, j_k)} = (j_1, j_2, \cdots, j_k, j_1, j_2, \cdots, j_k, j_1, j_2, \cdots)$, then $\{W^n_\tau \mid U\}$ is normal. Hence $\xi \in F(\varphi)$, the classical Fatou set of φ, which is a contradiction.

Corollary 2 *If z is a repelling periodic point, then*

$$J(R) = \text{closure}\left\{ \bigcup_{\sigma\in\Sigma_M} \bigcup_{n\geqslant 0} W^{-n}_\sigma(z) \right\}.$$

Proof Corollary 2 follows directly from Theorem 5 and Proposition 2.

For any rational function R, the classical Julia set $J(R)$ contains no interior points if it is not the whole complex plane $\overline{\mathbb{C}}$. However, this property does not hold for $J(R)$.

Example 2 Let $R_1(z) = 2z^2$ and $R_2(z) = z^2$. Then $J(R) = \left\{z \in \overline{\mathbb{C}} \,\middle|\, \frac{1}{2} < |z| < 1\right\}$, and $R_i(J(R)) \not\subset J(R)$, $i=1, 2$.

Proof Since $J(R_1)=\left\{z \in \overline{\mathbb{C}} \,\middle|\, |z| = \frac{1}{2}\right\}$, $J(R_2)=\{z \in \overline{\mathbb{C}} \mid |z|=1\}$. By Proposition 3, $J(R) \supset J(R_1) \cup J(R_2) = \left\{z \in \overline{\mathbb{C}} \,\middle|\, |z| = \frac{1}{2}\right\} \cup \{z \in \overline{\mathbb{C}} \mid |z| = 1\}$. For every $z \in \left\{z \in \overline{\mathbb{C}} \,\middle|\, \frac{1}{2} < |z| < 1\right\}$, let U be any neighborhood of z. Then $R_1^n(U) \to \{\infty\}$ and $R_2^n(U) \to \{0\}$. Therefore, we can choose a sequence $\{j_1, j_2, \cdots, j_k\}$ where $j_i \in \{1, 2\}$ for $i=1, 2, \cdots, k$, such that $R_{j_k} \circ R_{j_{k-1}} \circ \cdots \circ R_{j_1}(U) \cap \{z \in \overline{\mathbb{C}} \mid |z|=1\} \neq \varnothing$. Since U can be arbitrarily chosen, it follows that z is an accumulation point of inverse images of $\{z \in \overline{\mathbb{C}} \mid |z|=1\}$. Since $J(R)$ is backward invariant and closed, we have $z \in J(R)$.

Therefore, we have the following proposition.

Proposition 4 *If $R = \{R_1, R_2, \cdots, R_M\}$, then $J(R)$ may contain interior points and does not equal to the whole complex plane $\overline{\mathbb{C}}$ for $M \geqslant 2$.*

Theorem 6 *The Julia set $J(R)$ is contained in the closure of the set of periodic points.*

Proof Let

$$K = J(R) \setminus \{\infty, \text{ the critical points of } R_i^2, \text{ the poles of } R_i^2, \ i=1, 2, \cdots, M\};$$

then $J(R) = \overline{K}$. Suppose that $\xi \in K$ with $\xi \notin \overline{\{\text{periodic points}\}}$. Therefore, there is a neighborhood U of ξ containing no periodic points of W_σ^n for every $\sigma \in \Sigma_M$ and every $n > 0$. We may choose the inverse functions $\varphi_1(z)$, $\varphi_2(z)$ and $\varphi_3(z)$ of some $R_i^2(z)$, such that they are distinct.

For every $\sigma \in \Sigma_M$, we define

$$g_\sigma^n(z) = \frac{W_\sigma^n(z) - \varphi_1}{W_\sigma^n(z) - \varphi_2} \cdot \frac{\varphi_3 - \varphi_2}{\varphi_3 - \varphi_1}.$$

Then

$$W_\sigma^n(z)=\varphi_2+Q\frac{\varphi_2-\varphi_1}{g_\sigma^n(z)-Q}, \text{ where } Q=\frac{\varphi_3-\varphi_2}{\varphi_3-\varphi_1}.$$

It is clear that $\{W_\sigma^n(z)|U\}$ and $\{g_\sigma^n(z)|U\}$ are normal simultaneously. Since $g_\sigma^n(z)\neq 0, 1, \infty$, we conclude that $\{g_\sigma^n(z)|U\}$ is normal for every $\sigma\in\Sigma_M$. Therefore, $\{W_\sigma^n(z)|U\}$ is also normal, which is a contradiction. This completes the proof.

Ahlfors' Five Domains Theorem [15] *Let E_1, E_2, E_3, E_4 and E_5 be bounded simplyconnected domains which are bounded by sectional analytic Jordan curves with disjoint closures, and let $r>0$. Then there exists a constant C which depends only on the regions E_i and r with the following property:*

For every function f meromorphic in the disk U_r satisfying

$$\frac{|f'(0)|}{(1+|f(0)|^2)}>C,$$

there exists a domain $D\subset U_r$ and with closure mapped univalently by f onto one of the domain E_k.

By this theorem and Marty's theorem, we can now prove the following theorem.

Theorem 7 *The Julia set $J(R)$ is the closure of the set of repelling periodic points.*

Proof By Theorem 5 and the fact that $J(R)$ is backward invariant, it suffices to show that, if $\xi\in J(R)$, then every neighborhood U_0 of ξ contains a repelling periodic point of R.

Since $J(R)$ is perfect, we can find five distinct points $z_j\in J(R)\cap U_0$, $j=1, 2, 3, 4, 5$. Take disjoint disks $E_j=\{z \mid |z-z_j|<2r\}\subset U_0$. Since $z_j\in J(R)$, for $B(z_j, r)=\{z \mid |z-z_j|<r\}$, there exist some $\tau_j\in\Sigma_M$, such that $\{W_{\tau_j}^n|B(z_j, r)\}$ is not normal for $j=1, 2, 3, 4, 5$, respectively. By Marty's theorem, there are points $z_j'\in B(z_j, r)$ and integers m_j for $j=1, 2, 3, 4, 5$, such that

$$\frac{|(W_{\tau_j}^{m_j})'(z_j')|}{(1+|W_{\tau_j}^{m_j}(z_j')|^2)}>C,$$

where C is the constant in Ahlfors' five domains theorem. Let $g_j(t)=t+z_j'$ be an affine map which maps the disk U_r to $B(z_j', r)$ and maps 0 to z_j', then $W_{\tau_j}^{m_j}\circ g_j(t)$ is analytic in the disk U_r and satisfies the condition of Ahlfors' five domains theorem. By this theorem, for every j there exists a domain $\widetilde{D}_j\subset U_r$ which is mapped univalently by $W_{\tau_j}^{m_j}\circ g_j(t)$ onto one of the disks E_1, E_2, E_3, E_4, or E_5. Therefore, $W_{\tau_j}^{m_j}(z)$ maps the domain $D_j=g_j(\widetilde{D}_j)$ which is contained in $B(z_j', r)$ univalently onto one of disks E_1, E_2, E_3, E_4, or E_5. There are only the following five possibilities:

(i) $W_{\tau_i}^{m_i}(D_i)=E_i$, for some integer $i\in\{1, 2, 3, 4, 5\}$;

(ii) $W_{\tau_j}^{m_j}(D_j)=E_i$, and $W_{\tau_i}^{m_i}(D_i)=E_j$, for some distinct integers $i, j\in\{1, 2, 3, 4, 5\}$;

(iii) $W_{\tau_i}^{m_i}(D_i)=E_j$, $W_{\tau_j}^{m_j}(D_j)=E_k$ and $W_{\tau_k}^{m_k}(D_k)=E_i$, where $i, j, k\in\{1, 2, 3, 4, 5\}$ are distinct integers;

(iv) $W_{\tau_i}^{m_i}(D_i)=E_j$, $W_{\tau_j}^{m_j}(D_j)=E_k$, $W_{\tau_k}^{m_k}(D_k)=E_l$, and $W_{\tau_l}^{m_l}(D_l)=E_i$, where $i, j, k, l\in\{1, 2, 3, 4, 5\}$ are distinct integers;

(v) $W_{\tau_i}^{m_i}(D_i)=E_j$, $W_{\tau_j}^{m_j}(D_j)=E_k$, $W_{\tau_k}^{m_k}(D_k)=E_l$, $W_{\tau_l}^{m_l}(D_l)=E_n$, and $W_{\tau_n}^{m_n}(D_n)=E_i$, where $i, j, k, l, n\in\{1, 2, 3, 4, 5\}$ are distinct integers.

For case (i), the inverse map $\varphi(z)=(W_{\tau_i}^{m_i})^{-1}(z)$: $E_i\rightarrow D_i$ is a conformal contraction. Therefore, $\varphi(z)$ has an attracting fixed point in E_i, i. e., E_i contains a repelling fixed point of $W_{\tau_i}^{m_i}(z)$.

For case (ii), since $W_{\tau_i}^{m_i}(z)$: $D_i\rightarrow E_j$ is univalent, let $\Omega=(W_{\tau_i}^{m_i})^{-1}(D_j)$. Then $\Omega\subset D_i$ and $W_{\tau_j}^{m_j}\circ W_{\tau_i}^{m_i}(z)$: $\Omega\rightarrow E_i$ is univalent. It follows from the same argument as in case (i) that E_i contains a repelling fixed point of $W_{\tau_j}^{m_j}\circ W_{\tau_i}^{m_i}(z)$. Assume that $\tau_i=(i_1, i_2, \cdots, i_{m_i}, \cdots)$ and $\tau_j=(j_1, j_2, \cdots, j_{m_j}, \cdots)$, and let $\tau=(i_1, i_2, \cdots, i_{m_i}, j_1, j_2, \cdots, j_{m_j}, \cdots)\in\Sigma_M$, then we have $W_\tau^{m_i+m_j}(z)=W_{\tau_j}^{m_j}\circ W_{\tau_i}^{m_i}(z)$. Thus E_i contains a repelling fixed point of $W_\tau^{m_i+m_j}(z)$.

By the above argument, it is not hard to show that for cases (iii), (iv) and (v), there is some E_i which contains a repelling fixed point of $W_\tau^n(z)$ for some $\tau \in \Sigma_M$ and some $n \geqslant 0$.

Therefore, U_0 contains a repelling periodic point of some $W_\tau^n(z)$. Hence the Julia set is the closure of the set of repelling periodic points.

Acknowledgements

We would like to express our thanks to the Editor, Professor H. M. Srivastava, for having spent a lot of his invaluable time in preparing this thoroughly revised version of our submissions for publication.

This work was supported by the Qimingxing Project of Shanghai and the National Natural Science Foundation of China.

◇ References ◇

[1] L. V. Ahlfors, *Complex Analysis*. McGraw-Hill, New York, 1979.

[2] M. F. Barnsley, Fractal modeling of real world images, in *The Science of Fractal Images* (H.-O. Peitgen and D. Sauper, Editors), Springer-Verlag, New York, 1988.

[3] M. F. Barnsley, Leture notes on iteration function systems, *Proc. Symp. Appl. Math.* **39** (1989), 127 - 144.

[4] M. F. Barnsley and S. Demko, Iterated function systems and the global construction of fractals, *Proc. Roy. Soc. London Ser A* **399** (1985), 243 - 275.

[5] P. Blanchard, Complex analytic dynamics on the Riemann sphere, *Bull. Amer. Math. Soc.* (N. S.) **11** (1984), 85 - 141.

[6] H. Brolin, Invariant sets under iteration of rational functions, *Ark. Mat.* **6** (1965), 103 - 144.

[7] A. Douady and J. H. Hubbard, Itération des polynômes quadratiques complexes, *C. R. Acad. Sci. Paris Sér. I. Math.* **294** (1982), 123 - 126.

[8] A. Douady and J. H. Hubbard, Étude dynamique des polynômes complexes Part I, *Publ. Math. Orsay* **84 - 02** (1984).

[9] A. Douady and J. H. Hubbard, Étude dynamique des polynômes complexes Part II, *Publ. Math. Orsay* **85 - 02** (1985).

[10] A. Douady, Systèmes dynamiques holomorphes, *Astérisque* **105 - 106** (1983), 39 - 63.

[11] A. É. Eremenko and M. Yu. Lyubich, The dynamics of analytic transformations, *Leningrad Math. J.* **1** (1990), 563 - 634.

[12] P. Fatou, Sur les équations fonctionnelles, *Bull. Soc. Math. France* **47** (1919), 161 -

271.

[13] P. Fatou, Sur les équations fonctionnelles, *Bull. Soc. Math. France* **48** (1920), 33-94 and 208-314.

[14] J. Guckenheimer, Endomorphisms of the Riemann sphere, in *Global Analysis*, *Proc. Symp. Pure Math.* **48** (1970), 95-124.

[15] W. K. Hayman, *Meromorphic Functions*, Clarendon Press, Oxford, 1964.

[16] M. V. Jakobson, On the problem of the classification of polynomial endomorphisms of the plane, *Mat. Sb.* **80** (1969), 365-387; English transl in *Math. USSR Sb.* **9** (1969), 345-364.

[17] G. Julia, Mémoire sur l'iteration des fonctions rationelles, *J. Math.* **8** (1918), 47-245.

[18] J. Milnor, Dynamics in one complex variable: introductory lectures, Institute for Mathematical Science, SUNY at Stony Brook, Preprint No. 1990/5.

[19] D. Sullivan, Itération des fonctions analytiques complexes, *C. R. Acad. Sci. Paris Sér. I Math.* **294** (1982), 301-303.

[20] D. Sullivan, Conformal dynamical systems, in *Geometric Dynamics*, Lecture Notes in Mathematics **1007** (1983), 725-752.

[21] D. Sullivan, Quasiconformal homeomorphisms and dynamics. I: Solution of the Fatou-Julia problem on wandering domains, *Ann. of Math.* **122** (1985), 401-418.

[22] D. Sullivan, Quasiconformal homeomorphisms and dynamics. III: Topological conjugacy classes of analytic endomorphisms. Preprint 1983.

Wei-Min Zhou
Department of Fundamental Science
Shanghai Maritime University
Shanghai 200135
People's Republic of China

Fu-Yao Ren
Department of Mathematics
Fudan University
Shanghai 200433
People's Republic of China

超越函数随机迭代系统的 Julia 集*

周维民　任福尧

（复旦大学数学系）

关键词：迭代，Julia 集，超越整函数，超越亚纯函数

单个超越整函数自身迭代生成的动力系统的研究始于 1926 年 Fatou 的工作[1]，后来主要是 Baker 等人继续了这方面的研究[2-4]，但近年来，这一领域又得到了飞速发展[5-7]. 在本文中，我们将研究由有限多个超越整函数和超越亚纯函数生成的随机迭代系统，可以说这一工作既是 Fatou 等人的工作的推广，又是 Barnsley 等人在迭代函数系统方面工作的推广[8, 9]. 由于在动力系统的研究中，最基本的对象就是 Julia 集，所以我们首先研究了随机迭代系统的 Julia 集，下面就是我们在这方面得到的主要结果.

设 $\mathscr{F}=\{f_1, f_2, \cdots, f_M\}$ 是由有限多个超越整函数和超越亚纯函数所组成的集合，记 $\Sigma_M=\prod_{k=1}^{\infty}\{1, 2, \cdots, M\}$.

对于 Σ_M 中任一元素 $\sigma=(j_1, j_2, \cdots, j_n, \cdots)$，我们定义 $\mathscr{F}$ 关于 σ 的随机迭代序列 $\{W_\sigma^n\}$ 如下：

$$W_\sigma^1(z)=f_{j_1}(z),$$

$$W_\sigma^2(z)=f_{j_2}\circ f_{j_1}(z),$$

$$\cdots\cdots$$

$$W_\sigma^{n+1}(z)=f_{j_{n+1}}\circ W_\sigma^n(z)=f_{j_{n+1}}\circ f_{j_n}\circ\cdots\circ f_{j_1}(z),$$

$$\cdots\cdots$$

对于 $n\geqslant 1$，我们定义 $\{W_\sigma^{-n}\}$ 如下：

---

* 国家自然科学基金资助项目. 原载《科学通报》，1993 年，第 38 卷，第 4 期，289 - 290.

$$W_\sigma^{-n}(z)=(W_\sigma^n)^{-1}(z)=f_{j_1}^{-1}\circ f_{j_2}^{-1}\circ\cdots f_{j_n}^{-1}(z).$$

定义 1 复平面$\mathbb{C}$上一点 z_0 称为是$\mathscr{F}$的一个正规点，如果存在 z_0 的一个邻域U，使得对任何 $\sigma\in\Sigma_M$ 及任何 $n\geqslant 1$，$W_\sigma^n(z)$在U上均有定义，且迭代序列$\{W_\sigma^n(z)|U\}$是一个 Montel 意义下的正规族.

由全体正规点组成的集合称为 Fatou 集，记为 $F(\mathscr{F})$，它的全集称为 Julia 集，记为 $J(\mathscr{F})$.

定义 2 复平面$\overline{\mathbb{C}}$ 上一点 z_0 称为是系统 $\mathscr{F}$ 的例外点，如果对任何 $f_i\in\mathscr{F}$，$i=1, 2, \cdots, M$，z_0 都是 f_i 的 Picard 例外值.

定义 3 设 $z_0\in\overline{\mathbb{C}}$，$z_0$ 称为是系统 $\mathscr{F}=\{f_1, f_2, \cdots, f_M\}$ 的一个周期点，如果存在 $\sigma\in\Sigma_M$ 以及正整数 n 使得 $W_\sigma^n(z_0)=z_0$. 我们称 $\lambda=(W_\sigma^n)'(z_0)$ 为 z_0 的特征值. 周期点 z_0 称为是排斥的、中性的、吸引的，如果它的特征值分别满足 $|\lambda|>1$，$|\lambda|=1$ 或 $|\lambda|<1$.

命题 1 Fatou 集 $F(\mathscr{F})$是前向不变的开集，而 Julia 集 $J(\mathscr{F})$是后向不变的闭集.

定理 1 若 z_0 是一个非例外点，则

$$J(\mathscr{F})\subset\{z\mid z\in\overline{\mathbb{C}},\ z\text{ 为}[\bigcup_{\sigma\in\Sigma_M}\bigcup_{n\geqslant 1}W_\sigma^{-n}(z_0)]\text{的凝聚点}\}.$$

推论 1 若 $z_0\in J(\mathscr{F})$，且 z_0 不是 $\mathscr{F}$ 的例外点，则

$$J(\mathscr{F})=\text{closure}\left\{\bigcup_{\sigma\in\Sigma_M}\bigcup_{n\geqslant 0}W_\sigma^{-n}(z)\right\}.$$

推论 2 若 p 是 $f_{j_0}\in\mathscr{F}$ 的一个极点，且 p 不是 $\mathscr{F}$ 的例外点，则

$$J(\mathscr{F})=\text{closure}\left\{\bigcup_{\sigma\in\Sigma_M}\bigcup_{n\geqslant 0}W_\sigma^{-n}(p)\right\}.$$

命题 2 设 $J_i=J(f_i)$ 是 f_i 的 Julia 集，其中 $f_i\in\mathscr{F}$，$i=1, 2, \cdots, M$，则

$$J(\mathscr{F})=\text{closure}\{\bigcup_{\sigma\in\Sigma_M}\bigcup_{n\geqslant 0}W_\sigma^{-n}(\bigcup_{i=1}^{M}J_i)\},$$

因而有

$$J(\mathscr{F})\supset\bigcup_{i=1}^{M}J_i.$$

命题 3 Julia 集 $J(\mathscr{F})$是一个非空完全集.

定理 2 设 $\mathscr{F}=\{f_1, f_2, \cdots, f_M\}$，则

$$J(\mathscr{F})=\bigcup_{i=1}^{M} f_i^{-1}(J(\mathscr{F})).$$

定理 3 所有排斥周期点均属于 Julia 集.

推论 3 若 z_0 是 $\mathscr{F}$ 的一个排斥周期点，而且 z_0 不是 $\mathscr{F}$ 的例外点，则

$$J(\mathscr{F})=\text{closure}\{\bigcup_{\sigma\in\Sigma_M}\bigcup_{n\geqslant 0} W_\sigma^{-n}(z_0)\}.$$

定理 4 Julia 集 $J(\mathscr{F})$ 等于 $\mathscr{F}$ 的排斥周期点集的闭包.

注记 若 $\mathscr{F}=\{f\}$，则当 f 是超越整函数时，Baker[2] 得到了定理 4 的结果；当 f 是超越亚纯函数，且 f 至少有一个非 Picard 例外值的极点时，吕以辇①得到了上述结果。

◇参考文献◇

[1] Fatou, P., *Acta Math.*, 1926, 47: 337 - 370.

[2] Baker, I. N., *Math Z.*, 1963, 81: 296 - 214; 1968, 104: 252 - 265.

[3] Baker, I. N., *Proc. London Math. Soc.*, 1984, 49: 563 - 576.

[4] Baker, I. N., Rippon, P. J., *Ann. Acad. Sci. Fenn. Ser. A I Math.*, 1983, 8: 179 -186; 1984, 9: 49 - 77.

[5] Devaney, R. L., Krych, M., *Erg. Th. & Dyn Sys.*, 1984, 4: 35 - 52.

[6] Devaney, R. L., *Proc. Amer. Math. Soc.*, 1985, 94: 545 - 548.

[7] Lyubich, M. Yu., *Sibirsk Mat. Zh.*, 1987, 28 (5): 111 - 127.

[8] Barnsley, M. F., *Proc. of Symposia in Appl. Math.*, 1989, 39: 127 - 144.

[9] Barnsley, M. F., Demko, S., *Proc. R. Soc. London*, 1985, A399: 243 - 275.

① Lü Yi-nian, Iterations of meromorphic functions. Institute of Math., Academia Sinica, Beijing, China, August 1988 (preprint).

H^p Multipliers on Bounded Symmetric Domains*

REN Fu-Yao(任福尧)
(*Department of Mathematics, Fudan University, Shanghai* 200433, *PRC*)
and XIAO Jian-Bin(肖建斌)
(*Department of Mathematics, Hunan Normal University, Changsha* 410006, *PRC*)

Abstract: The following results are obtained in this paper: (i) if $0 < p \leqslant 2 \leqslant q < \infty$, $\alpha = \dfrac{1}{p} - \dfrac{1}{q}$, and $\lambda_k = O(k^{-n\alpha})$. then $\{\lambda_k\}$ is a multiplier of $H^p(\Omega)$ into $H^q(\Omega)$; (ii) if $0 < p < q < 2$ and $\alpha = \dfrac{1}{p} - \dfrac{1}{q}$, then there is a sequence $\{\lambda_k\}$ with $\lambda_k = O(k^{-n\alpha})$ which is not a multiplier of $H^p(B)$ into $H^q(B)$; (iii) if $2 < p < q < \infty$ and $\alpha = \dfrac{1}{p} - \dfrac{1}{q}$, then there is a sequence $\{\lambda_k\}$ with $\lambda_k = O(k^{-n\alpha})$ which is not a multiplier of $H^p(B)$ into $H^q(B)$.

Keywords: bounded symmetric domain, multiplier, Hardy space.

1. Introduction and Early Results

Let Ω be a bounded symmetric domain in $\mathbb{C}^n$ which contains the origin. By b we denote its Silov boundary, by Γ the holomorphic automorphism group of

---

* Project supported by the Tianyuan Foundation of China. Originally published in *Science in China* (*Ser. A*), Vol. 37, No. 3, (1994), 257 - 264.

Ω, and by Γ_0 the isotropy group of Γ at the origin. There exists a unique Γ_0-invariant measure σ on b such that $\sigma(b)=1$.

By the group representation method, Hua[1] constructed a system of polynomials

$$\{\Phi_{k,v}\},\ k=0,1,\cdots,\ v=1,2,\cdots,m_k,\ m_k=C_{n+k-1}^{k},$$

which are completely orthogonal on Ω and orthonormal on b.

Denote the set of all holomorphic functions on Ω by $H(\Omega)$. Hardy spaces $H^p=H^p(\Omega)$ $(0<p<\infty)$ on Ω are defined as

$$H^p(\Omega)=\{f \mid f\in H(\Omega),\ \|f\|_p<\infty\},$$

where

$$\|f\|_p=\sup_{0<r<1}M_p(r,f),\ M_p(r,f)=\left\{\int_b |f(r\xi)|^p \mathrm{d}\sigma(\xi)\right\}^{1/p}.$$

Each $f\in H(\Omega)$ possesses a series of expansion[2]

$$f(z)=\sum_{k,v}a_{k,r}\Phi_{k,r}(z),\ a_{k,v}=\lim_{r\to 1}\int_b f(r\xi)\overline{\Phi_{k,v}(\xi)}\mathrm{d}\sigma(\xi),$$

where $\sum\limits_{k,v}=\sum\limits_{k=0}^{\infty}\sum\limits_{v=1}^{m_k}$ and the series is uniformly convergent on any compact set of Ω.

A complex sequence $\{\lambda_k\}_{k=0}^{\infty}$ is called a multiplier of H^p into H^q if $\sum\limits_{k,v}\lambda_k a_{k,v}\Phi_{k,v}(z)$ is in H^q whenever $\sum\limits_{k,v}a_{k,v}\Phi_{k,v}(z)$ is in H^p.

When Ω is the unit disk U, there are many known results on $H^p(U)$-multipliers (see the survey[3] for reference). For example. Duren[4] obtained the following three theorems.

Theorem A *If* $0<p\leqslant 2\leqslant q<\infty$, $\alpha=\frac{1}{p}-\frac{1}{q}$, *and* $\lambda_k=O(k^{-\alpha})$, *then* $\{\lambda_k\}$ *is a multiplier of* $H^p(U)$ *into* $H^q(U)$. *The number is best possible. For each* $\beta<\alpha$, *there is a sequence* $\{\lambda_k\}$ *with* $\lambda_k=O(k^{-\beta})$ *which is not a multiplier of* $H^p(U)$ *into* $H^q(U)$.

Theorem B *If* $0<p<q<2$ *and* $\alpha=\frac{1}{p}-\frac{1}{q}$, *the condition* $\lambda_k=O(k^{-\alpha})$

does not imply that $\{\lambda_k\}$ *is a multiplier of* $H^p(U)$ *into* $H^q(U)$. *In fact, for each number* $\beta < \frac{1}{p} - \frac{1}{2}$, *there is a sequence* $\{\lambda_k\}$ *with* $\lambda_k = O(k^{-\beta})$ *which is not a multiplier of* $H^p(U)$ *into* $H^q(U)$ *for any* $q > 0$.

Theorem C *If* $2 < p < q \leqslant \infty$ *and* $\alpha = \frac{1}{p} - \frac{1}{q}$, *the condition* $\lambda_k = O(k^{-\alpha})$ *does not imply that* $\{\lambda_k\}$ *is a multiplier of* $H^p(U)$ *into* $H^q(U)$. *In fact, for each number* $\beta < \frac{1}{2} - \frac{1}{q}$, *there is a sequence* $\{\lambda_k\}$ *with* $\lambda_k = O(k^{-\beta})$ *which is not a multiplier of* $H^p(U)$ *into* $H^p(U)$ *for any* $p < \infty$.

Shi[5] generalized Theorem A to the case with the bounded symmetric domain and obtained the following theorem.

Theorem D *If* $0 < p \leqslant 1$ *and* $2 \leqslant q < \infty$, $\alpha = \frac{1}{p} - \frac{1}{q}$. *If*

$$\lambda_k = O(k^{-n\alpha - n + 1}),\ n > 1,$$

then $\{\lambda_k\}$ *is a multiplier of* $H^p(\Omega)$ *into* $H^q(\Omega)$.

Obviously, Theorem D is not the best form of the generalization of Theorem A.

In this paper, the first part of Theorem A is generalized to the case with the bounded symmetric domains, and the second part, i. e. to explain that the number α is best possible, is illustrated by the unit ball B. In addition, the generalizations of Theorem B and Theorem C to the case with the unit ball B are obtained. The method we use is completely different from Duren's and Shi's.

We always denote the bounded symmetric domain on $\mathbb{C}^n$ by Ω, the unit ball on $\mathbb{C}^n$ by B and the unit disk on $\mathbb{C}$ by U throughout the paper.

2. Theorems and Proofs

Theorem 1 *If* $0 < p \leqslant 2 \leqslant q < \infty$, $\alpha = \frac{1}{p} - \frac{1}{q}$, *and*

$$\lambda_k = O(k^{-n\alpha}),\ n > 1, \tag{1}$$

then $\{\lambda_k\}$ *is a multiplier of* $H^p(\Omega)$ *into* $H^q(\Omega)$. *When* $\Omega=B$, *the number* α *is best possible, i.e. for each* $\beta<\alpha$, *there is a sequence* $\{\lambda_k\}$ *with* $\lambda_k=O(k^{-n\beta})$ *which is not a multiplier of* $H^p(B)$ *into* $H^q(B)$.

Proof First we prove the case with $q=2$. If $p=2$, then $\lambda_k=O(1)$. From the definition of $H^2(\Omega)$, it is obvious that $\{\lambda_k\}$ is a multiplier of $H^2(\Omega)$ into $H^2(\Omega)$. If $p<2$, then for $f(z)=\sum_{k,v}a_{k,v}\Phi_{k,v}(z)\in H^p(\Omega)$, by Theorem 4 in Ref. [6], we have

$$\int_0^1(1-r)^{2n\left(\frac{1}{p}-\frac{1}{2}\right)-1}M_2(r,f)^2\mathrm{d}r<\infty,$$

i. e.

$$\int_0^1(1-r)^{2n\left(\frac{1}{p}-\frac{1}{2}\right)-1}\sum_{k,v}|a_{k,v}|^2r^{2k}\mathrm{d}r<\infty. \tag{2}$$

Lemma 4 in Ref. [7] leads to

$$\int_0^1(1-r)^{2n\left(\frac{1}{p}-\frac{1}{2}\right)-1}r^{2k}\mathrm{d}r=\frac{\Gamma(2k+1)\Gamma\left(\frac{2n}{p}-n\right)}{\Gamma\left(2k+1+\frac{2n}{p}-n\right)},$$

where Γ is the gamma function. By Stirling formula we know

$$\int_0^1(1-r)^{2n\left(\frac{1}{p}-\frac{1}{2}\right)-1}r^{2k}\mathrm{d}r\sim k^{-2n\left(\frac{1}{p}-\frac{1}{2}\right)}.$$

Hence Eq. (2) is equivalent to

$$\sum_{k=1}^{n}k^{-2n\left(\frac{1}{p}-\frac{1}{2}\right)}\sum_{v=1}^{m_k}|a_{k,v}|^2<\infty. \tag{3}$$

So when $\{\lambda_k\}$ satisfies $\lambda_n=O(k^{-n\left(\frac{1}{p}-\frac{1}{2}\right)})$, we have

$$\sum_{k=1}^{\infty}|\lambda_k|^2\sum_{v=1}^{m_k}|a_{k,v}|^2<\infty$$

for each $f(z)=\sum_{k,v}a_{k,v}\Phi_{k,v}(z)\in H^p(\Omega)$, i. e. $\sum_{k,v}\lambda_k a_{k,v}\Phi_{k,v}(z)\in H^2(\Omega)$, or $\{\lambda_k\}$ is a multiplier of $H^p(\Omega)$ into $H^2(\Omega)$.

Next we prove the case with $q>2$. Let $\mu_k=\dfrac{\Gamma\left(k+n\left(\frac{1}{2}-\frac{1}{q}\right)+1\right)}{k!}\lambda_k$. By Stirling formula and condition (1), we know $\mu_k=O(k^{-n\left(\frac{1}{p}-\frac{1}{2}\right)})$. It follows from the conclusion proved above that $\{\mu_k\}$ is a multiplier of $H^p(\Omega)$ into $H^2(\Omega)$. On the other hand, by Theorem 5 in Ref. [5], we have known that $\left\{\dfrac{k!}{\Gamma\left(k+n\left(\frac{1}{2}-\frac{1}{q}\right)+1\right)}\right\}$ is a multiplier of $H^2(\Omega)$ into $H^q(\Omega)$. So $\{\lambda_k\}$ is a multiplier of $H^p(\Omega)$ into $H^q(\Omega)$.

Now we turn to the second part of the theorem, i. e. to explain that the number α is best possible when $\Omega=B$. For each $\beta<\alpha$, take $c=\dfrac{np(\beta-\alpha)}{2}$, and $\zeta\in\partial B$, and put

$$f(z)=\frac{1}{(1-\langle z, \zeta\rangle)^{(n+c)/p}}=\sum_{k,v}a_{k,v}\Phi_{k,v}(z),$$

where $\langle z, \zeta\rangle$ denotes the inner product of points z and ζ. Since $c<0$, it is easy to see that $f(z)\in H^p(B)$. Let

$$\lambda_k=\frac{\Gamma(k+n)}{\Gamma(k+n+n\beta)}=O(k^{-n\beta}).$$

If $\{\lambda_k\}$ is a multiplier of $H^p(B)$ into $H^q(B)$, then $h(z)=\sum_{k,v}\lambda_k a_{k,v}\Phi_{k,v}(z)\in H^q(B)$. By Theorem 1 in Ref. [8], there is a positive number C_1 independent of f and r such that

$$M_q(r, f)\leqslant C_1\|h\|_q(1-r)^{-n\beta}. \tag{4}$$

From the definition of $f(z)$, however, Proposition 1. 4. 10 in Ref. [9] implies that there is a positive number C_2 independent of f and r such that

$$M_q(r, f)\geqslant C_2(1-r)^{-\left(\frac{n+c}{p}-\frac{n}{q}\right)}. \tag{5}$$

Combining Eq. (4) and Eq. (5), we have $n\beta\geqslant\dfrac{n+c}{p}-\dfrac{n}{q}$, i. e. $\beta\geqslant\alpha$. This contradicts the assumption of $\beta<\alpha$. The proof of Theorem 1 is completed.

From the proof of Corollary in Ref. [4], using Theorem 1 and Theorem 5 in Ref. [5], we can deduce the following corollary.

Corollary *If* $0 < p < q < \infty$ *and* $\alpha = \frac{1}{p} - \frac{1}{q}$, *then* $\{k^{-n\alpha}\}$ *is a multiplier of* $H^p(\Omega)$ *into* $H^q(\Omega)$.

Theorem 2 *If* $0<p<q<2$ *and* $\alpha=\frac{1}{p}-\frac{1}{q}$, *the condition* $\lambda_k=O(k^{-n\alpha})$ *does not imply that* $\{\lambda_k\}$ *is a multiplier of* $H^p(B)$ *into* $H^q(B)$. *In fact, for each positive number* $\beta < \frac{1}{p} - \frac{1}{2}$, *there is a sequence* $\{\lambda_k\}$ *of complex number with* $\lambda_k = O(k^{-n\beta})$ *which is not a multiplier of* $H^p(B)$ *into* $H^q(B)$ *for any* $q > 0$.

Proof Let $H_0^p(B) = \{f \in H^p(B) \mid f$ depend only on one-dimensional complex variable $z_1\}$. It is not hard to prove that $H_0^p(B)$ is a closed subspace of $H^p(B)$, so it is a complete metric space. If $\{\lambda_k\}$ is a multiplier of $H_0^p(B)$ into $H^2(B)$, we define a linear operator Λ on $H_0^p(B)$:

$$\Lambda(f)=\sum_{k,v}\lambda_k a_{k,v}\Phi_{k,v}(z), \quad \forall f(z)=\sum_{k,v}a_{k,v}\Phi_{k,v}(z) \in H_0^p(B).$$

It follows from the closed graph theorem that Λ is a bounded operator of $H_0^p(B)$ into $H^2(B)$, i. e.

$$\|\Lambda(f)\|_2 \leqslant \|\Lambda\| \|f\|_p. \tag{6}$$

For $\rho < 1$, take $\alpha = n-1+\frac{p(n+1)}{2}$ and define

$$f(z)=(1-\rho z_1)^{-\frac{1+\alpha}{p}}=\sum_{k=0}^{\infty}\frac{\Gamma\left(k+\frac{1+\alpha}{p}\right)}{k!\ \Gamma\left(\frac{1+\alpha}{p}\right)}\rho^k z_1^k=\sum_{k=0}^{\infty}\sum_{v=1}^{m_k}a_{k,v}\Phi_{k,v}(z).$$

Here, if we put

$$\omega_k=\sqrt{\frac{k!\ (n-1)!}{(n+k-1)!}}, \tag{7}$$

then

$$a_{k,v}=\begin{cases}\dfrac{\Gamma\left(k+\dfrac{1+\alpha}{p}\right)}{k!\ \Gamma\left(\dfrac{1+\alpha}{p}\right)}\omega_k\rho^k, & \text{if } \Phi_{k,v}(z)=\omega_k^{-1}z_1^k,\\ 0, & \text{otherwise.}\end{cases}$$

Thus, from the second formula of Sec. 1. 4. 5 in Ref. [9],

$$\| f \|_p^p=\frac{n-1}{\pi}\int_0^1\int_0^{2\pi}(1-r^2)^{n-2}\frac{r\mathrm{d}\theta\mathrm{d}r}{|1-\rho r\mathrm{e}^{i\theta}|^{1+\alpha}}.$$

According to Lemma 5 and Lemma 6 in Ref. [7], the right-hand side in the above expression is dominated by $O((1-\rho)^{-\frac{p(n+1)}{2}})$, and it follows from Ref. [7] that there exists a constant A such that

$$\sum_{k=0}^{\infty}\sum_{v=1}^{m_k}|\lambda_k a_{k,v}|^2\leqslant A(1-\rho)^{-(n+1)}.$$

By Stirling formula, it is equivalent to

$$\sum_{k=1}^{\infty}k^{\frac{2n}{p}}|\lambda_k|^2\rho^{2k}\leqslant A(1-\rho)^{-(n+1)}. \tag{8}$$

Putting $\rho^2=1-\dfrac{1}{N}$, we can easily obtain

$$\sum_{k=1}^{N}k^{\frac{2n}{p}}|\lambda_k|^2=O(N^{n+1}). \tag{9}$$

So expression (9) is a necessary condition for $\{\lambda_k\}$ being a multiplier of $H_0^p(B)$ into $H^2(B)$. For any $\beta<\dfrac{1}{p}-\dfrac{1}{2}$, not satisfying (9), $\{k^{-n\beta}\}$ is not a multiplier of $H_0^p(B)$ into $H^2(B)$, i. e. there is

$$f(z_1)=\sum_{k=0}^{\infty}a_k z_1^k=\sum_{k=0}^{\infty}\sum_{v=1}^{m_k}a_{k,v}\Phi_{k,v}(z)\in H_0^p(B)$$

such that

$$\sum_{k=1}^{\infty}\sum_{v=1}^{m_k}|k^{-n\beta}a_{k,v}|^2=\infty, \tag{10}$$

where

$$a_{k,v}=\begin{cases}\omega_k a_k, & \text{if } \Phi_{k,v}(z)=\omega_k^{-1}z_1^k,\\ 0, & \text{otherwise.}\end{cases}$$

Hence (10) turns into

$$\sum_{k=1}^{\infty}|k^{-n\beta}\omega_k a_k|^2=\infty \tag{11}$$

and it follows from Theorem 4. 5 in Ref. [11] that, for almost all choices of signs $\{\varepsilon_k\}$ ($\varepsilon_k=+1$ or -1), the function

$$g(z_1)=\sum_{k=1}^{\infty}\varepsilon_k k^{-n\beta}\omega_k a_k z_1^k$$

has no radial limit almost everywhere. So $g(z_1)$ does not belong to any $H^q(B)$. Fix such a sequence $\{\varepsilon_k\}$ and put $\lambda_k=\varepsilon_k k^{-n\beta}$. Clearly, $\lambda_k=O(k^{-n\beta})$ and $g(z_1)$ may be written as

$$g(z_1)=\sum_{k=1}^{\infty}\sum_{v=1}^{m_k}\lambda_k a_{k,v}\Phi_{k,v}(z).$$

It is obvious that $\{\lambda_k\}$ is not a multiplier of $H_0^p(B)$ into any $H^q(B)$. Thus, $\{\lambda_k\}$ is not a multiplier of $H^p(B)$ into $H^q(B)$. This completes the proof.

In order to obtain the conclusion corresponding to Theorem C with B in place of U, we have to introduce the following definition with a lemma. Define

$$(H^p)^a \triangleq \Big\{y(z)=\sum_{k,v}b_{k,v}\Phi_{k,v}(z)\,\Big|\,\lim_{r\to 1}\sum_{k,v}a_{k,v}b_{k,v}r^k$$
$$\text{exists for all } f(z)=\sum_{k,v}a_{k,v}\Phi_{k,v}(z)\in H^p\Big\}.$$

Lemma *Let $1<p<\infty$, and let $1<q<\infty$. Then $\{\lambda_k\}$ is a multiplier of H^p into H^q if and only if it is a multiplier of $H^{q'}$ into $H^{p'}$, where $\frac{1}{p}+\frac{1}{p'}=1$, $\frac{1}{q}+\frac{1}{q'}=1$.*

Proof By the symmetry, we need only to prove that if $\{\lambda_k\}$ is a multiplier of H^p into H^q, it is also a multiplier of $H^{q'}$ into $H^{p'}$. From

Theorem 1 in Ref. [6], it is easy to see that for $1 < p < \infty$,

$$(H^q)^a = H^q,\ (H^p)^a = H^{p'}.$$

If $\{\lambda_k\}$ is a multiplier of H^p into H^q, then for any given $g(z) = \sum_{k,v} b_{k,v}\Phi_{k,v}(z) \in H^{q'}$ and $f(z) = \sum_{k,v} a_{k,v}\Phi_{k,v}(z) \in H^p$,

$$\lim_{r\to 1}\sum_{k,v}(\lambda_k b_{k,v})a_{k,v}r^k = \lim_{r\to 1}\sum_{k,v}(\lambda_k a_{k,v})b_{k,v}r^k$$

exists. i. e. $\{\lambda_k\}$ is a multiplier of $H^{q'}$ into $H^{p'}$. This completes the proof of the lemma.

Remark In the above lemma, $H^p = H^p(\Omega)$.

Theorem 3 *If $2 < p < q < \infty$ and $\alpha = \frac{1}{p} - \frac{1}{q}$, the condition $\lambda_k = O(k^{-n\alpha})$ does not imply that $\{\lambda_k\}$ is a multiplier of $H^p(B)$ into $H^q(B)$. In fact, for each number $\beta < \frac{1}{2} - \frac{1}{q}$, there is a sequence $\{\lambda_k\}$ with $\lambda_k = O(k^{-n\beta})$ which is not a multiplier of $H^p(B)$ into $H^q(B)$ for any $p < \infty$.*

Proof For given $p < \infty$, $p_1 > p_2$ implies $H^{p_1}(B) \subset H^{p_2}(B)$. Hence we may assume $p > 1$ in the proof. For any $\beta < \frac{1}{2} - \frac{1}{q} = \frac{1}{q'} - \frac{1}{2}$, it follows from Theorem 2 that there is a sequence $\{\lambda_k\}$ with $\lambda_k = O(k^{-n\beta})$ which is not a multiplier of $H^{q'}(B)$ into $H^p(B)$. Hence it follows immediately from the above lemma that $\{\lambda_k\}$ is also not a multiplier of $H^p(B)$ into $H^q(B)$. The proof is completed.

◇ References ◇

[1] Hua Lo-Geng. *Harmonic Analysis on Canonical Domains in the Theory of Several Complex Variables* rev. ed., Science Press, Beijing, 1965.

[2] Hahn, K. T. Mitchell, J., *Ann. Polon. Math.*, 1973, **28**: 89.

[3] Campbell, D. M. & Leach, R. J., *Complex Variable*, 1984, **3**: 85.

[4] Duren, P. L., *Proc. Amer. Math. Soc.*, 1969, **22**: 24.

[5] Shi Ji-Huai, *Scientia Sinica*, 1988, **B4**: 366.

[6] Mitchell, J. & Kahn, K. T., *J. Math. Anal. Appl.*, 1976, **56**: 379.

[7] Shields, A. L. & Williams, D. L., *Trans. Amer. Math Soc.*, 1971, **162**: 287.

[8] Stoll, M., *J. Math. Anal. Appl.*, 1983, **93**: 109.

[9] Rudin, W., *Function Theory in the Unit Ball of C^n*, Springer-Verlag, New York, 1980.

[10] Zygmund, A., *Trigonometric Series*. 2nd ed., Cambridge Univ. Press, Cambridge, 1959.

[11] Duren. P. L., *Theory of H^p Spaces*, Academic Press, New York, 1970.

On Taylor's Conjecture about the Packing Measures of Cartesian Product Sets*

Xu You Ren Fuyao

(*Institute of Mathematics*, *Fudan University*, *Shanghai*, 200433, *China*)

Abstract: It is proved that if $E \subset \mathbb{R}$, $F \subset \mathbb{R}^n$, then $\mathscr{P}(E\times F,\ \varphi_1\varphi_2) \leqslant c \cdot \mathscr{P}(E,\ \varphi_1)\mathscr{P}(E,\ \varphi_2)$, where $\mathscr{P}(\cdot,\ \varphi)$ denotes the φ-packing measure, φ belongs to a class of Hausdorff functions, the positive constant c deponds only on φ_1, φ_2 and n.

Keywords: packing measure, Hausdorff measure, Cartesian product set

1. Introduction

In the geometry of fractals, Hausdorff measure and dimension play a very important role. On the other hand, the recent introduction of packing measures has led to a greater understanding of the geometric theory of fractals, as packing measures behave in a way that is "dual" to Hausdorff measures in many respects[2]. For example, denoting Hausdorff dimension and packing dimension by dim and Dim respectively, we have $\dim(E\times F) \geqslant \dim E + \dim F$, while $\mathrm{Dim}\,(E\times F) \leqslant \mathrm{Dim}\,E + \mathrm{Dim}\,F$. It is well-known that if $E \subset \mathbb{R}^m$, $F \subset \mathbb{R}^n$, then

$$\mathscr{H}(E\times F,\ \varphi_1\varphi_2) \geqslant b \cdot \mathscr{H}(E,\ \varphi_1)\mathscr{H}(F,\ \varphi_2)$$

* Originally published in *Chinese Ann. Math. Ser. B*, Vol. 17, No. 1, (1996), 121 - 126.

for some Hausdorff functions and constant b, where $\mathscr{H}(\cdot, \varphi)$ denotes the φ-Hausdorff measure. Taylor conjectures that we should have

$$\mathscr{P}(E\times F, \varphi_1\varphi_2)\leqslant c\cdot \mathscr{P}(E, \varphi_1)\mathscr{P}(F, \varphi_2).$$

In this paper, it is shown that if E or F is a subset of $\mathbb{R}$, then the conjecture is correct.

2. Packing Premeasure

We restrict our attention to subsets of Euclidean space $\mathbb{R}^d (d\geqslant 1)$. The Cartesian product of sets $E\subset\mathbb{R}^m$ and $F\subset\mathbb{R}^n$ is denoted by $E\times F$. We use $|E|$ to denote the diameter of E and $\| x \|$ to denote the distance from 0 to $x\in\mathbb{R}^n$. The open ball with center at x and radius $r>0$ is denoted by

$$B_r(x)=\{y\in\mathbb{R}^n: \| x-y\| < r\}.$$

Ω stands for the class of balls:

$$\Omega(E)=\{B_r(x): r>0, x\in E\}.$$

Γ^* stands for the class of dyadic cubes in $\mathbb{R}^d$, $C\in\Gamma^*$ if it has side length 2^{-n}, $n\in\mathbb{N}$, and each of its projections $\text{proj}_i C$ on the i-th axis is a half-open interval of the form $[k_i 2^{-n}, (k_i+1)2^{-n})$ with $k_i\in\mathbb{Z}$. $u_n(x)$ denotes the unique cube which is in Γ^* and contains x with side length 2^{-n}.

$$\Gamma^*(E)=\{u_n(x): x\in E, n\in\mathbb{N}\}.$$

Γ^{**} stands for the class of semidyadic cubes in $\mathbb{R}^d$, $C\in\Gamma^{**}$ if it has side length 2^{-n} and $\text{proj}_i C=\left[\frac{1}{2}k_i 2^{-n}, \left(\frac{1}{2}k_i+1\right)2^{-n}\right)$ with $k_i\in\mathbb{Z}$. $v_n(x)$ is the unique cube in Γ^{**} of side length 2^{-n} such that on the i-th axis the complement of $\text{proj}_i C$ is at distance 2^{-n-2} from $u_{n+2}(\text{proj}_i x)\subset\mathbb{R}$. It is not difficult to see that if $x\in\mathbb{R}^{m+n}$ and $n\in\mathbb{N}$, then $\text{proj}_{\mathbb{R}^n}(v_n(x))=v_n(\text{proj}_{\mathbb{R}^n}x)$, where $v_n(\text{proj}_{\mathbb{R}^n}x)$ is in $\mathbb{R}^n$.

$$\Gamma^{**}=\{v_n(x): x\in E, n\in\mathbb{N}\}.$$

Φ denotes the class of functions φ: $[0, +\infty) \to \mathbb{R}$ which are increasing, continous with $\varphi(0)=0$ and

$$\varphi(2x) < c_0\varphi(x) \quad \text{for some } c_0 > 0 \quad \text{and } 0 < x < \frac{1}{2}. \tag{2.1}$$

We use $\mathscr{B}(\mathbb{R}^n)$ to denote the family of bounded subsets of $\mathbb{R}^n$. For $\mathscr{R} \subset \mathscr{B}(\mathbb{R}^n)$, put $\|\mathscr{R}\| = \sup\{|E| : E \in \mathscr{R}\}$ and

$$\varphi(R) = \sum_{R \in \mathscr{R}} \varphi(|E|). \tag{2.2}$$

We say $R \subset \mathscr{B}(\mathbb{R}^n)$ is a packing of E if for all $F \in R$, $\bar{E} \cap \bar{F} \neq \varnothing$, and the sets in $\mathscr{R}$ are disjoint. Put

$$\tau(E, \varphi, \varepsilon) = \sup\{\varphi(\mathscr{R}): \|\mathscr{R}\| \leqslant \varepsilon, \mathscr{R} \text{ is a packing of } E\}. \tag{2.3}$$

Particularly, if $\mathscr{R} \subset \Omega(E)$ or $\mathscr{R} \subset \Gamma^{**}(E)$, the corresponding $\tau(E, \varphi, \varepsilon)$ is denoted by $P(E, \varphi, \varepsilon)$ or $P^{**}(E, \varphi, \varepsilon)$.

Obviously $\tau(E, \varphi, \varepsilon)$ is an increasing function of ε. Let

$$\begin{aligned} \tau(E, \varphi) &= \lim_{\varepsilon \to 0} \tau(E, \varphi, \varepsilon), \\ P(E, \varphi) &= \lim_{\varepsilon \to 0} P(E, \varphi, \varepsilon), \\ P^{**}(E, \varphi) &= \lim_{\varepsilon \to 0} P^{**}(E, \varphi, \varepsilon). \end{aligned} \tag{2.4}$$

3. Packing Measure

For $E \subset \mathbb{R}^n$, let

$$\mathscr{P}(E, \phi) = \inf\left\{\sum P(E_i, \varphi): E_i \in \mathscr{B}(\mathbb{R}^n), E \subset \bigcup E_i\right\}, \tag{3.1}$$

$$\mathscr{P}^{**}(E, \phi) = \inf\left\{\sum P^{**}(E_i, \varphi): E_i \in \mathscr{B}(\mathbb{R}^n), E \subset \bigcup E_i\right\}. \tag{3.2}$$

Then they are two outer measures. We call $\mathscr{P}(E, \varphi)$ the φ-packing measure of E.

4. Packing Measures of Cartesian Product Sets

Lemma 4.1[3] *Let $E \subset \mathbb{R}^n$. Then there exist $0 < c_1 \leqslant c_2 < +\infty$ such* *that*

$$c_1 P(E, \varphi) \leqslant P^{**}(E, \varphi) \leqslant c_2 P(E, \varphi). \tag{4.1}$$

c_1 and c_2 depend only on φ and n.

Proof From the definition of $v_i(x)$, we can get $B_{2^{-i-2}}(x) \subset v_i(x) \subset B_{\rho, 2^{-i}}(x)$, where $i \in N$, and $\rho = n^{\frac{1}{2}}$. So according to (2.3) and (2.4), the result is obvious.

Corollary 4.1[3] *Let $E \subset \mathbb{R}^n$. Then there exist $0 < c_1 \leqslant c_2 < +\infty$ such that*

$$c_1 \mathscr{P}(E, \varphi) \leqslant \mathscr{P}^{**}(E, \varphi) \leqslant c_2 \mathscr{P}(E, \varphi).$$

c_1 and c_2 depend only on φ and n.

Proof Use (3.1), (3.2) and Lemma 4.1.

Lemma 4.2 *Let $E \subset [a, b]$, $-\infty < a \leqslant b < +\infty$, $u = \{U_i, i = 1, 2, 3, \cdots\} \subset \Gamma^{**}(E)$. U_i and U_j may be the same set when $i \neq j$, $q > 0$, $\| u \| \leqslant q$. For all $x \in [a, b]$,*

$$\sum_{U_i \in u} \chi_{U_i}(x) \leqslant n, \quad n \in \mathbb{N}, \tag{4.2}$$

where $\chi_{U_i}(x)$ is the characteristic function of U_i. Then

$$\sum_{U_i \in u} \varphi(|U_i|) \leqslant n \cdot P^{**}(E, \varphi, q). \tag{4.3}$$

Proof Use mathematical induction.

If $n = 1$, then from (4.2) we know that u is a packing of E, so

$$\sum_{U_i \in u} \varphi(|U_i|) \leqslant P^{**}(E, \varphi, q).$$

Suppose that the lemma is true when $n = k - 1$. Let $n = k$. Let $u' = \{U_1, U_2, \cdots, U_N\}$. Then

$$\sum_{U_i \in u'} \chi_{U_i}(x) \leqslant \sum_{U_i \in u} \chi_{U_i}(x) \leqslant k, \quad x \in [a, b]. \tag{4.4}$$

Let $U_i = [a_i, b_i)$, $i = 1, 2, \cdots, N$. We can assume that $a_1 \leqslant a_2 \leqslant \cdots \leqslant a_{N-1} \leqslant a_N$. Let $U_{r_1} = [a_1, b_1)$, $U_{r_2} = [a_{r_2}, b_{r_2})$, where r_2 is the smallest number which satisfies $a_{r_2} \geqslant b_1$. Also we can get $U_{r_3} = [a_{r_3}, b_{r_3})$, where r_3 is the smallest number such that $a_{r_3} \geqslant b_{r_2}$. In such a way, we can get

$$U_{r_1}, U_{r_2}, \cdots, U_{r_l}, 1 = r_1 < r_2 < \cdots \leqslant r_l \leqslant N.$$

Let $\widetilde{u} = \{U_{r_1}, U_{r_2}, \cdots, U_{r_l}\}$, $u'' = u' \setminus \widetilde{u}$. Then $\widetilde{u}$ is a packing of E, so

$$\sum_{i=1}^{l} \varphi(|U_{r_i}|) \leqslant P^{**}(E, \varphi, q). \tag{4.5}$$

We need to prove

$$\sum_{U \in u''} \chi_{U_i}(x) \leqslant k - 1, \quad x \in [a, b]. \tag{4.6}$$

If $x \in \bigcup_{i=1}^{l} U_{r_i}$, then (4.6) is obviously correct.

If $x \notin \bigcup_{i=1}^{l} U_{r_i}$, then there must exist r_i such that $x \in [b_{r_i}, a_{r_{i+1}})$ (If $i = l$, then let $a_{r_{i+1}} = b$). So if $x \in U_i \in u''$, then U_i must satisfy $U_i \cap [a_{r_i}, b_{r_i}) \neq \varnothing$; otherwise U_i should have been selected into $\widetilde{u}$ before $[a_{r_{i+1}}, b_{r_{i+1}})$. So if there are more than $k-1$ sets containing x in u'', we can find a point b'_{r_i} in the left neighborhood of b_{r_i} such that

$$\sum_{U_i \in u'} \chi_{U_i}(b'_{r_i}) = \sum_{U_i \in \widetilde{u}} \chi_{U_i}(b'_{r_i}) + \sum_{U_i \in u''} \chi_{U_i}(b'_{r_i}) > 1 + (k-1) = k, \tag{4.7}$$

which contradicts (4.4). So we have

$$\sum_{U_i \in u''} \chi_{U_i}(x) \leqslant k - 1$$

and (4.6) is correct. So

$$\sum_{U_i \in u''} \varphi(|U_i|) \leqslant (k-1) \cdot P^{**}(E, \varphi, q),$$

and

$$\sum_{U_i \in u'} \varphi(|U_i|) = \sum_{U_i \in \widetilde{u}} \varphi(|U_i|) + \sum_{U_i \in u''} \varphi(|U_i|) \leqslant k \cdot P^{**}(E, \varphi, q).$$

Letting $N \to +\infty$, we complete the proof.

Lemma 4.3 *If $E \subset \mathbb{R}$, $F \subset \mathbb{R}^n$, then*

$$P^{**}(E\times F,\ \varphi_1\varphi_2)\leqslant c\cdot P^{**}(E,\ \varphi_1)\cdot P^{**}(F,\ \varphi_2),\tag{4.8}$$

where $0<c<+\infty$. *c depends only on* φ_1, φ_2 *and* n.

Proof If E or F is an unbounded set, then $P^{**}(E,\ \varphi_1)=+\infty$ or $P^{**}(F,\ \varphi_2)=+\infty$ and (4.8) holds. So we need only to consider the case that both E and F are bounded sets.

Let $u=\{U_i\}\subset\Gamma^{**}(E\times F)$, $\|u\|\leqslant q$ and u be a packing of $E\times F$. Put

$$P_1(U_i)=\text{proj}_{\mathbf{R}}(U_i),\ P_2(U_i)=\text{proj}_{\mathbf{R}^n}(U_i),\ u_1=\{P_1(U_i):\ U_i\in u\}$$

and $u_2=\{P_2(U_i):\ U_i\in u\}$. Then

$$u_1\subset\Gamma^{**}(E),\ u_2\subset\Gamma^{**}(F),\ \|u_1\|\leqslant q\text{ and }\|u_2\|\leqslant q.$$

Suppose $E\subset[a,\ b]$, $-\infty<a\leqslant b<+\infty$. For any fixed $x\in[a,\ b]$, $\{P_2(U_i):\ x\in P_1(U_i)\}$ is a packing of F. So

$$\sum_{U_i\in u}\varphi_2(|P_2(U_i)|)\cdot\chi_{P_1(U_i)}(x)\leqslant P^{**}(F,\ \varphi_2,\ q),\quad x\in[a,\ b].\tag{4.9}$$

For u we have

$$\begin{aligned}\sum_{U_i\in u}\varphi_1\varphi_2(|U_i|)&=\sum_{U_i\in u}\varphi_1(|U_i|)\cdot\varphi_2(|U_i|)\\&=\sum_{U_i\in u}\varphi_1(\sqrt{n+1}\,|P_1(U_i)|)\cdot\varphi_2\left(\frac{\sqrt{n+1}}{\sqrt{n}}|P_2(U_i)|\right).\end{aligned}\tag{4.10}$$

Let $u'=\{U_1,\ U_2,\ \cdots,\ U_N\}$. Then

$$\sum_{U_i\in u'}\varphi_2(|P_2(U_i)|)\cdot\chi_{P_1(U_i)}(x)\leqslant P^{**}(F,\ \varphi_2,\ q).\tag{4.11}$$

Let $\varphi_2(|P_2(U_i)|)=f_i$, $i=1,\ 2,\ \cdots,\ N$, and $P^{**}(F,\ \varphi_2,\ q)=g$. f_i and g can be approximated by rational numbers d_i and h so that

$$\frac{h}{1+\varepsilon}\leqslant g\leqslant h,\ \frac{d_i}{1+\varepsilon}\leqslant f_i\leqslant d_i,\ i=1,\ 2,\ \cdots,\ N,$$

where ε is also a rational number. Then

$$\sum_{U_i \in u'} d_i \cdot \chi_{P_1(U_i)}(x) \leqslant (1+\varepsilon)h.$$

Let M be the common demoninator of ε, h and d_i, $i=1, 2, \cdots, N$, $d_i = \frac{k_i}{M}$. Then

$$\sum_{U_i \in u'} Mk_i \cdot \chi_{P_1(U_i)}(x) \leqslant (1+\varepsilon)hM^2.$$

Put $(1+\varepsilon)hM^2 = K$. Then $K \in N$. Using Lemma 4.2 we get

$$\sum_{U_i \in u'} Mk_i \cdot \varphi_1(|P_1(U_i)|) \leqslant K \cdot P^{**}(E, \varphi_1, q).$$

So

$$\sum_{U_i \in u'} d_i \cdot \varphi_1(|P_1(U_i)|) \leqslant (1+\varepsilon)h \cdot P^{**}(E, \varphi_1, q),$$

$$\sum_{U_i \in u'} f_i \cdot \varphi_1(|P_1(U_i)|) \leqslant (1+\varepsilon)^2 g \cdot P^{**}(E, \varphi_1, q).$$

$$\sum_{U_i \in u'} \varphi_2(|P_2(U_i)|) \cdot \varphi_1(|P_1(U_i)|)$$
$$\leqslant (1+\varepsilon)^2 P^{**}(F, \varphi_2, q) \cdot P^{**}(E, \varphi_1, q).$$

Let $\varepsilon \to 0$ and then $N \to +\infty$. We get

$$\sum_{U_i \in u} \varphi_2(|P_2(U_i)|) \cdot \varphi_1(|P_1(U_i)|) \leqslant P^{**}(F, \varphi_2, q) \cdot P^{**}(E, \varphi_1, q). \tag{4.12}$$

From (4.12), (4.10) and (2.1) we get

$$\begin{aligned}\sum_{U_i \in u} \varphi_1\varphi_2(|U_i|) &\leqslant \sum_{U_i \in u} \varphi_1(2^n |P_1(U_i)|) \cdot \varphi_2(2|P_2(U_i)|) \\ &\leqslant \sum_{U_i \in u} c_1{}^n \varphi_1(|P_1(U_i)|) \cdot c_2\varphi_2(|P_2(U_i)|) \\ &\leqslant c \cdot P^{**}(E, \varphi_1, q) \cdot P^{**}(F, \varphi_2, q),\end{aligned} \tag{4.13}$$

where $c = c_1{}^n \cdot c_2$ depends only on φ_1, φ_2 and n. (4.13) is valid for any packing u of $E \times F$ on the condition that $u \in \Gamma^{**}(E \times F)$, $\| u \| \leqslant q$ and q is small enough. So we have

$$P^{**}(E\times F,\ \varphi_1\varphi_2,\ q)\leqslant c\cdot P^{**}(E,\ \varphi_1,\ q)\cdot P^{**}(F,\ \varphi_2,\ q). \tag{4.14}$$

Let $q\to 0$. We get

$$P^{**}(E\times F,\ \varphi_1\varphi_2)\leqslant c\cdot P^{**}(E,\ \varphi_1)\cdot P^{**}(F,\ \varphi_2).$$

Corollary 4.2 *If $E\subset\mathbb{R}$, $F\subset\mathbb{R}^n$, then*

$$P(E\times F,\ \varphi_1\varphi_2)\leqslant c'\cdot P(E,\ \varphi_1)\cdot P(F,\ \varphi_2), \tag{4.15}$$

where $0<c'<+\infty$, c' depends only on φ_1, φ_2 and n.

Proof Use Lemma 4.1 and Lemma 4.3.

Now we can prove the main result.

Theorem 4.1 *If $E\subset\mathbb{R}$, $F\subset\mathbb{R}^n$, then*

$$\mathscr{P}^{**}(E\times F,\ \varphi_1\varphi_2)\leqslant c\cdot\mathscr{P}^{**}(E,\ \varphi_1)\cdot\mathscr{P}^{**}(F,\ \varphi_2), \tag{4.16}$$

where $0<c<+\infty$. c depends only on φ_1, φ_2 and n.

Proof According to (3.2), for any $\varepsilon>0$ there exist $\{E_i,\ i=1,\ 2,\ \cdots\}$ such that $E_i\subset\mathscr{B}(\mathbb{R})$, $E\subset\bigcup E_i$ and

$$\mathscr{P}^{**}(E,\ \varphi_1)\leqslant\sum P^{**}(E_i)\leqslant\mathscr{P}^{**}(E,\ \varphi_1)+\varepsilon.$$

We can also get $\{F_i,\ i=1,\ 2,\ \cdots\}$ so that $F_i\subset\mathscr{B}(\mathbb{R}^n)$, $F\subset\bigcup F_i$ and

$$\mathscr{P}^{**}(F,\ \varphi_2)\leqslant\sum P^{**}(F_i)\leqslant\mathscr{P}^{**}(F,\ \varphi_2)+\varepsilon.$$

Let $u=\{E_i\times F_j,\ i,\ j=1,\ 2,\ \cdots\}$. Then $E_i\times F_j\subset\mathscr{B}(\mathbb{R}\times\mathbb{R}^n)$ and $E\times F\subset\bigcup_i\bigcup_j E_i\times F_j$. So

$$\mathscr{P}^{**}(E\times F,\ \varphi_1\varphi_2)\leqslant\sum_i\sum_j P^{**}(E_i\times F_j,\ \varphi_1\varphi_2).$$

From Lemma 4.3, we have

$$P^{**}(E_i\times F_j,\ \varphi_1\varphi_2)\leqslant c\cdot P^{**}(E_i,\ \varphi_1)\cdot P^{**}(F_j,\ \varphi_2).$$

So

$$\mathscr{P}^{**}(E\times F,\varphi_1\varphi_2)\leqslant\sum_i\sum_j c\cdot P^{**}(E_i,\varphi_1)\cdot P^{**}(F_j,\varphi_2)$$
$$\leqslant c\cdot(\mathscr{P}^{**}(E,\varphi_1)+\varepsilon)\cdot(\mathscr{P}^{**}(F,\varphi_2)+\varepsilon).$$

Let $\varepsilon\to 0$. We get

$$\mathscr{P}^{**}(E\times F,\varphi_1\varphi_2)\leqslant c\cdot\mathscr{P}^{**}(E,\varphi_1)\cdot\mathscr{P}^{**}(F,\varphi_2).$$

Theorem 4.2 *If* $E\subset\mathbb{R}$, $F\subset\mathbb{R}^n$, *then*

$$\mathscr{H}(E\times F,\varphi_1\varphi_2)\leqslant c'\cdot\mathscr{H}(E,\varphi_1)\cdot\mathscr{H}(F,\varphi_2),$$

where $0<c'<+\infty$. c' *depends only on* φ_1, φ_2 *and* n.

Proof Use Theorem 4.1 and Corollary 4.1.

Acknowledgment We would like to thank Dr. Lu Jin for his help.

◇ References ◇

[1] Falconer, K. J., *The Geometry of Fractal Sets*, Cambridge University Press, New York, 1985.

[2] Falconer, K. J., *Fractal Geometry*, Wiley, New York (1990).

[3] Taylor, S. J. & Tricot, C., Packing measure, and its evaluation for a Brownian path, *Trans. Amer. Math., Soc.*, 288 (1985), 679 - 699.

[4] Tricot, C., Two definitions of fractional dimension, *Math. Proc. Cambridge Philos. Soc.*, 91 (1982), 57 - 74.

[5] Wegmann, H., Die Hausdorff-dimension von kartesischen Producten metrischer Räume, *J. Reine Angew Math.*, 246 (1971), 46 - 75.

[6] Xu You, The equivalence of packing dimension and metric dimension in Euclidean space, *Chinese Journal of Contemporary Mathematics*, 13 (1992), 73 - 77.

Quasiconformal Extension of Biholomorphic Mappings of Several Complex Variables*

Ren Fuyao
(*Fudan University*)

Ma Jianguo
(*ZhengZhou University*)

Abstract: In this paper, the authors obtain two univalence criteria and conditions for quasiconformal extension of local biholomorphic mappings defined on B^n by the method of subordination chain generalized to $\mathbb{C}^n$ by Pfaltzgraff.

Keywords: biholomorphic mapping, subordination chain, quasiconformal extension

Let $H(B^n)$ be the set of holomorphic mappings of $B^n=\{z\in\mathbb{C}^n:\ \|z\|<1\}$ into $\mathbb{C}^n$, that is, $f(z)\in H(B^n)$ if $f(z)=(f_n(z),\ \cdots,\ f_n(z))^t$ and that $f_j:\ B^n\to\mathbb{C}$ is holomorphic functions of n complex variables, $j=1,\ \cdots,\ n$. Let Df be the Jacobian matrix of mapping $f(z)\in H(B^n)$[1]. The second derivative of $f(z)$ is a symmetric bilinear operator $D^2f(z)(\cdot,\cdot)$ on $\mathbb{C}^n\times\mathbb{C}^n$ and $D^2f(z)(w,\cdot)$ is the linear operator obtained by restricting $D^2f(z)(\cdot,\cdot)$ to $w\times\mathbb{C}^n$ and $D^2f(z)(z,\cdot)$ has the matrix representation

$$D^2f(z)(z,\cdot)=\left(\sum_{m=1}^{n}\frac{\partial^2\hat{T}_R(z)}{\partial z_j\partial z_m}z_m\right),\quad 1\leqslant j,k\leqslant n.$$

If $A\in L(\mathbb{C}^n)$ is a continuous linear operator acting on $\mathbb{C}^n$, then we denote by $\|A\|=\sup\{\|Az\|,\ \|z\|\leqslant 1\}$ the norm of A. Assume that $f(z)\in$

* Originally published in *J. Fudan Univ.* (*Nat. Sci.*), Vol. 34, No. 5, (1995), 545 - 556.

$H(B^n)$, we say that $f(z)$ is local biholomorphic if Df is nonsingular everywhere in B^n.

In this paper, we shall prove the following theorems.

Theorem 1 *Let $f(z)=z+\cdots$ be a holomorphic mapping of B^n satisfying*

(1) $\| Df(z)\|^n \leqslant K|\det Df(z)|$, $z\in B^n$;

(2) $\| c\| z\|^2 I(1-\| z\|^2)Df^{-1}(z)D^2 f(z)(z,\cdot)\| \leqslant k$, $z\in B^n$;

where $c\in\mathbb{C}$, $|c|<1$, $k<1$. Then $f(z)$ can be extended quasiconformally onto $\mathbb{R}^{2n}$.

Theorem 2 *Let $f(z)=z+\cdots$ be a holomorphic mapping of B^n, $g(z)$ be a nonsingular holomorphic matrix on B^n and satisfy the following conditions:*

(1) $\| Df\|^n \leqslant K|\det Df|$;

(2) $g(0)=I$;

(3) $\| I-g^{-1}(z)Df\| \leqslant k_1$;

(4) $\|\,\| z\|^2(I-g^{-1}(z)Df(z))-(1-\| z\|^2)g^{-1}(z)Dg(z)(z,\cdot)\| \leqslant k_2$,

where k_1, $k_1<1$, $z\in B^n$. Then f can be extended quasiconformally onto $\mathbb{R}^{2n}$.

1. Quasiconformal Mappings

Let $\Omega\subset\mathbb{C}^n$ be a domain. We consider a homeomorphic mapping f of Ω onto $f(\Omega)$, $f(z)=(f_1(z),\cdots,f_n(z))^t$. Assume that the partial derivatives of f exist almost everywhere. We define the Jacobian matrix of f as

$$\frac{\partial(f,\bar f)}{\partial(z,\bar z)}=\begin{pmatrix} D_z f & D_{\bar z} f\\ D_z\bar f & D_{\bar z}\bar f\end{pmatrix}\qquad \text{where } D_z f=\left(\frac{\partial f_k}{\partial z_j}\right)_{n\times n}$$

and its norms as
$$\left\|\frac{\partial(f,\bar f)}{\partial(z,\bar z)}\right\|=\max_{\left\|\binom{\xi}{\bar\xi}\right\|=1}\left\|\begin{pmatrix} D_z f & D_{\bar z} f\\ D_z\bar f & D_{\bar z}\bar f\end{pmatrix}\begin{pmatrix}\xi\\ \bar\xi\end{pmatrix}\right\|$$

where $\left\|\begin{pmatrix}\xi\\ \bar\xi\end{pmatrix}\right\|=(\bar\xi^t\cdot\xi+\xi^t\bar\xi)^{\frac{1}{2}}=\sqrt{2}\,\|\xi\|$, $\xi\in\mathbb{C}^n$.

Definition 1 A homeomorphic mapping f of Ω onto $f(\Omega)$ is said to be K-quasiconformal if

(1) f is ACL in Ω;

(2) f is differentiable almost everywhere in Ω;

(3) $\operatorname{ess\,sup}_{z\in\Omega} \left| \left\| \left(\frac{\partial(f, \bar{f})}{\partial(z, \bar{z})} \right) \right\|^{2n} \Big/ \det \frac{\partial(f, \bar{f})}{\partial(z, \bar{z})} \right| = K < \infty.$

Here K is a constant called the dilatation of f.

Remark $f(z)$ is differentiable at z_0 if the following holds:

$$f(z) = f(z_0) D_z f(z_0)(z - z_0) + D_{\bar{z}} f(z_0)(\bar{z} - \bar{z}_0) + o(\| z - z_0 \|).$$

Remark We refer ACL property to Väisälä[2].

By a linear transformation of matrix, we can show that our definition of quasiconformal mapping is the same as in Väisälä[2]. In fact, if u_k, v_k denote respectively the real part and the image part of $f_k(z)$, and denote

$$\begin{pmatrix} D_x u & D_y u \\ D_x v & D_y v \end{pmatrix} = \frac{\partial(u_1, \cdots, u_n, v_1, \cdots, v_n)}{\partial(x_1, \cdots, x_n, y_1, \cdots, y_n)},$$

where $D_x u = \frac{\partial(u_1, \cdots, u_n)}{\partial(x_1, \cdots, x_n)}$, $D_x v = \frac{\partial(v_1, \cdots, v_n)}{\partial(x_1, \cdots, x_n)}$, then we have

$$\frac{1}{\sqrt{2}} \begin{pmatrix} I & \mathrm{i}I \\ I & -\mathrm{i}I \end{pmatrix} \begin{pmatrix} D_x U & D_y U \\ D_x V & D_y V \end{pmatrix} \frac{1}{\sqrt{2}} \begin{pmatrix} I & I \\ -\mathrm{i}I & \mathrm{i}I \end{pmatrix} = \begin{pmatrix} D_z f & D_{\bar{z}} f \\ D_z \bar{f} & D_{\bar{z}} \bar{f} \end{pmatrix}$$

where $Q = \begin{pmatrix} I & \mathrm{i}I \\ I & -\mathrm{i}I \end{pmatrix}$ is a unitary matrix; $Q^{-t} Q = I$,

hence $$\det \frac{\partial(u, v)}{\partial(x, y)} = \det \frac{\partial(f, \bar{f})}{\partial(z, \bar{z})},$$

on the other hand, writing $\xi_R = s_k + \mathrm{i}t_k$, $k = 1, 2, \cdots, n$;

we have $\begin{pmatrix} \xi \\ \bar{\xi} \end{pmatrix} = \begin{pmatrix} I & \mathrm{i}I \\ I & -\mathrm{i}I \end{pmatrix} = \begin{pmatrix} S \\ t \end{pmatrix}$, where $\xi = \begin{pmatrix} \xi_1 \\ \vdots \\ \xi_n \end{pmatrix}$, $S = \begin{pmatrix} S_1 \\ \vdots \\ S_n \end{pmatrix}$, $t = \begin{pmatrix} t_1 \\ \vdots \\ t_n \end{pmatrix}$.

A simple calculus shows that the quadratic form

$$\overline{(\xi, \bar{\xi})\left[\frac{\partial(f, \bar{f})}{\partial(z, \bar{z})}\right]^t}\left[\frac{\partial(f, \bar{f})}{\partial(z, \bar{z})}\right]\binom{\xi}{\bar{\xi}} = 2(s, t)\left(\frac{\partial(u, v)}{\partial(x, y)}\right)^t\left(\frac{\partial(u, v)}{\partial(x, y)}\right)\binom{s}{t},$$

hence

$$\max_{\left\|\binom{\xi}{\bar{\xi}}\right\|=1}\left\|\frac{\partial(f, \bar{f})}{\partial(x, \bar{z})}\binom{\xi}{\bar{\xi}}\right\| = \max_{\left\|\binom{\xi}{\xi}\right\|=1}\left\|\frac{\partial(u, v)}{\partial(x, y)}\binom{s}{t}\right\|$$

and finally

$$\left\|\frac{\partial(f, \bar{f})}{\partial(z, \bar{z})}\right\|^{2n} \Big/ \left|\det\frac{\partial(f, \bar{f})}{\partial(z, \bar{z})}\right| = \left\|\frac{\partial(u, v)}{\partial(x, y)}\right|^{2n} \Big/ \left|\det\frac{\partial(u, v)}{\partial(x, y)}\right|.$$

2. Subordination Chains and Univalence Criteria

A mapping $v(z) \in H(B^n)$ is called a Schwarz function if $\| v(z) \| \leqslant \| z \|$ for all $z \in B^n$. Now assume that $\{f(z, t)\}$ $(t \geqslant 0)$ is a one parameter function family, i. e., for every $t \geqslant 0$, $f_t(z) = f(z, t) \in H(B^n)$ and $f(t) = 0$. If for any $0 \leqslant s \leqslant t < \infty$, there exists a Schwarz function $v(z, s, t) = v_{s,t}(z)$ such that

$$f(z, s) = f(v(z, s, t), t), \quad z \in B^n,$$

then we call $\{f(z, t)\}$ a subordination chain.

The existence of such subordination chains is guaranteed by the following lemmas.

Lemma 1 *Assume that* $h_t(z) = h(z, t)$ *is a mapping from* $B^n \times [0, \infty)$ *into* $\mathbb{C}^n$ *satisfying*

(1) *for each* $t \geqslant 0$, $h_t(z) \in M = \{h \in H(B^n): h(0) = 0, Dh(0) = I, \mathrm{Re}\langle h, z\rangle \geqslant 0\}$;

(2) *for each* $z \in B^n$, $h(z, t)$ *is a measurable function of* t;

(3) *for any* $T > 0$, $r \in (0, 1)$, *there exists a constant* $K = K(r, T)$ *such that*

$$\| h(z, t) \| \leqslant K(r, T) \quad \textit{whenever} \quad \| z \| \leqslant r \textit{ and } 0 \leqslant t \leqslant T.$$

Then for each $s \geqslant 0$, $z \in B^n$, *the initial value problem*

$$\frac{\partial v}{\partial t} = -h(v, t) \quad (\text{a. e. } t \geqslant s),\ v(s) = z,$$

has a unique solution $v(t) = v(z, s, t) = e^{s-t} I z + \cdots$; *furthermore, for fixed* s *and* $t (0 \leqslant s \leqslant t)$, $v_{s,t}(z) = v(z, s, t)$ *is a univalent Schwarz function on* B^n.

The proof is a standard application of the method of successive approximation[1].

Lemma 2 *Let* $f_t(z) = f(z, t) = e^t z + \cdots$ *be a mapping from* $B^n \times [0, \infty)$ *into* $\mathbb{C}^n$ *and satisfy the following conditions:*

(1) *For each* $t \geqslant 0$, $f_t(z) \in H(B^n)$;

(2) $f(z, t)$ *is a locally absolutely continuous function of* t *and locally uniformly with respect to* $z \in B^n$;

(3) *There exists an* $h(z, t)$ *satisfying the conditions of lemma* 1 *such that*

$$\frac{\partial}{\partial t} f(z, t) = Df(z, t) h(z, t), \quad \text{a. e. } t \geqslant 0,\ z \in B^n.$$

Then $\{f(z, t) : t \geqslant 0\}$ *is a subordination chain; furthermore, if there is a sequence* $\{t_m\}$, $t_m > 0$, *increasing to* ∞ *such that*

$$\lim_{m \to \infty} e^{-t_m} f(z, t_m) = F(z),$$

locally uniformaly in B^n, *then for each* $t \geqslant 0$, (f, z, t) *is a mapping that is univalent and holomorphic on* B^n.

Example 1 Let $f(z) = z + \cdots$ be a locally biholomorphic mapping on B^n, $g(z)$ a holomorphic non-singular matrix on B^n satisfying

(1) $g(0) = I$;

(2) $\| I - g^{-1}(z) Df(z) \| < 1$;

(3) $\| \|z\|^2 \cdot (I - g^{-1}(z) Df(z)) - (1 - \|z\|^2) g^{-1}(z) Dg(z, \cdot) \| < 1$.

Then $f(z)$ is univalent on B^n. In fact, if we construct

$$f(z, t) = f(z e^{-t}) + (e^t - e^{-t}) g(z e^{-t})(z), \tag{1}$$

then by simple calculation, we have

$$\frac{\partial f(z,t)}{\partial t}=e^t g(ze^{-t})[I+E(z,t)](z)$$

and
$$Df(z,t)=e^t g(ze^{-t})[I-E(z,t)],$$
$$E(z,t)=e^{-2t}(I-g(ze^{-t})^{-1}Df(ze^{-t}))$$
$$-(1-e^{-2t})g(ze^{-t})^{-1}Dg(ze^{-t})(ze^{-t},\bullet). \tag{2}$$

It can be showed that $E(z,t)$ is holomorphic with respect to z, and $E(z,t)=0$ for $z=0$. Furthermore, by Schwarz lemma, $\| E(z,t)\| <1$ for $z\in B^n$.

If $$h(z,t)=(I-E(z,t))^{-1}(I+E(z,t))(Z),$$

then $h(z,t)$ satisfies the conditions of lemma 1 and the equation

$$\frac{\partial f(z,t)}{\partial t}=Df(z,t)h(z,t),\quad z\in B^n,$$

holds on the other hand, it can be verified that $f(z,t)$ satisfies the assumption of lemma 2, hence from the conclusion of this lemma, for each $t\geqslant 0$, $f(z,t)$ is univalent, and in particular, $f(z)=f(z,0)$ is univalent on B^n.

We may also modify lemma 2 to obtain the following univalence criterion.

Example 2 Let $f(z)=z+\cdots$ be a locally biholomorphic mapping on B^n satisfying

$$\| c\| z\|^2 I-(I-\| z\|^2)Df^{-1}(z)D^2 f(z)(z,\bullet)\| <1,\text{ where } c\in\mathbb{C},\ |c|<1,$$

then $f(z)$ is univalent on B^n.

The proof is something like in example 1. Let

$$f(z,t)=f(ze^{-t})+\left(\frac{e^t-ce^{-t}}{1-c}-e^{-t}\right)Df(ze^{-t})(z).$$

A simple computation yields

$$\frac{\partial f(z,t)}{\partial t}=\frac{e^t}{a-c}Df(ze^{-t})(I+E(z,t))(Z)$$

and
$$Df(z,t)=\frac{e^t}{1-c}Df(ze^{-t})(I-E(z,t)),$$

where $E(z,t)=ce^{-2t}I-(1-e^{-2t})Df^{-1}(ze^{-t})D^2 f(ze^{-t})(ze^{-t},\bullet)$.

3. Biholomorphic Mapping of B^n Which Can Be Extended Quasi-conformally onto $\mathbb{R}^{2n}$

In this section, we shall prove theorem 1 and theorem 2 in section 1.

From example 1 and 2 we know that if $f(z)$ satisfies the conditions of theorem 1 or theorem 2, then $f(z)$ is a biholomorphic mapping of B^n onto $f(B^n)$ and is the first element of a univalent subordination chain, $f(z)=f(z, 0)$.

Next, we shall prove that $f(z)$ can be extended continuously to B^{-n}.

Lemma 3 *Let $g(z)$ be a holomorphic matrix on B^n satisfying*

(1) *there exists a positive constant $K<\infty$ such that*

$$\| g(z) \|^n \leqslant K|\det g|, \quad z \in B^n;$$

(2) *there exists a positive constant $k<1$ such that*

$$\| g(z^{-1}) Dg(z)(z, \cdot) \| \leqslant \frac{2R}{1-\| z \|^2},$$

holds for all $z \in B^n$. Then, for $z \in B^n$, we have

$$\| g(z) \| = O(1/(1-\| z \|^R)).$$

Proof Fix $z \in \bar{B}^n$ and let $H(\xi)=g(z\xi)$, $\xi \in B^1$. By simple calculation, we obtain

$$\frac{\mathrm{d}H(\xi)}{\mathrm{d}\xi}=\left(\sum_{R=1}^{n} \frac{\partial g_{ij}(z\xi)}{\partial w_R} z_R\right)_{n\times n}, \quad w_R=z_R\xi,$$

or

$$\xi \frac{\mathrm{d}H(\xi)}{\mathrm{d}\xi}=Dg(z\xi)(z\xi, \cdot). \tag{3}$$

Further, we have
$$\frac{\mathrm{d}\det H(\xi)}{\mathrm{d}\xi}=\sum_{i=1}^{n} A_{i1} \frac{\mathrm{d}h_{i1}}{\mathrm{d}\xi}+\cdots+\sum_{i=1}^{n} A_{in} \frac{\mathrm{d}h_{in}}{\mathrm{d}\xi},$$

where A_{ij} is the cofactor of h_{ij} in $\det(h_{ij})\det H(\xi)$.

On the other hand

$$H(\xi)^{-1}=\det H(\xi)^{-1}(A_{ij})_{n\times n}^{t},$$

thus

$$\det H(\xi)^{-1}\cdot\frac{\mathrm{d}\det H(\xi)}{\mathrm{d}\xi}=\operatorname{trace}H(\xi)^{-1}\frac{\mathrm{d}H(\xi)}{\mathrm{d}\xi}. \tag{4}$$

Combining (4) and (5) yields

$$\xi\frac{\mathrm{d}\det H(\xi)}{\mathrm{d}\xi}/\det H(\xi)=\operatorname{trace}(g^{-1}(z\xi)Dg(z\xi)(z\xi,\cdot)). \tag{5}$$

Recall that if $A=(a_{ij})_{n\times n}$, and $\lambda_1\geqslant\cdots\geqslant\lambda_n\geqslant 0$ are n eigenvalues of $\overline{A}^tA$, then

$$\|A\|^2=\lambda_1 \quad\text{and}\quad \sum_{j,k=1}^{n}|a_{jk}|^2=\sum_{j=1}^{n}\leqslant n\lambda_1,$$

hence

$$|\operatorname{trace}A|\leqslant\sum_{j=1}^{n}|a_{jj}|\leqslant\left\{n\sum_{j=1}^{n}|a_{jj}|^2\right\}^{1/2}\leqslant n\lambda_1^{1/2}=n\|A\|.$$

Applying this to (5), we have

$$\left|\xi\cdot\frac{\mathrm{d}\det H(\xi)}{\mathrm{d}\xi}/\det H(\xi)\right|=|\operatorname{trace}(g^{-1}(z\xi)Dg(z\xi)(z\xi,\cdot))|$$

$$\leqslant n\|g^{-1}(z\xi)Dg(z\xi)(z\xi,\cdot)\|\leqslant\frac{2nk}{1-\|z\|^2|z|^2}\quad\text{(by condition (2)).}$$

Taking integration will yield

$$|\det H(\xi)|=O\left(\frac{1}{1-|\xi|}\right)^{nk},$$

by condition (1), we obtain

$$\|g(z\xi)\|=O\left(\frac{1}{1-1|\xi|}\right)^{k}.$$

For any $w\in B^n$, let $\xi=\|w\|$, $z=w/\|w\|$, we finally obtain

$$\|g(w)\|=O\left(\frac{1}{1-\|w\|}\right)^{k}.$$

For $f(z)$ in theorem 1, making use of above lemma and noticing that

$$\| Df^{-1}(z)D^2 f(z)(z, \cdot) \| \leqslant \frac{|c|+k}{1-\|z\|^2},$$

we have

$$\| Df(z) \| = O\left(\frac{1}{1-\|z\|}\right)^{\frac{k+|c|}{2}}.$$

For $f(z)$ in theorem 2, if we can deduce from conditions (1~3) and (4) that $g(z)$ satisfies

$$\| g(z) \|^n \leqslant K' |\det g(z)| \text{ for some positive constant } K' < \infty, \tag{6}$$

then

$$\begin{aligned} \| Df(z) \| &\leqslant \| g(z) \| \cdot \| g^{-1}(z) Df(z) \| \\ &\leqslant \| g(z) \| \cdot (k_1+1) = O\left(\frac{1}{1-\|z\|}\right)^{\frac{R_1+R_2}{2}}. \end{aligned}$$

The last equality holds because of condition (4).

Now we prove inequality (6). First of all, we recall that inequality

$$\| E^{-1} \| \leqslant \| E \|^{n-1} / |\det E|, \tag{7}$$

holds for any nonsingular square matrix E. Indeed, let $\lambda_1 \geqslant \lambda_2 \geqslant \cdots \geqslant \lambda_n > 0$ are n eigenvalues of $E^t E$, then $\| E \|^2 = \lambda_1$, $\| E^{-1} \|^2 = \lambda_n^{-1}$ and $(\det E)^2 = \lambda_1 \lambda_2 \cdots \lambda_n$, thus

$$\| E^{-1} \| = \left(\frac{\lambda_1 \lambda_2 \cdots \lambda_{n-1}}{\lambda_1 \lambda_2 \cdots \lambda_n}\right)^{\frac{1}{2}} \leqslant \| E \|^{n-1} / |\det E|.$$

Furthermore, by (3) of theorem 2, we get

$$1 - k_1 \leqslant \| g^{-1}(z) Df(z) \| \leqslant k_1 + 1. \tag{8}$$

Applying (7) and (8), we have the following inequalities

$$\begin{aligned} \| g(z) \| &\leqslant \| Df(z) \| \, \| Df^{-1}(z) g(z) \| \leqslant \| Df(z) \| \frac{\| g^{-1}(z) Df(z) \|^{n-1}}{|\det(g^{-1}(z) Df(z))|} \\ &\leqslant \| Df(z) \| \frac{(R_1+1)^{n-1}}{|\det(g^{-1}(z) Df(z))|}. \end{aligned}$$

Hence

$$\| g(z) \|^n / |\det g(z)| \leqslant \frac{\| Df \|^n}{|\det Df(z)|}(R_1+1)^{n(n-1)} \frac{1}{|\det(g^{-1}(z)Df(z))|^{n-1}},$$

notice that

$$|\det g^{-1}Df(z)| \geqslant (1-k_1)^n \text{ and } \| Df(z) \|^n / \det Df(z) \geqslant K,$$

thus

$$\| g(z) \|^n / \det g(z) \leqslant K\left(\frac{1+K_1}{1-K_1}\right)^{n(n-1)} = K' < \infty.$$

Now, we can deduce a Lipschitz condition for $f(z)$ satisfying

$$\| Df(z) \| = O\left(\frac{1}{1-\| z \|}\right)^s, \quad 0 < s < 1.$$

For any given two points w_1, $w_2 \in B^n$, let $\rho = 1 - \| w_2 - w_1 \| /3$, we have that

$$f(w_2) - f(w_1) = \int_{w_1}^{\rho w_1} Df(w)\mathrm{d}w + \int_{\rho w_1}^{\rho w_2} Df(w)\mathrm{d}w + \int_{\rho w_2}^{w_2} Df(w)\mathrm{d}w$$

and

$$\left\| \int_{w_1}^{\rho w_1} Df(w)\mathrm{d}w \right\| \leqslant \left| \int_1^{\rho} \frac{M_1}{(1-\| w_1 \| t)^s}\mathrm{d}t \right|$$

$$\leqslant M'_1(1-\rho)^{1-s} = \frac{M'_1}{3^{1-s}} \| w_2 - w_1 \|^{1-s},$$

$$\left\| \int_{\rho w_2}^{w_2} Df(w)\mathrm{d}w \right\| \leqslant \left| \int_{\rho}^{1} \frac{M_2}{(1-\| w_2 \| t)^s}\mathrm{d}t \right| \leqslant \frac{M'_2}{3^{1-s}} \| w_2 - w_1 \|^{1-s},$$

$$\left\| \int_{\rho w_1}^{\rho w_2} Df(w)\mathrm{d}w \right\| \leqslant \int_0^1 \frac{M_3}{(1-\rho)^s} \| w_2 - w_1 \| \mathrm{d}t = \frac{M_3}{3^{1-s}} \| w_2 - w_1 \|^{1-s},$$

hence

$$\| f(w_2) - f(w_1) \| \leqslant M(\| w_2 - w_1 \|)^{1-s},$$

and the continuous extension of $f(z)$ to ∂B^n thus follows.

Let $f(z, t)$ be the subordination chain appearing in example 1 or 2. We also need a Lipschitz inequality for $f(z, t)$. First, if

$$f(z, t)=f(ze^{-t})+(e^{t}-e^{-t})g(ze^{-t})(z),$$

where $f(z)$, $g(z)$ satisfy the conditions of theorem 2, then

$$Df(z, t)=e^{t}g(ze^{-t})(I-E(z, t)),$$

where

$$\| E(z, t) \| \leqslant k_2 < 1 \quad \text{(see example 1)}.$$

For fixed $t \geqslant 0$ and by lemma 3, we have

$$e^{-t} \| Df(z, t) \| \leqslant \| g(ze^{-t}) \| (1+k_2)=O\left(\frac{1}{1-\| z \|}\right)^{k}. \tag{9}$$

Hence

$$\| f(z_2, t)-f(z_1, t) \| \leqslant e^{t}M \| z_2 - z_1 \|^{1-k}, \tag{10}$$

holds for any z_1, $z_2 \in B^n$ and M is a positive constant.

Employing the same method, we can obtain Lipschitz inequality (10) for

$$f(z, t)=f(ze^{-t})+\left(\frac{e^{t}-ce^{-t}}{1-c}-e^{-t}\right)Df(ze^{-t})(z)$$

where $f(z)$ satisfies the conditions of theorem 1, and we have

$$e^{-t} \| Df(z, t) \| \leqslant \frac{1}{|1-c|} \| Df(ze^{-t}) \| (1+k)=O\left(\frac{1}{1-\| z \|}\right)^{k}. \tag{11}$$

Next, we shall construct a quasiconformal mapping $F: \mathbb{R}^{2n} \to \mathbb{R}^{2n}$ such that $F|_{B^n}=f$.

For $f(z)$ in theorem 1 and theorem 2, we define mapping

$$F_\rho(z)=\begin{cases} \dfrac{1}{\rho}f(\rho z, 0), & \| z \| < 1, \\ \dfrac{1}{\rho}f\left(\dfrac{\rho z}{\| z \|}, \ln \| z \|\right), & \| z \| \geqslant 1, \end{cases}$$

where $\rho < 1$ is a positive number.

Since $f(w, t)$ is a univalent subordination chain on B^n, it is easy to show that $F_\rho(z)$ is a homeomorphic mapping of $\mathbb{R}^{2n}$, and more details can be found in Becker's[3]. Further by (10), we can show that $E_\rho(z)$ converge to a mapping F uniformly on compact in $\mathbb{R}^{2n}$ as ρ increases to 1, in fact, let $\{\rho_i\}$ $i=1, 2, \cdots$ be a sequence increasing to 1,

$$F_{\rho_n}(z)-F_{\rho_m}(z)=\frac{1}{\rho_n}f\left(\frac{\rho_n z}{\|z\|}, \ln\|z\|\right)-\frac{1}{\rho_m}f\left(\frac{\rho_m z}{\|z\|}, \ln\|z\|\right)$$

$$=\frac{1}{\rho_n}\left[f\left(\frac{\rho_m z}{\|z\|}, \ln\|z\|\right)-f\left(\frac{\rho_m z}{\|z\|}, \ln\|z\|\right)\right]$$

$$+\left(\frac{1}{\rho_n}-\frac{1}{\rho_m}\right)f\left(\frac{\rho_m z}{\|z\|}\ln\|z\|\right),$$

by (10),

$$\left\|f\left(\frac{\rho_n z}{\|z\|}, \ln\|z\|\right)-f\left(\frac{\rho_m z}{\|z\|}, \ln\|z\|\right)\right\|\leqslant\|z\|M(\rho_n-\rho_m)^{1-k}.$$

Taking $\frac{\rho_m z}{\|z\|}$ for z_1, 0 for z_2 in (10) and noticing that $f(0, t)=0$, we have

$$\left\|f\left(\frac{\rho_m z}{\|z\|}, \ln\|z\|\right)\right\|\leqslant\|z\|M\rho_m^{1-k}.$$

Hence $F_{\rho_i}(z)$ converge to $F(z)=\begin{cases} f(z, 0), & \|z\|<1, \\ f\left(\frac{z}{\|z\|}, \ln\|z\|\right), & \|z\|\geqslant 1 \end{cases}$

uniformaly on compact in $\mathbb{R}^{2n}$ when $i\to\infty$.

Now, if we can show that $F_\rho(z)$ is a quasiconformal mapping of $\mathbb{R}^{2n}$ with dilatation bounded above by a constant K independent of ρ, then Väisälä's[6] told us that $F(z)$ must be a quasiconformal mapping of the same dilatation.

To prove that $F_\rho(z)$ is ACL and differentiable a. e., we fix $\rho<1$ and consider $f_\rho(z, t)=\frac{1}{\rho}f(\rho z, t)$, $z\in\overline{B}^n$. Since

$$e^t\|Df_\rho(z, t)\|=e^t\left\|\frac{1}{\rho}Df(\rho z, t)\right\|=O\left(\frac{1}{1-\rho\|z\|}\right)^k$$

(the last equality holds because of (9) and (11)),

we know that $\|Df_\rho(z, t)\|$ is bounded above by a constant depending only upon ρ. Then, taking integration for $e^t Df_\rho(z, t)$, we obtain that

$$e^{-t} \| f_\rho(z, t) - f_\rho(w, t) \| \leqslant M_\rho \| z - w \|, \quad z, w \in \overline{B}^n, \tag{12}$$

where $M_\rho > 0$ depends only upon $\rho > 1$. Next, for $0 \leqslant s < t < \infty$, fix $z \in \overline{B}^n$ and make use of (12), we have

$$\begin{aligned} e^{-t} \| f_\rho(z, t) - f_\rho(z, s) \| &= e^{-t} \| f_\rho(z, t) - f_\rho(v(z, s, t,), t) \| \\ &\leqslant M_\rho \| z - v(z, s, t) \| . \end{aligned}$$

By Pfaltzgraff's[4]

$$\| z - v(z, s, t) \| = \| \int_s^t h(v(z, s, t), \tau) d\tau \| ,$$

where

$$\| h(v, \tau) \| = \| (I - E)^{-1}(I + E)v \| \leqslant (1+k)/(1-k).$$

Hence

$$\| z - v(z, s, t) \| \leqslant C(t - s)$$

and

$$e^{-1} \| f_\rho(z, t) - f_\rho(z, s) \| \leqslant L_\rho(t - s), \tag{13}$$

where L_ρ is a positive constant depending only upon ρ.

(12) and (13) imply that $F_\rho(z)$ is ACL and

$$L(z, F_\rho(z)) = \limsup_{h \to 0} \frac{| F_\rho(z + h) - F_\rho(z) |}{h} < \infty.$$

Then by the theorem of Rademacher and Stepanov (see Väisälä[6]), $E_\rho(z)$ is differentiable a. e. the following is devoted to a proof that the dilatation of $F_\rho(z)$ is bounded above by a constant.

Let

$$f(z, t) = f(ze^{-t}) + \left(\frac{e^t - Ce^{-t}}{1 - C} - e^{-t} \right) Df(ze^{-t})(z).$$

We have shown that

$$Df(z, t)=\frac{e^t}{1-C}Df(ze^{-t})(I-E(z, t))$$

and

$$\frac{\partial f(z, t)}{\partial t}=\frac{e^t}{1-c}Df(Ze^{-t})(I+E(z, t))(z),$$

where $\| E(z, t) \| \leqslant k < 1$. If we denote w for $\frac{\rho z}{\| z \|}$ and t for $\ln \| z \|$ and notice that by the above equality, $Df(w, t) = Df(w, \ln \| z \|) = \frac{\| z \|}{1-C}Df\left(\frac{w}{\| z \|}, 0\right)[I-E(w, t)]$, then we have

$$\begin{aligned}
D_z F_\rho(z) &= \frac{1}{\rho}D_z f\left(\frac{\rho z}{\| z \|}, \ln \| z \|\right) \\
&= \frac{1}{\rho}\left[Df(w, t)\frac{\partial}{\partial z}\left(\rho\frac{z}{\| z \|}\right)+\frac{\partial f(w, t)}{\partial t}\frac{\partial \ln \| z \|}{\partial z}\right] \\
&= \frac{1}{\rho}\left[Df(w, t)\left(\rho \| z \|^{-1} I-\frac{1}{2}\frac{\rho}{\| z \|^3}z\bar{z}^{-t}\right)+\frac{\partial f(w, t)}{\partial t}\frac{1}{2\| z \|}\bar{z}^t\right] \\
&= \| z \|^{-1}Df(w, t)\left[I+\frac{1}{2\rho^2}(h(w, t)-w)\bar{w}^{-t}\right] \\
&= (1-c)^{-1}Df\left(\frac{w}{\| z \|, 0}\right)\left(I-E(w, t)+\frac{1}{\rho^2}E(w, t)w\bar{w}^{-t}\right),
\end{aligned}$$

where we recall that $h(w, t)=(I-E(w, t))^{-1}(I+E(w, t))(w)$ and

$$\frac{\partial f(w, t)}{\partial t}=Df(w, t)h(w, t).$$

By a similar computation, we have

$$D_{\bar{z}}F_\rho(z)=(1-c)^{-1}\frac{1}{\rho^2}\cdot Df\left(\frac{w}{\| z \|}, 0\right)E(w, t)w\bar{w}^{-t}.$$

Hence the Jacobi matrix

$$\frac{\partial(F_\rho(z), \bar{F}_\rho(z))}{\partial(z, \bar{z})}=(1-c)^{-1}\begin{pmatrix} Df\left(\frac{w}{\| z \|}, 0\right)A & Df\left(\frac{w}{\| z \|}, 0\right)B \\ Df\left(\frac{w}{\| z \|}, 0\right)\bar{B} & Df\left(\frac{w}{\| z \|}, 0\right)\bar{A} \end{pmatrix}$$

$$=(1-c)^{-1}\begin{pmatrix} Df\left(\frac{w}{\| z \|},\ 0\right) & 0 \\ 0 & Df\left(\frac{w}{\| z \|},\ 0\right) \end{pmatrix}\begin{pmatrix} A & B \\ \bar{B} & \bar{A} \end{pmatrix},$$

where

$$A=I-E(w,\ t)+\frac{1}{\rho^2}E(w,\ t)ww^{-t},$$

$$B=\frac{1}{\rho^2}E(w,\ t)ww^t.$$

The norm of the Jacobi matrix

$$\left\| \frac{\partial(F_\rho,\ \bar{F}_\rho)}{\partial(z,\ \bar{z})}\begin{pmatrix} \xi \\ \bar{\xi} \end{pmatrix} \right\| = |1-c|^{-1}\left\| \begin{pmatrix} \overline{Df\left(\frac{w}{\| z \|},\ 0\right) A\xi + Df\left(\frac{w}{\| z \|},\ 0\right) B\bar{\xi}} \\ Df\left(\frac{w}{\| z \|},\ 0\right) B\xi + Df\left(\frac{w}{\| z \|},\ 0\right) A\bar{\xi} \end{pmatrix} \right\|$$

$$=|1-c|^{-1}\sqrt{2\left\| Df\left(\frac{w}{\| z \|},\ 0\right) A\xi + Df\left(\frac{w}{\| z \|},\ 0\right) B\bar{\xi} \right\|^2}$$

$$\leqslant |1-c|^{-1}\left\| Df\left(\frac{w}{\| z \|},\ 0\right) \right\| \left\| A\sqrt{2}\xi + B\sqrt{2}\bar{\xi} \right\|.$$

Let $\xi'=\sqrt{2}$. Then

$$\left\| \frac{\partial(F_\rho,\ \bar{F}_\rho)}{\partial(z,\ \bar{z})} \right\| \leqslant |1-c|^{-1}\left\| Df\left(\frac{w}{\| z \|},\ 0\right) \right\| \cdot \max_{\| \xi=1 \|}\| A\xi' + B\bar{\xi}' \|,$$

but

$$\| A\xi' + B\bar{\xi}' \| = \left\| \xi' - E\xi' + \frac{<\xi',\ w>}{\rho^2}Ew + \frac{\overline{<\xi',\ w>}}{\rho^2}Ew \right\|$$

$$= \left\| \xi' - E\left(\xi' - \frac{<\xi',\ w> + \overline{<\xi',\ w>}}{\rho^2}\cdot w\right) \right\|.$$

It is easy to vertify that

$$\left\| \xi' - \frac{<\xi,\ w> + \overline{<\xi',\ w>}}{\rho^2}w \right\| = 1.$$

Hence

$$\max_{\|\xi'\|=1} \| A\xi' + B\bar{\xi}' \| \leqslant \| \xi' \| + \| E \| \leqslant 1+k$$

and

$$\left\| \frac{\partial(F_\rho, \bar{F}_\rho)}{\partial(z, \bar{z})} \right\|^{2n} \leqslant |1-c|^{-2n} (1+k)^{2n-1} \left\| Df\left(\frac{w}{\|z\|}\right), 0 \right\|^{2n} \left\| \begin{vmatrix} A & B \\ \bar{B} & \bar{A} \end{vmatrix} \right\|.$$

Here we have used the fact that $\left\| \begin{vmatrix} A & B \\ \bar{B} & \bar{A} \end{vmatrix} \right\| = \max_{\|\xi'\|=1} \| A\xi' + B\bar{\xi}' \|.$

On the other hand

$$\left| \det \frac{\partial(F_\rho, \bar{F}_\rho)}{\partial(z, \bar{z})} \right| = (1-c)^{-2n} \left| \det Df\left(\frac{w}{\|z\|}, 0\right) \right|^2 \left| \det\begin{pmatrix} A & B \\ \bar{B} & \bar{A} \end{pmatrix} \right|.$$

But

$$\left| \det\begin{pmatrix} A & B \\ \bar{B} & \bar{A} \end{pmatrix} \right| \geqslant \left\| \begin{vmatrix} A & B \\ \bar{B} & \bar{A} \end{vmatrix} \right\| \begin{pmatrix} A & B \\ \bar{B} & \bar{A} \end{pmatrix}^{n-1},$$

where we denote $1\begin{pmatrix} A & B \\ \bar{B} & \bar{A} \end{pmatrix}$ for $\min_{\|\xi'\|=1} \| A\xi' + B\bar{\xi}' \| \geqslant 1-k.$

Hence

$$\left| \det \frac{\partial(F_\rho, \bar{F}_\rho)}{\partial(z, \bar{z})} \right| \geqslant |1-c|^{-2n} \left| \det Df\left(\frac{w}{\|z\|}, 0\right) \right|^2 \cdot \left\| \begin{vmatrix} A & B \\ \bar{B} & \bar{A} \end{vmatrix} \right\| \cdot (1-k)^{2n-1}.$$

Finally

$$\left\| \frac{\partial(F_\rho, \bar{F}_\rho)}{\partial(z, \bar{z})} \right\|^{2n} \Big/ \left| \det \frac{\partial(F_\rho, \bar{F}_\rho)}{\partial(z, \bar{z})} \right| \leqslant \frac{\left\| Df\left(\frac{w}{\|z\|}, 0\right) \right\|^{2n}}{\left| \det Df\left(\frac{w}{\|z\|}, 0\right) \right|^2} \left(\frac{1+k}{1-k}\right)^{2n-1}$$

$$\leqslant K^2 \left(\frac{1+k}{(1-k)}\right)^{2n-1}.$$

Remark 1 When $n=1$ and $c=0$, $K=1$, we obtain Becker's result[3]. For $f(z)$ satisfying the conditions of theorem 2, and

$$f(z, t) = f(ze^{-t}) + (e^t - e^{-t}) g(ze^{-t})(z),$$

by the same argument, we can show that there is a quasiconformal extension to $\mathbb{R}^{2n}$ for $f(z)$ with dilatation $\leqslant K^2\left(\frac{1+k_1}{1-k_1}\right)^{2n(n-1)}\left(\frac{1+k_2}{1-k^2}\right)^{2n-1}$.

Remark 2 Taking $g(z)$ in theorem 2 to be special matrix I or $Df(z)$, we obtain respectively Brodskii's and Pfaltzgraff's results[4, 5].

◇ References ◇

[1] Becker J. Lownersche differentialglei chung und quasikonfor fortsetzbare schlichte funktionen. *J Reine Angew Math*, 1972, 255: 23 - 43.

[2] Brodskii A A, Quasiconformal extension of biholomorphic mappings (Russian), Theory of mappings and approximation of functions, 30 - 34 Kiev: Naukona Dumka, 1983.

[3] Pfaltzgraff J A. Subordination chains and univalence of holomorphic mappings in C^n. *Math Ann*, 1974, 210: 55 - 68.

[4] Pfaltzgraff J A. Subordination chains and quasi conformal extension of holomorphic maps in C^n. *Ann Acad Sci Fenn Ser A I Math*, 1975 1: 13 - 25.

[5] Väisälä J. Lecture on n-dimensional quasi conformal mappings. Lecture Notes in Mathematics 229. Berlin-Heidelberg-New York: Springer-Verlag, 1971.

多复变双全纯映照的拟共形扩张

任福尧　　　　马建国
(复旦大学)　　(郑州大学)

提　要: 利用 Pfaltzgraff 推广到$\mathbb{C}^n$空间的从属链的方法,获得两个 B^n 上双全纯映照拟共形扩张到 $\mathbb{R}^{2n}$ 上的条件.

关键词: 双全纯映照,从属链,拟共形扩张

Bounded Projections and Duality on Spaces of Holomorphic Functions in the Unit Ball of $\mathbb{C}^{n*}$

Ren Fuyao & Xiao Jianbin

Abstract: In this paper three Banach spaces $A_0(\varphi)$, $A_\infty(\varphi)$ and $A^1(\varphi)$ of functions holomorphic in the unit ball B of $\mathbb{C}^n$ are defined. We exhibit bounded projections from $C_0(B)$ onto $A_0(\varphi)$, from $L^1(B)$ onto $A^1(\varphi)$, and from $L^\infty(B)$ onto $A_\infty(\varphi)$. Using these projections, we show that $A_0(\varphi)^* \cong A^1(\varphi)$ and $A^1(\varphi)^* \cong A_\infty(\varphi)$.

1. Introduction

Throughout this paper, n will be a fixed positive integer, and $\mathbb{C}^n$ will be the vector space of all ordered n-tuples $z=(z_1, \cdots, z_n)$ of complex numbers, with inner product

$$\langle z, w\rangle = z_1\overline{w}_1 + z_2\overline{w}_2 + \cdots + z_n\overline{w}_n$$

and norm

$$|z| = \langle z, z\rangle^{1/2}.$$

Let B denote the unit ball of $\mathbb{C}^n$, the letter v denote the Lebesgue measure on $\mathbb{C}^n$, normalized so that $v(B)=1$. And let S be the boundary of B, the letter

---

* Supported in part by the National Natural Science Foundation of China. Originally published in *Acta Math. Sinica* (*N. S.*), Vol. 11, No. 1, (1995), 29–36.

σ denote the positive rotation-invariant measure on S with $\sigma(S)=1$. The class of all holomorphic functions with domain B will be denoted by $H(B)$.

Let $\varphi(t)$ be positive and continuous in $(0, 1]$, with

(A) $\varphi(t)$ is increasing;

(B) $\int_\delta^1 \varphi(t)t^{-2}\mathrm{d}t = O(\varphi(\delta)/\delta)$ for $\delta > 0$.

For $f \in H(B)$, we define

$$\| f \|_\varphi = \sup_{z\in B} \frac{1-|z|^2}{\varphi(1-|z|)}|f(z)| = \sup_{0\leqslant r<1} \frac{1-r^2}{\varphi(1-r)} M_\infty(r, f),$$

$$\| f \|_{1,\varphi} = \int_B \varphi(1-|z|)|f(z)|\mathrm{d}v(z),$$

where

$$M_\infty(r, f) = \max_{|z|=r} |f(z)|.$$

We shall be concerned with the following spaces of holomorphic functions.

$$A_\infty = A_\infty(\varphi) = \{f \in H(B): \| f \|_\varphi < \infty\},$$

$$A_0 = A_0(\varphi) = \left\{f \in A_\infty(\varphi): \lim_{r\to 1} M_\infty(r, f)\frac{1-r^2}{\varphi(1-r)} = 0\right\},$$

$$A^1 = A^1(\varphi) = \{f \in H(B): \| f \|_{1,\varphi} < \infty\}.$$

In the case $\varphi(t) = t^\alpha (0 < \alpha < 1)$, Duren-Romberg-Shields[1] studied the representation of linear functionals on A^1 spaces in the unit disk U of $\mathbb{C}$. In [2], Chen generalized Duren-Romberg-Shields' results to the unit ball B of $\mathbb{C}^n$. Furthermore, if φ satisfies (A) and (B) and

(C) $t/\varphi(t)$ is increasing and $t/\varphi(t)\to 0$ as $t\to 0$;

(D) $\int_0^\delta \varphi(t)/t\,\mathrm{d}t = O(\varphi(\delta))$ for $\delta > 0$.

Lyevshina[3] defined the spaces $\Lambda_\varphi(U)$, $\lambda_\varphi(U)$, $\Gamma_\varphi(U)$ in the unit disk U of $\mathbb{C}$, which are similar to $A_\infty(\varphi)$, $A_0(\varphi)$, $A^1(\varphi)$, respectively. And he obtained that $\lambda_\varphi(U)^* \cong \Gamma_\varphi(U)$ and $\Gamma_\varphi(U)^* \cong \Lambda_\varphi(U)$, where X^* denotes the dual of X, and $\cong$ denotes topological isomorphism. In [4], Shi extended Lyevshina's results to the unit ball B of $\mathbb{C}^n$. Shi's proof was based on fractional derivative and fractional integration.

In this paper, for the class φ which satisfies (A) and (B), we exhibit bounded projections from $C_0(B)$ onto $A_0(\varphi)$, from $L^1(B)$ onto $A^1(\varphi)$, and from $L^\infty(B)$ onto $A_\infty(\varphi)$. Using these projections, we show that $A_0(\varphi)^* \cong A^1(\varphi)$ and $A^1(\varphi)^* \cong A_\infty(\varphi)$. Our methods used here are different from Shi's.

The letter c denotes a finite constant, not necessarily the same at each appearance.

2. Some Lemmas

We can show that $A_0(\varphi)$ contains a large class of holomorphic functions, for example, the bounded holomorphic functions belong to $A_0(\varphi)$. It suffices to show that $\lim\limits_{r\to 1}\dfrac{1-r^2}{\varphi(1-r)}=0$, which can be proved by the following lemma.

Lemma 1 *If φ satisfies (A) and (B), then there exists a number s, $0<s<1$, such that $\varphi(t)\geqslant ct^s$ for $t\in(0, 1]$.*

Proof Define a function $\psi(t)$ by the equation

$$\psi(t)=t\int_t^1 r^{-2}\varphi(r)\mathrm{d}r. \tag{1}$$

Note that $\psi(t)\leqslant \alpha\varphi(t)$ for some constant α, and we may assume $\alpha>1$. Differentiating (1) gives

$$\psi'(t)=t^{-1}(\psi(t)-\varphi(t)),$$

which leads to

$$\psi'(t)/\psi(t)\leqslant t^{-1}(1-1/\alpha).$$

Let $s=1-1/\alpha$ and $c=\psi(1/2)>0$. If $t\leqslant 1/2$, then integrating from t to $1/2$ gives

$$\ln(c/\psi(t))\leqslant -s\ln t+s\ln(1/2),$$

or

$$\varphi(t)\geqslant \alpha^{-1}\psi(t)\geqslant c\alpha^{-1}2^s t^s. \tag{2}$$

If $1/2 < t \leqslant 1$, then $\varphi(t) \geqslant \varphi(1/2) \geqslant \varphi(1/2)t^s$, which proves the lemma.

The following formula will be needed in the sequel:

$$\int_B F(z)\mathrm{d}v(z) = 2n\int_0^1 r^{2n-1}M_1(r, F)\mathrm{d}r, \quad F \in L^1(B), \tag{3}$$

where

$$M_1(r, F) = \int_S |F(r\zeta)|\mathrm{d}\sigma(\zeta).$$

(3) is just the integration in polar coordinates.

Lemma 2 *Let A denote any one of the three spaces A_∞, A_0, A^1.*

(i) *If D is a bounded subset of A, then the functions in D are uniformly bounded on each compact subset of B;*

(ii) *If $\{f_k\}$ is a Cauchy sequence in A, then it converges uniformly on each compact subset of B;*

(iii) *A point evaluation at any point of B is a bounded linear functional on A;*

(iv) *A is a Banach space;*

(v) *A_0 is a closed subspace of A_∞.*

Proof (i) This is obvious for A_∞ and A_0. In A^1 case, suppose D is a bounded subset of A^1, $f \in D$. For $|z| \leqslant R < 1$, the Cauchy integral formula gives

$$f(z) = \int_S f(\rho\zeta)(1-\rho^{-1}\langle z, \zeta\rangle)^{-n}\mathrm{d}\sigma(\zeta),$$

where $\rho = (1+R)/2$. Hence

$$|f(z)| \leqslant ((1+R)/(1-R))^n M_1(\rho, f).$$

Also formula (3) gives

$$\| f \|_{1,\varphi} \geqslant \int_\rho^1 \varphi(1-r)M_1(r, f)\mathrm{d}r \geqslant M_1(\rho, f)\int_0^{1-\rho}\varphi(r)\mathrm{d}r,$$

since $M_1(r, f)$ is an increasing function on r. The result follows.

(ii) and (iii) follow from the above estimate for $f(z)$.

(iv) It is only necessary to establish completeness, and this is easy for A_∞

and A_0. If $\{f_k\}$ is a Cauchy sequence in A^1, then it converges uniformly on compact sets to the holomorphic function f, by (ii). Also, $f \in A^1$ by Fatou's lemma. Thus A^1 is complete.

(v) This follows from (iv).

Define $f_r(z)=f(rz)$, $0 \leqslant r < 1$, for $f \in H(B)$.

Lemma 3 (i) *For $f \in A^1(\varphi)$ or $A_0(\varphi)$, $f_\rho \to f$ in norm as $\rho \to 1$.*

(ii) *The polynomials in $z_1, \cdots, z_n$ are dense in $A^1(\varphi)$ and in $A_0(\varphi)$.*

Proof (i) This is obvious for A_0. For $f \in A^1$ and $\varepsilon > 0$ choose $R < 1$ so that

$$\int_{|z|>R} |f(z)| \varphi(1-|z|) \mathrm{d}v(z) = 2n \int_R^1 r^{2n-1} \varphi(1-r) M_1(r, f) \mathrm{d}r < \varepsilon.$$

Since $M_1(r, f_\rho) = M_1(\rho r, f) \leqslant M_1(r, f)$, we have

$$\int_{|z|>R} |f_\rho(z)| \varphi(1-|z|) \mathrm{d}v(z) < \varepsilon.$$

Choose ρ so that $|f(z) - f_\rho(z)| < \varepsilon$ on $|z| \leqslant R$. Then

$$\int_B |f(z) - f_\rho(z)| \varphi(1-|z|) \mathrm{d}v(z)$$

$$< \varepsilon \int_{|z| \leqslant R} \varphi(1-|z|) \mathrm{d}v(z) + 2\varepsilon \leqslant \varepsilon(\varphi(1)+2).$$

So $f_\rho \to f$ in norm as $\rho \to 1$.

(ii) In either A_0 or A^1, if $\varepsilon > 0$ is given, choose ρ so that $\| f - f_\rho \| < \varepsilon$, which is possible by (i). Since the multiple power series of f converges uniformly to f on every compact subset of B, we may choose a polynomial P so that $|f_\rho(z) - P(z)| < \varepsilon$ on all of B. The result follows from $\| f - P \| \leqslant \| f - f_\rho \| + \| f_\rho - P \|$.

Lemma 4 *Let φ satisfies* (A) *and* (B). *Then*

$$\int_0^1 (1-rR)^{-2} \varphi(1-r) \mathrm{d}r \leqslant c(1-R)^{-1} \varphi(1-R), \quad 0 \leqslant R < 1.$$

Proof Since φ satisfies (A) and (B), then

$$\int_0^1 (1-rR)^{-2} \varphi(1-r) \mathrm{d}r \leqslant \int_0^R (1-r)^{-2} \varphi(1-r) \mathrm{d}r + \varphi(1-R) \int_R^1 (1-rR)^{-2} \mathrm{d}r$$

$$\leqslant c(1-R)^{-1} \varphi(1-R).$$

3. Duality

We shall use the following pairing between $A_\infty(\varphi)$ and $A^1(\varphi)$:

$$(f, g)=\int_B f(z)g(\bar{z})(1-|z|^2)\mathrm{d}v(z), \quad f\in A_\infty(\varphi), g\in A^1(\varphi). \tag{4}$$

Note that (f, g) is unchanged if $f(z)g(\bar{z})$ is replaced by $f(\bar{z})g(z)$.

Lemma 5 *Let*

$$K_w(z)=(n+1)(1-\langle w, \bar{z}\rangle)^{-n-2}, \ z, w\in B. \tag{5}$$

Then

(i) K_w *is in both* $A_0(\varphi)$ *and* $A^1(\varphi)$;

(ii) $g(w)=(K_w, g)$, *for all* $g\in A^1(\varphi)$;

(iii) $f(w)=(f, K_w)$, *for all* $f\in A_0(\varphi)$.

Proof (i) K_w is holomorphic for $|z|<|w|^{-1}$ and so is in both A_0 and A^1.

(ii) The result follows for polynomials in $z_1, \cdots, z_n$ from Theorem (i) of [5]. The general case follows from Lemmas 2(iii) and 3 (ii) (if the two bounded linear functionals agree on a dense set then they agree identically).

(iii) Let $f\in A_\infty$. It is easily known that $\int_B |f(z)|(1-|z|^2)\mathrm{d}v(z)<\infty$. By the Lebesgue dominated convergence theorem,

$$(f, K_w)=(n+1)\lim_{t\to 1}\int_{|z|<t} f(z)(t^2-|z|^2)(1-\langle w, z\rangle)^{-n-2}\mathrm{d}v(z). \tag{6}$$

Now in the integral on the right hand side of the above equality replace $(1-\langle w, z\rangle)^{-n-2}$ by its multiple power series

$$\begin{aligned}(1-\langle w, z\rangle)^{-n-2}&=\sum_{k=0}^{\infty}\frac{(k+n+1)!}{k!\,(n+1)!}\langle w, z\rangle^k\\&=\sum_{k=0}^{\infty}\sum_{|\alpha|=k}\frac{(k+n+1)!}{\alpha!\,(n+1)!}w^\alpha\bar{z}^\alpha,\end{aligned} \tag{7}$$

replace f by its multiple power series $f(z)=\sum_{k=0}^{\infty}\sum_{|\alpha|=k} a_\alpha z^\alpha$, and then write z in polar coordinates. Then computation by means of Propositions 1. 4. 8 and 1. 4. 9 in [6] yields that the right hand side of (6) equals

$$\lim_{t\to 1} t^{2n+2} f(t^2 w).$$

Since the above limit equals $f(w)$, the proof is complete.

Let $C_0(B)$ denote the Banach space of continuous functions on the closed ball that vanish on the boundary, with the supremum norm. Also let $L^1(B)$ and $L^\infty(B)$ denote, respectively, the usual Banach spaces of integrable and essentially bounded measurable functions associated with Lebesgue measure on B. The maps

$$T_0: A_0(\varphi)\to C_0(B),\quad T_\infty: A_\infty(\varphi)\to L^\infty(B),\quad T_1: A^1(\varphi)\to L^1(B),$$

defined by $(T_0 f)(z)=f(z)(1-|z|^2)/\varphi(1-|z|)$, $(T_\infty f)(z)=f(z)(1-|z|^2)/\varphi(1-|z|)$, $(T_{1g})(z)=g(z)\varphi(1-|z|)$ are isometries. We use the following notation for the ranges of these operators.

Notation. $TA_0=T_0A_0(\varphi)$, $TA^1=T_1A^1(\varphi)$, $TA_\infty=T_\infty A_\infty(\varphi)$.

Thus TA_0 is a subspace of $C_0(B)$, TA^1 is a subspace of $L^1(B)$, and TA_∞ is a subspace of $L^\infty(B)$. These subspaces are closed by Lemma 2.

Let $M(B)$ denote the Banach space of complex-valued, bounded Borel measures on B with the variation norm. The map

$$M: A^1(\varphi)\to M(B)$$

defined by $(Mg)(z)=g(z)\varphi(1-|z|)dv(z)$ is an isometry of $A^1(\varphi)$ onto a closed subspace of $M(B)$, which we denote by MA^1.

We now come to one of our main results.

Theorem 1 (i) *The transformation P defined by*

$$(Ph)(w)=\int_B K_w(\bar z)h(z)\varphi(1-|z|)dv(z),\quad h\in L^\infty(B),\quad w\in B, \tag{8}$$

is a bounded operator mapping $L^\infty(B)$ onto $A_\infty(\varphi)$; the operator $T_\infty P$ is a *bounded projection of* $L^{\infty}(B)$ *onto the subspace* TA_{∞}.

(ii) *The transformation* $P_0 = P \mid C_0(B)$ *is a bounded operator mapping* $C_0(B)$ *onto* $A_0(\varphi)$; *the operator* T_0P_0 *is a bounded projection of* $C_0(B)$ *onto the subspace* TA_0.

(iii) *The transformation* Q *defined by*

$$(Q\mu)(w) = \int_B K_w(\bar{z})(1-|z|^2)/\varphi(1-|z|)\mathrm{d}\mu(z), \quad \mu \in M(B), \quad w \in B, \tag{9}$$

is a bounded operator mapping $M(B)$ *onto* $A^1(\varphi)$; *the operator* MQ *is a bounded projection of* $M(B)$ *onto the subspace* MA^1.

(iv) *The transformation* $Q_1 = Q \mid L^1(B)$ *is a bounded operator mapping* $L^1(B)$ *onto* $A^1(\varphi)$; *the operator* T_1Q_1 *is a bounded projection of* $L^1(B)$ *onto the subspace* TA^1.

Proof (i) From Formula (3) and Proposition 1.4.10 of [6], we have

$$|(Ph)(w)| \leqslant \|h\|_{\infty}\int_0^1 2n\varphi(1-r)M_1(r, K_w)\mathrm{d}r$$
$$\leqslant c\|h\|_{\infty}\int_0^1 (1-r|w|)^{-2}\varphi(1-r)\mathrm{d}r.$$

It follows from Lemma 4 that

$$|(Ph)(w)| \leqslant c\|h\|_{\infty}\varphi(1-|w|)/(1-|w|^2). \tag{10}$$

Thus P is a bounded operator mapping $L^{\infty}(B)$ into $A_{\infty}(\varphi)$. Now let $f \in A_{\infty}(\varphi)$ be given. Then from Lemma 5 (iii),

$$(P(T_{\infty}f))(w) = (f, K_w) = f(w).$$

Thus $PT_{\infty} = I$, the identity on $A_{\infty}(\varphi)$, and so P is onto and $T_{\infty}P$ is a bounded projection of $L^{\infty}(B)$ onto the subspace TA_{∞}. This proves (i).

(ii) This will follow from (i) if it can be shown that $h \in C_0(B)$ implies $P_0h \in A_0(\varphi)$. Given $\varepsilon > 0$, choose R, $0 < R < 1$, such that $|h(z)| < \varepsilon$ for $|z| > R$. Then

$$|(Ph)(w)| \leqslant \left(\int_{|z|\leqslant R} + \int_{|z|>R}\right)|K_w(z)h(z)|\varphi(1-|z|)\mathrm{d}v(z) = I_1 + I_2.$$

From (10), $I_2 \leqslant c\varepsilon\varphi(1-|w|)/(1-|w|^2)$. Also, $I_1 \leqslant c_R \| h \|_\infty$, where c_R is a constant depending on R. Hence

$$|(Ph)(w)|(1-|w|^2)/\varphi(1-|w|)$$
$$\leqslant c_R \| h \|_\infty (1-|w|^2)/\varphi(1-|w|)+c\varepsilon, \quad w \in B.$$

It follows from Lemma 1 that

$$\limsup_{|w|\to 1} |(Ph)(w)|(1-|w|^2)/\varphi(1-|w|) \leqslant c\varepsilon,$$

and the result follows.

(iii) For $\mu \in M(B)$,

$$\int_B |(Q\mu)(w)| \varphi(1-|w|)\mathrm{d}v(w)$$
$$\leqslant \int_B\int_B |K_w(\bar{z})|(1-|z|^2)\varphi(1-|w|)/\varphi(1-|z|)\mathrm{d}v(w)\mathrm{d}|\mu|(z).$$

By a procedure like that in (10), we have

$$\int_B |K_w(\bar{z})| \varphi(1-|w|)\mathrm{d}v(w) \leqslant c\varphi(1-|z|)/(1-|z|^2).$$

Thus Q is a bounded operator from $M(B)$ into $A^1(\varphi)$. Now let $g \in A^1(\varphi)$. From Lemma 5(ii) we have $(Q(T_1 g))(w)=(K_w, g)=g(w)$. Thus $QT_1=I$, the identity on $A^1(\varphi)$, and so Q is onto and $T_1 Q$ is a projection. Hence MQ is a bounded projection of $M(B)$ onto the subspace MA^1.

(iv) This follows from (iii).

Let $\langle K_w \rangle$ denote the vector subspace spanned by the functions K_w, $w \in B$.

Lemma 6 *$\langle K_w \rangle$ is dense in $A^1(\varphi)$ and in $A_0(\varphi)$.*

Proof Consider first $A^1(\varphi)$. It is equivalent to showing that $\langle T_1 K_w \rangle$ is dense in TA^1. By the Hahn-Banach theorem and the Riesz representation theorem it suffices to show that if $h \in L^\infty(B)$ and if

$$\int_B K_w(z)\varphi(1-|z|)h(\bar{z})\mathrm{d}v(z)=0 \tag{11}$$

for all $w \in B$, then h annihilates all of TA^1. From (11) and (7) we have

$$0=\sum_{k=0}^{\infty}\sum_{|\alpha|=k}\frac{\Gamma(k+n+2)}{\alpha!\ \Gamma(n+2)}w^{\alpha}\int_{B}z^{\alpha}\varphi(1-|z|)h(\bar{z})\mathrm{d}v(z),$$

for all $w\in B$. Thus h annihilates polynomials in $z_1, z_2, \cdots, z_n$ and the result follows from Lemma 3(ii).

The proof for $A_0(\varphi)$ is similar, by using the duality between $C_0(B)$ and $M(B)$.

We come now to our second main result.

Theorem 2 *Using the pairing given by* (4) *we have*

(i) $A_0(\varphi)^*\cong A^1(\varphi)$;

(ii) $A^1(\varphi)^*\cong A_\infty(\varphi)$.

More precisely, if $g\in A^1(\varphi)$ and if we define $\lambda_g(f)=(f, g)$, $f\in A_0(\varphi)$, then $\lambda_g\in A_0(\varphi)^*$ and $\|\lambda_g\|\leqslant\|g\|_{1,\varphi}$. Conversely, given $\lambda\in A_0(\varphi)^*$ then there is a unique $g\in A^1(\varphi)$ such that $\lambda=\lambda_g$. Also, $\|g\|_{1,\varphi}\leqslant\|Q\|\|\lambda\|$.

Furthermore, if $f\in A_\infty(\varphi)$ and if we define $\lambda_{f(g)}=(f, g)$, $g\in A^1(\varphi)$, then $\lambda_f\in A^1(\varphi)^*$ and $\|\lambda_f\|\leqslant\|f\|_{\varphi}$. Conversely, given $\lambda\in A^1(\varphi)^*$ then there is a unqiue $f\in A_\infty(\varphi)$ such that $\lambda=\lambda_f$. Also, $\|f\|_{\varphi}\leqslant\|P\|\|\lambda\|$.

Proof (i) It is trivial that if $g\in A^1$ then $\lambda_g\in A_0^*$ and $\|\lambda_g\|\leqslant\|g\|_{1,\varphi}$. We also have the uniqueness: if $\lambda_g(f)=0$ for all $f\in A_0$, then $g=0$. Indeed, from Lemma 5(ii), $g(w)=\lambda_g(K_w)$, and from Lemma 5(i), $K_w\in A_0$.

Now let $\lambda\in A_0^*$ be given. Since T_0 is an isometric embedding of A_0 into $C_0(B)$, there exists $\mu\in M(B)$ with $\|\mu\|=\|\lambda\|$ and

$$\lambda(f)=\int_B f(\bar{z})(1-|z|^2)/\varphi(1-|z|)\mathrm{d}\mu(z)\tag{12}$$

for all $f\in A_0$, by the Riesz representation theorem. Let $g(w)=\lambda(K_w)$. Then

$$g(w)=\int_B K_w(\bar{z})(1-|z|^2)/\varphi(1-|z|)\mathrm{d}\mu(z)=(Q\mu)(w).$$

By Theorem 1(iii), $g\in A^1$ and $\|g\|_{1,\varphi}\leqslant\|Q\|\|\mu\|=\|Q\|\|\lambda\|$. From Lemma 5(ii) we see that $\lambda_g(K_w)=g(w)$ for $w\in B$. Hence $\lambda=\lambda_g$ on $\langle K_w\rangle$, and hence also on A_0 by Lemma 6.

(ii) The proof of the first part and the proof of the uniqueness of f are the same as in the proof of (i).

Now let $\lambda \in (A^1)^*$. T_1 is an isometric embedding of A^1 into $L^1(B)$. There exists $h \in L^\infty(B)$ with $\| h \| = \| \lambda \|$ and

$$\lambda(g) = \int_B g(\bar{z}) h(z) \varphi(1 - |z|) dv(z)$$

for all $g \in A^1$, by the Riesz representation theorem. Let $f(w) = \lambda(K_w)$. Then

$$f(w) = \int_B K_w(\bar{z}) h(z) \varphi(1 - |z|) dv(z) = (Ph)(w).$$

By Theorem 1(i), $f \in A_\infty$ and $\| f \|_\varphi \leqslant \| P \| \| h \| = \| P \| \| \lambda \|$. From Lemma 5(iii) we see that $\lambda_f(K_w) = f(w)$ for $w \in B$. Hence $\lambda = \lambda_f$ on $\langle K_w \rangle$, and hence also on A^1 by Lemma 6.

◇ References ◇

[1] Duren, P. L., Romberg, B. W. and Shields, A. L., Linear functionals on H^p spaces with $0 < p < 1$, *J. Reine Angew. Math.*, 238(1969), 32 - 60.

[2] Chen W. Y., Lipschitz spaces of holomorphic functions over bounded symmetric domains in $\mathbb{C}^n$, *J. Math. Anal. Appl.*, 81(1981), 63 - 87.

[3] Lyevshina, G. D., Linear functionals on Lipschitz spaces of holomorphic functions in the unit disk, *Mathematical Notice*, 33(1981), 679 - 688.

[4] Shi Jihuai, Representation of linear functionals on Lipschitz spaces Λ_φ of functions holomorphic in the unit ball of $\mathbb{C}^n$, *Chin. Ann. of Math.*, 8B(1987), 189 - 198.

[5] Forelli, F. and Rudin, W., Projections on spaces of holomorphic functions in balls, *Indiana Univ. Math. J.*, 24(1974), 593 - 602.

[6] Rudin, W., *Function Theory in the Unit Ball of* $\mathbb{C}^n$, Springer-Verlag, New York, 1980.

Ren Fuyao

Department of Mathematics

Fudan University

Shanghai, 200433

China

Xiao Jianbin
Department of Mathematics
Hunan Normal University
Changsha，410081
China

Local Fractional Brownian Motions and Gaussian Noises and Application*

Ren Fuyao, Zhao Xingqiu, Jiang Feng, Qiu Weiyuan, Su Feng
(Fudan University)
Qian Shaoxing, Shen Pinpin
(Scientific Research Institute of Petroleum Exploration & Development)

Abstract: In this paper, the models of local fractional Brownian motion and the local fractional Gaussian noise are studied. Apply the theory to analysing the sequence of seismic data in Jilake area, and obtain that the seismic trace is approximately a fractal curve of local fGn type with Hausdorff dimension $2-H_k$.

Keywords: local fractional Brownian motion and Gaussian noise, Hausdorff dimension, seismic trace, fractal

Statistical analyses of vertical sequences of property variations in sedimentary environments indicate that they have characteristics similar to fractal Gaussian noise, fGn[1]. This is consistent with the temporal behavior of meteorological and hydrological records. In [2], statistical analysis of sequence of reflection variations and velocity variation from a well log indicate that reflection curve has characteristics similar to fGn with $H<0.5$, and velocity curve has characteristics similar to fGn with $H>0.7$. Hence it has seen that the sequences of porosity, velocity, reflection variations have fractal character. Nevertheless, whether seismic trace is a fractal is still and open problem in the

* Originally published in *J. Fudan Univ.* (*Nat. Sci.*), Vol. 35, No. 4, (1996), 361 - 372.

world. By our calculation and analyses, we know that most of the seismic records are not suit to the model of fBm or fGn. In this paper, we establish the local fBm and local fGn models, modeling seismic signals, and obtain that seismic trace is a fractal curve.

Now we simply recall some definitions of Hausdorff and packing measure and Hausdorff, boxing and packing dimension. More detailed is seen in [3].

1. Local Fractional Brownian Motion and Gaussian Noise

It is well known that Brownian motion trace is a typical example of random fractals. The mathematical model of Brownian motion was applied widely to model many natural phenomena.

Since we require that the increment of Brownian motion be independent, Brownian motion is, for many purposes, too restrictive. However, random functions of other form are required for a variety of modeling purposes. Mandelbrot and Van Ness proposed the mathematical model and introduced the notion of fractional Brownian motion in [4]. It has been used to model a wide variety of phenomena. In general, the exponent H of fractional Brownian motion can be estimated by the R/S analysis. Obviously, if a natural phenomenon has characteristics similar to fBm or fGn, then values of H respectively obtained by the variogram analysis and the R/S analysis ought be equal approximately. If values of H obtained by the two different methods have obvious difference, fBm and fGn cannot be suited to modeling such phenomena. Therefore, we must establish more general models.

1.1 On local fractional Brownian motion

Definition 1 Let $\{Z(x); x \in \mathbb{R}\}$ be real-valued Gaussian process. let $0 < H < 1$. If it satisfies the following two conditions:

(i) $Z(0)=0$;

(ii) Local statistical self-similarity;

there exists $L>0$, such that for any real number $x \in \mathbb{R}$ and for $0<l<L$, $Z(x+l)-Z(x)$ and $l^H Z(1)$ have the same distribution.

Then we call $\{Z(x); x \in \mathbb{R}\}$ a local fractional Brownian motion with the parameter H. It is abbreviated by local fBm.

Remark Existence of the local fractional Brownian motion can be proved. Define a family of finite dimensional distribution function

$$\{F_{x_1, x_2, \cdots, x_n}(z_1, z_2, \cdots, z_n), n \in \mathbb{N}, x_i \in \mathbb{R}-\{0\}, i=1, 2, \cdots, n\}$$

as following:

(i) $F_{x_1}(z_1)$ is a normal distribution function $N(0, \sigma^2|x_1|^{2H})$, where $\sigma > 0$, $0 < H < 1$.

(ii) $F_{x_1, x_2, \cdots, x_n}(z_1, z_2, \cdots, z_n)$ is a n-dimensional normal distribution function $N(0, \Sigma)$ where $\Sigma=(\sigma_{ij})_{n\times n}$ with

$$\sigma_H = \sigma^2|x_i|^{2H},$$
$$\sigma_{ij} = r(x_i, x_j)\sigma^2(|x_i x_j|)^H, \quad \text{for } i \neq j,$$

where

$$0 < r(x, y) < 1, \ r(x, y) = r(y, x)$$

and

$$r(x, y) = \frac{1}{2(|xy|)^{2H}}(|x|^{2H} + |y|^{2H} - |x-y|^{2H}) \quad \text{for} \quad |x-y| < L,$$

therefore by[5], there exists a probability space $(\Omega, \mathscr{F}, P)$ and a stochastic process $\{Z_x, x \in \mathbb{R}-\{0\}\}$ on $(\Omega, \mathscr{F}, P)$ such that $\{F_{x_1, x_2, \cdots, x_n}(z_1, z_2, \cdots, z_n), n \in \mathbb{N}, x_i \in R-\{0\}, i=1, 2, \cdots, n\}$ is just its family of finite dimensional distribution functions.

Now let $Z_0=0$, then $\{Z_x, x \in \mathbb{R}\}$ is a local fBm.

Theorem 1 *Let Z_H be a local fractional Brownian motion. Then*

(1) *The variogram is given as*

$$2\gamma(l) = E|Z_H(x+l) - Z_H(x)|^2 = V_H l^{2H} \quad (0 < l < L), \tag{1}$$

where $V_H = E[Z_H(1)^2]$.

(ii) *The autocovariance is given as*

$$C(x, l)=EZ_H(x+l)Z_H(x)=\frac{V_H}{2}[(x+l)^{2H}+x^{2H}-l^{2H}] \quad (0<1<L). \tag{2}$$

(iii) *Its modulus of continuity is given as*

$$\sup_{\substack{s,\, t\in k \\ |s-t|\leqslant l}} |Z_H(s)-Z_H(t)|=O\left(l^H\left(\ln\frac{1}{l}\right)^{\frac{1}{2}}\right) \tag{3}$$

with probability one for any compact set K.

(iv) Z_H *is almost surely not differentiale; in fact,*

$$\lim \sup_{t\to t_0}\left|\frac{Z_H(t)-Z_H(t_0)}{t-t_0}\right|=\infty \tag{4}$$

with probability one.

(v) *The correlation of successive increments is given as*

$$\frac{E[-Z_H(-l)Z_H(l)]}{E[Z_H(l)^2]}=2^{2H-1}-1 \quad (0<l<L). \tag{5}$$

When $H=\frac{1}{2}$, *the correlation is zero, as expected for the independent increments of classical Brownian motion. When* $H<\frac{1}{2}$, *the correlation is negative. When* $H>\frac{1}{2}$, *the correlation is positive.*

(vi) *Interpolation of* $Z_H(x)$ *is given as*

$$\frac{E[Z_H(x)\mid Z_H(l)]}{Z_H(l)}=\frac{1}{2}\left[\left(\frac{x}{l}\right)^{2H}+1-\left|\frac{x}{l}-1\right|^{2H}\right], \tag{6}$$

for $0<x<1<L$.

(vii) *The graph range of* Z_H *is a fractal curve with Hausdorff dimension* $2-H$, *and*

$$\begin{aligned}\dim GrZ_H\mid[a, b]&=\underline{\dim}_{MB}GrZ_H\mid[a, b]\\&=\overline{\dim}_{MB}GrZ_H\mid[a, b]=\dim_P GrZ_H[a, b]=2-H,\end{aligned}$$

with probability one, where $[a, b]\subset\mathbb{R}$, $\dim$, $\dim_{MB}$, $\dim_P$ *is Hausdorff, modify box and packing dimension respectively and*

$$GrZ_H \mid [a, b]=\{(x, Z_H(x)); x \in [a, b]\}.$$

Proof (i), (ii), (v) are obtained by definition of local fBm. We argue that (iii), (iv), (vi) hold by the similar method in [6, 4], respectively. Now, we turn to prove (vii). First, by the similar method in [7], we have

$$\dim GrZ_H \mid [a, b]=2-H.$$

Now, we only need to prove

$$\overline{\dim}_B GrZ_H \mid [a, b] \leqslant 2-H.$$

By (iii), $\forall \eta>0$, there exists a positive number ε_0 such that $\forall l<\varepsilon_0$,

$$\sup_{\substack{s, t\in[a, b] \\ |s-t|<l}} |Z_H(s)-Z_H(t)| \leqslant t^{H-\eta}, \tag{7}$$

$\forall 0<\varepsilon<\varepsilon_0$, let $A_1(\varepsilon), A_2(\varepsilon), \cdots, A_N(\varepsilon)$ be minimal ε-cover of $[a, b]\left(N=N\varepsilon([a, b])=O\left(\frac{1}{\varepsilon}\right)\right)$. Then $\{Z'_N(A_i(\varepsilon), \delta)\}$ is a cover of $Z'_H([a, b], \delta)$. Now, $Z'_H(A_i(\varepsilon), \delta)$ is covered by N_i intervals of length ε, $A_{i,j}$ ($j=1, 2, \cdots, N_i$). So by (8), we have $\forall \eta>0$, $N_i=O(\varepsilon^{H-\eta-1})$.

Since $\{A_i(\varepsilon)\times A_{i,j}\}$ is a $\sqrt{2\varepsilon}$-cover of $GrZ'_H(\delta)|[a, b]$,

$$N_{\sqrt{2t}}(GrZ'_H(\delta) \mid [a, b]) \leqslant O(N_g([a, b]\varepsilon^{H-\eta-1}))=O(\varepsilon^{H-\eta-2}),$$

therefore

$$\overline{\dim} GrZ'_H(\delta) \mid [a, b] \leqslant 2-H.$$

1.2 On local fractional Gaussian noise

The local fBm does not have a derivative. Now, we introduce the local fractional Gaussian noise.

Definition 2 Let Z_H be a local fBm and let $0<\delta<L$. Define

$$Z_H(x, \delta)=\delta^{-1}\int_x^{x+\delta} Z_H(u)\mathrm{d}u, \tag{8}$$

$$Z'_H(x, \delta)=\delta^{-1}[Z_H(x+\delta)-Z_H(x)]. \tag{9}$$

Then we call $\{Z'_H(x, \delta), x \in \mathbb{R}\}$ a local fractional Gaussian noise. It is abbreviated by local fGn.

Theorem 2 *Suppose* $Z'_H(x, \delta)$ *be a local fGn, then*

(i) *The autocovariance is given as*

$$C(l, \delta)=EZ'_H(x+l, \delta)Z'_H(x, \delta)=$$
$$\frac{V_H}{2}\delta^{2H-2}\left[\left(\frac{l}{\delta}+l\right)^{2H}+\left|\frac{l}{\delta}-1\right|^{2H}-2\left(\frac{l}{\delta}\right)^{2H}\right]\quad(0<l<L). \tag{10}$$

(ii) *The variogram is given as*

$$2\gamma(l, \delta)=E|Z'_H(x+l, \delta)-Z'_H(x, \delta)|^2=$$
$$V_H\delta^{2H-2}\left[2-\left(\frac{l}{\delta}+l\right)^{2H}-\left|\frac{l}{\delta}-1\right|^{2H}+2\left(\frac{l}{\delta}\right)^{2H}\right] \tag{11}$$

and

$$2\gamma(l, \delta)=O(l^{2H})\quad(as\ l\to 0). \tag{12}$$

(iii) $(H-\eta)$*-order Hölder condition for any* $\eta>0$ *hold. That is*

$$\sup_{\substack{s, t\in K\\ |s-t|\leqslant l}}|Z'_H(s, \delta)-Z'_H(t, \delta)|\leqslant l^{H-\eta} \tag{13}$$

with probability one for any compact K.

(iv) $Z'_H(x, \delta)$ *is almost surely not differentiable.*

(v) $\{Z'_H(x, \delta);\ x\in\mathbb{R}\}$ *is a stationary process identified distribution with* $N(0, V_H\delta^{2H-2})$.

(vi) *The graph range of* $Z'_H(x, \delta)$ *is a fractal curve with the Hausdorff dimension* $2-H$ *and*

$$\begin{aligned}\dim GrZ'_H(\delta)\mid[a, b]&=\dim_B GrZ'_H(\delta)\mid[a, b]\\&=\dim_{MB} GrZ'_H(\delta)\mid[a, b]\\&=\dim GrZ'_H(\delta)\mid[a, b]=2-H\end{aligned} \tag{14}$$

with probability one, where $[a, b]\subset\mathbb{R}$,

$$GrZ'_H(\delta)\mid[a, b]=\{(x, Z'_H(x, \delta));\ x\in[a, b]\}.$$

Proof We obtain (i)~(v) by definition of local fGn and Theorem 2. The proof of (vi) is divided into two parts as follows:

1) We at first prove

$$\overline{\dim}_B GrZ'_H(\delta) \mid [a, b] \leqslant 2-H,$$

$\forall \varepsilon > 0$, let $A_1(\varepsilon)$, $A_2(\varepsilon)$, $\cdots$, $A_N(\varepsilon)$ be minimal ε-cover of $[a, b]\left(N = N_\varepsilon([a, b]) = O\left(\frac{1}{\varepsilon}\right)\right)$. Then $\{Z'_H(A_i(\varepsilon), \delta)\}$ is a cover of $Z'_H([a, b], \delta)$. Now, $Z'_H(A_i(\varepsilon), \delta)$ is covered by N_i intervals of length ε, $A_{i,j}(j = 1, 2, \cdots, N_i)$. So according to (iii), we have $\forall \eta > 0$, $N_i = O(\varepsilon^{H-\eta-1})$. Since $\{A_i(\varepsilon) \times A_{i,j}\}$ is a $\sqrt{2\varepsilon}$ -cover of $GrZ'_H(\delta) \mid [a, b]$,

$$N_{\sqrt{2t}}(GrZ'_H(\delta) \mid [a, b]) \leqslant O(N_\varepsilon([a, b]\varepsilon^{H-\eta-2})).$$

Therefore

$$\overline{\dim}_B GrZ'_H(\delta) \mid [a, b] \leqslant 2-H.$$

2) We now turn to prove

$$\dim GrZ'_H(\delta) \mid [a, b] \geqslant 2-H.$$

Noticed that there exists $\varepsilon > 0$, such that for $l < \varepsilon$,

$$c_1^2 l^{2H} \leqslant 2\gamma(l, \delta) \leqslant c_2^2 l^{2H},$$

where c_1, c_2 are two finite positive constants, it is sufficient to show that

$$\dim GrZ'_H(\delta) \mid [a, b] \geqslant 2-H, \quad \text{for } |b-a| \leqslant \varepsilon.$$

Choose $1 < \alpha < 2-H$, let $\mu = \lambda(G^{-1})$, λ denotes 1-dimensional Lebesgue measure, and let $G(x) = (x, Z'_H(x, \delta))$, $x \in [a, b]$, then μ is a finite measure supported on $GrZ'_H(\delta) \mid [a, b]$. Since

$$\int_{G[a, b]\times G[a, b]} \frac{\mu(\mathrm{d}x)\mu(\mathrm{d}y)}{|x-y|^\alpha}$$

$$= \int_{G[a, b]\times G[a, b]} \frac{\mathrm{d}t\,\mathrm{d}s}{(|t-s|^2 + |Z'_H(t, \delta) - Z'_H(s, \delta)|^2)^{\frac{\alpha}{2}}},$$

we have

$$E\left(\int_{G[a, b]\times G[a, b]} \frac{\mu(\mathrm{d}x)\mu(\mathrm{d}y)}{|x-y|^\alpha}\right) \tag{15}$$

$$= \int_{[a, b]\times[a, b]} \int_{-\infty}^{+\infty} ((t-s)^2 + \sigma_{t,s}^2 u^2)^{-\frac{\alpha}{2}} \frac{1}{\sqrt{2\pi}} \mathrm{e}^{-\frac{u^2}{2}} \mathrm{d}u\,\mathrm{d}t\,\mathrm{d}s \tag{16}$$

$$\leqslant c_3 \int_{[a,b]\times[a,b]} |t-s|^{-\alpha+1-H} \mathrm{d}t\,\mathrm{d}s \int_0^{+\infty} (1+u^2)^{-\frac{\alpha}{2}} \mathrm{d}u \tag{17}$$

$$< \infty. \tag{18}$$

Hence according to Theorem 4.13 in [2],

$$\dim GrZ'_H(\delta) \mid [a, b] \geqslant 2-H,$$

with probability one.

2. The Determining of Parameter *H*

2.1 On variogram analysis

Let Z_H be a local fBm, then, when $0<l<L$,

$$2\gamma_{Z_H}(l) = E|Z_H(x+l) - Z_H(x)|^2 = V_H l^{2H}.$$

The experimental variogram is defined as

$$2\gamma^*_{Z_H}(l) = \frac{1}{N_l}\sum_{i=1}^{N_l} [Z_H(x_i+l) - Z_H(x_i)]^2, \tag{19}$$

where N_l is the number of pairs of points in the data that are separated by a distance l, and $|x_i - x_j| < L$ $(i, j = 1, 2, \cdots, N_l)$. We have

Theorem 3 $$E\{[\gamma_{Z_H}(l) - \gamma^*_{Z_H}(l)]^2\} \leqslant \begin{cases} c_1 \dfrac{1}{N_l} l^{4H}, & as\ H < \dfrac{3}{4}, \\ c_2 \left(\dfrac{1}{N_l}\right)^{4-4H} l^{4H}, & as\ H > \dfrac{3}{4}, \\ c_3 \left(\dfrac{\ln N_l}{N_l}\right) l^{4H}, & as\ H = \dfrac{3}{4}. \end{cases}$$

Proof Let $I = E[\gamma_z(l) - \gamma^* z(l)]^2 = \frac{1}{4N^2}\sum_{i=1}^{N_l} [E(Z(x_i+l) - Z(x_i))^4 - (2\gamma z(l))^2] + \frac{2}{4N^2}\sum_{1<i<j<N_l} [E(Z(x_j+l) - Z(x_j))^2 (Z(x_i+l) - Z(x_i))^2 - (2\gamma z(l))^2] = I_1 + I_2$,

where $x_i = (i-1)l \quad (i = 1, 2, \cdots, N_l)$.

For convenience, let $Z(1) \sim N(0, 1)$. Then

$$2\gamma z(l)=E(Z(x_i+l)-Z(x_i))^2=l^{2H},$$
$$E(Z(x_i+l)-Z(x_i))^4=3l^{4H},$$

so $I_1=\frac{l^{4H}}{2N_l}$.

For $i\neq j$, let joint distribution density of $\frac{Z(x_i+l)-Z(x_i)}{l^H}$ and $\frac{Z(x_j+l)-Z(x_j)}{l^H}$ be

$$f_{i,j}(x,y)=\frac{1}{2\pi\sqrt{1-r_{i,j}^2}}\exp\left\{-\frac{1}{2(1-r_{i,j}^2)}(x^2-2r_{i,j}xy+y^2)\right\},$$

where

$$\begin{aligned} r_{i,j}&=E\left[\frac{Z(x_i+l)-Z(x_i)}{l^H}\right]\left[\frac{Z(x_j+l)-Z(x_j)}{l^H}\right]\\ &=\frac{1}{2}[(j-i+1)^{2H}+(j-i-1)^{2H}-2(j-i)^{2H}]. \end{aligned}$$

We calculate

$$\begin{aligned} &E[(Z(x_i+l)-Z(x_i))^2(Z(x_j+l)-Z(x_j))^2]\\ =&\,l^{4H}\int_{-\infty}^{+\infty}\int_{-\infty}^{+\infty}\frac{x^2y^2}{2\pi\sqrt{(1-r_{i,j}^2)}}\exp\left\{-\frac{1}{2(1-r_{i,j}^2)}(x^2-2r_{i,j}xy+y^2)\right\}dx\,dy\\ =&\,l^{4H}(1+2r_{i,j}^2), \end{aligned}$$

then

$$I_2=\frac{l^{4H}}{N_l^2}\sum_{1<i<j\leqslant N_l}r_{i,j}^2.$$

Now, we estimate $r_{i,j}$. Define

$$g(x)=\frac{1}{2x^{2H}}[(1+x)^{2H}+(1-x)^{2H}-2].$$

Since

$$(1+x)^{2H}=1+2Hx+\frac{2H(2H-1)}{2}x^2+o(x^2), \tag{20}$$

we have

$$g(x)=\frac{1}{2}x^{2-2H}(2H(2H-1)+o(1)).$$

Hence there exists a positive integer N_0, such that for any $k>N_0$,

$$g\left(\frac{1}{k}\right)\leqslant c_0\left\{\frac{1}{k}\right)^{2-2H},$$

where c_0 is a finite positive constant.

So when $j-i\geqslant N_0$,

$$r_{j,i}\leqslant c_0\left(\frac{1}{j-i}\right)^{2-2H}.$$

Therefore

$$\frac{1}{2N_l^2}\sum_{\substack{1<i<j<N_l\\ j-i>N_0}} r_{i,j}^2\leqslant\frac{1}{2N_l^2}c_0^2\sum_{1\leqslant i<j\leqslant N_l}\sum_{i+N_0\leqslant j\leqslant N_l}\left(\frac{1}{j-i}\right)^{4-4H}$$

$$\leqslant\frac{c_0^2}{2N_l^2}N_l\int_1^{N_l}\left(\frac{1}{x}\right)^{4-4H}\mathrm{d}x\leqslant\begin{cases}c_1'\left(\dfrac{1}{N_l}\right), & \text{as } H<\dfrac{3}{4},\\ c_2'\left(\dfrac{1}{N_l}\right)^{4-4H}, & \text{as } H>\dfrac{3}{4},\\ c_3'\left(\dfrac{\ln N_l}{N_l}\right), & \text{as } H=\dfrac{3}{4},\end{cases}$$

where c_1', c_2', c_3' are three finite positive constants.

But

$$\frac{1}{2N_l^2}\sum_{\substack{1<i<j\leqslant N_l\\ 1\leqslant j-i\leqslant N_0}} r_{i,j}^2\leqslant\frac{1}{2N_l^2}\max_{\substack{1\leqslant i<j\leqslant N_l\\ j-i\leqslant N_0}} r_{i,j}^2 N_0(N_l-N_0)\leqslant c_4\left(\frac{1}{N_l}\right)$$

where c_4 is a finite positive constant.

Thus

$$I\leqslant\begin{cases}c_1\left(\dfrac{1}{N_l}\right)l^{4H}, & \text{as } H<\dfrac{3}{4},\\ c_2\left(\dfrac{1}{N_l}\right)^{4-4H}l^{4H}, & \text{as } H>\dfrac{3}{4},\\ c_3\left(\dfrac{\ln N_l}{N_l}\right)l^{4H}, & \text{as } H=\dfrac{3}{4},\end{cases}$$

where c_1, c_2, c_3 are three finite positive constants.

When l is small enough, N_l will often be large and the local variogram can be accurately estimated with the experimental variogram.

Since the graph of theoretical variogram $\gamma_{Z_H}(l)$ on logarithmic coordinate is a line with slope $2H$, the graph of experimental variogram $\gamma^*_{Z_H}(l)$ on logarithmic coordinate is approximated by a line with slope $2H$. Then the value of H can be determined from the least squares fit.

2.2 On R/S analysis

The R/S analysis is a usual method determining parameter H of fGn or fBm. Let $\{Z(x);\ x \in \mathbb{R}\}$ be fGn, let

$$Z_R(x) = \sum_{u=1}^{x} Z(x), \tag{21}$$

$$R(l) = \max_{0 \leqslant u \leqslant 1} - \min \left\{ Z_R(u) - \frac{u}{l} Z_R(l) \right\}, \tag{22}$$

$$S^2(l) = l^{-1} \sum_{u=1}^{l} Z(u)^2 - \left[l^{-1} \sum_{u=1}^{l} Z(u) \right]^2, \tag{23}$$

we have

$$\frac{R(l)}{S(l)} = O(l^H) \quad (\text{as } l \to \infty). \tag{24}$$

Therefore when l is sufficiently large, the graph range of $\frac{R(l)}{S(l)}$ on logarithmic coordinate ought to be approximated by a line with slope H. The value of H can be determined from the least squares fit.

If $\{Z(x);\ x \in \mathbb{R}\}$ is an fBm, then value of H can be determined by R/S analysis of increment of Z.

Discussion (i) If values of H obtained by above two different methods are approximately equal, then we think that the natural phenomena suit the model of fBm or fGn.

(ii) If values of H obtained by above two different methods have obvious difference, the fBm and fGn cannot be applied to modeling the phenomena.

Likely, the phenomena are well approximated by the models of local fBm or local fGn.

3. Application

We consider the sequences of seismic date CDPs 698～702, 705, 710, 715, 720, 725, 730, 735, 740, 741, denoted as $Z_k(x)$, $x=1, 2, \cdots, N$; $N=1\,002$. We will analyze and manage them according to the following steps for every k. The figures in the paper originate from CDP 700.

Step 1 We calculate probability density of Z_k, shown in Fig. 1. We observe that Z_k has approximately a Gaussian distribution $N(a_k, \sigma_k^2)$.

Step 2 Z_k is normalized by formula:

$$\overline{Z}_k = \frac{Z_k - a_k}{\sigma_k}, \tag{25}$$

the normalized sequences $\overline{Z}_k$ are shown in Fig. 2.

Step 3 Let

$$\overline{Z}_{k,g}(x) = \sum_{u=1}^{x} \overline{Z}_k(u), \tag{26}$$

the graded normalized sequences $\overline{Z}_{k,g}$ are shown in Fig. 3.

Step 4 Comparing respectively Fig. 2 and Fig. 3 with Fig. 4 and Fig. 5, we find that Fig. 2 and Fig. 3 are similar somewhat to traces of fGn and fBm, respectively.

Is it true? We need to do further investigation.

Step 5 The variograms of the graded sequences are shown on logarithmic coordinates in Fig. 6.

The values of H determined from the least squares fit of $\gamma^*_{\overline{Z}_{k,g}}(l)$ for l small enough is denoted by H_k, seen in Table 1.

Step 6 The values of H determined from the least squares fit of $\frac{R(l)}{S(l)}$ for l large enough is denoted by H'_k, seen in Table 1.

We notice that when $k \neq 720$, H_k is obviously different with H'_k. It means that those seismic traces are suited to the model of local fGn rather than the models of fBm and fGn.

When $k = 720$, $H_k = H'_k$, it means that the seismic trace CDP 720 is approximated by the model of fGn.

Now, we divide each CDP into 5 parts, denoted by

$$\{Z_{ij}(x)\},\ i = 698 \sim 702,\ 705,\ 710,\ 715,\ 720,\ 725,\ 730,\ 735,\ 740,\ 741,\ j = 1,\ 2,\ 3,\ 4,\ 5.$$

We checked that these sequences are stationary.

For each sequence Z_{ij}, the value of H determined from the variogram analysis is denoted by H_{ij}, seen in Table 2.

Summarizing above, we obtain each seismic trace Z_{ij} is approximately fractal curve with Hausdorff dimension

$$D_{ij} = 2 - H_{ij}. \tag{27}$$

Discussion (i) We guess that seismic trace in other area has also the similar characteristics to the seismic trace in Jilake area from above detailed analysis.

(ii) The models established in the paper are expected further generalization and application.

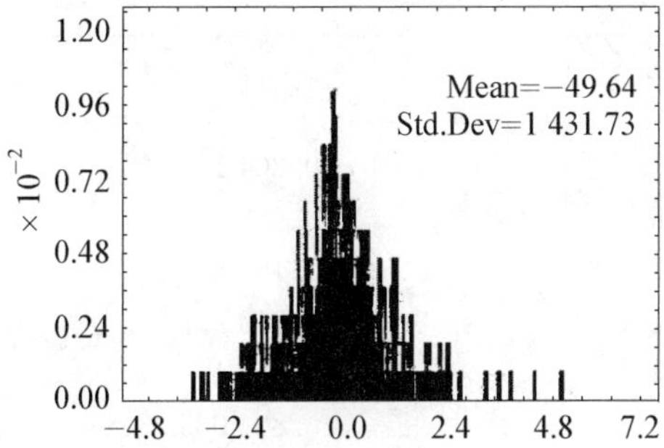

Fig. 1 Probability Density of CDP 700

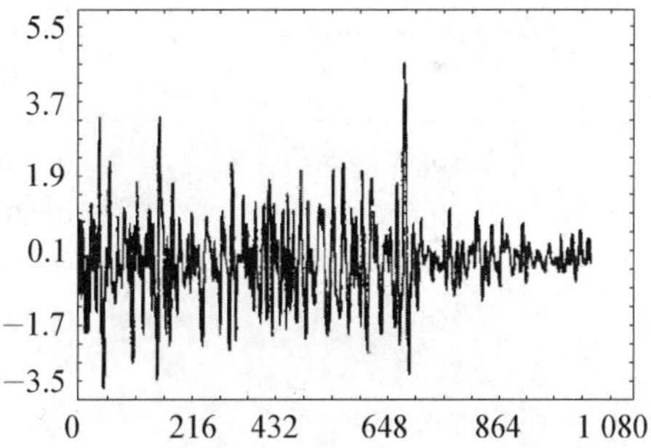

Fig. 2 Normalized Sequence of CDP 700

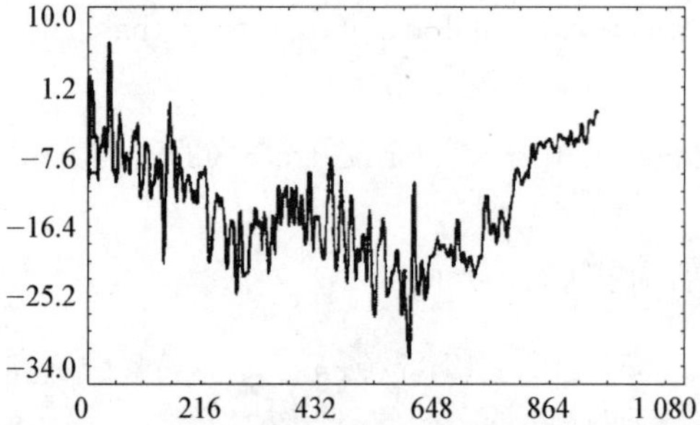

Fig. 3 Graded Normalized Sequence of CDP 700

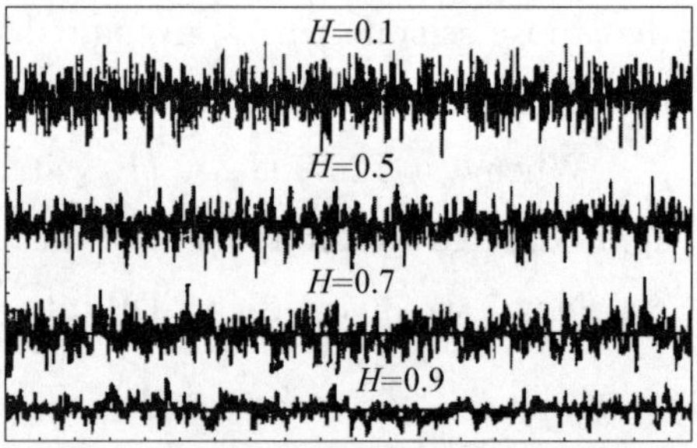

Fig. 4 Traces of Fractional Gaussian Noises, CDP

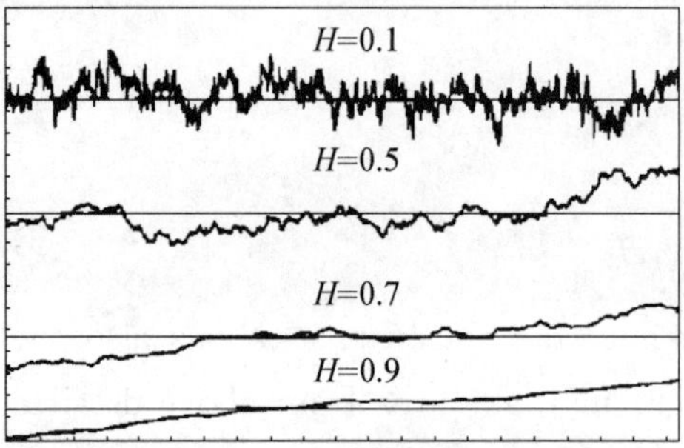

Fig. 5 Traces of Fractional Brownian Motion, fBm

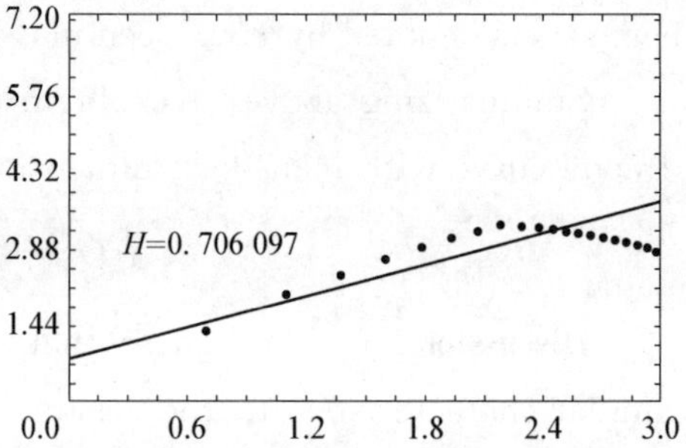

Fig. 6 Variogram Analysis of Graded Normalized CDP 700

Table 1 Values of H_k obtained by the variogram and $\frac{R}{S}$ analysis for seismic signals

CDP_k	H_k	H_k'
698	0. 696 685	0. 204 169
699	0. 694 225	0. 499 900
700	0. 706 097	1. 162 873
701	0. 698 642	0. 345 267
705	0. 702 754	0. 258 189
710	0. 715 100	0. 834 063
715	0. 692 552	0. 108 988
720	0. 682 856	0. 684 736
725	0. 710 224	0. 792 607
730	0. 704 148	0. 283 385

续 表

CDP_k	H_k	H_k'
735	0.718 881	0.178 857
740	0.721 249	0.412 784
741	0.727 655	0.352 688

Table 2 Values of H_{ij} obtained by the variogram analysis for seismic signals

CDP_i	H_{i1}	H_{i2}	H_{i3}	H_{i4}	H_{i5}
698	0.550 661	0.443 691	0.570 697	0.450 370	0.684 292
699	0.535 512	0.435 825	0.549 908	0.524 345	0.629 287
700	0.480 447	0.443 612	0.424 430	0.438 311	0.515 633
701	0.503 995	0.488 988	0.546 744	0.481 131	0.649 324
705	0.486 957	0.478 405	0.591 148	0.537 535	0.709 150
710	0.551 514	0.475 712	0.596 642	0.542 340	0.657 466
715	0.526 425	0.432 824	0.542 570	0.487 262	0.622 188
720	0.534 210	0.436 416	0.575 402	0.646 429	0.535 760
725	0.569 872	0.454 637	0.554 585	0.546 989	0.548 774
730	0.509 508	0.507 094	0.534 858	0.547 982	0.550 061
735	0.484 910	0.484 875	0.580 060	0.572 396	0.568 944
740	0.507 144	0.416 107	0.565 930	0.538 880	0.623 672
741	0.541 869	0.403 879	0.587 938	0.559 884	0.635 317

◇ References ◇

[1] Hewett T A. Fractal distributions of reservoir heterogeneity and their influence on fluid transport. US: Chevron oil field research Co SPE 15386, 1986.

[2] Ren Fuyao, Zhao Xingqiu, Jiang Feng, et al. On the Fractal properties of the reflection curve and velocity curve from a well log. *Journal of Fudan University (Natural Sience)*, 1997, 36 (to appear).

[3] Falconer K J. Fractal geometry. New York: John Wiley & Sons, 1990.

[4] Mandelbrot B B, Van Ness J W. Fractional Brownian motions, fractional noises, and application. *SIAM Rev*, 1968, 10(4): 442 - 437.

[5] Wang Zikun. Introduction to stochastic process. Beijing: Scientific Press, 1965.

[6] Kahane J P. Some random series of functions. 2nd ed. Cambridge: Cambridge University Press, 1985.
[7] Adler R J. Hausdorff dimension and Gaussian fields. *Ann Probab*, 1977, 5: 145 - 151.
[8] Feder J. Fractals. New York: Plenum Press, 1988.

局部分式 Brown 运动和局部分式 Gauss 噪声及应用

任福尧　赵兴球　姜　峰　邱维元　苏　峰
（复旦大学）
钱绍新　沈平平
（北京石油勘探开发科学研究院）

提　要：研究了局部分式 Brown 运动和局部分式 Gauss 噪声的基本性质，并用于分析吉拉克地区地震数据序列的分形性质，得到各地震道 CDP 近似地是 Hausdorff 维数为 $2-H_k$ 的局部分式 Gauss 噪声曲线.

关键词：局部分式 Brown 运动，局部分式 Gauss 噪声，Hausdorff 维数，地震道，分形

Fractional Integral Associated to the Self-similar Set or the Generalized Self-similar Set and Its Physical Interpretation*

Fu-Yao Ren, Zu-Guo Yu, Feng Su

Institute of Mathematics, Fudan University, Shanghai 200433, China

Abstract: This paper is based on a study of Nigmatullin [Teor. Mat. Fiz. 90 (1992) 354]. When the "residual" memory set is a self-similar set which is generated by similarities $S_j x = \xi_j x + b_j$ $(0 < \xi_j < 1,\ b_1 = 0 < b_2 < \cdots < b_K = t(1-\xi_K),\ j = 1, 2, \cdots, K)$ on $[0, t]$ or a generalized self-similar set which is generated by a family of similarities $\{S_{n,j}(x) = \xi_{n,j} x + b_{n,j} : 0 < \xi_{n,j} < 1,\ b_{n,j} \in \mathbb{R},\ j = 1, 2, \cdots, K_n\}_{n \in Z^+}$ on $[0, t]$, we prove that the fractional exponent of the fractional integral is not uniquely determined by the fractal dimension of the self-similar set or generalized self-similar set, it is determined by $\ln P_1 / \ln \xi_1$ of the self-similar measure $\mu = \sum_{j=1}^{K} P_j \mu \circ S_j^{-1}$, $0 < P_j < 1$, $\sum_{j=1}^{K} P_j = 1$ on this self-similar set or of the generalized self-similar measure $\mu' = \sum_{j=1}^{\infty} P_j \mu' \circ S_j^{-1}$, $0 < P_j < 1$, $\sum_{j=1}^{\infty} P_j = 1$ on the generalized self-similar set, and it can have the value of all positive real numbers. Our results generalize and extend the results of Nigmatullin.

1. Introduction

When we describe a structure of the evolution of a physical system far

---

* Project partially supported by the Tianyuan Foundation of China and PhD station Foundation of the State Education Committee. Originally published in *Physics Letters A*, Vol. 219, (1996), 59 - 68.

from thermodynamic equilibrium, in amorphous materials [1～3], in the description of structural relaxation of high-T_c oxide superconductors [4], in the process of plastic deformation [5] and fracture of solids [6], in the description of solid solutions [7] and the macrostructure of martensite [8], and so on, the medium exhibits memory. The existence of memory means that if at time τ a force $f(\tau)$ acts on the system, then there arises a flux J whose magnitude at time $t > \tau$ is given through a memory function $m(\tau)$ by the equation

$$J(t) = \int_0^t m(t-\tau) f(\tau) \mathrm{d}\tau. \tag{1}$$

If the "residual" memory set is a Cantor fractal set (or Cantor's k-bars) generated by $S_1 = \xi x$, $S_2 = \xi x + (1-\xi)t$ (or $S_j = \xi x + (j-1)\xi t + (j-1)(1-k\xi)t/(k-1)$, $j = 1, 2, \cdots, k$), and if the total number of remaining states in each stage of the division of this set is normalized to unity, Nigmatullin [9] obtained the following result,

$$J(t) = A_v t^{-v} [\Gamma(v)]^{-1} \int_0^t (t-t')^{v-1} f(t') \mathrm{d}t' = A_v D^{-v} f(t), \tag{2}$$

where $v = \ln 2/\ln(1/\xi)$ (or $v = \ln k/\ln(1/\xi)$) is the fractal dimension of the Cantor set (or Cantor's k-bars), $A_v = [\sqrt{2}(1-\xi)]^{-v}$ for the Cantor set, and $A_v = \exp[-I_1/\ln(1/\xi)]$ where

$$I_1 = \int_0^\infty \frac{f'(y)}{f(y)} \ln y \mathrm{d}y, \quad f(y) = \frac{1-\mathrm{e}^{-ky/(k-1)}}{k(1-\mathrm{e}^{-y/(k-1)})}$$

for Cantor's k-bars, $\Gamma(v)$ is the gamma function, and where the fractional exponent of the fractional integral

$$D^{-v} f(t) = [\Gamma(v)]^{-1} \int_0^1 (1-u)^{v-1} f(tu) \mathrm{d}u \tag{3}$$

$$= [v\Gamma(v)]^{-1} \int_0^1 f((1-u_1)t) \mathrm{d}u_1^v, \tag{4}$$

is equal to the fractal dimension of the Cantor set $s = \ln 2/\ln\xi$ (or Cantor's k-bar $s = \ln k/\ln\xi$), furthermore, the physical interpretation of the fractional integral is given. We naturally raise the following three problems:

(1) Is the fractional exponent of the fractional integral v uniquely determined by the fractal dimension of this Cantor set (or Cantor's k-bar)?

(2) If v is not uniquely determined by the fractal dimension of this Cantor set (or Cantor's k-bar), what is it determined by?

(3) Which values can v have?

These problems are all basic problems of the fractal space of spatial fractional (non-integer) integrals. In these problems, in this paper, we do not only consider the Cantor set or Cantor's k-bar, we consider a family of more extensive sets, i. e. self-similar sets or generalized self-similar sets. We give an approximate solution of (1), and prove that the fractional exponent v of the fractional integral is not uniquely determined by the fractal dimension of the self-similar set E_t or generalized self-similar set E_t', it is determined by the self-similar measure or the generalized self-similar measure, and it can have the value of all positive real numbers. This paper provides a rigorous mathematical frame for the statement made by Nigmatullin in Ref. [9].

In this paper, we denote by $\mathbb{R}$ the set of all real numbers, by $\mathbb{C}$ the set of all complex numbers, and by $\mathbb{Z}^+$ the set of all positive integer numbers.

2. Construction and Fractal Dimension of Self-similar Sets

For any given $t \in (0, \infty)$, let

$$S_j(x) = \xi_j(x) + b_j, \quad 0 < \xi_j < 1, \quad b_j \in [0, t], \quad j = 1, \cdots, K$$

be similarities on $E_0 = [0, t]$, denote E_t the *self-similar set* generated by $\{S_j(x)\}_{j=1}^K$, and assume $\{S_j(x)\}_{j=1}^K$ satisfy the *open set condition*, i. e. there exists an open set $U \subset [0, \infty)$ such that $S_j(U) \subset U$ for all $j = 1, \cdots, K$ and $S_j(U) \cap S_i(U) = \varnothing$ for $i \neq j$. We may assume $E_0 \subset U$, we have $S_j(E_0) \cap S_i(E_0) = \varnothing$ for $j \neq i$. For convenience, we may as well assume

$$b_1 = 0 < b_2 < \cdots < b_K = t(1 - \xi_K). \tag{5}$$

For natural numbers n, let $I = \{1, 2, \cdots, K\}$ and

$$E_{j_1 j_2 \cdots j_n} = S_{j_n} \circ \cdots \circ S_{j_2} \circ S_{j_1}(E_0) = [b_{j_n \cdots j_1},\ b_{j_n \cdots j_1} + \Delta_{j_n \cdots j_1}], \tag{6}$$

$$E(n) = \bigcup_{j_i \in I} E_{j_1 \cdots j_n},$$

where $[a,\ b]$ is an interval, i. e. $a \leqslant x \leqslant b$, $j_i \in I\ (i=1,\ 2,\ \cdots,\ n)$, and

$$\Delta_{j_n \cdots j_1} = t \prod_{i=1}^{n} \xi_{j_i},\quad b_{j_n \cdots j_1} = b_{j_n} + \xi_{j_n} b_{j_{n-1} \cdots j_1}. \tag{7}$$

It is obvious that

$$E_{j_1 \cdots j_n} \subset E_{j_1 \cdots j_{n-1}},\quad E(n) \subset E(n-1)$$

and

$$E_t = \bigcap_{n=1}^{\infty} E(n).$$

From Ref. [10], the fractal dimension s of E is the solution of the equation $\sum_{j=1}^{K} \xi_j^s = 1$.

The Cantor set and Cantor's k-bars are self-similar sets, the Von Koch curve is also a self-similar set.

3. Memory Measures and Self-similar Measures

Let real numbers $P_1, P_2, \cdots, P_K$ satisfy

$$0 < P_j < 1,\quad j \in I,\quad \sum_{j=1}^{K} P_j = 1. \tag{8}$$

A probability measure μ is called a *self-similar measure* if it satisfies

$$\mu(\cdot) = \sum_{j=1}^{K} P_j \mu \circ S_j^{-1}(\cdot), \tag{9}$$

where P_j are called the *weights*. In Ref. [10], Hutchinson proved the existence of a self-similar measure with operator theory. Now we concretely construct the self-similar measure on E_t.

Let the *memory measure* on $E(n)$ be defined by $d\mu_n(\tau) = m_n(\tau)d\tau$, where

$$m_n(\tau) = \sum_{j_1, \cdots, j_n \in I} P_{j_1} \cdots P_{j_n} \eta(b_{j_n \cdots j_1} < \tau < b_{j_n \cdots j_1} + \Delta_{j_n \cdots j_1}) / \Delta_{j_n \cdots j_1} \tag{10}$$

and $\eta(a<\tau<b)=\eta(\tau-a)-\eta(\tau-b)$,

$$\eta(\tau)=\begin{cases}1, & \tau>0,\\ 0, & \tau<0,\end{cases}$$

then the support set $\mathrm{supp}(\mu_n)$ of the memory measure μ_n is $E(n)$ and

$$\int_0^t \mathrm{d}\mu_n(\tau)=\int_{E(n)}\mathrm{d}\mu_n(\tau)=1,$$

and for any $A\subset E_0$, we have

$$\int_A \mathrm{d}\mu_n(\tau)=\sum_{i=1}^{K}P_j\int_A \mathrm{d}\mu_{n-1}\circ S_j^{-1}(\tau), \tag{11}$$

i. e.

$$\mu_n=\sum_{j=1}^{K}P_j\mu_{n-1}\circ S_j^{-1}.$$

We also have $\mathrm{supp}(\mu_0)\supset\mathrm{supp}(\mu_1)\supset\cdots$.

For any $E_{j_1\cdots j_k}$ and natural number l, we have $\mu_{k+l}(E_{j_1\cdots j_k})=\mu_k(E_{j_1\cdots j_k})$. For any continuous real function $f(\tau)$ on $\mathbb{R}$, f is uniformly continuous on $\overline{E}_0$, hence for any $\varepsilon>0$, there exists a $\delta>0$ such that if $|U'|<\delta$, $U'\subset\mathbb{R}$, we have

$$\max_{\tau\in U'}f(\tau)-\min_{\tau\in U'}f(\tau)<\varepsilon.$$

We want to prove that $\left\{\int_{\mathbb{R}}f\mathrm{d}\mu_k\right\}_{k=1}^{\infty}$ is a Cauchy sequence. Let k be large enough such that all $|E_{j_1\cdots j_k}|<\delta$. Then

$$\begin{aligned}\left|\int_{\mathbb{R}}f\mathrm{d}\mu_k-\int_{\mathbb{R}}f\mathrm{d}\mu_{k+l}\right| &\leqslant \sum_{j_1,\cdots,j_k\in I}\left|\int_{E_{j_1\cdots j_k}}f\mathrm{d}\mu_k-\int_{E_{j_1\cdots j_k}}f\mathrm{d}\mu_{k+l}\right|\\ &\leqslant \sum_{j_1,\cdots,j_k\in I}\left(\left|\int_{E_{j_1\cdots j_k}}f\mathrm{d}\mu_k-m_{j_1\cdots j_k}\mu_k(E_{j_1\cdots j_k})\right|\right.\\ &\quad\left.+\left|m_{j_1\cdots j_k}\mu_{k+l}(E_{j_1\cdots j_k})-\int_{E_{j_1\cdots j_k}}f\mathrm{d}\mu_{k+l}\right|\right)\\ &\leqslant \sum_{j_1,\cdots,j_k\in I}2[M_{j_1\cdots j_k}\mu_k(E_{j_1\cdots j_k})-m_{j_1\cdots j_k}\mu_{k+1}(E_{j_1\cdots j_k})]\end{aligned}$$

$$= \sum_{j_1, \cdots, j_k \in I} 2(M_{j_1\cdots j_k} - m_{j_1\cdots j_k})\mu_k(E_{j_1\cdots j_k}) \leqslant 2\varepsilon\mu_0(E_0),$$

where $M_{j_1\cdots j_k} = \max_{\tau \in E_{j_1\cdots j_k}} f(\tau)$, $m_{j_1\cdots j_k} = \min_{\tau \in E_{j_1\cdots j_k}} f(\tau)$, hence $\left\{\int_{\mathbb{R}} f \mathrm{d}\mu_k\right\}$ converges as $k \to \infty$. It is easy to see that $\lim_{k\to\infty}\int_{\mathbb{R}} f \mathrm{d}\mu_k$ is a continuous linear functional on the space of continuous functions, from the Riesz representation theorem, there exists a measure μ satisfying $\int_{E_t} \mathrm{d}\mu(\tau) = 1$, $\mathrm{supp}(\mu) = E_t$ such that

$$\lim_{k\to\infty}\int_{\mathbb{R}} f \mathrm{d}\mu_k = \int_{\mathbb{R}} f \mathrm{d}\mu, \tag{12}$$

i. e.

$$\mu_n \to \mu \text{ (weak convergence)}. \tag{13}$$

For any continuous complex function $g(\tau)$, we can write $g(\tau) = u(\tau) + \mathrm{i}v(\tau)$, where $u(\tau)$, $v(\tau)$ are continuous real functions, and hence

$$\lim_{k\to\infty}\int_{\mathbb{R}} u \mathrm{d}\mu_k = \int_{\mathbb{R}} u \mathrm{d}\mu, \quad \lim_{k\to\infty}\int_{\mathbb{R}} v \mathrm{d}\mu_k = \int_{\mathbb{R}} v \mathrm{d}\mu.$$

Hence $\lim_{k\to\infty}\int_{\mathbb{R}} g \mathrm{d}\mu_k = \int_{\mathbb{R}} g \mathrm{d}\mu$. In particular, for $f(\tau) = \mathrm{e}^{-p\tau}$, $p \in \mathbb{C}$, we have

$$\lim_{n\to\infty}\int_0^\infty \mathrm{e}^{-p\tau} \mathrm{d}\mu_n(\tau) = \int_0^\infty \mathrm{e}^{-p\tau} \mathrm{d}\mu(\tau). \tag{14}$$

We call μ the *memory measure* on the self-similar set E_t. We obtain from (11)

$$\int_A \mathrm{d}\mu(\tau) = \sum_{j=1}^K P_j \int_A \mathrm{d}\mu \circ S_j^{-1}(\tau),$$

i. e.

$$\mu(\cdot) = \sum_{j=1}^K P_j \mu \circ S_j^{-1}(\cdot),$$

hence the memory measure μ is a self-similar measure. From Ref. [10], μ is the unique self-similar mearure satisfying (9) for given $\{P_j\}_{j=1}^K$ and $\{S_j\}_{j=1}^K$.

4. Flux Function and Memory Function

We consider

$$J_n(t') = \int_0^{t'} m_n(t' - \tau) f(\tau) \mathrm{d}\tau. \tag{15}$$

If $f(t)$ is a generating function, i. e. $f(t) = 0$ for $t < 0$, $f(t)$ has only finite many first class discontinuous points on any $[a, b] \subset [0, \infty)$ and $|f(t)| \leqslant M\mathrm{e}^{s_0 t}$ for $t \in [0, \infty)$, $s_0 \geqslant 0$. Acting the Laplace transformation on both sides of (15), from the product theorem of the Laplace transformation, we have

$$\mathscr{J}_n(p) = M_n(p) F(p), \quad p \in \mathbb{C}, \tag{16}$$

where

$$\mathscr{J}_n(p) = \int_0^{\infty} \exp(-pt') J_n(t') \mathrm{d}t',$$

$$M_n(p) = \int_0^{\infty} \exp(-pt') m_n(t') \mathrm{d}t' = \sum_{j_1, \cdots, j_n \in I} \frac{P_{j_1} \cdots P_{j_n}}{p \Delta_{j_n \cdots j_1}} \mathrm{e}^{-pb_{j_n \cdots j_1}} (1 - \mathrm{e}^{-p\Delta_{j_n \cdots j_1}}). \tag{17}$$

From (17) and (7), we obtain

$$\begin{aligned} M_{n+1}(p) &= \sum_{j_{n+1}=1}^{K} P_{j_{n+1}} \sum_{j_1, \cdots, j_n \in I} \frac{P_{j_1} \cdots P_{j_n}}{p \Delta_{j_{n+1} \cdots j_1}} \mathrm{e}^{-pb_{j_{n+1} \cdots j_1}} (1 - \mathrm{e}^{-p\Delta_{j_{n+1} \cdots j_1}}) \\ &= \sum_{j_{n+1}=1}^{K} P_{j_{n+1}} \mathrm{e}^{-pb_{j_{n+1}}} \sum_{j_1, \cdots, j_n \in I} \frac{P_{j_1} \cdots P_{j_n}}{p \xi_{j_{n+1}} \Delta_{j_n \cdots j_1}} \mathrm{e}^{-p\xi_{j_{n+1}} b_{j_n \cdots j_1}} (1 - \mathrm{e}^{-p\xi_{j_{n+1}} \Delta_{j_n \cdots j_1}}), \end{aligned}$$

hence we obtain the function equation

$$M_{n+1}(p) = \sum_{j=1}^{K} P_j \mathrm{e}^{-pb_j} M_n(\xi_j p). \tag{18}$$

In particular, when $K = 2$, $\xi_1 = \xi_2 = \xi$, $b_1 = 0$, $b_2 = (1 - \xi)t$, and when $K = k$, $\xi_1 = \xi_2 = \cdots = \xi_k = \xi$, $k\xi < 1$, $b_j = (j-1)\xi t + (j-1)(1 - k\xi)l/(k-1)$, $1 \leqslant j \leqslant k$, (18) becomes the function equation of the Cantor set and Cantor's k-bars which Nigmatullin obtained.

Noting that

$$M_n(p)=\int_0^{\infty} e^{-p\tau}\,d\mu_n(\tau), \tag{19}$$

$$M(p)=\int_0^{\infty} e^{-p\tau}\,d\mu(\tau), \tag{20}$$

we have from (14)

$$\lim_{n\to\infty} M_n(p)=M(p). \tag{21}$$

Hence from (18) and (16), we have

$$M(p)=\sum_{j=1}^{K} P_j e^{-pb_j} M(\xi_j p), \tag{22}$$

$$\mathscr{J}(p)=M(p)F(p), \tag{23}$$

where $\mathscr{J}(p)=\lim\limits_{n\to\infty}\mathscr{P}_n(p)$.

Now we want to obtain the approximate solution of the function equation (22).

From(22), noting that $b_1=0$, $b_j\neq 0$ $(j=2,\cdots,K)$ and $M(p)$ is bounded, when $\mathrm{Re}(p)$ is large enough, we have

$$M(p)=P_1 M(\xi_1 p)+o(1),\quad |p|\to\infty. \tag{24}$$

The unique solution of the function equation

$$\overline{M}(p)=P_1\overline{M}(\xi_1 p) \tag{25}$$

has the form

$$\overline{M}(p)=Ap^{-v}, \tag{26}$$

where A is a constant depending only on $\{P_j\}_{j=1}^{K}$ and $\{S_j\}_{j=1}^{K}$, and

$$v=\ln P_1/\ln\xi_1. \tag{27}$$

Hence when $\mathrm{Re}(p)$ is large enough,

$$M(p)\approx Ap^{-v}, \tag{28}$$

and

$$\mathscr{J}(p) \approx A p^{-v} F(p). \tag{29}$$

Since μ has a compact support set and $M(p)$ is the Laplace transform of μ, $M(p)$ is an entire analytic function. From (21) and (26), when $|p| \to \infty$, we obtain that $M(p)$ with respect to $\arg(M(p))$ uniformly converges to 0 and $\int_{a-i\infty}^{a+i\infty} M(p)\mathrm{d}p$ absolutely converges. Hence we can make the Laplace inversion transformation for (28) and (29) and obtain

$$m(\tau) \approx A(\Gamma(v))^{-1}\tau^{v-1}, \tag{30}$$

$$\begin{aligned} J(t) &\approx A(\Gamma(v))^{-1}\int_0^t (t-\tau)^{v-1} f(\tau)\mathrm{d}\tau \\ &= At^v(\Gamma(v))^{-1}\int_0^1 (1-u)^{v-1} f(ut)\mathrm{d}u = At^v D^{-v} f, \end{aligned} \tag{31}$$

where

$$D^{-v} f = (\Gamma(v))^{-1}\int_0^1 (1-u)^{v-1} f(ut)\mathrm{d}u = (v\Gamma(v))^{-1}\int_0^1 f((1-u_1)t)\mathrm{d}u_1^v. \tag{32}$$

(32) is the fractional integral. Then from (31), we establish the connection between the fractional integral and the flux. In particular, when $P_1 = \xi_1^s$ and $\sum_{j=1}^K P_j = 1$, where s is the fractal dimension of the self-similar set E_t, i. e. $\sum_{j=1}^K \xi_j^s = 1$, from (27), we have $v = s$, thus we establish the connection between the fractional integral and the dimension of the self-similar set.

5. Generalized Self-similar Sets and Generalized Self-similar Measures

Let $\{S_{n,j}(x) = \xi_{n,j}x + b_{n,j}:\ 0 < \xi_{n,j} < 1,\ b_{n,j} \in \mathbb{R},\ j = 1, 2, \cdots, K_n\}_{n\in Z^+}$ be a family of similarities, and for each n, $\{S_{n,j}(x)\}_{j=1}^{K_n}$ satisfies the open set condition. For convenience, we assume

$$b_{n,1} = 0 < b_{n,2} < \cdots < b_{n,K_n} = t(1-\xi_{n,K_n}),\quad n \in \mathbb{Z}^+,$$

Let $E_0 = [0, t]$, $I_N = \{1, 2, \cdots, K_N\}$, $\sigma = (j_1, j_2, \cdots, j_N)$, $\Lambda_N = I_1 \times I_2 \times \cdots \times I_N$,

$$E'_{j_1j_2\cdots j_N}=S_{N,\,j_N}\circ S_{N-1,\,j_{N-1}}\circ\cdots\circ S_{1,\,j_1}(E_0)=[b'_{j_N\cdots j_1},\,b'_{j_N\cdots j_1}+\Delta'_{j_N\cdots j_1}],$$
$$E'(N)=\bigcup_{\sigma\in\Lambda_N}E'_{j_1j_2\cdots j_N},$$

where

$$\Delta'_{j_N\cdots j_1}=t\prod_{i=1}^{N}\xi_{i,\,j_i},\quad b'_{j_N\cdots j_1}=b_{N,\,j_N}+\xi_{N,\,j_N}b'_{j_{N-1}\cdots j_1}.$$

Then $E'_t=\bigcap_{N=1}^{\infty}E'(N)$ is called a *generalized self-similar set*. If $c=\inf_{N\in\mathbb{Z}^+,\,j\in I_N}\xi_{N,\,j}>0$ and $\lim_{N\to\infty}\sup_{\sigma\in\Lambda_N}|E'_{j_1j_2\cdots j_N}|=0$, then from Refs. [11] or [12], the fractal dimension s of E'_t is $\lim\inf_{N\to\infty}\beta_N$, where β_N is the solution of $\prod_{n=1}^{N}(\sum_{j=1}^{K_n}\xi_{n,\,j}^{\beta_N})=1$. Even if $\lim_{n\to\infty}K_n=\infty$, the conclusion of the fractal dimension of E'_t holds.

Let the family of real numbers $\{P_{n,\,j}:\,j=1,\,2,\,\cdots,\,K_n\}_{n\in\mathbb{Z}^+}$ satisfy

$$0<P_{n,\,j}<1,\ j\in I_n,\quad \sum_{j=1}^{K_n}P_{n,\,j}=1.$$

Let the memory measure on $E'(N)$ be defined by $d\mu'_N(\tau)=m'_N(\tau)d\tau$, where

$$m'_N(\tau)=\sum_{\sigma\in\Lambda_N}P_{1,\,j_1}P_{2,\,j_2}\cdots P_{N,\,j_N}\eta(b'_{j_N\cdots j_1}<\tau<b'_{j_N\cdots j_1}+\Delta'_{j_N\cdots j_1})/\Delta'_{j_N\cdots j_1}. \tag{33}$$

Similar to the above, there exists a measure μ' on E'_t satisfying

$$\lim_{N\to\infty}\int_{\mathbb{R}}f\,d\mu'_N=\int_{\mathbb{R}}f\,d\mu'$$

for every continuous complex function $f(\tau)$. In particular, for $f(\tau)=e^{-p\tau}$, we have

$$\lim_{N\to\infty}M'_N(p)=\lim_{N\to\infty}\int_0^{\infty}e^{-p\tau}d\mu'_N=\int_0^{\infty}e^{-p\tau}d\mu'=M'(p), \tag{34}$$

where $M'_N(p)=\int_0^{\infty}e^{-p\tau}d\mu'_N$ and $M'(p)=\int_0^{\infty}e^{-p\tau}d\mu'$.

We also have

$$\mu'_N(\cdot)=\sum_{j=1}^{K_N}P_{N,\,j}\mu'_{N-1}\circ S_{N,\,j}^{-1}(\cdot) \tag{35}$$

and

$$M'_N(p)=\sum_{j=1}^{K_N} P_{N,j}\mathrm{e}^{-pb_{N,j}}M'_{N-1}(\xi_{N,j}p). \tag{36}$$

If $\lim_{N\to\infty}P_{N,j}=P_j$, $\lim_{N\to\infty}b_{N,j}=b_j$, $\lim_{N\to\infty}\xi_{N,j}=\xi_j$, and $\lim_{N\to\infty}K_N=K<+\infty$, we obtain from (35)

$$u'(\cdot)=\sum_{j=1}^{K} P_j\mu'\circ S_j^{-1}(\cdot), \tag{37}$$

where $S_j(x)=\xi_j x+b_j(j=1, 2, \cdots, K)$. Hence μ' is a self-similar measure on E'_t. From (34) and (36), we have

$$M(p)=\sum_{j=1}^{K} P_j\mathrm{e}^{-pb_j}M(\xi_j p). \tag{38}$$

When $\lim_{N\to\infty}K_N=\infty$, we obtain from (34)~(36)

$$M(p)=\sum_{j=1}^{\infty} P_j\mathrm{e}^{-pb_j}M(\xi_j p), \tag{39}$$

and

$$\mu'(\cdot)=\sum_{j=1}^{\infty} P_j\mu'\circ S_j^{-1}(\cdot), \tag{40}$$

where $S_j(x)=\xi_j+b_j(j=1, 2, \cdots)$. We call μ' a *generalized self-similar measure* on E'_t. From Ref. [13], μ' is the unique generalized self-similar measure satisfying (40) for given $\{P_j\}_{j=1}^{\infty}$ and $\{S_j\}_{j=1}^{\infty}$.

Since $b_{N,1}=0$, $N\in\mathbb{Z}^+$, we have $b_1=0$. When $\mathrm{Re}(p)$ is large enough, as previously we also obtain (30)~(32).

6. Physical Interpretations

If the similarities $\{S_j\}$ satisfy $\bigcup_{j=1}^{K}S_j(E_0)=E_0$ and $P_1=\xi_1$, and the fractional exponent $v=1$ it follows from (31) that $J(t)$ is related to $f(\tau)$ through the complete integral and corresponds to the case of complete memory. If all $\xi_j\to 0$ $(v\to 0)$, then from (22) or (31), it follows that

$$M(p)=\sum_{j=1}^{K}P_j e^{-pb_j}, \quad b_1=0, \quad b_K=t, \tag{41}$$

where K may be equal to infinity. In the t representation, expression (41) corresponds to ($m(t)=\sum_{j=1}^{K}P_j\delta(t-b_j)$) a linear combination of K delta functions of P_j intensity localized at the ends of the chosen interval $[0, t]$ and the point b_2, b_3, …, b_{K-1}, this case corresponds to complete absence of memory. Thus, it also follows from the above analysis that the exponent v of the fractional integral corresponds to the fraction of the preserved states in the process of evolution of the considered physical system and encompasses the cases of the completely closed ($v=1$) and Markov ($v=0$) system when all states degenerate into finitely many or countable infinitely many with infinitely high density. An interesting case for analysis is $P_1=\xi_1^{1/2}$, in this case $v=\frac{1}{2}$, which also corresponds to classical diffusion in quasi-one-dimensional semi-infinite systems, in which the connection between the concentration and flux is always expressed through an integral or derivative of only half order [14~16].

From these arguments, some physical systems that can be described by equations in fractional derivatives must contain channels belonging to some branching fractal structure. This was confirmed in Ref. [16], in which an "ultraslow" diffusion equation of the following type was obtained for the main channel,

$$\frac{\partial^\alpha c}{\partial t^\alpha}=\mathscr{D}_x\frac{\partial^\alpha c}{\partial x^\alpha}, \quad 0<\alpha<1. \tag{42}$$

The structure of the channels may differ and be generated by the definite fractal structure of the medium. In Refs. [17 ~ 19], such processes were classified as processes with "residual" memory. A process with "residual" memory corresponds to the energy principle formulated by Jonscher for dielectric relaxation [20] in the frequency domain.

From this point of view, transport processes in percolation clusters, fractal trees, and porous systems really must be reanalyzed in order to obtain the correct transport equations for such systems.

Another large class of physical systems in which one can expect the appearance of equations in fractional derivatives is represented by processes with loss due to collisions. We write Newton's equation in the form

$$\Delta v_i = \frac{1}{m_i}\int_0^t F_i(r, p, \tau)\mathrm{d}\tau = \frac{t}{m_i}\int_0^1 F_i(r, p, ut)\mathrm{d}u, \tag{43}$$

where m_i is the mass of particle i, and F_i is the force of the interaction of particle i with the medium. If the interaction with the medium with self-similar fractal structure or generalized self-similar fractal structure is collisional in nature, then the force can be expressed in the form

$$F_i(r, p, \tau) = F_i(r, p, \tau)\sum_k \eta(t_k < \tau < t_{k+1})\rho_k, \tag{44}$$

where ρ_k is the density of state, and $\eta(t_k < \tau < t_{k+1})$ is the step function defined as previously. For a force acting for only a definite time, we obtain, repeating the arguments of the previous section,

$$m_i \Delta V_i(t) = A(\Gamma(v))^{-1}\int_0^t (t-\tau)^{v-1} m_i F_i(r, p, \tau)\mathrm{d}\tau. \tag{45}$$

Using the Leibniz formula, we can rewrite Eq. (45) in the more elegant form

$$m_i \frac{\mathrm{d}^v(\Delta V_i)}{\mathrm{d}t^v} = AF_i \frac{(v-1)!}{\Gamma(v)}. \tag{46}$$

This equation can be used to describe Brownian motion and loss due to collisions.

Similar to Ref. [9], the results of the previous section can also be applied to the Liouville equation.

7. Conclusions

(1) When the self-similar set E_t or generalized self-similar set E_t' on $[0, t]$ is given, no matter which self-similar measure satisfying (9) or generalized self-similar measure satisfying (40) for the memory measure on E_t is taken, the approximate expressions (30) and (31) of the memory function and flux

function are invariable.

(2) From (27), no matter which self-similar measure or generalized self-similar measure for the memory measure is taken, the fractional exponent v of the fractional integral is determined only by $\ln P_1/\ln \xi_1$, while it does not depend on another weight P_j of the self-similar measure or generalized self-similar measure and the similar coefficient ξ_j of other similarities $\{S_j\}$.

(3) The fractional exponent v of the fractional integral is equal to the fractal dimension s of the self-similar set E_t or generalized self-similar set E_t' if and only if $P_1 = \xi_1^s$. In particular, $v = \ln 2/\ln(1/\xi)$ when E_t is the Cantor set, and $v = \ln k/\ln(1/\xi)$ when E_t is Cantor's k-bar, this is Nigmatullin's result [9].

(4) When P_1 changes from 0 to 1, v can have all positive real values.

(5) Since the structure of the self-similar set or generalized self-similar set is similar to the Cantor set (or Cantor's k-bar), its corresponding fractional integral also has physical interpretations.

◇ References ◇

[1] K. Binder and A. P. Joung, Rev. Mod. Phys. 58 (1986) 801.

[2] S. L. Ginzbug, Irreversible phenomena in spin glasses (Nauka, Moscow, 1989) [in Russian].

[3] A. I. Olemskoi and E. A. Toropov, Fiz. Met. Metalloved 9 (1991) 5.

[4] A. I. Olemskoi and E. A. Toropov, Fiz. Met. Metalloved 7 (1991) 32.

[5] A. I. Olemskoi and I. A. Sklyar, Usp. Fiz. Nauk. 162 (1992) 29.

[6] Synergetics and fatigue fracture of metals (Nauka, Moscow, 1989) [in Russian].

[7] A. I. Olemskoi, Fiz. Met. Metalloved 68 (1989) 56.

[8] A. I. Olemskoi and Yu. I. Paskal, preprint No. 30, Institute of Physics and Applied Mathematics, Tomsk Affiliate of the Siberian Branch of the USSR Academy of Sciences, Tomsk (1988).

[9] R. R. Nigmatullin, Teor. Mat. Fiz. 90 (1992) 354.

[10] J. P. Hutchinson, Indiana Univ. Math. J. 30 (1981) 713.

[11] Ren Fu-Yao and Liang Jin-Rong, Chin. Ann. Math. B. 16 (1995) 153.

[12] Hua-Su, Acta. Math. Appl. Sinica China 17 (1994) 551.

[13] R. H. Riedi and B. B. Mandelbrot, Appl. Math. 16 (1995) 132.

[14] Yu. I. Babenko, Heat and mass transfer. A method of calculating thermal and diffusion fluxes (Khimiya, Leningrad, 1986) [in Russian].

[15] R. R. Nigmatullin and B. A. Belavin, Tr. KAI 82 (1964).

[16] R. R. Nigmatullin, Phys. Status Solidi B 133 (1986) 425.

[17] R. R. Nigmatullin, Phys. Status Solidi B 124 (1984) 389.

[18] L. A. Dissado, R. R. Nigmatullin and R. M. Hill, in: Dynamical processes in condensed matter, Vol. 63, ed. M. Evans (1985) p. 253.

[19] R. R. Nigmatullin, Fiz. Tverd. Tela (Leningrad) 27 (1985) 1583.

[20] A. K. Jonscher, Dielectric relaxation in solids (Chelsea Dielectric Press, London, 1983).

The Relationship between the Fractional Integral and the Fractal Structure of a Memory Set*

Fu-Yao Ren[a], Zu-Guo Yu[b], Ji Zhou[a],
Alain Le Mehaute[c], Raoul R. Nigmatullin[d]

[a] *Institute of Mathematics, Fudan University, Shanghai 200433, People's Republic of China*

[b] *Institute of Theoretical Physics, Academia Sinica, P.O. Box 2375, Beijing 100080, People's Republic of China*

[c] *Institut Supérieur des Materiaux, du Mans, 72000 - Le Mans, France*

[d] *Department of Theoretical Physics, Kazan State University, 420008, Kazan, Tatarstan, Russia*

Abstract: It is shown that there is no direct relation between the fractional exponent v of the fractional integral and the fractal structure of the memory set considered, v depends only on the first contraction coefficient ξ_1 and the first weight P_1 of the self-similar measure (or infinite self-similar measure) μ on the memory set. If and only if $P_1 = \xi_1^\beta$ (where $\beta \in (0, 1)$ is the fractal dimension of the memory set), v is equal to the fractal dimension of the memory set. It is also true that v is continuous about ξ_1 and P_1.

Keywords: flux, memory measure, Laplace transform, memory set, self-similar (or infinite self-similar) measure

---

* Project partially supported by the Tianyuan Foundation of China and Ph. D. station Foundation of the State Education Committee. Originally published in *Physica A*, Vol. 246, (1997), 419 - 429.

1. Introduction

When we describe the structure of evolution of a physical system far from thermodynamic equilibrium, e. g. , in amorphous materials [1~3], in the description of structural relaxation of high-T_c oxide superconductors [4], in the process of plastic deformation [5], fracture of solids [6], in the description of solid solutions [7], the macrostructure of martensite [8], and so on, it has been found that the medium exhibits memory. The existence of memory means that if at time τ a force $f(\tau)$ acts on the system, then there arises a flux J whose magnitude at time $t > \tau$ is given through a memory function $m(\tau)$ by the equation

$$J(t) = \int_0^t m(t-\tau) f(\tau) d\tau. \tag{1}$$

For any given $T \in (0, \infty)$, if the "residual" memory set is a Cantor's fractal set (or Cantor's k-bars) in $[0, T]$ generated by $\varphi_1 = \xi x$, $\varphi_2 = \xi x + (1-\xi)T$ (or $\varphi_j = \xi x + (j-1)\xi T + (j-1)[(1-k\xi)T/(k-1)]$, $j = 1, 2, \cdots, k$), and if the total number of remaining states in each stage of the division of this set is normalized to unity, Nigmatullin [9], in 1992, obtained the folllowing result:

$$\begin{aligned} J(t) &\backsimeq A_v T^{-v} [\Gamma(v)]^{-1} \int_0^t (t-t')^{v-1} f(t') dt' \\ &= A_v T^{-v} t^v \hat{D}^{-v} f(t), \end{aligned} \tag{2}$$

where $v = \ln 2/\ln(1/\xi)$ (or $v = \ln k/\ln(1/\xi)$) is the fractal dimension of Cantor's set (or Cantor's k-bars), $A_v = [\sqrt{2}(1-\xi)T]^{-v}$ for Cantor's set, and $A_v = \exp[-I_1/\ln(1/\xi)]$ where $I_1 = \int_0^\infty [f'(y)/f(y)] \ln y dy$ and $f(y) = \{1 - \exp[-ky/(k-1)]\}/\{k(1-\exp[-y/(k-1)])\}$ for Cantor's k-bars, $\Gamma(v)$ is the gamma function, and the fractional exponent v of the fractional integral is defined by

$$\hat{D}^{-v}f(t)=(\Gamma(v))^{-1}\int_0^1(1-u)^{v-1}f(tu)\mathrm{d}u \tag{3}$$

$$=[v\Gamma(v)]^{-1}\int_0^1 f((1-u_1)t)\mathrm{d}u_1^v \tag{4}$$

is equal to the fractal dimension of the Cantor set $s=\ln 2/\ln\xi$ (or Cantor's k-bar $s=\ln k/\ln\xi$); furthermore, the physical interpretation of the fractional integral is also given in Ref. [9].

In this paper, we denote by $\mathbb{R}$ the set of all real numbers, $\mathbb{C}$ the set of all complex numbers and $\mathbb{Z}^+$ the set of all positive integer numbers.

Mehaute, Nigmatullin and Nivanen in their book "*Irreversibilité Temporel et Geometrie Fractale*" proposed the following question:

Is there any relationship between the fractional integral written in the form of Eq. (2) *and the geometrical characteristics of the fractal structure considered?*

In fact, the relationship between the fractional exponent v of the fractional integral and the geometrical characteristics of the fractal structure of the memory set considered needs to be studied.

In this paper, we want to solve this problem.

2. Fractal Constrution of Self-similar Sets

2.1 Self-similar set

For any given $T\in(0,\infty)$, fixed $K<\infty$ or $K=\infty$, let

$$\varphi_j(x)=\xi_j x+b_j,\quad 0<\xi_j<1,\quad b_j\in[0,T],\quad j\in I=\{1,2,\cdots,K\}$$

be similarities on $E_0=[0,T]$, and assume that $\mathrm{Int}(\varphi_j(E_0))\cap\mathrm{Int}(\varphi_i(E_0))=\varnothing$ for $j\neq i$ and $b_1=0<b_2<\cdots<b_K=T(1-\xi_K)$.

For natural numbers n, let

$$E^{(1)}_{j_1j_2\cdots j_n}=\varphi_{j_1}\circ\cdots\circ\varphi_{j_n}(E_0)=[b^{(1)}_{j_1\cdots j_n},a^{(1)}_{j_1\cdots j_n}],$$

$$E^{(1)}(n)=\overline{\bigcup_{j_1\in I}E^{(1)}_{j_1\cdots j_n}},$$

where $[a, b]$ is an interval, i. e., $a \leqslant x \leqslant b$, $j_i \in I(i=1, 2, \cdots, n)$, $b_{j_1}^{(1)} = b_{j_1}$, $j_1 \in I$ and

$$b_{j_1\cdots j_n}^{(1)} = b_{j_n} \prod_{i=1}^{n-1} \xi_{j_i} + b_{j_1\cdots j_{n-1}}^{(1)}, \quad \Delta_{j_1\cdots j_n}^{(1)} = T \prod_{i=1}^{n} \xi_{j_i}, \quad a_{j_1\cdots j_n}^{(1)} = b_{j_1\cdots j_n}^{(1)} + \Delta_{j_1\cdots j_n}^{(1)}.$$

It is obvious that

$$E_{j_1\cdots j_n}^{(1)} \subset E_{j_1\cdots j_{n-1}}^{(1)}, \quad E^{(1)}(n) \subset E^{(1)}(n-1).$$

Then

$$E_T^{(1)} = \bigcap_{n=1}^{\infty} E^{(1)}(n)$$

is called a *self-similar set* (if $K < \infty$) or an *infinite self-similar set* (if $K = \infty$).

Now $E_j^{(1)} = [b_j^{(1)}, a_j^{(1)}]$, $j \in I$ and $\mathrm{Int}(E_j^{(1)}) \cap \mathrm{Int}(E_{j'}^{(1)}) = \varnothing$, $j \neq j'$. Denote $O^{(1)}(1) = E_0 \backslash E^{(1)}(1)$. If $E_j^{(1)} \cap E_{j'}^{(1)} = \varnothing$, $j \neq j'$, $j, j' \in I$, then

$$O^{(1)}(1) = \bigcup_{j=1}^{K-1} (a_j^{(1)}, b_{j+1}^{(1)}), \quad (a_j^{(1)}, b_{j+1}^{(1)}) \cap (a_{j'}^{(1)}, b_{j'+1}^{(1)}) = \varnothing, \quad j \neq j'. \tag{5}$$

If there exists an $a_j^{(1)} = b_{j+1}^{(1)}$, we assume $\#\{j \in I: a_j^{(1)} = b_{j+1}^{(1)}\} = k$ $(k < K)$ and $\{j \in I: a_j^{(1)}, b_{j+1}^{(1)}\} = \{i_1, i_2, \cdots, i_k\}$, denote $I_1 = \{j \in I: j \neq i_1, i_2, \cdots, i_k\}$, then

$$O^{(1)}(1) = \bigcup_{j \in I_1} (a_j^{(1)}, b_{j+1}^{(1)}), \quad (a_j^{(1)}, b_{j+1}^{(1)}) \cap (a_{j'}^{(1)}, b_{j'+1}^{(1)}) = \varnothing, \quad j, j' \in I_1, j \neq j'. \tag{5'}$$

Hence

$$E_0 = E^{(1)}(1) \cup O^{(1)}(1), \quad E^{(1)}(1) \cap O^{(1)}(1) = \varnothing. \tag{6}$$

For any $j_1, j_2 \in I$, let $F_{j_1}^{(1)} = \bigcup_{j_2 \in I} E_{j_1 j_2}^{(1)}$, $O_{j_1}^{(1)} = E_{j_1}^{(1)} \backslash F_{j_1}^{(1)}$, then

$$O_{j_1}^{(1)} = \bigcup_{j_2=1}^{K-1} (a_{j_1 j_2}^{(1)}, b_{j_1 j_2+1}^{(1)}), \quad \text{or} \quad O_{j_1}^{(1)} = \bigcup_{j_2 \in I_1} (a_{j_1 j_2}^{(1)}, b_{j_1 j_2+1}^{(1)}),$$
$$(a_{j_1 j_2}^{(1)}, b_{j_1 j_2+1}^{(1)}) \cap (a_{j_1 j_2'}^{(1)}, b_{j_1 j_2'+1}^{(1)}) = \varnothing, \ j_2 \neq j_2'$$

and $E_{j_1}^{(1)} = F_{j_1}^{(1)} \cup O_{j_1}^{(1)}$. Denote

$$O^{(1)}(2)=\bigcup_{j_1\in I,\ j_2\in I_1}(a^{(1)}_{j_1j_2},\ b^{(1)}_{j_1j_2+1}),\tag{7}$$

then

$$E^{(1)}(1)=E^{(1)}(2)\cup O^{(1)}(2),\quad E^{(1)}(2)\cap O^{(1)}(2)=\varnothing.\tag{8}$$

Similarly, let

$$O^{(1)}(n)=\bigcup_{j_1,\cdots,j_{n-1}\in I,\ j_n\in I_1}(a^{(1)}_{j_1\cdots j_{n-1}j_n},\ b^{(1)}_{j_1\cdots j_{n-1}j_{n+1}}),$$

then

$$E^{(1)}(n-1)=E^{(1)}(n)\cup O^{(1)}(n),\quad E^{(1)}(n)\cap O^{(1)}(n)=\varnothing.\tag{9}$$

The self-similar set

$$E^{(1)}_T\overset{\text{denote}}{=\!=\!=}E^{(1)}_T(\{(b_j,\ \xi_j)\}^K_1)=\lim_{n\to\infty}E^{(1)}(n).\tag{10}$$

From the above we can see

$$E_0=E^{(1)}(n)\cup(\bigcup_{i=1}^{n}O^{(1)}(i)).\tag{11}$$

We denote by $m(E)$ the Lebesgue measure of set E, since

$$m(E^{(1)}(n))=\sum_{j_1\cdots j_n\in I}(a^{(1)}_{j_1\cdots j_n}-b^{(1)}_{j_1\cdots j_n})=\sum_{j_1\cdots j_n\in I}(\xi_{j_1}\cdots\xi_{j_n})T=T(\sum_{j=1}^{K}\xi_j)^n,\tag{12}$$

$$m(O^{(1)}(n))=\sum_{j_1\cdots j_{n-1}\in I,\ j_n\in I_1}(b^{(1)}_{j_1\cdots j_{n-1}j_{n+1}}-a^{(1)}_{j_1\cdots j_n}).\tag{13}$$

From Eqs. (11)~(13), we have

$$m(E^{(1)}_T)=\lim_{n\to\infty}m(E^{(1)}(n))=0,\tag{14}$$

$$\sum_{n=1}^{\infty}m(O^{(1)}(n))=T-m(E^{(1)}_T)=T.\tag{15}$$

So the self-similar (or infinite self-similar) set $E^{(1)}_T=E^{(1)}_T(\{(b_j,\ \xi_j)\}^K_1)$ is of the following geometrical characteristics:

(i) $E^{(1)}_T$ is obtained from the closed interval $[0,\ T]$ cutting out countable infinite disjoint open intervals with form $(a^{(1)}_{j_1\cdots j_n},\ b^{(1)}_{j_1\cdots j_{n-1}j_{n+1}})$. The sum of

Lebesgue measures of all these open intervals is equal to T.

(ii) $E_T^{(1)}$ is a nowhere dense set, i. e. it is not dense in any neighbourhood of any point x in $\mathbb{R}$.

(iii) $E_T^{(1)}$ is stationary about $\{b_j\}_{j=1}^K$ and $\{\xi_j\}_{j=1}^K$, i. e. if $b_j \to c_j$, $\xi_j \to \eta_j$, $\sum_{j=1}^K \eta_j < 1$, then $E_T^{(1)}(\{(b_j, \xi_j)\}_1^K)$ converges to $E_T^{(1)}(\{(c_j, \eta_j)\}_1^K)$ under the Hausdorff metric d_H.

(iv) From Ref. [10] or Refs. [11, 12], the fractal dimension s of $E_T^{(1)}$ is the solution of equation $\sum_{j=1}^K \xi_j^s = 1$.

Cantor's set and Cantor's k-bars are self-similar sets, and the Von Koch curve is also a self-similar set.

2.2 Generalized self-similar set

Let $\{\varphi_{n,j}(x) = \xi_{n,j}x + b_{n,j}: 0 < \xi_{n,j} < 1,\ b_{n,j} \in \mathbb{R},\ j = 1, 2, \cdots, K_n < \infty\}_{n \in \mathbb{Z}^+}$ be a family of similarities, and assume that for each n, $\mathrm{Int}(\varphi_{n,j}(E_0)) \cap \mathrm{Int}(\varphi_{n,i}(E_0)) = \varnothing$ and

$$b_{n,1} = 0 < b_{n,2} < \cdots < b_{n,K_n} = T(1 - \xi_{n,K_n}), \quad n \in \mathbb{Z}^+.$$

Let $I_n = \{1, 2, \cdots, K_n\}$, $\sigma = (j_1, j_2, \cdots, j_n)$, $\Lambda_n = I_1 \times I_2 \times \cdots \times I_n$,

$$E_{j_1 j_2 \cdots j_n}^{(2)} = \varphi_{1,j_1} \circ \cdots \circ \varphi_{n,j_n}(E_0) = [b_{j_1 \cdots j_n}^{(2)}, a_{j_1 \cdots j_n}^{(2)}],$$

$$E^{(2)}(n) = \bigcup_{\sigma \in \Lambda_n} E_{j_1 j_2 \cdots j_n}^{(2)},$$

where $b_{j_1}^{(2)} = b_{1,j_1}$, $1 \leqslant j_1 \leqslant K_1$ and

$$b_{j_1 \cdots j_n}^{(2)} = b_{n,j_n} \prod_{i=1}^{n-1} \xi_{i,j_i} + b_{j_1 \cdots j_{n-1}}^{(2)}, \quad \Delta_{j_1 \cdots j_n}^{(2)} = T \prod_{i=1}^{n} \xi_{i,j_i},$$

$$a_{j_1 \cdots j_n}^{(2)} = b_{j_1 \cdots j_n}^{(2)} + \Delta_{j_1 \cdots j_n}^{(2)}.$$

Then

$$E_T^{(2)} = \bigcap_{n=1}^{\infty} E^{(2)}(n)$$

is called a *generalized self-similar set*. Let

$$O^{(2)}(n) = E^{(2)}(n-1) \backslash E^{(2)}(n) = \bigcup_{j_n=1}^{K_{n-1}} \bigcup_{(j_1, \cdots, j_{n-1}) \in \Lambda_{n-1}} (a_{j_1 \cdots j_{n-1} j_n}^{(2)}, b_{j_1 \cdots j_{n-1} j_{n+1}}^{(2)}),$$

then the generalized self-similar set is given by

$$E_T^{(2)} \xlongequal{\text{denote}} E_T^{(2)}(\{b_{n,j}, \xi_{n,j})\}_{j=1}^{K_n}) = \lim_{n\to\infty} E^{(2)}(n),$$

$$E_0 = E^{(2)}(n) \cup (\bigcup_{i=1}^{n} O^{(2)}(i)) = E_T^{(2)} \cup (\bigcup_{n=1}^{\infty} O^{(2)}(n)).$$

Similarly the generalized self-similar set $E_T^{(2)} = E_T^{(2)}(\{(b_{n,j}, \xi_{n,j})\}_{j=1}^{K_n})$ has the following geometrical characteristics:

(i) $E_T^{(2)}$ is obtained from the closed interval $[0, T]$ cutting out countable infinite disjoint open intervals with form $(a_{j_1\cdots j_n}^{(2)}, b_{j_1\cdots j_{n-1}j_{n+1}}^{(2)})$. The sum of Lebesgue measures of all these open intervals is equal to T.

(ii) $E_T^{(2)}$ is a nowhere dense set.

(iii) $E_T^{(2)}$ is stationary about $\{b_{n,j}: 1 \leqslant j \leqslant K_n\}_{n=1}^{\infty}$ and $\{\xi_{n,j}: 1 \leqslant j \leqslant K_n\}_{n=1}^{\infty}$.

(iv) If $c = \inf_{n\in Z^+, j\in I_n} \xi_{n,j} > 0$ and $\lim_{n\to\infty} \sup_{\sigma\in A_n} |E_{j_1j_2\cdots j_n}^{(2)}| = 0$, then from Ref. [13] or [14], the fractal dimension s'' of $E_T^{(2)}$ is $\liminf_{n\to\infty} \beta_n$, where β_n is the solution of $\prod_{i=1}^{n} (\sum_{j=1}^{K_i} \xi_{i,j}^{\beta_n}) = 1$. Even if $\lim_{n\to\infty} K_n = \infty$, the conclusion of fractal dimension of $E_T^{(2)}$ also holds.

3. Fractal Construction of Cookie-Cutter Sets

3.1 Cookie-cutter set

For any given $T \in (0, \infty)$, fixed $K < \infty$ or $K = \infty$, denote $E_0 = [0, T]$, $E_1, E_2, \cdots, E_K$ are closed subintervals of E_0, and $\mathrm{Int}(E_i) \cap \mathrm{Int}(E_j) = \varnothing$ $(i \neq j)$. A *cookie-cutter map* (if $K < \infty$) or *infinite cookie-cutter map* (if $K = \infty$) is a mapping $S: \bigcup_{j=1}^{K} E_j \to E_0$ with the properties that

(1) $S|E_j$ is a $1-1$ mapping onto E_0;

(2) S is $C^{1+\gamma}$ differentiable, i. e. differentiable with a Hölder continuous derivative DS satisfying $|DS(x) - DS(y)| < c|x-y|^{\gamma}$ for some $c > 0$, and $|DS(x)| > 1$ for all $x \in \bigcup_{j=1}^{K} E_j$.

The *cookie-cutter set* (if $K < \infty$) or *infinite cookie-cutter set* (if $K = \infty$) associated to S is the set

$$E_T^{(3)} = \{x \in \overline{\bigcup_{j=1}^{K} E_j} : S^n(x) \in \overline{\bigcup_{j=1}^{K} E_j}, \ \forall n \geq 0\}.$$

Examples If $K=2$, $E_1 = \left[0, \frac{1}{3}T\right]$, $E_2 = \left[\frac{2}{3}T, T\right]$ and $S(x) \equiv 3x$ mod T, then $E_T^{(3)}$ is a Cantor set. More generally, Cantor k-bars and self-similar sets are all cookie-cutter sets.

Now if we denote $S|_{E_j} = \varphi_j^{-1}$, then $|D\varphi_j(x)| < 1$, $\forall x \in E_0$. For natural numbers n, let $I = \{1, 2, \cdots, K\}$,

$$E_{j_1 j_2 \cdots j_n}^{(3)} = \varphi_{j_1} \circ \varphi_{j_2} \circ \cdots \circ \varphi_{j_n}(E_0), \quad E^{(3)}(n) \overline{\bigcup_{j_i \in I} E_{j_1 j_2 \cdots j_n}^{(3)}}.$$

It is obvious that

$$E_T^{(3)} = \bigcap_{n \geq 1} E^{(3)}(n) = \bigcap_{n \geq 1} \bigcup_{j_i \in I} E_{j_1 j_2 \cdots j_n}^{(3)}. \tag{16}$$

We can see that $E_{j_1 j_2 \cdots j_n}^{(3)} = [b_{j_1 j_2 \cdots j_n}^{(3)}, a_{j_1 j_2 \cdots j_n}^{(3)}]$, where $b_{j_1 j_2 \cdots j_n}^{(3)} = \varphi_{j_1} \circ \varphi_{j_2} \circ \cdots \circ \varphi_{j_n}(0)$, $a_{j_1 j_2 \cdots j_n}^{(3)} = \varphi_{j_1} \circ \varphi_{j_2} \circ \cdots \circ \varphi_{j_n}(T)$. Similar to the discussion in Section 2. 1, let

$$O^{(3)}(n) = \bigcup_{j_1, \cdots, j_{n-1} \in I, \, j_n \in I_1} (a_{j_1 \cdots j_{n-1} j_n}^{(3)}, b_{j_1 \cdots j_{n-1} j_{n+1}}^{(3)}), \tag{17}$$

then

$$E^{(3)}(n-1) = E^{(3)}(n) \cup O^{(3)}(n), \quad E^{(3)}(n) \cap O^{(3)}(n) = \varnothing. \tag{18}$$

The cookie-cutter (or infinite cookie-cutter) set

$$E_T^{(3)} \xlongequal{\text{denote}} E_T^{(3)}(\{\varphi_j\}_1^K) = \lim_{n \to \infty} E^{(3)}(n). \tag{19}$$

From the above we can see

$$E_0 = E^{(3)}(n) \cup (\bigcup_{i=1}^{n} O^{(3)}(i)) = E_T^{(3)} \cup (\bigcup_{n=1}^{\infty} O^{(3)}(n)), \tag{20}$$

$$T = \lim_{n \to \infty} m(E^{(3)}(n)) + \sum_{n=1}^{\infty} m(O^{(3)}(n)).$$

From Ref. [15], $m(E_T^{(3)}) = \lim_{n \to \infty} m(E^{(3)}(n)) = 0$. So the cookie-cutter (or infinite cookie-cutter) set $E_T^{(3)} = E_T^{(3)}(\{\varphi_j\}_1^K)$ has the following geometrical characteristics:

(i) $E_T^{(3)}$ is obtained from the closed interval $[0, T]$ cutting out countable infinite disjoint open intervals with form $(\varphi_{j_1}\circ\varphi_{j_2}\circ\cdots\circ\varphi_{j_n}(T), \varphi_{j_1}\circ\cdots\circ\varphi_{j_{n-1}}\circ\varphi_{j_{n+1}}(0))$. The sum of Lebesgue measures of all these open intervals is equal to T.

(ii) $E_T^{(3)}$ is a nowhere dense set.

(iii) $E_T^{(3)}$ is stationary about $\{\varphi_j\}_{j=1}^K$.

(iv) For the fractal dimension formula of a cookie-cutter set $E_T^{(3)}$ (when $K<\infty$) see Ref. [15]. The fractal dimension s of an infinite cookie-cutter set $E_T^{(3)}$ (when $K=\infty$) is estimated in Ref. [16]. (Let $\bar{r}_j=\sup_{x\in E_0}|D\varphi_j(x)|$, $\underline{r}_j=\inf_{x\in E_0}|D\varphi_j(x)|$, then $\min\{1, l\}\leqslant s\leqslant u$, where l, u satisfy equations $\sum_{j=1}^{\infty}\underline{r}_j^l=1$ and $\sum_{j=1}^{\infty}\bar{r}_j^u=1$.)

3.2 Generalized cookie-cutter set

Let $\{\varphi_{n,j}(x): E_0\to E_0, j=1, 2, \cdots, K_n<\infty\}_{n\in\mathbb{Z}^+}$ be a family of functions satisfying

(1) $\varphi_{n,j}: E_0\to\varphi_{n,j}(E_0)$ is a $1-1$ mapping and $\mathrm{Int}(\varphi_{n,i}(E_0))\cap\mathrm{Int}(\varphi_{n,j}(E_0))=\varnothing$ $(i\neq j)$ for any n and $1\leqslant i, j\leqslant K_n$.

(2) For all n, the mapping $S_n: \bigcup_{j=1}^{K_n}\varphi_{n,j}(E_0)\to E_0$, defined by $S_n|_{\varphi_{n,j}(E_0)}=(\varphi_{n,j})^{-1}$ is $C^{1+\gamma}$ differentiable, i. e. differentiable with a Hölder continuous derivative DS satisfying $|DS_n(x)-DS_n(y)|<c_n|x-y|^\gamma$ for some $c_n>0$, and $|DS_n(x)|>c_0>1$ for some constant number c_0 and all $x\in\bigcup_{j=1}^{K_n}\varphi_{n,j}(E_0)$.

Then $\{S_n\}_{n\in\mathbb{Z}^+}$ is called a sequence of *cookie-cutter maps*. Now $|D\varphi_{n,j}(x)|<c_0^{-1}<1$, $\forall x\in E_0$. For natural numbers n, let $I_n=\{1, 2, \cdots, K_n\}$, $\Lambda_n=I_1\times I_2\times\cdots\times I_n$,

$$E_{j_1j_2\cdots j_n}^{(4)}=\varphi_{1,j_1}\circ\varphi_{2,j_2}\circ\cdots\circ\varphi_{n,j_n}(E_0),\quad E^{(4)}(n)=\bigcup_{j_1j_2\cdots j_n\in\Lambda_n}E_{j_1j_2\cdots j_n}^{(4)}.$$

Then

$$E_T^{(4)}\xlongequal{\text{denote}}E_T^{(4)}(\{\varphi_{n,j}, 1\leqslant j\leqslant K_n\}_{n=1}^{\infty})=\bigcap_{n\geqslant1}E^{(4)}(n)=\bigcap_{n\geqslant1}\bigcup_{j_1j_2\cdots j_n\in\Lambda_n}E_{j_1j_2\cdots j_n}^{(4)}$$

is called a *generalized cookie-cutter set* associated to $\{S_n\}$.

Similar to the discussion in Section 3.1, the generalized cookie-cutter set $E_T^{(4)}$ is of the same geometrical characteristics (i) and (ii) of the cookie-cutter set $E_T^{(3)}$. Moreover, $E_T^{(4)}$ is stationary about $\{\varphi_{n,j}, 1 \leqslant j \leqslant K_n\}_{n=1}^{\infty}$, and from Theorems 3 and 4 of Ref. [14], we can estimate its fractal dimension s. (Let $a_{n,j} = \sup_{x \in E_0} |D\varphi_j(x)|$, $b_{n,j} = \inf_{x \in E_0} |D\varphi_j(x)|$, $\bar{\beta}_n$ and $\underline{\beta}_n$ satisfy equations $\prod_{i=1}^{n}(\sum_{j=1}^{K_i} a_{i,j}^{\bar{\beta}_n}) = 1$ and $\prod_{i=1}^{n}(\sum_{j=1}^{K_i} b_{i,j}^{\underline{\beta}_n}) = 1$, then $\liminf_{n\to\infty} \underline{\beta}_n \leqslant s \leqslant \liminf_{n\to\infty} \bar{\beta}_n$.)

4. Memory Measures, (Infinite) Self-similar Measures and The Flux Function

For given probabilities $(P_1, P_2, \cdots, P_K)$ satisfying $\sum_{j=1}^{K} P_j = 1$, $K < \infty$ or $K = \infty$, let the memory measure on $E^{(i)}(n)(i = 1, 3)$ is defined by $d\mu_n^{(i)}(\tau) = m_n^{(i)}(\tau)d\tau$, where

$$m_n^{(i)}(\tau) = \sum_{j_1, \cdots, j_n \in I} P_{j_1} \cdots P_{j_n} \chi_{E^{(i)}_{j_1 \cdots j_n}}(\tau) / |E^{(i)}_{j_1 \cdots j_n}|, \quad i = 1, 3 \tag{21}$$

and

$$\chi_{E^{(i)}_{j_1 \cdots j_n}}(\tau) = \begin{cases} 1 & \text{if } \tau \in E^{(i)}_{j_1 \cdots j_n}, \\ 0 & \text{otherwise.} \end{cases}$$

For a given family of probabilities $\{P_{n,j}\}$ satisfying $\sum_{j=1}^{K_n} P_{n,j} = 1$, let the memory measure on $E^{(i)}(n)(i = 2, 4)$ be defined by $du_n^{(i)}(\tau) = m_n^{(i)}(\tau)d\tau$, where

$$m_n^{(i)}(\tau) = \sum_{j_1 \cdots j_n \in \Lambda_n} P_{1,j_1} \cdots P_{n,j_n} \chi_{E^{(i)}_{j_1 \cdots j_n}}(\tau) / |E^{(i)}_{j_1 \cdots j_n}|, \quad i = 2, 4. \tag{22}$$

Then we proved elsewhere [16~18] that there exists a measure $\mu^{(i)}$ $(i = 1, 3)$ satisfying $\int_{E_T^{(i)}} d\mu^{(i)}(\tau) = 1$, $\mathrm{supp}(\mu^{(i)}) = E_T^{(i)}$ such that

$$\mu_n^{(i)} \to \mu^{(i)} \quad \text{(weakly converges)}, \ i = 1, 3. \tag{23}$$

We call $\mu^{(i)}$ the *memory measure* on the (infinite) self-similar (or (infinite) cookie-cutter) set $E_T^{(i)}$ $(i=1, 3)$. We also have

$$\int_A \mathrm{d}\mu^{(i)}(\tau)=\sum_{j=1}^{K} P_j \int_A \mathrm{d}\mu^{(i)} \circ \varphi_j^{-1}(\tau), \tag{24}$$

i. e.,

$$\mu^{(i)}(\cdot)=\sum_{j=1}^{K} P_j \mu^{(i)} \circ \varphi_j^{-1}(\cdot),\ i=1,\ 3. \tag{25}$$

A probability measure satisfying Eq. (25) is called a *self-similar measure* (when $K<\infty$) or *infinite self-similar measure* (when $K=\infty$) and $\{P_j\}$ are called *weights*. Hence the memory measure $\mu^{(i)}$ is a self-similar (or infinite self-similar) measure. From Ref. [10], $\mu^{(i)}$ is the unique self-similar (or infinite self-similar) measure satisfying Eq. (25) for given $\{P_j\}_{j=1}^{K}$ and $\{\varphi_j\}_{j=1}^{K}$.

We proved elsewhere [17, 19] that there exists a measure $\mu^{(i)}$ $(i=2, 4)$ satisfying $\int_{E_T^{(i)}} \mathrm{d}\mu^{(i)}(\tau)=1$, $\operatorname{supp}(\mu^{(i)})=E_T^{(i)}$ such that

$$\mu_n^{(i)} \rightarrow \mu^{(i)} \quad \text{(weakly converges)},\ i=2,\ 4. \tag{26}$$

We call $\mu^{(i)}$ the *memory measure* on the generalized self-similar (or generalized cookie-cutter) set $E_T^{(i)}$ $(i=2, 4)$. If $\lim_{n\to\infty} K_n=K<\infty$, $\lim_{n\to\infty} P_{n,j}=P_j$ and $\lim_{n\to\infty} \varphi_{n,j}=\varphi_j$, we have

$$\mu^{(i)}(\cdot)=\sum_{j=1}^{K} P_j \mu^{(i)} \circ \varphi_j^{-1}(\cdot),\ i=2,\ 4. \tag{27}$$

Hence by uniqueness, the memory measure $\mu^{(i)}$ is the self-similar measure corresponding to weights $\{P_j\}_{j=1}^{K}$. If $\lim_{n\to\infty} K_n=\infty$, $\lim_{n\to\infty} P_{n,j}=P_j$ and $\lim_{n\to\infty} \varphi_{n,j}=\varphi_j$, we have

$$\mu^{(i)}(\cdot)=\sum_{j=1}^{\infty} P_j \mu^{(i)} \circ \varphi_j^{-1}(\cdot),\ i=2,\ 4. \tag{28}$$

Hence by uniqueness, the memory measure $\mu^{(i)}$ $(i=2, 4)$ is the infinite self-similar measure corresponding to weights $\{P_j\}_{j=1}^{\infty}$.

Let

$$M^{(i)}(p)=\int_0^\infty e^{-p\tau}\mathrm{d}\mu^{(i)}(\tau),\ i=1,\cdots,4 \tag{29}$$

be the Laplace transform of $\mu^{(i)}$. When $i = 3, 4$, we assume $\varphi_1(x)=\xi_1 x$, $0<\xi_1<1$. In Refs. [16~19], we proved that when $\mathrm{Re}(p)$ is large enough, we have

$$M^{(i)}(p)=P_1 M^{(i)}(\xi_1 p)+o(1),\quad i=1,\cdots,4\quad (\text{as } \mathrm{Re}(p)\to+\infty). \tag{30}$$

The unique solution of equation

$$\overline{M^{(i)}}(p)=P_1\overline{M^{(i)}}(\xi_1 p) \tag{31}$$

has the form

$$\overline{M^{(i)}}(p)=A^{(i)}p^{-v}, \tag{32}$$

where $A^{(i)}$ is a constant depending only on $\{P_j\}_{j=1}^K$ and $\{\varphi_j\}_{j=1}^K$ (or $\{P_j\}_{j=1}^\infty$ and $\{\varphi_j\}_{j=1}^\infty$), and

$$v=\ln P_1/\ln\xi_1. \tag{33}$$

Hence when $\mathrm{Re}(p)$ is large enough,

$$M^{(i)}(p)\approx A^{(i)}p^{-v}. \tag{34}$$

We apply inverse Laplace transform to Eq. (34) and obtain

$$\mathrm{d}\mu^{(i)}(\tau)\approx A^{(i)}(\Gamma(v))^{-1}\tau^{v-1}\mathrm{d}\tau \tag{35}$$

$$\begin{aligned} J^{(i)}(t) &\approx A^{(i)}(\Gamma(v))^{-1}\int_0^t (t-\tau)^{v-1}f(\tau)\mathrm{d}\tau \\ &= A^{(i)}t^v(\Gamma(v))^{-1}\int_0^1 (1-u)^{v-1}f(ut)\mathrm{d}u \\ &= A^{(i)}t^v D^{-v}f,\quad i=1,\cdots,4, \end{aligned} \tag{36}$$

where

$$D^{-v}f=(\Gamma(v))^{-1}\int_0^1 (1-u)^{v-1}f(ut)\mathrm{d}u$$

$$=(v\Gamma(v))^{-1}\int_0^1 f((1-u_1)t)\mathrm{d}u_1^v \tag{37}$$

is the fractional integral. Then from Eq. (36), we establish the connection between the fractional integral and the flux. In particular, if we assume $\beta \in (0, \infty)$ is the fractal dimension of memory set $E_T^{(i)}$ $(i=1, \cdots, 4)$, when $P_1 = \xi_1^\beta$ and $\sum_{j=1}^K P_j = 1$ (or $P_1 = \xi_1^\beta$ and $\sum_{j=1}^\infty P_j = 1$), from Eq. (33) we have $v = \beta$; thus we establish the connection between the fractional integral and the dimension of the memory set.

5. Conclusions

(1) From Eq. (33) $v = \ln P_1/\ln \xi_1$ and from Eq. (36) we can see that there is no direct relation between the fractional integral written in the form of Eq. (2) and the geometrical characteristics of the fractal structure of memory set $E_T^{(i)}$ $(i=1, \cdots, 4)$ considered, v only depends on ξ_1 and the first weight P_1 of the self-similar (or infinite self-similar) measure.

(2) If and only if $P_1 = \xi_1^\beta$ (where $\beta \in (0, 1)$ is the fractal dimension of the memory set), v is equal to the fractal dimension of the memory set.

(3) From Eq. (33), we can see that v is continuous about P_1 and ξ_1.

(4) In Eqs. (32), (34)~(36) the constant $A^{(i)}$ depends on $\{P_j\}_{j=1}^K$ and $\{\varphi_j\}_{j=1}^K$ (or $\{P_j\}_{j=1}^\infty$ and $\{\varphi_j\}_{j=1}^\infty$), i. e. it depends on $\{P_j\}$ and the geometrical characteristics of the fractal structure of memory sets $E_T^{(i)}$ $(i = 1, \cdots, 4)$ considered.

◇ References ◇

[1] K. Binder, A. P. Joung, Rev. Mod. Phys. 58 (1986) 801.

[2] S. L. Ginzbug, Irreversible Phenomena in Spin Glasses, Nauka, Moscow, 1989, (in Russian).

[3] A. I. Olemskoi, E. A. Toropov, Fiz. Met. Metalloved 9 (1991) 5.

[4] A. I. Olemskoi, E. A. Toropov, Fiz. Met. Metalloved 7 (1991) 32.

[5] A. I. Olemskoi, I. A. Sklyar, Usp. Fiz. Nauk. 162 (1992) 29.

[6] A. I. Olemskoi, A. Ya. Flat, Physics-Uspekhi 36 (12) (1993) 1087.

[7] A. I. Olemskoi, Fiz. Met. Metalloved 68 (1989) 56.

[8] A. I. Olemskoi, Yu. I. Paskal, Preprint No. 30, Institute of Physics and Applied Mathemtatics, Tomsk Affiliate of the Siberian Branch of the USSR Academy of Sciences, Tomsk, 1988.

[9] R. R. Nigmatullin, Fractional integral and its physical interpretation, Teor. Matem. Fiz. 90 (1992) 354.

[10] J. P. Hutchinson, Fractals and self-similarity, Indiana Univ. Math. J. 30 (1981) 713 - 747.

[11] R. H. Riedi, B. B. Mandelbrot, Multifractal formalism for infinite multinomial measures, Adv. Appl. Math. 16 (1995) 132 - 150.

[12] Z. G. Yu, F. Y. Ren, J. R. Liang, Some results on infinite similarity, Bull. Hongkong Math. Soc., submitted.

[13] F. Y. Ren, J. R. Liang, The Hausdorff dimension and measure of generalized Moran fractals and Fourier series, Chin. Ann. Math. B 16 (1995) 153 - 162.

[14] H. Su, On the dimension of generalized self-similar sets (in Chinese), Acta. Math. App. Sinica China 17 (1994) 551 - 558.

[15] T. Bedford, Applications of dynamical systems theory of fractals — a study of cookie-cutter Cantor sets, in: Proc. Séminaire de Mathématiques "Fractal geometry and analysis" Université de Montréal, NATO Series, Huwer, Amsterdam, 1989.

[16] Z. G. Yu, F. Y. Ren, Fractional integral associated to infinite cookie-cutter set and its physical interpretation, Chinese Science Bull., submitted.

[17] F. Y. Ren, Z. G. Yu, F. Su, Fractional integral associated to self-similar set or generalized self-similar set and its physical interpretation, Phys. Lett. A 219 (1996) 59 - 68.

[18] F. Y. Ren, Z. G. Yu, Fractional integral associated to cookie-cutter set and its physical interpretation, Progress in Natural Science 7 (1997) 422 - 428.

[19] Z. G. Yu, F. Y. Ren, J. Zhou, Fractional integral associated to generalized cookie-cutter set and its physical interpretation, J. Phys. A 30 (1997) 5569 - 5577.

Dynamics of Periodically Random Orbits*

REN Fuyao, ZHOU Ji and QIU Weiyuan

(*Department of Mathematics, Fudan University, Shanghai* 200433, *China*)

Abstract: This paper discusses the dynamics of iteration of periodically random orbits formed by a collection of rational functions. Some important results, such as the density theorem, the eventual periodicity theorem and the classification theorem are obtained. The dynamics of iteration of non-periodically random orbits is also investigated and a few problems are posed.

Keywords: Fatou sets, Julia sets, normal family.

Let $\mathscr{R}=\{R_i(z): i\in M\}$, where M is a finite index set, being a collection of rational functions, and suppose that $\Sigma_M=\Pi_1^\infty M=\{\sigma=(j_1, j_2, \cdots): j_i\in M, i=1, 2, \cdots\}$. A metric on Σ_M is given by

$$d'(\{j_i\}, \{j'_i\})=\sum_{i=1}^{\infty}\frac{|j_i-j'_i|}{2^i},$$

where $\{j_i\}$ and $\{j'_i\}$ are in Σ_M. For each random orbit $\sigma=(j_1, j_2, \cdots)\in\Sigma_M$, we define a sequence $\{W_\sigma^n(z)\}$ of the iteration formed by $\mathscr{R}$ with respect to orbit σ,

$$W_\sigma^n(z)=R_{j_n}\circ R_{j_{n-1}}\circ\cdots\circ R_{j_1}(z),$$

and its inverse $W_\sigma^{-n}(z)$,

$$W_\sigma^{-n}(z)=(W_\sigma^n)^{-1}(z)=R_{j_1}^{-1}\circ R_{j_2}^{-1}\circ\cdots\circ R_{j_n}^{-1}(z),\ n=1, 2, \cdots.$$

---

* Project supported by the National Natural Science Foundation of China (Grant No. 19771023) and the Tianyuan Foundation of China. Originally published in *Progress in Natural Science*, Vol. 9, No. 4, (1999), 248-255.

In refs. [1] and [2], the authors discussed the dynamics of sequence $\{W_\sigma^n(z)\}$ with respect to all orbits in Σ_M. Recently, Hinkkanen and Martin[3] investigated the dynamics of rational semigroup G generated by $\mathscr{R}$, $G=\langle\mathscr{R}\rangle$.

For any random orbit $\sigma=(j_1, j_2, \cdots)$, it is very interesting to make a study on the dynamics of the iteration sequence $\{W_\sigma^n(z)\}$ with respect to σ. However this is very complicated. As far as we know, there are only a few articles about it. Fornaess and Sibony[4], and Rohde[5] discussed the iteration of random perturbation of one rational function and more.

For any random orbit $\sigma=(j_1, j_2, \cdots)\in\Sigma_M$, we define the one-sided shift S of Σ_M to itself as follows: $S(\sigma)=(j_2, j_3, \cdots)$. If there is a positive integer p such that $S^p(\sigma)=\sigma$, where $S^p=S\circ S^{p-1}$ (henceforth assume that S denotes the one-sided shift), we say σ is a periodic orbit, and the minimal number p satisfying the above equation is called its period. In this situation, $j_{np+i}=j_i$ for $n\in\mathbb{N}^+$, $i=1, 2, \cdots, p$; that is,

$$\sigma=(j_1, j_2, \cdots, j_p, j_1, j_2, \cdots, j_p, \cdots).$$

Let

$$\Sigma_{\text{prd}}=\{\sigma\in\Sigma_M:\ S^p(\sigma)=\sigma,\ \text{for some } p\in\mathbb{N}^+\}.$$

It is known that Σ_{prd} is dense in Σ_M (see ref. [6]), thus it is significant to make researches on the dynamics of the iteration of periodically random orbits. The principal aim of this paper is to discuss the dynamics of the iteration of periodically random orbits and the density theorem about the repelling periodic points. The eventual periodicity theorem and the classification theorem are obtained. Finally we investigate the dynamics of the iteration of non-periodically random orbits and pose a few problems.

1. Periodically Random Orbits

Assume that $\sigma\in\Sigma_M$, and let $J(\sigma)$ denote the Julia set of σ, which is the set of points of non-normality of the sequence $\{W_\sigma^n(z)\}$, and its complement is called the Fatou set of σ, denoted by $F(\sigma)$. If $\sigma(j_1, j_2, \cdots)\in\Sigma_M$ is a periodic orbit with period p, we call $g(z)=R_{j_p}\circ R_{j_{p+1}}\circ\cdots\circ R_{j_1}(z)$ the characteristic function of σ, and assume always that its degree $\deg g\geqslant 2$. Here we have the following theorem.

Theorem 1 *Suppose that $\sigma=(j_1, j_2, \cdots)\in\Sigma_M$ is a periodic orbit of period p and $g(z)$ is its characteristic function, then $J(\sigma)=J(g)$ and $F(\sigma)=F(g)$, where $F(g)$ and $J(g)$ denote the Fatou set and the Julia set of g respectively.*

Proof From the definitions of $F(g)$ and $F(\sigma)$, it is clear that $F(\sigma)\subset F(g)$. So we show next that $F(g)\subset F(\sigma)$. First we note that R_i and $i\in M$ are rational functions, and for each R_i there exists a constant $K_i>0$ such that, for any x, $y\in\overline{C}$,

$$|R_i(x)-R_i(y)|_\rho<K_i|x-y|_\rho, \tag{1}$$

where $|\cdot|_\rho$ denotes the spherical metric. We let $K=\max_i\{K_i, 1\}$.

If $z_0\in F(g)$, then there exists a neighborhood V of z_0 and for any $\varepsilon>0$ there is $\delta>0$ such that for every $n\in\mathbb{N}^+$ and z, $z_0\in V$ with $|z-z_0|_\rho<\delta$,

$$|g^n(z)-g^n(z_0)|_\rho<\frac{\varepsilon}{K^{p-1}}. \tag{2}$$

For each positive integer m, it can be represented as $m=pn+l$, $0\leqslant l<p$. Hence from eqs. (1) and (2) we can get

$$\begin{aligned}|W_\sigma^m(z)-W_\sigma^m(z_0)|_\rho&=|W^l_{S^{pn}(\sigma)}\circ W_\sigma^{pn}(z)-W^l_{S^{pn}(\sigma)}\circ W_\sigma^{pn}(z)|_\rho\\&<K^l|W_\sigma^{pn}(z)-W_\sigma^{pn}(z_0)|_\rho\\&<\varepsilon.\end{aligned}$$

Therefore the sequence $\{W_\sigma^m(z)\}$ is equicontinuous in V and $z\in F(\sigma)$. Thus the proof is completed.

Corollary 1 *$J(\sigma)$ and $F(\sigma)$ are completely invariant under g; that is*

$$g^{-1}(J(\sigma))=g(J(\sigma))=J(\sigma) \text{ and } g^{-1}(F(\sigma))=g(F(\sigma))=F(\sigma).$$

From Theorem 1 and ref. [7], we get

Corollary 2 *$J(\sigma)$ is a nonempty and uniformly perfect set, and if $J(\sigma)\neq\overline{C}$, then it does not contain any interior point.*

Given $z_0 \in \overline{C}$, we call $O_\sigma^+(z_0) = (z_0, W_\sigma^1(z_0), W_\sigma^2(z_0), \cdots)$ the forward orbit of σ at z_0. z_0 is a periodic point of σ if there exists a positive integer q such that $S^q(O_\sigma^+(z_0)) = O_\sigma^+(z_0)$, moreover the minimal number q satisfying the above equation is called the period of σ at z_0. In this situation, $W_\sigma^{nq}(z_0) = z_0$, $W_\sigma^{nq+i}(z_0) = W_\sigma^i(z_0)$, $n \in \mathbb{N}^+$, $i = 1, 2, \cdots, q-1$.

Suppose that z_0 is a periodic point of period q with respect to periodic orbit σ of period p. We call $\lambda = (W_\sigma^l)'(z_0)$ its multiplier, where l is the least common multiple of p and q. When $p=1$, $\lambda = (W_\sigma^q)'(z_0)$ is the multiplier in classical sense.

Next we give a classification of periodic points.

Definition 1 Assume that z_0 is a periodic point of period p with respect to periodic orbit σ of period q, and its multiplier is λ. Then

(i) z_0 is called an attracting periodic point of σ if $0 < |\lambda| < 1$.

(ii) z_0 is called a superattracting periodic point of σ if $\lambda = 0$.

(iii) z_0 is called a repelling periodic point of σ if $|\lambda| > 1$.

(iv) z_0 is called an indifferent periodic point of σ if $|\lambda| = 1$, moreover we let $\lambda = e^{2\pi\theta i}$, if θ is a rational number, z_0 is called a rationally indifferent periodic point of σ and if θ is an irrational number, z_0 is called an irrationally indifferent periodic point of σ. The corresponding orbit $O_\sigma^+(z_0)$ is called to be attracting, superattracting, repelling, rationally indifferent, and irrationally indifferent, respectively.

According to the above classification, we have

Theorem 2 *The (super)attracting periodic points of σ belong to the Fatou set $F(\sigma)$ and the repelling periodic points of σ belong to the Julia set $J(\sigma)$.*

Theorem 3 *The Julia set $J(\sigma)$ is equal to the closure of all of the repelling periodic points of σ. Hence the repelling periodic points of σ are dense in $J(\sigma)$.*

To prove the above Theorems 2 and 3, we need the following lemma.

Lemma 1 *z_0 is a periodic point of period q with respect to the periodic orbit σ of period p if and only if z_0 is also a periodic point of the characteristic* *function* $g(z)=W_\sigma^p(z)$. *Moreover* z_0 *is of the same type as periodic points of* σ *and* $g(z)=W_\sigma^p(z)$.

Proof If z_0 is a periodic point of period q with respect to the periodic orbit σ of period p, then for every $n\in\mathbb{N}^+$, $W_\sigma^{nq}(z_0)=z_0$ and $W_\sigma^{nq+i}(z_0)=W_\sigma^i(z_0)$, $i=1, 2, \cdots, q-1$. Since l is the least common multiple of p and q, we let $m=l/p$ and $k=l/q$, and get

$$g^m(z_0)=\overbrace{W_\sigma^p\circ\cdots\circ W_\sigma^p}^{m}(z_0)=W_\sigma^l(z_0)=W_\sigma^{kq}(z_0)=z_0; \tag{3}$$

that is, z_0 is a periodic point of $g(z)$. Conversely, if z_0 is a periodic point of $g(z)$ with period r, that is, $g^r(z_0)=z_0$ and $W_\sigma^{rp}(z_0)=z_0$, note that σ is a periodic orbit of period p, hence the orbit $O_\sigma^+(z_0)$ is periodic and z_0 is a periodic point with respect to σ. Now we observe multiplier of z_0. Assume that the period of g at z_0 is r, from eq. (3) it follows

$$\lambda=(W_\sigma^l)'(z_0)=(g^m)'(z_0)=((g^r)^{n_1})'(z_0)=((g^r)'(z_0))^{n_1},$$

where $n_1=m/r$ is a positive integer. This implies that the lemma holds.

Applying Theorem 1 and Lemma 1, it is easy to verify that Theorems 2 and 3 are true.

Now we investigate the components of Fatou set. From Corollary 1, it follows that the Fatou set and Julia set of σ are completely invariant under $g(z)=W_\sigma^p(z)$ and hence if D is a maximal connected component of $F(\sigma)$, i. e. a Fatou component, $g(D)$ is also a Fatou component. We let $\mathscr{F}_\sigma$ denote the collection of all the components of $F(\sigma)$. Therefore, by the above discussion, g induces a map $\hat{g}$ from $\mathscr{F}_0$ to itself, $\hat{g}: D\to g(D)$. A Fatou component $D\in\mathscr{F}_\sigma$ is called periodic if D is a periodic point of $\hat{g}$, i. e. , there is a positive integer n such that $g^n(D)=D$. We say a Fatou component D is eventually periodic if there exists a positive integer k such that $g^k(D)$ is periodic. If a Fatou component D is not eventually periodic we say it is wandering, that is to say, for any positive integers n, m, $n\neq m$, we have $g^n(D)\cap g^m(D)=\varnothing$. Since $F(\sigma)=F(g)$, from the Sullivan's eventual periodicity theorem[8], we have

Theorem 4 (*Eventual Periodicity Theorem*) *Let* $\sigma=(j_1, j_2, \cdots)\in\Sigma_M$

be a periodically random orbit of period p and its characteristic function $g(z) = W_\sigma^p(z) = R_{j_p} \circ R_{j_{p-1}} \circ \cdots \circ R_{j_1}(z)$ is a rational function of degree $d \geqslant 2$, then each component of the Fatou set $F(\sigma)$ is eventually periodic, that is, $F(\sigma)$ has no wandering domains under g.

Applying the Sullivan's classification theorem[8] and Lemma 1, note that $F(\sigma) = F(g)$, we get

Theorem 5 *(Classification Theorem) Let $\sigma = (j_1, j_2, \cdots) \in \Sigma_M$ be a periodically random orbit of period p and its characteristic function $g(z) = W_\sigma^p(z)$ is a rational function of degree $d \geqslant 2$. If $D \in \mathcal{F}_\sigma$ is a forward invariant component of the Fatou set under g, i.e., $g(D) = D$, then D is*

(i) *attracting. There is an attracting periodic point $z_0 \in D$ and g^n converges locally uniformly to z_0 in D, or*

(ii) *parabolic. There is a rationally indifferent periodic point $z_0 \in \overline{D}$ and g^n converges locally uniformly to z_0 in D, or*

(iii) *superattracting. There is a superattracting periodic point $z_0 \in D$ and g^n converges locally uniformly to z_0 in D, or*

(iv) *Siegel disk. D is conformally equivalent to the unit disk Δ and g is conformally conjugate to $z \to e^{2\pi\theta i} z$, $z \in \Delta$, θ is an irrational number, or*

(v) *Herman ring. D is conformally equivalent to a ring $A(\delta, 1) = \{z \mid 0 < \delta < |z| < 1\}$ and g is conformally conjugate to $z \to e^{2\pi\theta i} z$, $z \in A(\delta, 1)$, θ is an irrational number.*

We next introduce the exceptional set of σ. If $z \in J(\sigma)$, and V is a neighborhood of z. Set

$$E_V = \overline{C} - \bigcup_{n>0} W_\sigma^n(V).$$

By Montel theorem, E_V contains at most two points. Put $E_z = \bigcup_V E_V$ and $E(\sigma) = \bigcup_{z \in J(\sigma)} E_z$. We call $E(\sigma)$ the exceptional set of σ and each element of $E(\sigma)$ is called an exceptional point. We get

Lemma 2 *The exceptional set contains at most two points and the exceptional points belong to the Fatou set $F(\sigma)$.*

Proof It is clear that for any neighborhood V of $z \in J(\sigma)$,

$$\bigcup_{n>0} W_\sigma^n(V) \supset \bigcup_{n>0} g^n(V)$$

and hence $E_V = \overline{C} - \bigcup_{n>0} W_\sigma^n(V) \subset \overline{C} \bigcup_{n>0} g^n(V)$. Noting that $J(\sigma) = J(g)$, we get $E(\sigma) \subset E(g)$, where $E(g)$ denotes the exceptional set of g[9]. From the classical Fatou-Julia theory[9, 10], we see that $E(g)$ contains at most two points and both of them belong to $F(g)$. Note that $F(\sigma) = F(g)$, therefore the lemma holds.

Applying the same way of proving Theorem 2 in ref. [1], from Lemma 2 we get

Theorem 6 *If* $z \in \overline{C} - E(\sigma)$, *then*

$$J(\sigma) \subset \{\textit{accumulation points of } \bigcup_{n>0} W_\sigma^{-n}(z)\}.$$

And hence if $z \in J(\sigma)$,

$$J(\sigma) \subset \overline{\{\bigcup_{n>0} W_\sigma^{-n}(z)\}}. \tag{4}$$

We notice that $J(\sigma) = \overline{\{\bigcup_{n>0} W_\sigma^{-n}(z)\}}$ does not hold. Here is an example, if we take $R_1(z) = z+1$, $R_2(z) = (z-1)^2$ and $\sigma = (1, 2, 1, 2, \cdots)$. Clearly, $W_\sigma^1(z) = z+1$, $W_\sigma^2(z) = z^2$, $W_\sigma^3(z) = z^2+1$ and $J(\sigma) = J(z^2) = \{z: |z| = 1\}$. Since $1 \in J(\sigma)$, it follows that the origin $O \in W_\sigma^{-3}(1)$, but $O \notin J(\sigma)$.

2. Non-periodically Random Orbits

Since Σ_{prd} is dense in Σ_M, this implies that for any $\sigma \in \Sigma_M$, there exists a sequence $\{\sigma_n\}$ of Σ_{prd} such that σ_n converges to σ in the metric d'. We notice that $J(\sigma_n)$ and $n \in \mathbb{N}^+$ are closed subsets of $\overline{C}$, and are also compact. Hence a problem is to ask whether $\{J(\sigma_n)\}$ is convergent with respect to the Hausdorff metric d_H when σ_n converges to σ. Clearly, (Σ_M, d') is a compact metric space. From ref. [11], if $\mathscr{F}$ stands for the collection of all compact subsets of $\overline{C}$, then $(\mathscr{F}, d_{\mathrm{H}})$ is a complete metric space. Using the way to prove Theorem 3.16 in ref. [12], note that $\overline{C}$ is compact, we obtain

Lemma 3 *If* $\mathscr{G}$ *is an infinite collection of nonempty compact subsets of* $\overline{C}$, *then there exists a sequence* $\{A_i\}$ *of* $\mathscr{G}$ *such that* $\{A_i\}$ *converges in the*

Hausdorff metric d_H to a nonempty compact subset of $\overline{\mathbb{C}}$. Hence $(\mathscr{F}, d_H)$ is compact.

Proof Let $\{A_{1,i}\}_i$ by any infinite sequence of $\mathscr{F}$. For each $k>1$, define an infinite subsequence $\{A_{k,i}\}_i$ of $\{A_{k-1,i}\}_i$ as follows. Let $\mathscr{B}_k$ be the finite cover of $\overline{\mathbb{C}}$ by closed disks with their diameters less than $1/k$ in the spherical metric, since $\overline{\mathbb{C}}$ is compact. For any nonempty compact $A \subset \overline{\mathbb{C}}$. Let $\mathscr{B}_k(A)$ be the subcollection of $\mathscr{B}_k$ consists of those disks which intersect with A. By the pigeon-hole principle, there are only finite many combinations of those disks in $\mathscr{B}_k$. Hence there must exist an infinite subsequence $\{A_{k,i}\}_i$ of $\{A_{k-1,i}\}_i$ such that $\mathscr{B}_k(A_{k,i}) = \mathscr{B}_k(A_{k,j})$, this implies that $d_H(A_{k,i}, A_{k,j}) \leqslant 2/k$. Let $A_i = A_{i,i}$. Hence this sequence $\{A_i\}_i$ is a Cauchy sequence in the Hausdorff metric. Again since $(\mathscr{B}, d_H)$ is a complete metric space, it follows that $\{A_i\}_i$ converges to a nonempty compact. Therefore the lemma holds.

From the above lemma, we see that there must be a subsequence of $J(\sigma_n)$ which is convergent with respect to d_H. Therefore we ask naturally: for any $\sigma \in \Sigma_M$, if there is a sequence $\{\sigma_n\}$ of Σ_{prd} satisfying $\sigma_n \to \sigma$, as $n \to \infty$ and $\{J(\sigma_n)\}$ is a convergent sequence with respect to the Hausdorff metric d_H, does $J(\sigma_n)$ converge to $J(\sigma)$ with respect to the Hausdorff metric, as $\eta \to \infty$? In general, the answer is negative. For example, we choose rational functions $R_1(z)$ and $R_2(z)$ with z_0 as their fixed point, and the multipliers of R_1 and R_2 at z_0 are 3 and 1/2 respectively, and their degrees are both more than one. Meanwhile we take the sequence $\{\sigma_n\}$ of random orbits and an orbit σ as follows:

$$\sigma_n = (\overbrace{2, \cdots, 2}^{n}, \overbrace{1, \cdots, 1}^{n}, \overbrace{2, \cdots, 2}^{n}, \overbrace{1, \cdots, 1}^{n}, \cdots),$$
$$n = 1, 2, \cdots \text{ and } \sigma = (2, 2, \cdots, 2, \cdots).$$

Clearly, $\sigma_n \to \sigma$, as $n \to \infty$ in the metric d'. From Theorems 1 and 2 it is clear that $z_0 \in J(\sigma_n)$, $n \in \mathbb{N}^+$ and $z_0 \notin J(\sigma)$, and that $J(\sigma_n)$ does not converge to $J(\sigma)$ with respect to the Hausdorff metric d_H[11]. Now we ask

Problem 1 *For any $\sigma \in \Sigma_M \backslash \Sigma_{\text{prd}}$, is there a sequence $\{\sigma_n\}$ of Σ_{prd} satisfying $\sigma_n \to \sigma$, as $n \to \infty$ and $\{J(\sigma_n)\}$ is convergent with respect to the*

Hausdorff metric d_H, such that $J(\sigma_n)$ converges to $J(\sigma)$, as $n \to \infty$?

From the above discussions we see that if $\sigma \in \Sigma_{prd}$ and the degree of its characteristic function is at least 2, then the Julia set $J(\sigma)$ is nonempty and perfect. But if $\sigma \in \Sigma_M \backslash \Sigma_{prd}$, what can we say? Here we get

Theorem 7 *Let $\sigma \in \Sigma_M$, if there is a rational function R_j with degree at least 2 such that the corresponding j appears in σ infinitely often, then the Julia set $J(\sigma)$ is nonempty.*

Proof This proof is by contradiction. Assume that there is a rational function R_j with degree at least 2 such that the corresponding j appears in σ infinitely often and $J(\sigma)$ is vacuous. Then there is a subsequence $\{W_\sigma^{n_i}(z)\}$ of sequence $\{W_\sigma^n(z)\}$ that would converge uniformly on the entire sphere $\overline{C}$ to a function G. By Theorem 2. 8. 2 in ref. [9], G is a rational function and for sufficiently large n_i, $\deg(W_\sigma^{n_i}) = \deg G$. However, $\deg(W_\sigma^{n_i}) \to \infty$ as $n_i \to \infty$, since the degree of R_j is at least 2 and j appears in σ infinitely often, which is a contradiction. Hence $J(\sigma)$ is nonempty and the proof is completed.

Problem 2 *If $\sigma \in \Sigma_M \backslash \Sigma_{prd}$, is the Julia set $J(\sigma)$ of σ perfect?*

We see that the periodic points play an important role in classical dynamics. Now we give a classification of periodic points.

Definition 2 $z_0 \in \overline{C}$ is a periodic point of period q with respect to σ. Then z_0 is

(i) attracting if $\liminf_{n\to\infty} |(W_\sigma^{nq})'(z_0)| = 0$;

(ii) superattracting if there is a k such that $(W_\sigma^{kq})'(z_0) = 0$;

(iii) repelling if $\limsup_{n\to\infty} |(W_\sigma^{nq})'(z_0)| = \infty$;

(iv) indifferent if $0 < \liminf_{n\to\infty} |(W_\sigma^{nq})'(z_0)| \leqslant \limsup_{n\to\infty} |(W_\sigma^{nq})'(z_0)| < \infty$.

It is obvious that if $\mathcal{R}$ contains one element, the classification of periodic points as above is the same as that in classical case. Moreover we get

Theorem 8 *All of the repelling periodic points with respect to σ defined as above belong to the Julia set $J(\sigma)$.*

Proof Assume that z_0 is a repelling periodic point of degree q with respect to σ. We see that there is a sequence $\{\eta_k\}$ of positive integers such that

$\lim_{k\to\infty}|(W_{\sigma}^{n_k q})'(z_0)|=\infty$. Now suppose, by contradiction, that $z_0 \in F(\sigma)$. Hence there is a subsequence of sequence $\{\eta_k\}$, still say $\{\eta_k\}$, such that the iteration sequence $\{W_{\sigma}^{\eta_k q}(z)\}$ converges locally uniformly to some analytic function $h(z)$ on some neighborhood of z_0. Now $h(z_0)=z_0$, so $h'(z_0)$ is finite. On the other hand, the uniform convergence implies that for the given subsequence, $|h'(z_0)|=\lim_{k\to\infty}|(W_{\sigma}^{n_k q})'(z_0)|=\infty$, which is a contradiction. Therefore the proof of the theorem is completed.

Problem 3 *If $\sigma \in \Sigma_M \backslash \Sigma_{\text{prd}}$, are the repelling periodic points with respect to σ defined as above dense in the Julia set $J(\sigma)$?*

Problem 4 *If $\sigma \in \Sigma_M \backslash \Sigma_{\text{prd}}$, do the (super)attracting periodic points with respect to σ defined as above belong to the Fatou set $F(\sigma)$?*

It is well known that the Hausdorff dimension of Julia set of single rational function of degree at least 2 is more than zero. Hence we ask

Problem 5 *If $\sigma \in \Sigma_M \backslash \Sigma_{\text{prd}}$ and the Julia set $J(\sigma)$ is nonempty, is its Hausdorff dimension more than zero?*

◇ References ◇

[1] Zhou, W. M., Ren, F. Y., A dynamical system formed by a set of rational functions, *Current Topics in Analytic Function Theory* (eds: Srivastava, H. M., Owa, S.), Singapore: World Scientific, 1992.

[2] Gong, Z. M., Ren, F. Y., A random dynamical system formed by infinitely many functions, *J. Fudan Univ.*, (natural science), 1996, 35: 387.

[3] Hinkkanen, A., Martin, G. J., The dynamics of semigroups of rational functions I, *Proc. London Math. Soc.*, 1996, 73: 358.

[4] Fornaess, J. E., Sibony, N., Random iteration of rational function, *Ergod. Th. & Dymam. Sys.*, 1991, 11: 687.

[5] Rohde, S., Compositions of random rational functions, *Complex Variables*, 1996, 29: 1.

[6] Walters, P., *An Introduction to Ergodic Theory*, New York: Springer-Verleg, 1982.

[7] Mane, R., Da Rocha, L. F., Julia sets are uniformly perfect. *Proc. Amer. Math. Soc.*, 1997, 116: 251.

[8] Sullivan, D., Quasiconformal homeomorphisms and dynamics I: Solution of the Fatou-

Julia problem on wandering domain, *Ann, of Math.*, 1985, 122: 401.

[9] Beardon, A. F., *Iteration of Rational Functions*, New York: Springer, 1992.

[10] Milnor, J., *Dynamics in One Complex Variable: Introductory Lectures*, Institute for Methematical Science, SUNY at Stony Brook, Preprint No, 1990/5.

[11] Barnsley, H. F., *Fractals Everywhere*, Boston: Academic Press, 1988.

[12] Falconer, K. J., *Geometry of Fractal Sets*, Cambridge: Cambridge University Press, 1985.

The Determination of the Diffusion Kernel on Fractals and Fractional Diffusion Equation for Transport Phenomena in Random Media*

Fu-yao Ren[a], Jin-rong Liang[b], Xiao-Tian Wang[a]

[a] *Institute of Mathematics, Fudan University, Shanghai 200433, China*

[b] *Institute of Mathematics, Zhejiang University, Hangzhou 310027, China*

Abstract: This paper studies the diffusion kernel and fractional diffusion equation on fractals. The concrete expressions of the diffusion kernel and fractional diffusion equation on (generalized) Cantor type fractals are obtained. Particularly, the exact solution of the fractional diffusion equation is given.

Keywords: (generalized) Cantor type sets, diffusion kernel, diffusion measures, fractional diffusion equation

1. Introduction

In Refs. [1～5], the fractional integral associated to fractal memory sets and its physical interpretation were discussed systematically. Le Mehaute, Nigmatullin and Nivanen asked in Ref. [6] whether there are possible generalizations of the dynamical process simulated by the fractal set and whether the region of its application may be increased. Intensive analytical and numerical work has been performed in recent years to elucidate the unusual transport properties on fractal structures [7～12]. Much work has focused on

* Originally published in *Physics Letter A*, Vol. 252, (1999), 141 - 150.

understanding diffusion processes on such spatially correlated media.

In Ref. [13], Giona and Roman discussed the fractional diffusion equation for transport phenomena in random isotropic and homogeneous fractal structures media. Since they dealt with stationary stochastic processes, they had a restriction for the diffusion kernel. That is, for the radial probability current $i(r, t)$ leaving the spherical region of radius r at time t and the average probability density $P(r, t)$, they dealt with an integral relation of the form

$$\int_0^t i(r, \tau)\mathrm{d}\tau = r^{d_f - 1}\int_0^t K(t, \tau)P(r, \tau)\mathrm{d}\tau = r^{d_f - 1}\int_0^t K(t-\tau)P(r, \tau)\mathrm{d}\tau, \tag{1.1}$$

and with the constitutive local equation on fractals

$$i(r, t) = -Br^{d_f - 1}r^{-\theta'}\left(\frac{\partial P(r, t)}{\partial r} + \frac{\kappa}{r}P(r, t)\right), \tag{1.2}$$

where $K(t, \tau) = K(t-\tau)$ is the diffusion kernel (i. e. the diffusion intensity), d_f is the fractal dimension, $B > 0$ is the diffusion coefficient, the parameter θ' is necessarily positive, $\theta' \geqslant 0$, and κ remains to be determined.

Under the assumption that on fractals the diffusion kernel should behave as

$$K(t-\tau) = (t-\tau)^{-\gamma}, \tag{1.3}$$

where $0 < \gamma < \frac{1}{2}$, by using the Laplace transform and the normalization condition $\int_0^\infty r^{d_f - 1}P(r, t)\mathrm{d}t = 1$ they obtained the asymptotic solution of the fractional diffusion equations (1.1) and (1.2),

$$P(r, t) \sim t^{-d_f/d_w}\left(\frac{r}{R}\right)^{\alpha}\exp\left[-\mathrm{const}\cdot\left(\frac{r}{R}\right)^{u'}\right], \tag{1.4}$$

when $r/R \gg 1$ and $t \to \infty$, where $d_w > 2$ is the anomalous diffusion exponent, R is the diffusion length,

$$u' = \frac{d_w(1+\theta')}{d_w - (1+\theta')}, \quad \alpha = u'\left[\frac{1}{2}(d_s - 1) - \kappa\right], \quad d_s = 2d_f/d_w. \quad (1.5)$$

But they did not show the relation between the exponents γ in (1. 3) and fractals and how γ is determined. This paper will study and answer these problems. We obtain the solution of the fractional diffusion equations (1. 1) and (1. 2) and give a concrete expression for γ.

2. The Determination of the Diffusion Kernel Functions on Fractals

In this section, we obtain the concrete expressions for the diffusion kernel and the fractional diffusion equation on fractals generated by a family of contractive transformations.

For any $T \in (0, \infty)$, let $E_0 = [0, T]$ and $I = \{1, 2, \cdots, k\}$ and let $\{\phi_j(x)\}_{j=1}^{k}$, $k < \infty$ be the contractive transformations E_0, i. e.

$$|\phi_j(x) - \phi_j(y)| \leqslant b_j |x - y|, \quad x, y \in E_0, \quad (2.1)$$

where $0 < b_j < 1$. Suppose $\operatorname{int}(\phi_j(E_0)) \cap \operatorname{int}(\phi_i(E_0)) = \varnothing$, $i \neq j$. For any $n \in \mathbb{Z}^+$ ($\mathbb{Z}^+$ denotes the set of all positive integer numbers), let

$$E^{(1)}_{j_1 \cdots j_n} = \phi_{j_1} \circ \cdots \circ \phi_{j_n}(E_0), \quad E^{(1)}(n) = \bigcup_{j_1 \cdots j_n} E^{(1)}_{j_1 \cdots j_n}.$$

The nonempty compact set

$$E^{(1)}_T = \bigcap_{n=1}^{\infty} E^{(1)}(n)$$

is called a Cantor type set. According to the result in Ref. [14], one can obtain

$$\dim_H E^{(1)}_T \leqslant u,$$

where u is the solution of the equation $\sum_{j=1}^{k} b_j^u = 1$.

For any $n \in \mathbb{Z}^+$, let $I_n = \{1, 2, \cdots, k_n\}$ and $\{\phi_{nj}(x)\}_{j \in I_n}$ be a family of contractive transformations on E_0, i. e.

$$|\phi_{n,j}(x) - \phi_{n,j}(y)| \leqslant b_{n,j} |x - y|, \quad x, y \in E_0, \quad (2.2)$$

where $0 < b_{n,j} \leqslant b_0 < 1$ for some b_0. Suppose, for any $n \in \mathbb{Z}^+$,

$$\operatorname{int}(\phi_{n,j}(E_0)) \cap \operatorname{int}(\phi_{n,j}(E_0)) = \varnothing, \quad i \neq j, \quad i, j \in I_n.$$

Let

$$E^{(2)}_{j_1 \cdots j_n} = \phi_{1, j_1} \circ \cdots \circ \phi_{n, j_n}(E_0), \quad j_i \in I_i, \ i = 1, \cdots, n,$$
$$E^{(2)}(n) = \bigcup_{j_1 \cdots j_n} E^{(2)}_{j_1 \cdots j_n}.$$

The nonempty compact set $E_T^{(2)} = \bigcap_{n=1}^{\infty} E^{(2)}(n)$ is called a generalized Cantor type set. According to the result in Ref. [15] one can get

$$\dim_H E_T^{(2)} \leqslant \liminf_{n \to \infty} \overline{\beta_n},$$

where $\overline{\beta_n}$ is the solution of the equation $\prod_{i=1}^{n} \sum_{j=1}^{k_i} b_{i,j}^{\overline{\beta_n}} = 1$.

Assume that the diffusion set E_T on $[0, T]$ is a Cantor type set $E_T^{(1)}$ or a generalized Cantor type set $E_T^{(2)}$ and that $\phi_1(x) = \xi_1 x$, $0 < \xi_1 < 1$. Let d_{f} denote the fractal dimension of E_T. If the total number of remaining states in each stage of the division of this set is normalized to unity and if $P(r, t)$ is a generating function of Laplace transform, then using the same method as in Ref. [5], for any given probability vector $\boldsymbol{p} = (p_1, p_2, \cdots, p_k)$ (i. e. $p_i > 0$, $\sum_{i=1}^{k} p_i = 1$), there exists a unique self-similar measure μ on E_T, such that $\int_{E_T} \mathrm{d}\mu(\tau) = 1$, $\operatorname{supp}(\mu) = E_T$,

$$\mu(\cdot) = \sum_{j=1}^{k} p_j \mu \circ \varphi_j^{-1}(\cdot) \tag{2.3}$$

(where we take $\varphi_j(x) = \lim_{n \to \infty} \varphi_{n,j}(x)$ and $k = \lim_{n \to \infty} k_n$ associated to $E_T = E_T^{(2)}$) and

$$j(r, t) = \int_0^t i(r, \tau) \mathrm{d}\tau = r^{d_{\mathrm{f}} - 1} \int_0^t P(r, \tau) \mathrm{d}\mu(t - \tau), \tag{2.4}$$

where $\mathrm{d}\mu(t) = K(t)\mathrm{d}t$. Moreover, we also obtain the approximate expressions for the diffusion kernel and the asymptotic conservation equation on the diffusion set E_T respectively,

$$K(t) = \frac{A}{\Gamma(1 - \gamma)} t^{-\gamma}, \tag{2.5}$$

and

$$\int_0^t i(r, \tau)\mathrm{d}\tau = A[\Gamma(1-\gamma)]^{-1} r^{d_f-1} \int_0^t P(r, \tau)(t-\tau)^{-\gamma}\mathrm{d}\tau, \tag{2.6}$$

where

$$\gamma = 1 - v < 1, \quad v = \ln p_1 / \ln \xi_1, \quad 0 < v < \infty. \tag{2.7}$$

Since $\mathrm{supp}(\mu) = E_T \subset [0, T]$, (2.5) and (2.6) are true iff $t \in [0, T]$. The measure μ is called a diffusion measure on E_T corresponding to weights $\{p_j\}_{j=1}^k$. Eq. (2.6) is the conservation equation on the fractal diffusion sets discussed in Ref. [13]. γ is called fractional diffusion exponent.

Problem How to determine the constant A is still an open problem.

From the above results, we can summarize the following conclusions:

(1) When the memory set on $[0, T]$ is E_T, no matter which self-similar measure satisfying (2.3) for the memory measure on E_T is taken, the approximate expressions (2.5) and (2.6) for the diffusion kernel and the conservation equation are invariable.

(2) No matter which self-similar measure is taken, the fractional diffusion exponent γ of the diffusion kernel is determined only by $1 - \ln p_1 / \ln \xi_1$.

(3) $\gamma = 1 - d_f$ if and only if $p_1 = \xi_1^{d_f}$ and $\gamma = d_f$ if and only if $p_1 = \xi^{1-d_f}$.

(4) When p_1 changes from ξ_1 to 1, γ changes from 0 to 1.

(5) From the formula $\gamma = 1 - \ln p_1 / \xi_1$, we can see that there is no direct relationship between the fractional diffusion exponent γ in (2.7) and the geometrical characteristics of the fractal structure of the diffusion set E_T considered.

(6) When the diffusion set E_T is a Cantor type set or a generalized Cantor type set on $[0, T]$, the diffusion kernel function and the flux function $\int_0^t i(r, \tau)\mathrm{d}\tau$ of the radial probability current $i(r, t)$ are not explicitly related to the geometrical structure of the diffusion set E_T and the probability vector $\boldsymbol{p}$.

(7) From (2.6) and the constitutive local equation of fractal E_T we have

$$i(r, t) = -B r^{d_f-1} r^{-\theta'} \left(\frac{\partial P(r, t)}{\partial r} + \frac{\kappa}{r} P(r, t) \right), \tag{2.8}$$

and thus the fractional diffusion equation on fractals is

$$\frac{\partial^\gamma P(r,t)}{\partial t^\gamma}=-Gr^{-\theta'}\left(\frac{\partial P(r,t)}{\partial r}+\frac{\kappa}{r}P(r,t)\right), \tag{2.9}$$

where

$$\frac{\partial^\gamma P(r,t)}{\partial t^\gamma}=\frac{1}{\Gamma(1-\gamma)}\frac{\partial}{\partial t}\int_0^t\frac{P(r,\tau)}{(t-\tau)^\gamma}\mathrm{d}\tau,\quad \gamma<1. \tag{2.10}$$

As in Ref. [13], by simple scaling considerations, from (2.9) we see that $t^\gamma\sim r^{1+\theta'}$. According to the anomalous behavior of the mean square displacement of a random Brownian particle on all time scales

$$R^2\equiv\langle r^2(t)\rangle\sim t^{2/d_w}\quad[8], \tag{2.11}$$

we require that $t^{1/d_w}\sim r$, obtaining

$$\gamma=\frac{1+\theta'}{d_w}. \tag{2.12}$$

From $d_w>2$, $\theta'\geqslant 0$ and $\gamma<1$, we have

$$0<\gamma<1,\quad 0\leqslant\theta'<d_w-1. \tag{2.13}$$

If we take $\gamma=\frac{1}{2}$ in (2.9), i.e. $p_1=\xi_1^{1/2}$, $\theta'=0$ and $\kappa=0$, then

$$\frac{\partial^{1/2}P(r,t)}{\partial t^{1/2}}=-G\frac{\partial P(r,t)}{\partial r}. \tag{2.14}$$

This is the diffusion equation for the standard Brownian motion [16].

If we take $\theta'=0$, $\kappa=0$, $\gamma=1/d_w$, i.e. $p_1=\xi_1^{1-1/d_w}$, then

$$\frac{\partial^{1/d_w}P(r,t)}{\partial t^{1/d_w}}=-G\frac{\partial P(r,t)}{\partial r}. \tag{2.15}$$

This is the diffusion equation of the fractional Brownian motion (fBm) when its Hurst exponent is $0<H=1/d_w<\frac{1}{2}$ [16]. fBm has the anti-protracted property. In this case, a previous increasing tendency implies a future decreasing tendency, but a previous decreasing tendency may imply a future increasing tendency.

If we take $\theta' = \kappa = 0$ and $\gamma = 1 - 1/d_w$, i. e. $p_1 = \xi_1^{1/d_w}$, then we have

$$\frac{\partial^{1-1/d_w} P(r, t)}{\partial t^{1-1/d_w}} = -G \frac{\partial P(r, t)}{\partial t}. \tag{2.16}$$

This is the diffusion equation of the fractional Brownian motion when its Hurst exponent is $\frac{1}{2} < H = 1/d_w < 1$. fBm has the protracted property. In this case, a previous increasing tendency implies a future increasing tendency, and this is true for any time t; conversely, a previous decreasing tendency may imply a future continuous decreasing tendency.

Finally, we study the exact solution of the fractional diffusion equation (2.9). In Ref. [13], Giona and Roman obtained the asymptotic behavior of the solution of (2.9), i. e.

$$P(r, t) \sim t^{-d_f/d_w} \left(\frac{r}{R}\right)^{\alpha} \exp\left[-\text{const} \cdot \left(\frac{r}{R}\right)^{u'}\right], \tag{2.17}$$

when $r/R \gg 1$ and $t \to \infty$, and

$$u' = \frac{d_w(1+\theta')}{d_w - (1+\theta')}, \tag{2.18}$$

$$\alpha = u'\left[\frac{1}{2}(d_s - 1) - \kappa\right], \tag{2.19}$$

$$\kappa = \frac{1}{2}(d_s - 1), \tag{2.20}$$

where $d_s = 2d_f/d_w$ is called the fractional or spectral dimension [7, 17].

Using the Laplace transform we can get the exact solution of the fractional diffusion equation (2.9) on the fractal set,

$$P(r, t) = \frac{A}{\Gamma(1 - (d_f - \kappa)/d_w)} \int_0^t h(\tau)(t-\tau)^{-(d_f-\kappa)/d_w} d\tau, \tag{2.21}$$

where

$$h(t)=(1/\pi)\int_0^\infty e^{-tx}e^{-Bx^\gamma\cos(\gamma\pi)}\sin(Bx^\gamma\sin(\gamma\pi))dx,$$

see Appendix A.

3. The Diffusion Kernel for Diffusion Phenomena and Diffusion Equation on Self-similar Sets

In this section, we discuss diffusion kernels and diffusion equations when diffusion sets are self-similar fractals. Let the diffusion set E_T be a self-similar set generated by the similarities

$$\phi_j(x)=b_j+\xi_j x,\quad 0<\xi_j<1,\quad j=1,\cdots,k,$$

$$\sum_{j=1}^{k}\xi_j<1,\quad 0=b_0<b_1<\cdots<b_k=T(1-\xi_k),$$

then for any given probability vector $\boldsymbol{p}=(p_1,\cdots,p_k)$, there exists a unique self-similar measure μ satisfying (2.3). From Ref. [4] we get that $d\mu(\tau)=K(\tau)d\tau$ and approximately

$$K(\tau)=A'(\Gamma(1-\gamma))^{-1}\sum_{j=1}^{k}p_j\xi_j^{-v}(t-\tau-b_j)^{v-1}\eta(t-\tau-b_j),\tag{3.1}$$

where $\eta(x)=0$ for $x<0$ and 1 for $x>0$,

$$0<\gamma=1-v<1,\quad v=\ln p_1/\ln\xi_1.\tag{3.2}$$

Therefore,

$$\int_0^t i(r,\tau)d\tau=\frac{r^{d_f-1}A'}{\Gamma(1-\gamma)}\sum_{j=1}^{k}p_j\xi_j^{\gamma-1}\int_0^t P(r,\tau)(t-\tau-b_j)^{-\gamma}\eta(t-\tau-b_j)d\tau.\tag{3.3}$$

This is the conservation equation for the diffusion set E_T (i.e. a self-similar set).

From (3.1) and (3.3) we can further see that

(1) The expressions for the diffusion kernel function and the flux function $\int_0^t i(r,\tau)d\tau$ of the radial probability current $i(r,t)$ are finer than that of

(2.5) and (2.6) respectively.

(2) When the diffusion set E_T is a self-similar set on $[0, T]$, the diffusion kernel function and the flux function $\int_0^t i(r, \tau)\mathrm{d}\tau$ of the radial probability current $i(r, t)$ are explicitly related to the ξ_j, b_j and p_j, $j=1, 2, \cdots, k$, thus to the geometrical structure of the diffusion set E_T and the probability vector $\boldsymbol{p}$, because the geometric structure of the diffusion set E_T is completely determined by ξ_j and b_j.

From (3.3) we obtain that

$$i(r, t)=r^{d_f-1}A'\sum_{j=1}^{k}p_j\xi_j^{\gamma-1}\frac{\partial_j^\gamma P(r, t)}{\partial t^\gamma}, \tag{3.4}$$

where

$$\frac{\partial_j^\gamma P(r, t)}{\partial t^\gamma}=\frac{1}{\Gamma(1-\gamma)}\frac{\partial}{\partial t}\int_0^t P(r, \tau)(t-\tau-b_j)^{-\gamma}\eta(t-\tau-b_j)\mathrm{d}\tau. \tag{3.5}$$

From (1.2) and (3.4) one gets

$$\sum_{j=1}^{k}p_j\xi_j^{\gamma-1}\frac{\partial_j^\gamma P(r, t)}{\partial t^\gamma}=-Gr^{-\theta'}\left(\frac{\partial P(r, t)}{\partial r}+\frac{\kappa}{r}P(r, t)\right), \quad G>0. \tag{3.6}$$

Using the Laplace transform and the normalization condition $\int_0^\infty r^{d_f-1}P(r, t)\mathrm{d}t=1$, we obtain the approximate solution of the fractional diffusion equation on self-similar sets,

$$P(r, t)=\frac{G_3}{r^\kappa\Gamma(1-(d_f-\kappa)/d_w)p_1\xi_1}\int_0^t\sum_{j=1}^{k}p_j\xi_j(t-b_j-\tau)^{-(d_f-\kappa)/d_w}h(\tau)\mathrm{d}\tau, \tag{3.7}$$

where

$$h(t)=\frac{1}{\pi}\int_0^\infty\exp\left[-tx-B_1x^\gamma\left(1+\sum_{j=2}^{k}(p_j/p_1)(\xi_j/\xi_1)^{\gamma-1}\mathrm{e}^{b_jx}\right)\cos(\gamma\pi)\right]$$

$$\cdot\sin\left[B_1x^\gamma\left(1+\sum_{j=2}^{k}(p_j/p_1)(\xi_j/\xi_1)^{\gamma-1}\mathrm{e}^{b_jx}\right)\sin(\gamma\pi)\right]\mathrm{d}x,$$

and $B_1=(r/G_4)^{\theta'+1}$, see Appendix B.

Acknowledgement

We would like to express our thanks to the editor, Professor C. R. Doering, for his invaluable comments, suggestions and pointing out the mistakes in references. This project 19771023 was supported by NSFC.

Appendix A

Let $P(r,s)=\mathscr{L}(P(r,t))$ denote the Laplace transform of $P(r,t)$. Acting the Laplace transform for (2.9) we obtain

$$s^{\gamma}P(r,s)=-Gr^{-\theta'}\left(\frac{\partial P(r,s)}{\partial r}+\frac{\kappa}{r}P(r,s)\right), \tag{A.1}$$

since the Laplace transform of the fractional derivative $\mathscr{L}[\partial(r,t)/\partial^{\gamma}]=s^{\gamma}P(r,s)$. Using this result, a simple integration of (A.1) yields

$$P(r,s)=Q(s)\frac{1}{(rs^{1/d_{\mathrm{w}}})^{\kappa}}\exp[-(rs^{1/d_{\mathrm{w}}}/G)^{\theta'+1}],\quad d_{\mathrm{f}}-\kappa>0, \tag{A.2}$$

where $Q(s)=G's^{-(1-d_{\mathrm{f}}/d_{\mathrm{w}})}$, $G'>0$, ensures the normalization condition

$$\int_0^t r^{d_{\mathrm{f}}-1}P(r,t)\mathrm{d}r=1,$$

i.e.

$$\int_0^t r^{d_{\mathrm{f}}-1}P(r,s)\mathrm{d}r=1/s.$$

Then

$$P(r,t)=\mathscr{L}^{-1}[As^{-\beta}\exp(-Bs^{\gamma})], \tag{A.3}$$

where $A=Gr^{-\kappa}$, $B=(r/G')^{\theta'+1}$ and $\beta=1-(d_{\mathrm{f}}-\kappa)/d_{\mathrm{w}}$.

If $\beta>0$, it is easy to show that

$$\mathscr{L}^{-1}(s^{-\beta})=t^{-(d_{\mathrm{f}}-\kappa)/d_{\mathrm{w}}}/\Gamma(\beta). \tag{A.4}$$

The inverse Laplace transform of $\exp(-Bs^{\gamma})$

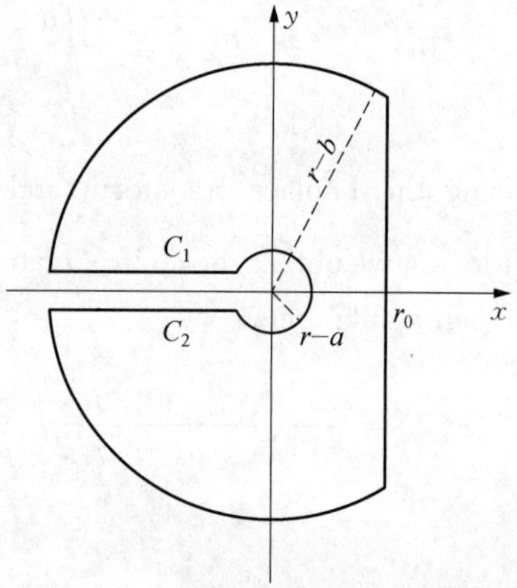

Fig. A. 1 Contour of integration in the complex plane for evaluating the inverse Laplace transform formula (A.5).

can be evaluated from the complex inversion formula

$$\mathscr{L}^{-1}[\exp(-Bs^{\gamma})]=\frac{1}{2\pi\mathrm{i}}\int_{\gamma_0-\mathrm{i}\infty}^{\gamma_0+\mathrm{i}\infty}\mathrm{e}^{xt-Bx^{\gamma}}\,\mathrm{d}x, \tag{A. 5}$$

following the integration path shown in Fig. A. 1.

It can be shown that the integrals along the circular paths vanish and applying the Cauchy theorem (A. 5) becomes

$$\begin{aligned}\mathscr{L}^{-1}[\exp(-Bx^{\gamma})]&=-\lim_{a\to 0,\, b\to+\infty}\frac{1}{2\pi\mathrm{i}}\left(\int_{C_1}\mathrm{e}^{tx-Bx^{\gamma}}\,\mathrm{d}x+\int_{C_2}\mathrm{e}^{tx-Bx^{\gamma}}\,\mathrm{d}x\right)\\&=-\frac{1}{2\pi\mathrm{i}}\left(\int_0^{\infty}\mathrm{e}^{-tx-Bx^{\gamma}\mathrm{e}^{\mathrm{i}\gamma\pi}}\,\mathrm{d}x+\int_{\infty}^{0}\mathrm{e}^{-tx-Bx^{\gamma}\mathrm{e}^{-\mathrm{i}\gamma\pi}}\,\mathrm{d}x\right)\\&=\frac{1}{\pi}\int_0^{\infty}\mathrm{e}^{-tx}\,\mathrm{e}^{-Bx^{\gamma}\cos(\gamma\pi)}\sin(Bx^{\gamma}\sin(\gamma\pi))\,\mathrm{d}x.\end{aligned} \tag{A. 6}$$

From (A. 3)~(A. 6) and by the product theorem of Laplace transform, we get the exact solution (2. 21) of the fractional diffusion equation (2. 9) on fractals.

Appendix B

Using the Laplace transform, from (3. 6) we get the transport equation for the fractional diffusion equation on self-similar sets,

$$\sum_{j=1}^{k}P_j\xi_j^{\gamma-1}\mathrm{e}^{-b_js}s^{\gamma}P(r,s)=-Gr^{-\theta'}\left(\frac{\partial P(r,s)}{\partial r}+\frac{\kappa}{r}P(r,s)\right). \tag{B. 1}$$

Since $P(r, t)$ satisfies the normalization condition of $P(r, s)$, i. e. $\int_0^{\infty}r^{d_f-1}P(r,s)\mathrm{d}r=1/s$, from (B. 1) we have

$$\begin{aligned}P(r,s)=&G_3s^{-1}r^{-\kappa}[1+g_1(s)]^{(d_f-\kappa)/(\theta'+1)}s^{(d_f-\kappa)/d_w}\\&\cdot\exp\{-[1+g_1(s)](rs^{1/d_w}/G_4)^{\theta'+1}\},\end{aligned} \tag{B. 2}$$

where

$$G_3=G_1(p_1\xi_1^{\gamma-1})^{(d_f-\kappa)/(\theta'+1)},\quad G_2=[(\theta'+1)G]^{1/(\theta'+1)},$$

$$G_4=G_2/(p_1\xi_1^{\gamma-1})^{1/(\theta'+1)},\quad G_1=C_1G_2^{-(d_f-\kappa)},$$

$$g_1(s)=\sum_{j=2}^{k}(p_j/p_1)(\xi_j/\xi_1)^{\gamma-1}\mathrm{e}^{-b_j s},\quad C_1^{-1}=\frac{1}{\theta'+1}\int_0^{\infty}\mathrm{e}^{-\chi}\chi^{(d_f-1-\kappa-\theta')/(\theta'+1)}\,\mathrm{d}\chi<\infty,$$

$$\chi=g(s)r^{\theta'+1},\ g(s)=\frac{1}{(\theta'+1)G}\Big(\sum_{j=1}^{k}p_j\xi_j^{\gamma-1}s^{\gamma}\mathrm{e}^{-b_j s}\Big).$$

Since $|g_1(s)|\ll 1$ when $\mathrm{Re}(s)\gg 1$, from (B. 2) we have approximately

$$P(r,s)=G_3r^{-\kappa}s^{-(1-(d_f-\kappa)/d_w)}\left(1+\frac{d_f-\kappa}{\theta'+1}g_1(s)\right)\exp\{-[1+g_1(s)](rs^{1/d_w}/G_4)^{\theta'+1}\}$$

$$\backsimeq G_3s^{-(1-d_f/d_w)}\frac{1}{(rs^{1/d_w})^{\kappa}}\exp[-(rs^{1/d_w}/G_4)^{\theta'+1}]. \tag{B. 3}$$

This is the result in Ref. [16].

Taking the inverse Laplace transform in (B. 3), we obtain

$$P(r,t)\backsimeq G_3r^{-\kappa}\mathscr{L}^{-1}\left\{s^{-(1-(d_f-\kappa)/d_w)}\left(1+\frac{d_f-\kappa}{\theta'+1}g_1(s)\right)\right.$$
$$\left.\cdot\exp\left[-(1+g_1(s))\left(\frac{rs^{1/d_w}}{G_4}\right)^{\theta'+1}\right]\right\}.$$

Similar to (A. 4) we have

$$\mathscr{L}^{-1}\left[s^{-(1-(d_f-\kappa)/d_w)}\left(1+\frac{d_f-\kappa}{\theta'+1}g_1(s)\right)\right]$$
$$=\frac{1}{\Gamma(\beta)p_1\xi_1}\sum_{j=1}^{k}(p_j/p_1)(\xi_j/\xi_1)^{\gamma-1}(t-b_j)^{-(d_f-\kappa)/d_w},$$

where $\beta=1-(d_f-\kappa)/d_w$. By using a method similar to the proof of (A. 6) (see Appendix A), we get

$$L^{-1}[\exp\{-[1+g_1(s)](rs^{1/d_w}/G_4)^{\theta'+1}\}]$$
$$=\mathscr{L}^{-1}\left[\exp\left\{-\Big(1+\sum_{j=2}^{k}(p_j/p_1)(\xi_j/\xi_1)^{\gamma-1}\mathrm{e}^{-b_j s}\Big)B_1s^{\gamma}\right\}\right]$$
$$=\frac{1}{\pi}\int_0^{\infty}\exp\left[-tx-B_1x^{\gamma}\Big(1+\sum_{j=2}^{k}(p_j/p_1)(\xi_j/\xi_1)^{\gamma-1}\mathrm{e}^{b_j x}\Big)\cos(\gamma\pi)\right]$$
$$\cdot\sin\left[B_1x^{\gamma}\Big(1+\sum_{j=2}^{k}(p_j/p_1)(\xi_j/\xi_1)^{\gamma-1}\mathrm{e}^{b_j x}\Big)\sin(\gamma\pi)\right]\mathrm{d}x.$$

Finally, by the properties of the Laplace transform we get (3. 7).

◇ References ◇

[1] F. Y. Ren, Z. G. Yu, F. Su, Phys. Lett. A 219 (1996) 59.

[2] Z. G. Yu, F. Y. Ren, J. Phys. A 30 (1997) 5569.

[3] F. Y. Ren, Z. G. Yu, J. Zhou, A. Le Mehaute, R. R. Nigmatullin, Physica A 246 (1997) 419.

[4] F. Y. Ren, Z. G. Yu, J. Zhou, Fine approximation of flux and fractional integral on (infinite) self-similar set or generalized self-similar set, Southeast Asian Bull. of Math. (to appear).

[5] F. Y. Ren, Z. G. Yu, Progress in Natural Science 7 (1997) 422.

[6] A. Le Mehaute, R. R. Nigmatullin, L. Nivanen, Irrveribilité Temporl et Geometrie Fractale, published in HERMES, 1998.

[7] S. Alexander, R. Orbach, J. Phys. (Paris) Lett. 43 (1982) 2625.

[8] S. Havlin. D. ben-Avraham, Adv. Phys. 36 (1987) 695.

[9] B. O'Shaughnessy, I. Procaccia, Phys. Rev. Lett. 54 (1985) 455.

[10] S. Havlin, B. Trus, G. H. Weiss, J. Phys. A 18 (1985) L1043.

[11] A. Brooks Harris, A. Aharony, Europhys. Lett. 4 (1987) 1355.

[12] A. Bunde, S. Havlin, eds., Fractals and Disordered Systems (Spinger, Heidelberg, 1991).

[13] M. Giona, H. E. Roman, Physica A 185 (1992) 87.

[14] K. J. Falconer, Fractal Geometry: Mathematics Foundations and Applications (Wiley, Chichester, 1990).

[15] S. Huan, Acta Math. Appl. Sin. 17 (1994) 551.

[16] M. Gioma, H. H. Roman, J. Phys. A 25 (1992) 2093.

[17] R. Rammal, G. Toulouse, J. Phys. (Paris) Lett. 44 (1983) 13.

Determination of Memory Function and Flux on Fractals*

Fu-Yao Ren[a], Wei-Yuan Qiu[a],
Jin-Rong Liang[b], Xiao-Tian Wang[a]
[a] *Institute of Mathematics, Fudan University, Shanghai 200433, PR China*
[b] *Department of Mathematics, East China Normal University, Shanghai 200062, PR China*

Abstract: When memory set is a fractal, whether the generating mappings of the memory set are linear, non-linear, monotone increasing, or monotone decreasing, the exact estimates of corresponding memory function and flux on the fractal can be obtained.

Keywords: flux, memory function, memory measure, Laplace transform

1. Introduction

When we describe a structure of the evolution of a physical system far from thermodynamic equilibrium — in amorphous materials [1～3], in the description of structural relaxation of high-T_c oxide superconductors [4], in the process of plastic deformation [5] and fracture of solids, in the description of solid solutions [6] and the macrostructure of martensite [7], and so on, it has been found that the medium exhibit memory. The existence of memory means that if at time τ a force $f(\tau)$ acts on the system, then there arises a flux

* Project partially supported by NSFC. Originally published in *Physics Letters A*, Vol. 288, (2001), 79-87.

J whose magnitude at time $t > \tau$ is given through a memory function $m(\tau)$ by the equation

$$J(t) = \int_0^t m(t-\tau) f(\tau) \mathrm{d}\tau. \tag{1}$$

The set with "residual" memory, i. e., the support of $m(t)$, is called the *memory set*. The procedure of calculation of (1) is rather complicate because the concrete form of the kernel $m(t-\tau)$ in the most cases is not known. Moreover, it is important to formulate the following questions [8]:

How to find the function $m(t-\tau)$ for a fractal structure and, in turn, how to find the averaged value of the function $f(\tau)$ distributed over a fractal structure of an arbitrary nature?

If memory sets are (generalized) self-similar sets or cookie-cutter sets or net fractals, the estimates of the flux and the fractional integral associated to these sets and its physical interpretation have been discussed systematically in Refs. [9~15]. That is, if the "residual" memory set is generated by a family of contractive mappings $\{\varphi_j(x)\}_{j=1}^k$ on $[0, T]$, where $\varphi_j(x)$ is a similarity or $\varphi_j(x)$ is $C^{1+\alpha}$-differentiable, and if the total number of remaining states in each stage of the division of this set is normalized to unity, then when $\varphi_1(0) = 0$, the approximation and fine approximation of the flux $J(t)$ have the following forms, respectively (cf. Refs. [10~14]):

$$J(t) \backsimeq A(\Gamma(v))^{-1} \int_0^t (t-\tau)^{v-1} f(\tau) \mathrm{d}\tau = A t^v D^{-v} f(t) \tag{2}$$

and

$$\begin{aligned} J(t) &\backsimeq A(\Gamma(v))^{-1} \left[\sum_{j=1}^k p_j \xi_j^{-v} \int_0^t (t-\tau-b_j)^{v-1} \eta(t-\tau-b_j) f(\tau) \mathrm{d}\tau \right] \\ &= A \left[p_1 \xi_1^{-v} t^v D_1^{-v} f + \sum_{j=2}^k p_j \xi_j^{-v} (t-b_j)^v D^{-v} f(t-b_j) \right], \end{aligned} \tag{3}$$

where $v = \ln p_1 / \ln \xi_1$ is the fractional exponent of the fractional integral,

$$D^{-v}f(t)=(\Gamma(v))^{-1}\int_0^1(1-u)^{v-1}f(ut)\mathrm{d}u=(v\Gamma(v))^{-1}\int_0^1 f((1-u_1)t)\mathrm{d}u_1^v \tag{4}$$

is a fractional integral, $D^{-v}f(t-b_j)$ is the translation of the fractional integral $D^{-v}f\ (t)$, i. e.,

$$D^{-v}f(t-b_j)=(v\Gamma(v))^{-1}\int_0^1 f[(1-u_1)(t-b_j)]\mathrm{d}u_1^v,$$

$b_j=\varphi_j(0)$, $\xi_j=\varphi_j'(0)>0$, $j=1,\cdots,k$, $\eta(x)=0$ for $x<0$ and 1 for $x>0$, and $\boldsymbol{p}=(p_1,p_2,\cdots,p_k)$ is any normalized probability vector (i. e., $\sum_{j=1}^k p_j=1$, $p_j>0$ for $j=1,\cdots,k$).

If $\varphi_1(0)\neq 0$, then

$$J(t)\backsimeq A\left(t-\frac{b_1}{1-\xi_1}\right)^v D^{-v}f\left(t-\frac{b_1}{1-\xi_1}\right) \tag{5}$$

and

$$J(t)\backsimeq A\sum_{j=1}^k p_j\xi_j^{-v}(t-a_j)^v D^{-v}f(t-a_j), \tag{6}$$

where

$$a_j=b_j+\frac{b_1}{1-\xi_1}\xi_j,\quad j=1,\cdots,k.$$

However, how to determine the constant A in (2)~(6) is still an open problem. Moreover, although the fractional exponent v of the fractional integral is determined by $v=\ln p_1/\ln\xi_1$, yet whether the exponent should satisfy other condition is not discussed.

In this Letter we will calculate this unknown constant and give the exact estimates of corresponding memory function and flux in terms of fractional integrals when memory sets are fractals, whether the generating mappings of the memory sets are linear, non-linear, monotone increasing, or monotone decreasing.

Remark $\varphi_j(x)$ is $C^{1+\alpha}$-differentiable means that $\varphi_j(x)$ is differentiable with a Hölder continuous derivative φ'_j satisfying $|\varphi'_j(x)-\varphi'_j(y)|<\beta_j|x-y|^{\alpha}$, where $\alpha>0$ and $\beta_j>0$ are constants.

2. Determination of Memory Function and Flux

For any given $T\in(0,\infty)$ and $k\in\{2,3,\cdots\}$, let $E_0=[0,T]$ and let $\{\varphi_j(x)\}_{j=1}^{k}$ be the contractive mappings on E_0, i. e.,

$$|\varphi_j(x)-\varphi_j(y)|\leqslant c_j|x-y|,\quad \forall x, y\in E_0, 0<c_j<1.$$

Suppose $\varphi_j(E_0)\cap\varphi_i(E_0)=\varnothing$, $i\neq j$. Then the nonempty compact set

$$E_T=\bigcap_{n=1}^{\infty}E(n) \tag{7}$$

is a net fractal, where

$$E(n)=\bigcup_{j_1,\cdots,j_n}E_{j_1\cdots j_n},\quad E_{j_1\cdots j_n}=\varphi_{j_n}\circ\cdots\circ\varphi_{j_1}(E_0).$$

For any given probability vector $\boldsymbol{p}=(p_1, p_2, \cdots, p_k)$, let the measure μ_n on $E(n)$ be a probability measure defined by $\mathrm{d}\mu_n(\tau)=m_n(\tau)\,\mathrm{d}\tau$:

$$m_n(\tau)=\sum_{j_1,\cdots,j_n\in I}p_{j_1}\cdots p_{j_n}\frac{\chi E_{j_1\cdots j_n}(\tau)}{|E_{j_1\cdots j_n}|},\quad I=\{1,\cdots,k\},$$

where

$$\chi E_{j_1\cdots j_n}(\tau)=\begin{cases}1, & \tau\in E_{j_1\cdots j_n},\\ 0, & \text{otherwise.}\end{cases}$$

Then, as shown in Refs. [10, 13], μ_n weakly converges to a unique probability measure μ as $n\to+\infty$ such that $\int_{E_T}\mathrm{d}\mu(\tau)=1$, $\mathrm{supp}(\mu)=E_T$, and

$$\mu(\cdot)=\sum_{j=1}^{k}p_j\mu\circ\varphi_j^{-1}(\cdot). \tag{8}$$

Let E_T be a memory set. The measure μ is called the *memory measure*. Write

$$\mathrm{d}\mu(\tau)\backsim m(\tau)\mathrm{d}\tau, \tag{9}$$

then $m(\tau)$ is the *memory function*. Let

$$M(p)=\int_0^{+\infty} e^{-p\tau}\,d\mu(\tau) \tag{10}$$

be the Laplace transform of the measure $\mu(\tau)$.

Assume that, for each $j \in I$, $\varphi_j(x)$ is a similarity or $\varphi_j(x)$ is a $C^{1+\alpha}$-differentiable mapping. Let

$$b_j=\varphi_j(0),\quad \xi_j=\varphi_j'(0)>0,\quad 0<\xi_j<1,\quad j=1,\cdots,k, \tag{11}$$

and $0=b_1<b_2<\cdots<b_k<\varphi_k(T)=T$. Then

$$M(p)=\sum_{j=1}^{k} p_j e^{-b_j p} M(\xi_j p)+o(1) \tag{12}$$

as Re $p \to +\infty$[14].

If $b_1=0$, it follows from (12) that

$$M(p) \backsimeq p_1 M(\xi_1 p)+o(1) \quad (\text{as Re } p \to +\infty).$$

Solving the function equation

$$\overline{M}(p)=p_1\overline{M}(\xi_1 p),$$

we have

$$\overline{M}(p)=Ap^{-v},$$

where A is a constant and $v = \ln p_1/\ln \xi_1$.

Note that $M(0)=\int_0^{+\infty} d\mu(\tau)=\int_0^{T} d\mu(\tau)=1$. But, $\overline{M}(0)\neq 1$. Thus we choose $\widetilde{M}(p)$ as an approximation of $M(p)$:

$$M(p) \backsimeq \widetilde{M}(p)=Ap^{-v}[1-\exp(-p^{v}/A)] \tag{13}$$

as Re $p \to +\infty$. Applying the theory of complex analysis we can obtain the inverse Laplace transform $L^{-1}[\widetilde{M}(p)]$ of $\widetilde{M}(p)$:

$$m(t)=L^{-1}[\widetilde{M}(p)]=\frac{1}{2\pi i}\int_{a-i\infty}^{a+i\infty} e^{pt}\widetilde{M}(p)\,dp=\frac{A\Gamma(1-v)\sin(v\pi)}{\pi}t^{v-1} \quad (a>0) \tag{14}$$

if and only if $0 < v < 1$. See Appendix A.

Noting that $\int_0^T \mathrm{d}\mu(t) = 1$ we obtain the estimates of the constant A, the memory function $m(t)$ and the flux $J(t)$ as follows:

$$A \backsimeq \frac{v\pi}{T^v \Gamma(1-v)\sin(v\pi)}, \tag{15}$$

$$m(t) = \frac{v}{T^v} t^{v-1}, \tag{16}$$

$$J(t) \backsimeq \frac{v}{T^v}\int_0^t (t-\tau)^{v-1} f(\tau)\mathrm{d}\tau = \frac{v\Gamma(v)}{T^v} t^v D^{-v} f(t), \tag{17}$$

where

$$0 < v = \ln p_1 / \ln \xi_1 < 1. \tag{18}$$

If we use $\widetilde{M}(\xi_j p)$ as an approximation of $M(\xi_j p)$, then

$$M(p) \backsimeq \sum_{j=1}^k p_j \mathrm{e}^{-b_j p} A(\xi_j p)^{-v}[1-\exp(-(\xi_j p)^v / A)]$$

as Re $p \to +\infty$. Therefore, we obtain the fine approximations as follows:

$$A \backsimeq \frac{\pi v}{\Gamma(1-v)[p_1 \xi_1^{-v} T^v + \sum_{j=2}^k p_j \xi_j^{-v}(T-b_j)^v]\sin(v\pi)}, \tag{19}$$

$$m(t) = \frac{v(p_1 \xi_1^{-v} t^{v-1} + \sum_{j=2}^k p_j \xi_j^{-v}(t-b_j)^{v-1}\eta(t-b_j))}{p_1 \xi_1^{-v} T^v + \sum_{j=2}^k p_j \xi_j^{-v}(T-b_j)^v}, \tag{20}$$

$$J(t) \backsimeq \frac{v\Gamma(v)[p_1 \xi_1^{-v} t^v D^{-v} f(t) + \sum_{j=2}^k p_j \xi_j^{-v}(t-b_j)^v D^{-v} f(t-b_j)]}{p_1 \xi_1^{-v} T^v + \sum_{j=2}^k p_j \xi_j^{-v}(T-b_j)^v}, \tag{21}$$

where $v = \ln p_1 / \ln \xi_1$, $0 < v < 1$.

If $b_1 \neq 0$, then

$$M(p) \backsimeq p_1 \mathrm{e}^{-b_1 p} M(\xi_1 p) + o(1) \quad (\text{as Re}\, p \to +\infty). \tag{22}$$

Note that the equation

$$\overline{M}(p)=p_1 e^{-b_1 p}\overline{M}(\xi_1 p)$$

has a unique solution of form

$$\overline{M}(p)=Ap^{-v}\exp\left(-\frac{b_1}{1-\xi_1}p\right),$$

where $v = \ln p_1/\ln \xi_1$ and A is a constant. We choose

$$M(p) \backsimeq \widetilde{M}(p)=Ap^{-v}\exp\left(-\frac{b_1}{1-\xi_1}p\right)(1-\exp(-p^v/A)) \tag{23}$$

as Re $p\to+\infty$, then we obtain the following approximations:

$$A \backsimeq \frac{\pi v}{\left(T-\frac{b_1}{1-\xi_1}\right)^v \Gamma(1-v)\sin(v\pi)}, \tag{24}$$

$$m(t)=\frac{v}{\left(T-\frac{b_1}{1-\xi_1}\right)^v}\left(t-\frac{b_1}{1-\xi_1}\right)^{v-1}\eta\left(t-\frac{b_1}{1-\xi_1}\right), \tag{25}$$

$$J(t) \backsimeq \frac{v\Gamma(v)}{\left(T-\frac{b_1}{1-\xi_1}\right)^v}\left(t-\frac{b_1}{1-\xi_1}\right)^v D^{-v} f\left(t-\frac{b_1}{1-\xi_1}\right). \tag{26}$$

If we choose

$$M(\xi_j p) \backsimeq \widetilde{M}(\xi_j p)=A\xi_j^{-v}p^{-v}\exp\left(-\frac{b_1\xi_j}{1-\xi_1}p\right)(1-\exp(-\xi_j^v p^v/A))$$

as Re $p\to+\infty$, then we obtain the fine estimates

$$A \backsimeq \frac{\pi v}{\Gamma(1-v)\left[\sum_{j=1}^k p_j\xi_j^{-v}(T-a_j)^v\right]\sin(v\pi)}, \tag{27}$$

$$m(t)=\frac{v\left[\sum_{j=1}^k p_j\xi_j^{-v}(t-a_j)^{v-1}\eta(t-a_j)\right]}{\sum_{j=1}^k p_j\xi_j^{-v}(T-a_j)^v}, \tag{28}$$

$$J(t) \backsimeq \frac{v\Gamma(v)\left[\sum_{j=1}^k p_j\xi_j^{-v}(t-a_j)^v D^{-v}f(t-a_j)\right]}{\sum_{j=1}^k p_j\xi_j^{-v}(T-a_j)^v}, \tag{29}$$

where

$$a_j = b_j + \frac{b_1}{1-\xi_1}\xi_1, \quad j = 1, \cdots, k, \tag{30}$$

and $v = \ln p_1 / \ln \xi_1$, $0 < v < 1$.

Remark (1) When $b_1 = 0$, (24)~(26) and (27)~(29) become (15)~(17) and (19)~(21), respectively.

(2) If $\varphi_j(0) \neq 0$ and $-1 < \xi_j < 0$ for all $j \in I$. By re-arranging the order of φ_j we can assume that

$$T \geqslant b_1 > b_2 > \cdots > b_k \geqslant 0. \tag{31}$$

For each $j \in I$, let $\widetilde{\varphi}_j(x) = T - \varphi_j(x)$, $x \in [0, T]$. Then $0 < \widetilde{\varphi}_j'(0) < 1$. Similar to (8),

$$\mu(\cdot) = \sum_{j=1}^{k} p_j \mu \circ \widetilde{\varphi}_j^{-1}(\cdot). \tag{32}$$

Thus

$$M(p) = \sum_{j=1}^{k} p_j \int_0^T \exp(-p\widetilde{\varphi}_j(\tau))\mathrm{d}\mu(\tau) = \sum_{j=1}^{k} p_j \mathrm{e}^{-p(T-b_j)} M(|\xi_j| p) + o(1) \tag{33}$$

as Re $p \to +\infty$. Applying the previous way and replacing (23) by

$$M(p) \backsim \widetilde{M}(p) = Ap^{-v}\exp\left(-\frac{T-b_1}{1-|\xi_1|}p\right)(1-\exp(-p^v/A)) \tag{34}$$

as Re $p \to +\infty$, we obtain the similar results to (24)~(26):

$$A \backsim \frac{\pi v}{\left(T - \frac{T-b_1}{1-|\xi_1|}\right)^v \Gamma(1-v)\sin(v\pi)}, \tag{35}$$

$$m(t) = \frac{v}{\left(T - \frac{T-b_1}{1-|\xi_1|}\right)^v}\left(t - \frac{T-b_1}{1-|\xi_1|}\right)^{v-1} \eta\left(t - \frac{T-b_1}{1-|\xi_1|}\right), \tag{36}$$

$$J(t) \backsim \frac{v\Gamma(v)}{\left(T - \frac{T-b_1}{1-|\xi_1|}\right)^v}\left(t - \frac{T-b_1}{1-|\xi_1|}\right)^v D^{-v} f\left(t - \frac{T-b_1}{1-|\xi_1|}\right). \tag{37}$$

Moreover, the fine estimates as follows:

$$A \backsimeq \frac{\pi v}{\Gamma(1-v)\left[\sum_{j=1}^{k} p_j |\xi_j|^{-v}(T-c_j)^v\right]\sin(v\pi)}, \tag{38}$$

$$m(t) \backsimeq \frac{v\left[\sum_{j=1}^{k} p_j |\xi_j|^{-v}(t-c_j)^{v-1}\eta(t-c_j)\right]}{\sum_{j=1}^{k} p_j |\xi_j|^{-v}(T-c_j)^v}, \tag{39}$$

$$J(t) \backsimeq \frac{v\Gamma(v)\left[\sum_{j=1}^{k} p_j |\xi_j|^{-v}(t-c_j)^v D^{-v} f(t-c_j)\right]}{\sum_{j=1}^{k} p_j |\xi_j|^{-v}(T-c_j)^v}, \tag{40}$$

where

$$c_j = T - b_j + \frac{T-b_1}{1-|\xi_1|}|\xi_1|,\ b_j = \varphi_j(0),\quad \xi_j = \varphi_j'(0),\quad j \in I, \tag{41}$$

and $v = \ln p_1 / \ln|\xi_1|$, $0 < v < 1$.

3. Conclusions

(1) In many problems of theoretical physics it is necessary to take into account the non-local effects caused by heterogeneity of a medium. In connection with penetration of fractal geometry ideas into modern physics it is important to calculate two types of heterogeneities: the heterogeneity caused by a fractal structure and the heterogeneity caused by physical value itself distributed over the fractal structure. Usually the relationship between two physical values $J(t)$ and $f(\tau)$ localized in the points t, τ accordingly is expressed in most cases by formula (1) [8] or similar formulas. For example, in the diffusion process discussed in [16] there exists the following relation between the total radial current $j(r, t)$ and diffusion kernel $K(t, \tau)$ with probability density $P(r,\tau)$:

$$j(r, t) = r^{d_f - 1}\int_0^t d\tau K(t, \tau) P(r, \tau).$$

Therefore, our discussion is valid not only in some memory processes and but also in other physical processes (e. g. , in some diffusion processes).

(2) When the memory set is E_T, whether the generating mappings $\{\varphi_j(x)\}_{j=1}^k$ of the memory set are linear, non-linear, increasing or decreasing and no matter which self-similar measure is taken, the fractional exponent v of the fractional integral is always determined only by $v = \ln p_1/\ln|\varphi_1'(0)|$ and v must satisfy $0 < v < 1$, where p_1 is the first weight of self-similar measure defined on the memory set and $\varphi_1'(0)$ is the derivative at 0 of the first generating mapping of the memory set, but it does not depend on the other weights of self-similar measure and other mappings.

(3) Whether the mappings $\{\varphi_j(x)\}_{j=1}^k$ are linear, non-linear, increasing or decreasing and no matter which self-similar measure is taken, estimates (17), (26) and (37) of the flux $J(t)$ depend only on $\ln p_1/\ln|\varphi_1'(0)|$, $\varphi_1(0)$ and T. In addition, the flux $J(t)$ has always the fine estimates (21), (29) and (40) and these estimates are determined not only by depend only on $v = \ln p_1/\ln|\varphi_1'(0)|$, but also by all $\{p_j\}_{j=1}^k$, $\{\varphi_j(0)\}_{j=1}^k$, $\{\varphi'_j(0)\}_{j=1}^k$ and T. Therefore, these fine estimates depend both on the generating mappings of the memory set and on the weights of self-similar measure. The fine estimates (21), (29) and (40) also give the relationship between the fractional integrals and the fractal structure of the memory set.

(4) When the memory set is E_T, whether the generating mappings of the memory set are linear or non-linear, so long as the generating mappings are monotone increasing, estimate (26) or (29) of the flux are invariable.

(5) The fractional exponent v is equal to the fractal dimension s of the self-similar set E_T if and only if $p_1 = \xi_1^s$. In particular, $v = \ln 2/\ln(1/\xi)$ when E_T is the Cantor set, and $v = \ln k/\ln(1/\xi)$ when E_T is Cantor's k-bar, this is Nigmatullin's result [9]. This results coincides with $0 < v < 1$ proved in this Letter.

(6) From the formula $v = \ln p_1/\ln|\varphi_1'(0)|$, there is no direct relationship between the fractional exponent v and the fractal structure of the memory set E_T. This is determined by the methodology.

(7) From (26) or (37), when $\varphi_1(0) \neq 0$ or $\varphi_1(0) \neq T$, the flux $J(t)$ has a delay of time $b_1/(1-\varphi_1'(0))$ or $(T-\varphi_1(0))/(1-|\varphi_1'(0)|)$ with respect to

the force $f(t)$.

(8) The fractal discussed in this Letter is only a net fractal generated by a family of contraction transformations. But, how to determine the memory function $m(\tau)$ and the flux $J(t)$ on general fractals is still an open problem.

Appendix A

The inverse Laplace transform of $L^{-1}[\widetilde{M}(p)]$ can be evaluated from the complex inversion formula

$$L^{-1}[\widetilde{M}(p)]=\frac{1}{2\pi \mathrm{i}}\int_{a-\mathrm{i}\infty}^{a+\mathrm{i}\infty} \mathrm{e}^{pt}\widetilde{M}(p)\mathrm{d}p \tag{A.1}$$

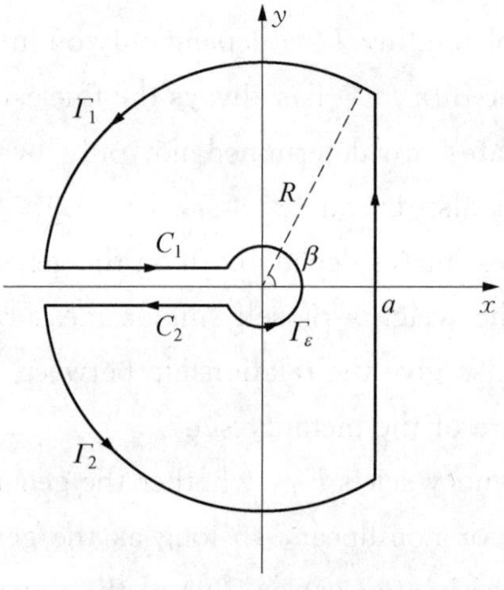

Fig. 1 Contour of integration in the complex plane for evaluating the inverse Laplace transform formula (A.1).

following the integration path shown in Fig. 1.

Applying the theorem of Cauchy (A.1) becomes

$$\begin{aligned} & L^{-1}[Ap^{-v}(1-\exp(-p^{v}/A))] \\ &= -\lim_{\substack{R\to+\infty \\ \varepsilon\to+0}} \frac{1}{2\pi \mathrm{i}} \int_{\Gamma_\varepsilon+\Gamma_1+\Gamma_2+C_1+C_2} A\mathrm{e}^{pt}p^{-v}(1-\exp(-p^{v}/A))\mathrm{d}p \qquad \text{(A. 2)} \\ &= -\lim_{\substack{R\to+\infty \\ \varepsilon\to+0}} \frac{1}{2\pi \mathrm{i}}(I_1+I_2+I_3+I_4), \end{aligned}$$

where

$$I_1=\int_{\Gamma_\varepsilon} A e^{pt} p^{-v}(1-\exp(-p^v/A))\mathrm{d}p, \quad I_2=\int_{\Gamma_1} A e^{pt} p^{-v}(1-\exp(-p^v/A))\mathrm{d}p,$$

$$I_3=\int_{\Gamma_2} A e^{pt} p^{-v}(1-\exp(-p^v/A))\mathrm{d}p, \quad I_4=\int_{C_1+C_2} A e^{pt} p^{-v}(1-\exp(-p^v/A))\mathrm{d}p.$$

Note that $\Gamma_\varepsilon: p=\varepsilon e^{i\theta}, -\pi<\theta<\pi, \mathrm{d}p=i\varepsilon e^{i\theta}\mathrm{d}\theta$. Then

$$|I_1|\leqslant 2\varepsilon^{1-v}\int_0^\pi \varepsilon^v(e-1)\exp(t\varepsilon\cos\theta)\mathrm{d}\theta<2\pi(e-1)\varepsilon e^{t\varepsilon}$$

When $\varepsilon \ll 1$ and $v>0$, and so

$$\lim_{\varepsilon\to+0}\left\{I_1=\int_{\Gamma_\varepsilon} A e^{pt} p^{-v}(1-\exp(-p^v/A))\mathrm{d}p\right\}=0. \tag{A. 3}$$

Using the monotonicity of the function $\exp(tR\cos\theta)$ and noting that Γ_1: $p=Re^{i\theta}$, $\beta<\theta<\pi$, $\mathrm{d}p=iRe^{i\theta}\mathrm{d}\theta$ we can prove that when $0<v<1$, $\lim_{R\to+\infty} I_2=0$. But when $v>1$, $\lim_{R\to+\infty} I_2\neq 0$. When $v=1$ and $t=1/A$ or $t<1/A$, $\lim_{R\to+\infty} I_2\neq 0$. Therefore,

$$\lim_{R\to+\infty}\left\{I_2=\int_{\Gamma_1} A e^{pt} p^{-v}(1-\exp(-p^v/A))\mathrm{d}p\right\}=0 \tag{A. 4}$$

if and only if $0<v<1$.

Similarly,

$$\lim_{R\to+\infty}\left\{I_3=\int_{\Gamma_2} A e^{pt} p^{-v}(1-\exp(-p^v/A))\mathrm{d}p\right\}=0 \tag{A. 5}$$

if and only if $0<v<1$.

Therefore, from (A. 2)～(A. 5) we obtain that, if and only if $0<v<1$,

$$L^{-1}[\widetilde{M}(p)]=-\lim_{\substack{R\to+\infty\\ \varepsilon\to+0}}\frac{1}{2\pi i}\int_{C_1+C_2} A e^{pt} p^{-v}(1-\exp(-p^v/A))\mathrm{d}p$$
$$=\frac{A\Gamma(1-v)\sin(v\pi)}{\pi}t^{v-1}.$$

◇ References ◇

[1] K. Binder, A. P. Joung, Rev. Mod. Phys. 58 (1986) 801.

[2] S. L. Ginzbug, Irreversible Phenomena in Spin Glasses, Nauka, Moscow, 1989, in Russian.

[3] A. I. Olemskoi, E. A. Toropov, Fiz. Met. Metalloved. 9 (1991) 5.

[4] A. I. Olemskoi, E. A. Toropov, Fiz. Met. Metalloved. 7 (1991) 32.

[5] A. I. Olemskoi, I. A. Sklyar, Usp. Fiz. Nauk 162 (1992) 29.

[6] A. I. Olemskoi, Fiz. Met. Metalloved. 68 (1989) 56.

[7] A. I. Olemskoi, Yu. I. Paskal, Preprint No. 30, Institute of Physics and Applied Mathematics, Tomsk Affiliate of the Siberian Branch of the USSR Academy of Sciences, Tomsk (1988).

[8] R. R. Nigmatullin, L. Nivanen, A. Le Mehaute, Time arrows and fractal geometry, published in HERMES.

[9] R. R. Nigmatullin, Teor. Mat. Fiz. (1992) 354.

[10] F.-Y. Ren, Z.-G. Yu, F. Su, Phys. Lett. A 219 (1996) 59.

[11] F.-Y. Ren, Z.-G. Yu, J. Zhou, Southeast Asian Bull. Math. 23 (1999) 497.

[12] F.-Y. Ren, Z.-G. Yu, Prog. Natural Sci. 7 (1997) 422.

[13] Z.-G. Yu, F.-Y. Ren, J. Phys. A: Math. Gen. 30 (1997) 5569.

[14] W.-Y. Qiu, J. Lü, Phys. Lett. A 272 (2000) 353.

[15] F.-Y. Ren, Z.-G. Yu, J. Zhou, A. Le Mehaute, R. R. Nigmatullin, Physica A 246 (1997) 419.

[16] M. Giona, H. E. Roman, Physica A 185 (1992) 87.

An Anomalous Diffusion Model in an External Force Fields on Fractals*

Fu-Yao Ren[a], Jin-Rong Liang[b], Wei-Yuan Qiu[a],
Xiao-Tian Wang[c], Y. Xu[a], R. R. Nigmatullin[d]

[a] *Department of Mathematics, Fudan University, Shanghai 200433, PR China*

[b] *Department of Mathematics, East China Normal University, Shanghai 200062, PR China*

[c] *Department of Mathematics, Wuhan University, Wuhan 430072, PR China*

[d] *Theoretical Physics Department, Kazan State University, Kazan 420008, Tatarstan, Russia*

Abstract: We present a fractional diffusion equation involving external force fields for transport phenomena in random media. It is shown that this fractional diffusion equation obey generalized Einstein relation, and its stationary solution is the Boltzmann distribution. It is proved that the asymptotic behavior of its solution is stretched Gaussian and that its solution can be expressed in the form of a function of a dimensionless similarity variable, not only for constant potentials but also for logarithm, harmonic, analytic and generic potentials. A comparison with the fractional Fokker-Planck equation is given.

1. Introduction

In connection with the growing interest in the physics of complex

* Project supported by NSFC (No. 10271031) and by the Shanghai Priority Academic Discipline. Originally published in *Physics Letters A*, Vol. 312, (2003), 187 - 197.

systems, anomalous transport properties and their description have received considerable interest. One found application in a wide field ranging from physics and chemistry to biology and medicine [1~5]. Anomalous diffusion in one dimension is characterized by the occurrence of a mean square displacement of the form

$$\langle X^2\rangle(t)=\langle\langle\Delta x\rangle^2\rangle(t)=\frac{2K_\gamma}{\Gamma(1+\gamma)}t^\gamma,\quad 0<\gamma<1 \tag{1.1}$$

which deviates from the linear Brownian dependence on time [3]. In Eq. (1.1), the anomalous diffusion coefficient K_γ is introduced, which has the dimension $[K_\gamma]=\mathrm{cm}^2\mathrm{s}^{-\gamma}$.

Diffusion on fractals exhibits many anomalous features due to geometrical constraints imposed by the complex structure on a diffusion process. These constraints represent spatial correlations of the structure (self-similarity) which persist on all length scales leading to an anomalous behavior of the mean square displacement of a random walker on all time scales.

Recently in the literature, in order to describe diffusion processes in disordered media, some authors proposed an extension of the Fokker-Planck equation, which is called fractional Fokker-Planck equation [6~12].

A fractional Fokker-Planck equation (FFPE) describing anomalous transport close to thermal equilibrium was presented recently [6]. Since it describes subdiffusion in the force-free case, it involves a strong, i. e., slowly decaying memory. In Ref. [8] Metzler et al. gave a seminal framework and investigated the anomalous diffusion and relaxation involving external fields with the one-dimensional fractional Fokker-Planck equation

$$\dot{W}(x,t)={}_0D_t^{1-\gamma}L_{\mathrm{FP}}W \tag{1.2}$$

in respect to its physical properties, where $W(x,t)$ is the probability density function at position x at time t and the FP operator

$$L_{\mathrm{FP}}=\frac{\partial}{\partial x}\left(\frac{V'(x)}{m\eta_\gamma}+K_\gamma\frac{\partial}{\partial x}\right), \tag{1.3}$$

with the external potential $V(x)$ [13], contains the anomalous diffusion constant K_γ and the anomalous friction coefficient η_γ with the dimension $[\eta_\gamma]=s^{\gamma-2}$, herein m denotes the mass of the diffusion particle, and

$$_0D_t^{1-\gamma}W=\frac{1}{\Gamma(\gamma)}\frac{\partial}{\partial t}\int_0^t d\tau\frac{W(x,\tau)}{(t-\tau)^{1-\gamma}},\quad 0<\gamma<1. \tag{1.4}$$

The right-hand side of the FFPE (1.2) is equivalent to the fractional expression $-{}_0D_t^{1-\gamma}\partial S(x,t)/\partial x$, where

$$S(x,t)=-\left(\frac{V'(x)}{m\eta_\gamma}+K_\gamma\frac{\partial}{\partial x}\right) \tag{1.5}$$

is the probability current.

It has been shown that as in the standard FPE, the FFPE (1.2) in the force-free case possesses a scaling variable, i. e., $W_{F=0}(x,t)=(K_\gamma t^\gamma)^{-1/2}f(z)$ with the dimensionless similarity variable $z=x/(K_\gamma t^\gamma)^{1/2}$ and that the asymptotic shape of $W_{F=0}$ is stretched Gaussian [14]. But for arbitrary external potentials $V(x)$, such simple scaling behavior is not found yet.

In Ref. [24], the fractional Fokker-Planck equation

$$_0D_t^\alpha W(x,t)=Gx^{-\theta'}L_{\mathrm{FP}}W(x,t) \tag{1.6}$$

has been presented, where $G>0$ is to be determined, $\theta'\geqslant 0$ is a parameter, $0<\alpha<1$, if $\theta'=0$ then $\alpha=\gamma$ and Eq. (1.6) reduces to

$$_0D_t^\gamma W(x,t)=GL_{\mathrm{FP}}W(x,t), \tag{1.7}$$

which leads to the FFPE (1.2). It is proved that for the FFPE (1.6), its solution has the asymptotic behavior

$$\ln W(x,t)\sim -c\xi^u, \tag{1.8}$$

where

$$\xi\equiv x/t^{\gamma/2}\gg 1,\quad u=1/\left(1-\frac{\gamma}{2}\right), \tag{1.9}$$

and possesses a scaling variable for constant potentials, linear potentials, logarithm potentials and harmonic potentials. We naturally ask the following problem:

Can we establish a reasonable model for describing anomalous transport processes in an external force fields on random fractal structure for which the stretched Gaussian asymptotic behavior and scaling variable are expected to be universal?

This question is interesting and important.

In this Letter, we generalize a fractional diffusion equation of Giona and Roman [15] to the case in external force fields and then discuss the diffusion in external force fields. Unlike fractional Fokker-Planck equation, such a fractional diffusion equation involves only a first order differential with respect to the space variable x. We prove that the solutions of the fractional diffusion equations in external force fields possess a scaling variable for arbitrary external potentials $V(x)$ and obtain that their asymptotic solutions are stretched Gaussian. In addition, it is proved that the necessary and sufficient condition of its stationary solution being the Boltzmann distribution is that the Stokes-Einstein-Smoluchowski relation holds.

The Letter is organized as follows. In Section 2 we derive fractional diffusion equations involving external potentials. In Section 3 we respectively give the solutions of fractional diffusion equations in the cases of constant potentials, logarithm potentials, harmonic potentials, analytic potentials and generic potentials.

2. Fractional Diffusion Equations Involving External Potentials

In this section, we will derive the fractional diffusion equation involving the external potential $V(x)$ by using a heuristic argument of Giona and Roman.

For the anomalous diffusion, we assume that diffusion sets are underlying fractals (underlying fractals denote self-similar sets in [19] or net fractals in [20, 21]). The relationship between the total probability current $S(x, t)$ up to time t and the average probability density $W(x, t)$ should be (cf. [15, 16, 18])

$$\int_0^t S(x,\tau)d\tau = x^{d_f-1}\int_0^t K(t-\tau)W(x,\tau)d\tau, \tag{2.1}$$

where d_f is the fractal dimension of the system considered, the diffusion kernel on the underlying fractal should behave as

$$K(t-\tau) = \frac{A_\alpha}{(t-\tau)^\alpha} \tag{2.2}$$

with $0<\alpha<1$, where α is a diffusion exponent and A_α is a constant that can be determined [21].

On the other hand, stimulated by the work of Metzler et al. on the probability current, the probability current on the underlying fractal structure is defined by

$$S(x,t) = -Bx^{d_f-1}x^{-\theta'}\left(\frac{V'(x)}{m\eta_\gamma} + K_\gamma\frac{\partial}{\partial x}\right)W(x,t), \tag{2.3}$$

where B is a positive constant to be determined, the parameter θ' is necessarily positive, $\theta'\geqslant 0$. If $B = d_f = 1$ and $\theta' = 0$, it reduces to (1.5).

Eq. (2.3) together with (2.1) and (2.2) yield a fractional diffusion equation on the fractal involving external potentials,

$${}_0D_t^\alpha W(x,t) = -Gx^{-\theta'}\left(\frac{V'(x)}{m\eta_\gamma} + K_\gamma\frac{\partial}{\partial x}\right)W(x,t), \tag{2.4}$$

where $G = \dfrac{B}{A_\alpha\Gamma(1-\alpha)} > 0$.

From (2.4) and (1.1), the simple scaling considerations give that

$$\alpha = \frac{\gamma}{2}(1+\theta'). \tag{2.5}$$

Furthermore, we assume that the probability density $W(x,t)$ satisfies the following normalization condition

$$\int_0^\infty dx\,x^{d_f-1}W(x,t) = 1, \tag{2.6}$$

and the extraction of moments $\langle(\Delta x)^n\rangle$ is defined by

$$\langle(\Delta x)^n\rangle = \int_0^\infty \mathrm{d}x x^{d_f - 1} x^n W(x, t). \qquad (2.7)$$

If $d_f = 1$, they reduce to the expressions of ordinary normalization condition and moments, respectively.

Before we discuss the solution of Eq. (2.4), let us consider its stationary solution.

It is readily seen that if a stationary state is reached, $S(x, t)$ in (2.3) must be a constant. Thus, if $S(x, t) = 0$ for any x, it vanished for all x [13], and the stationary solution is given by $V'(x)W_{st}/(m\eta_\gamma) + K_\gamma W'_{st} = 0$, i. e.,

$$W_{st}(x) = \exp[(V(0) - V(x))/(m\eta_\gamma K_\gamma)].$$

Comparing this expression with the required Boltzmann distribution $W_{st} \propto \exp\{-V(x)/(k_B T)\}$, we find a generalization of the Einstein relation, also referred to as Stokes-Einstein-Smoluchowski relation,

$$K_\gamma = \frac{k_B T}{m\eta_\gamma}, \qquad (2.8)$$

for the anomalous coefficients K_γ and η_γ. Thus, the process described by Eq. (2.4) fulfills the linear relation between anomalous friction and diffusion coefficients, reflecting the fluctuation-dissipation theorem. Contrarily, if (2.8) holds, then its stationary solution must be the Boltzmann distribution.

3. Solutions of Fractional Diffusion Equations

In this section we respectively give the solutions of the fractional diffusion equation (2.4) in the cases of constant potentials, logarithm potentials, harmonic potentials, analytic potentials and generic potentials.

It can be shown that the Laplace transform of the fractional derivative (1.4):

$$L[{}_0D_t^\alpha W(x, t)] = s^\alpha W(x, s).$$

Using this result, from (2.4) we get

$$s^{\alpha}W(x, s)=-Gx^{-\theta'}\left[\frac{V'(x)}{m\eta_{\gamma}}W(x, s)+K_{\gamma}\frac{\partial W(x, s)}{\partial x}\right], \tag{3.1}$$

where $W(x, s)$ denotes the Laplace transform of $W(x, t)$.

3.1 Case I: constant potentials

If the external field is force-free, i. e., $V(x) = \text{const}$, then (3.1) degenerates to

$$s^{\alpha}W(x, s)=-Gx^{-\theta'}K_{\gamma}\frac{\partial W(x, s)}{\partial x}, \tag{3.2}$$

which has the solution

$$W(x, s)=q_0(s)\exp\left\{-\frac{s^{\alpha}x^{\theta'+1}}{(\theta'+1)GK_{\gamma}}\right\}, \tag{3.3}$$

where $q_0(s)=G_1 s^{\xi}$ ensures the normalization condition. Herein

$$G_1=\frac{(\theta'+1)^{1-d_f/(\theta'+1)}}{\Gamma(d_f/(\theta'+1))(GK_{\gamma})^{d_f/(\theta'+1)}}, \tag{3.4}$$

and $\xi=\alpha d_f/(\theta'+1)-1$. It follows from (1.1) and (2.7) (when $n = 2$) that

$$\alpha=\frac{\gamma}{2}(1+\theta'), \quad B=g(d_f), \tag{3.5}$$

where

$$g(d_f)=A_{\alpha}\Gamma(1-\alpha)\frac{\gamma}{\alpha}(2K_{\gamma})^{\alpha/\gamma-1}\left[\frac{\Gamma(\gamma(d_f/2\alpha))}{\Gamma(\gamma(d_f+2)/2\alpha)}\right]^{\alpha/\gamma}. \tag{3.6}$$

Noting $0<\alpha<1$ and (3.7) we have $\alpha\geqslant\gamma/2$ and $\theta'<2/\gamma-1$.

Let us discuss the asymptotic behavior of $W(x, t)$ as predicted by (3.3). We expect that

$$W(x, t)\backsimeq t^{-\gamma d_f/2}\left(\frac{x}{X}\right)^{\delta'}\exp\left\{-b\left(\frac{x}{X}\right)^{u'}\right\}, \tag{3.7}$$

where $x/X\gg1$ and $t\to\infty$. The Laplace transform of (3.7) can be evaluated by applying the method of steepest descent and the result compared with (3.3). This yields (cf. Ref. [15])

$$u' = 2\alpha/\gamma(1-\alpha), \quad \delta' = \frac{u'}{2}(d_s - 1), \tag{3.8}$$

where $d_s = \gamma d_f$ is called the fraction or spectral dimension [22, 23]. Thus the asymptotic shape of $W(x, t)$ is stretched Gaussian. This result coincides with that in the force-free case [14].

Now we give the solution of the fractional diffusion equation (2.5) in Case I. Using the inversion theorem of Laplace transform, from (3.3) we have (see Appendix A)

$$W(x, t) = (K_\gamma t^{\gamma d_f})^{-1/2} f(z), \tag{3.9}$$

where

$$f(z) = \frac{G_1 K_\gamma^{1/2}}{\pi} \sum_{n=0}^{\infty} C_n(\xi) z^{n(2\alpha/\gamma)}, \quad z = x/t^{\gamma/2}, \quad \xi = \frac{1}{2}\gamma d_f - 1.$$

Putting $d_f = 1$ this result coincides with that in Ref. [14].

3.2 Case II: logarithm potentials

If the external field is a logarithm potential, i. e., $V(x) = b\ln x$, then (3.1) becomes

$$S^\alpha W(x, s) = -G' x^{-\theta'} \left[\frac{\partial W(x, s)}{\partial x} + \frac{\kappa}{x} W(x, s)\right], \tag{3.10}$$

where $G' = GK_\gamma > 0$, $\kappa = b/(m\eta_\gamma K_\gamma)$. Eq. (3.10) is the fractional diffusion equation on fractals in [15].

A simple integration of (3.10) yields

$$W(x, s) = \frac{q(s)}{x^k} \exp\left\{-\frac{s^\alpha x^{\theta'+1}}{(\theta'+1)G'}\right\}, \tag{3.11}$$

where

$$q(s) = \frac{\theta'+1}{\Gamma((d_f-\kappa)/(\theta'+1))} \frac{1}{s} \left[\frac{s^\alpha}{(\theta'+1)G'}\right]^{(d_f-\kappa)/(\theta'+1)}, \tag{3.12}$$

if $d_f - \kappa > 0$, ensures the normalization condition of $W(x, t)$. Thus, it follows from (3.11) that

$$W(x,s)=G_1 s^{\xi} x^{-\kappa}\exp\{-G_2 s^{\alpha}\}, \tag{3.13}$$

where

$$\xi=\alpha(d_f-\kappa)/(\theta'+1)-1 \tag{3.14}$$

and

$$G_1=\frac{(\theta'+1)^{1-(d_f-\kappa)/(\theta'+1)}}{\Gamma((d_f-\kappa)/(\theta'+1))(GK_\gamma)^{(d_f-\kappa)/(\theta'+1)}},\quad G_2=\frac{x^{\theta'+1}}{(\theta'+1)GK_\gamma}.$$

It follows from (1.1) and (2.7) (when $n=2$) that

$$\alpha=\frac{\gamma}{2}(1+\theta'),\quad B=g(d_f-\kappa).$$

Using the same methods in Ref. [15], from (3.18) we have

$$W(x,t)\backsimeq t^{-\gamma d_f/2}\left(\frac{x}{X}\right)^{\delta'}\exp\left\{-\text{const}\cdot\left(\frac{x}{X}\right)^{u'}\right\} \tag{3.15}$$

when $x/X\gg 1$ and $t\to\infty$, where

$$u'=2\alpha/\gamma(1-\alpha),\quad \delta'=u'\left(\frac{\gamma d_f-1}{2}-\kappa\right). \tag{3.16}$$

As in Case I, we can show that

$$W(x,t)=(K_\gamma t^{\gamma d_f})^{-1/2}\cdot z^{-\kappa}f(z), \tag{3.17}$$

where

$$f(z)=\frac{G_1K_\gamma^{1/2}}{\pi}\sum_{n=0}^{\infty}C_n(\xi)z^{n(2\alpha/\gamma)},$$

where $z=x/t^{\gamma/2}$ is a dimensionless similarity variable, $\{C_n(\xi)\}$ are the same as in (3.15) but herein $\xi=\frac{\gamma}{2}(d_f-\kappa)-1$.

3.3 Case III: harmonic potentials

In this paragraph we deal with the special and physically important case of a harmonic potential field, i.e., $V(x)=\frac{\lambda}{2}x^2$, leading to the Hookean force

$F(x) = -\lambda x$ directed at the original. Similar consideration may be found in Refs. [7,12]. In this case, Eq. (3.1) reduces to

$$s^{\alpha} W(x, s) = -Gx^{-\theta'} \left[\frac{\lambda x}{m\eta_{\gamma}} + K_{\gamma} \frac{\partial W(x, s)}{\partial x} \right]. \tag{3.18}$$

A simple integration of (3.18) yields

$$W(x, s) = q(s) \exp\left\{ -\frac{\kappa\lambda}{2} x^2 - [x s^{\alpha/(\theta'+1)} / G_3]^{(\theta'+1)} \right\} \tag{3.19}$$

where

$$q(s) = \frac{c_0 s^{\alpha z_0 - 1}}{1 + \sum_{n=1}^{\infty} r_{2n} s^{-2n\alpha/(\theta'+1)}},$$

ensures the normalization condition of $W(x, t)$. Herein,

$$\kappa = \frac{1}{m\eta_{\gamma} K_{\gamma}}, \quad G_3 = G_{\gamma}^{1/(\theta'+1)}, \quad G_{\gamma} = (\theta' + 1) G K_{\gamma}, \quad c_0 = \frac{\theta' + 1}{\Gamma(z_0) G_{\gamma}^{z_0}},$$

$$r_{2n} = \frac{c_{2n} \Gamma(z_{2n}) G_{\gamma}^{2n}}{\Gamma(z_0)}, \quad z_n = \frac{d_f + n}{\theta' + 1}, \quad c_n = \frac{1}{n!} \left(-\frac{\lambda\kappa}{2} \right)^n, \ n = 1, 2, \cdots. \tag{3.20}$$

It follows from (1.1) and (2.7) (when $n = 2$) that

$$\alpha = \frac{\gamma}{2} (1 + \theta'), \quad B = g(d_f). \tag{3.21}$$

Let us discuss the asymptotic behavior of $W(x, t)$ predicted by (3.25). Noting (3.27), (3.25) can be written as follows

$$W(x, s) = \phi_0(x) s^{-(1-\gamma d_f/2)} \exp\{ -(x s^{\gamma/2} / G_3)^{(\theta'+1)} \} \cdot \left(1 + \sum_{n=1}^{\infty} d_{2n} s^{-n\gamma}\right), \tag{3.22}$$

where

$$\phi_0(x) = c_0 \exp\left\{ -\frac{\kappa\lambda}{2} x^2 \right\}, \quad d_2 = -r_2,$$

$$d_{2n} = -(d_{2(n-1)} r_2 + d_{2(n-2)} r_4 + \cdots + d_2 r_{2(n-1)} + r_{2n}). \tag{3.23}$$

Thus

$$W(x, s) \approx \phi_0(x) s^{-(1-\gamma d_f/2)} \exp\{-(x s^{\gamma/2}/G_3)^{\theta'+1}\} \tag{3.24}$$

when $|s| \gg 1$. Using the same methods as in Refs. [15, 17], we have

$$W(x, t) \sim t^{-\gamma d_f/2} \phi_0(x) \left(\frac{x}{X}\right)^{\delta'} \exp\left\{-\text{const} \cdot \left(\frac{x}{X}\right)^{u'}\right\} \tag{3.25}$$

when $x/X \gg 1$ and $t \to \infty$, where δ' and u' are the same as in (3.8).

To determine the solution of (2.4) associated to Case III, taking the Laplacian inversion transform on both sides of (3.22) we have

$$W(x, t) = \phi_0(x) \sum_{n=0}^{\infty} d_{2n} f_{2n}(x, t), \tag{3.26}$$

where

$$d_0 = 1, \quad f_{2n}(x, t) = L^{-1}[F_{2n}(x, s)],$$
$$F_{2n}(x, s) = s^{\xi_{2n}} \exp\{-(x s^{\gamma/2}/G_\gamma)^{\theta'+1}\}, \ \xi_n = \frac{\gamma}{2}(d_f - n) - 1. \tag{3.27}$$

Since $\xi_{2n+1} + 1 < 0$ for $n \geqslant 1$, we can show that $f_{2n}(x, t) = 0$. Since $\xi_0 + 1 > 0$, then as in Case I we have

$$W(x, t) = (K_\gamma t^{\gamma d_f})^{-1/2} \phi_0(x) f(z), \tag{3.28}$$

where $z = x/t^{\gamma/2}$, $f(z) = \frac{G_1 K_\gamma^{1/2}}{\pi} \sum_{n=0}^{\infty} C_n(\xi_0) z^{n(2\alpha/\gamma)}$ and $C_n(\xi_0)$ is defined as in Appendix A.

3.4 Case IV: generic potentials

In this paragraph we discuss the generic external potential

$$V(x) = b \ln x + \sum_{k=1}^{\infty} a_k x^k. \tag{3.29}$$

In this case, the solution of the fractional diffusion equation (3.1) is as follows

$$W(x, s) = q(s) \exp\{-\kappa V(x)\} \cdot \exp\{-s^{\alpha} x^{\theta'+1}/G_\gamma\}, \tag{3.30}$$

where $\kappa = 1/(m\eta_\gamma K_\gamma)$, $G_\gamma = (\theta'+1) G K_\gamma$, and

$$q(s)=c_0 s^{\alpha z_0-1}/\left[1+\sum_{n=1}^{\infty}\eta_n s^{-\alpha n/(\theta'+1)}\right] \tag{3.31}$$

if $d_{\mathrm{f}}>b\kappa$, ensures the normalization condition of $W(x, t)$. Herein,

$$c_0=(\theta'+1)/\Gamma(z_0)G_\gamma^{z_0}, \quad \eta_n=\beta_n\Gamma(\xi_n)G_3^n, \quad G_3=G_\gamma^{1/(\theta'+1)},$$

$$\xi_n=(d_{\mathrm{f}}-b\kappa+n)/(\theta'+1), \quad n\geqslant 0,$$

$$\beta_0=1, \quad \beta_n=-\frac{\kappa}{n}\sum_{k=1}^{n}ka_k\beta_{n-k}, \quad n\geqslant 1.$$

It follows from (1.1) and (2.7) (when $n=2$) that

$$\alpha=\frac{\gamma}{2}(1+\theta'), \quad B=g(d_{\mathrm{f}}-b\kappa).$$

Thus, it follows from (3.30) that

$$W(x, s)\approx\phi_1(x)s^{-(1-\frac{1}{2}d_{\mathrm{f}})}\frac{\exp\{-(xs^{\gamma/2}/G_3)^{(\theta'+1)}\}}{(xs^{\gamma/2})^{b\kappa}}, \tag{3.32}$$

when $|s|\gg 1$, where

$$\phi_1(x)=c_0\exp\left\{-\kappa\sum_{n=1}^{\infty}a_n x^n\right\}, \quad G_3=G_\gamma^{1/(\theta'+1)}. \tag{3.33}$$

Using the same methods in Refs. [15, 17], from (3.32) we have

$$W(x, t)\sim t^{-\gamma d_{\mathrm{f}}/2}\phi_1(x)\left(\frac{x}{X}\right)^{\delta'}\exp\left\{-\mathrm{const}\cdot\left(\frac{x}{X}\right)^{u'}\right\} \tag{3.34}$$

when $x/X\gg 1$ and $t\to\infty$, where δ' and u' are the same as in (3.16).

We now turn to discuss the expression of dimensionless similarity variable of $W(x, t)$. It follows from (3.30) that

$$W(x, s)=c_0\exp\{-\kappa V(x)\}\sum_{n=0}^{\infty}d_n L^{-1}[F_n(s)], \tag{3.35}$$

where $d_0=1$, $d_1=\kappa a_1 G_3\Gamma(z_1)/\Gamma(z_0)$,

$$d_n=-(d_{n-1}\eta_1+d_{n-2}\eta_2+\cdots+d_1\eta_{n-1}+\eta_n) \quad (n=2, 3, \cdots), \tag{3.36}$$

$$F_n(s)=s^{\xi_n}\exp\{-G_2 s^\alpha\}, \ \xi_n=\frac{\gamma}{2}(d_{\mathrm{f}}-b\kappa-n)-1, \ n=0, 1, \cdots. \tag{3.37}$$

If $b \neq 0$ and $0 < d_f - b\kappa < 1$. Similar to Case III, since $\xi_0 + 1 > 0$ but $\xi_n + 1 < 0$ for $n \geqslant 1$, we have

$$W(x, t) = \phi_1(x)(K_\gamma t^{\gamma d_f})^{-1/2} z^{-b\kappa} f(z), \tag{3.38}$$

where $z = x/t^{\gamma/2}$, and $f(z) = \frac{G_1 K_\gamma^{1/2}}{\pi} \sum_{n=0}^{\infty} C_n(\xi_0) z^{n(2\alpha/\gamma)}$.

If $b \neq 0$ and $d_f - b\kappa > 1$, then $\xi_i + 1 > 0$ $(i = 0, 1)$, but $\xi_n + 1 < 0$ for $n \geqslant 2$. Thus

$$W(x, t) = \phi_1(x)(K_\gamma t^{\gamma d_f})^{-1/2} z^{-b\kappa} [f(z) + d_1 t^{\gamma/2} f_1(z)], \tag{3.39}$$

where $f_1(z) = \frac{G_1 K_\gamma^{1/2}}{\pi} \sum_{n=0}^{\infty} C_n(\xi_1) z^{n(2\alpha/\gamma)}$.

Furthermore, if $a_1 = 0$ we still have the same results as in case of $0 < d_f - b\kappa < 1$.

On the other hand, if $b = 0$, i.e., $V(x)$ is analytic with form

$$V(x) = \sum_{n=1}^{\infty} a_n x^n. \tag{3.40}$$

Similar consideration can be found in Ref. [7]. In this case, we have

$$W(x, t) \sim t^{-\gamma d_f/2} \phi_1(x) \left(\frac{x}{X}\right)^{\delta'} \exp\left\{- \text{const} \cdot \left(\frac{x}{X}\right)^{u'}\right\} \tag{3.41}$$

when $x/X \gg 1$ and $t \to \infty$, and

$$W(x, t) = \phi_1(x)(K_\gamma t^{\gamma d_f})^{-1/2} [f(z) + t^{\gamma/2} d_1 f_1(z)], \tag{3.42}$$

where δ' and u' are the same as in (3.8).

In particular, if $a_1 = 0$, then

$$W(x, t) = c_0 \exp\{-\kappa V(x)\} (K_\gamma t^{\gamma d_f})^{-1/2} f(z). \tag{3.43}$$

4. Conclusion and Discussion

In order to describe anomalous diffusion processes involving external potential fields on fractal structures, we introduce a fractional diffusion equation involving external potentials. This equation is different from the fractional Fokker-Planck equation. The solution of this fractional diffusion equation possesses following properties:

The necessary and sufficient condition of its stationary solution being the Boltzmann distribution $\exp\{-V(x)/(k_B T)\}$ is that the generalization of the Einstein relation, also preferred to as Stokes-Einstein-Smoluchowski relation holds

$$K_\gamma = k_B T / m\eta_\gamma,$$

for the anomalous coefficients K_γ and η_γ. This result is consistent with that in Ref. [8].

The asymptotic shape of solution of Eq. (2.5) is stretched Gaussian not only for constant potentials but also for logarithm, harmonic, analytic and generic potentials.

Solutions of (2.5) can be expressed in the form of

$$(K_\gamma t^{\gamma d_f})^{-1/2} z^{-\kappa} f(z) \quad \text{or} \quad (K_\gamma t^{\gamma d_f})^{-1/2} z^{-\kappa} \phi(x)[f(z) + t^{\gamma/2} d_1 f_1(z)]$$

with the dimensionless similarity variable $z = x/t^{\gamma/2}$ for any potentials mentioned above, where κ and d_1 are constants but may be zero, $f(z)$ and $f_1(z)$ are functions of z, and d_f is the fractal dimension of the underlying structure. It is surprising that the presence of $f_1(z)$ is due to the presence of term $a_1 x$ in the external potentials.

In each case (constant, logarithm, harmonic, analytic and generic potentials) there exists an intrinsic relationship, $2\alpha = \gamma(1+\theta')$. We know that α and γ are structural parameter of the underlying fractal structure and α can be determined explicitly (see Refs. [19~21]).

It is interesting that as same as the fractional Fokker-Planck equation (1.6), the solution of fractional diffusion equation (2.5) also has the asymptotic behavior $\ln W(x, t) \sim -c\xi^{u'}$ where $\xi \equiv x/t^{\gamma/2} \gg 1$, $u' = 2\alpha/\gamma(1-\alpha)$, only $u' > u = 1/(1-\gamma/2)$ of FFPE for $1/2 < \rho = \alpha/\gamma < 1/\gamma$ which means that the asymptotic behavior decays quickly than that ones of FFPE, and $u' = u$ if and only if $\rho = 1/2$ which means that both of FFPE and Eq. (2.4) have same decay.

Appendix A

Using the inversion theorem of Laplace transform, from (3.3) we have

$$W(x,t)=\frac{G_0}{\pi}\int_0^{\infty}\tau^{\xi}\exp\{-[t\tau+G_2\tau^{\alpha}\cos(\alpha\pi)]\}\sin[G_2\tau^{\alpha}\sin(\alpha\pi)-\xi\pi]\mathrm{d}\tau$$

$$=\frac{G_0}{\pi}t^{-\xi-1}\int_0^{\infty}y^{\xi}\exp\{-[y+G_2(y/t)^{\alpha}\cos(\alpha\pi)]\}\sin[G_2(y/t)^{\alpha}\sin(\alpha\pi)-\xi\pi]\mathrm{d}y, \tag{A.1}$$

where

$$G_2=\frac{x^{\theta'+1}}{(\theta'+1)GK_{\gamma}}.$$

Using the expressions of power series of entire-functions e^z, $\cos z$ and $\sin z$ and the integral representation of Γ function, a simple integration of (A.1) yields the following expression with the dimensionless similarity variable $z=x/t^{\gamma/2}$,

$$W(x,t)=(K_{\gamma}t^{\gamma d_f})^{-1/2}f(z), \tag{A.2}$$

where

$$f(z)=\frac{G_1K_{\gamma}^{1/2}}{\pi}\sum_{n=0}^{\infty}C_n(\xi)z^{n(2\alpha/\gamma)},$$

$$C_0=a'\Gamma(\gamma d_f/2)>0,\quad C_n(\xi)=a'b_n+b'a_{n-1},$$

$$a_n(\xi)=\sum_{k,\,m\geqslant 0,\,2k+m=n}(-1)^{k+m}\frac{a^{2k+1}b^m}{(2k+1)!\,m!}C^{n+1}\Gamma(\alpha(\xi+2+n)),$$

$$b_n(\xi)=\sum_{k,\,m\geqslant 0,\,2k+m=n}(-1)^{k+m}\frac{a^{2k}b^m}{(2k)!\,m!}C^{n}\Gamma(\alpha(\xi+1+n)),$$

$$a'=-\sin(\xi\pi)>0,\quad b'=\cos(\xi\pi),\quad a=\sin(\alpha\pi),\quad b=\cos(\alpha\pi),$$

$$C=\frac{\gamma}{2\alpha GK_{\gamma}},\quad G_1=\frac{(2\alpha/\gamma)^{1-\gamma d_f/2\alpha}}{(GK_{\gamma})^{\gamma d_f/2\alpha}\Gamma(\gamma d_f/2\alpha)}. \tag{A.3}$$

◇ References ◇

[1] S. Havlin, D. B. Avraham, Adv. Phys. 36 (1987) 695.

[2] M. B. Isichenko, Rev. Mod. Phys. 64 (1992) 961.

[3] J. P. Bouchaud, A. Georges, Phys. Rep. 195 (1990) 127.

[4] A. Blumen, J. Klafter, G. Zumofen, in: I. Zschokke (Ed.), Optical Spectres Copy of Glasses, Reidel, Dordrecht, 1986.

[5] G. A. Losa, E. R. Weibl, Fractals in Biology and Medicine, Birkhäuser, Basel, 1993.

[6] R. Metzler, J. Klafter, Phys. Rep. 339 (2000) 1.

[7] R. Metzler, J. Klafter, L. Sokolov, Phys. Rev. E 58 (1998) 1621.

[8] R. Metzler, E. Barkai, J. Klafter, Phys. Rev. Lett. 82 (1999) 3563.

[9] S. A. El-Wakil, A. Elhanbaly, M. A. Zahran, Chaos Solitons Fractals 12 (2001) 1035.

[10] G. Jumarie, Chaos Solitons Fractals 12 (2001) 1873.

[11] S. A. El-Wakil, M. A. Zahran, Chaos Solitons Fractals 12 (2001) 1929.

[12] S. Jesperson, R. Metzler, H. C. Fogedby, Phys. Rev. E 59 (3) (1999) 2736.

[13] H. Risken, The Fokker-Planck Equation, Springer-Verlag, Berlin, 1989.

[14] W. R. Schneider, W. Wyss, J. Math. Phys. 30 (1989) 134.

[15] M. Giona, H. E. Roman, Physica A 185 (1992) 87.

[16] M. Giona, H. E. Roman, J. Phys. A: Math. Gen. 25 (1992) 2093.

[17] H. E. Roman, M. Giona, J. Phys. A: Math. Gen. 25 (1992) 2107.

[18] A. Le Mehaute, J. Stat. Phys. 36 (5-6) (1984) 665.

[19] F.-Y. Ren, Z.-G. Yu, F. Su, Phys. Lett. A 219 (1996) 59.

[20] W.-Y. Qiu, J. Lü, Phys. Lett. A 272 (2000) 353.

[21] F.-Y. Ren, X.-T. Wang, J.-R. Liang, J. Phys A: Math. Gen. 34 (2001) 9815.

[22] S. Alexander, R. Orbach, J. Phys. (Paris) Lett. 43 (1982) 2625.

[23] R. Rammel, G. Toulouse, J. Phys. (Paris) Lett. 44 (1983) 12.

[24] F.-Y. Ren, W.-Y. Qiu, J.-R. Liang, Y. Xu, preprint.

Integrals and Derivatives on Net Fractals*

Fu-Yao Ren[a], Jin-Rong Liang[b],
Xiao-Tian Wang[a], Wei-Yuan Qiu[a]
[a] *Institute of Mathematics, Fudan University, Shanghai 200433, China*
[b] *Department of Mathematics, East China Normal University, Shanghai 200062, China*

Abstract: In this paper a framework of calculus on net fractals is built. Integrals and derivatives of functions in net measure are discussed. Approximate calculations of the integrals and derivatives and approximate solutions to the inverse problem of integrals in net measure are given. In addition, applications of the calculus in some physical systems, such as in diffusion processes and in memory processes are given.

Keywords: net fractals, net measures, integrals and derivatives

1. Introduction

In many problems of theoretical physics it is necessary to take into account the non-local effects caused by heterogeneity of a medium. In connection with penetration of fractal geometry ideas into modern physics it is important to calculate two types of heterogeneities: the heterogeneity caused by a fractal structure and the heterogeneity caused by physical value itself distributed over the fractal structure. Usually the relationship between two physical values $\Phi(x)$ and $f(\tau)$ localized in the points x, τ accordingly is expressed in most

* Originally published in *Chaos, Solitons and Fractals*, Vol. 16, (2003), 107 – 117.

cases by the formula:

$$\Phi(x)=\int_E K(x-\tau)f(\tau)\mathrm{d}\tau. \tag{1.1}$$

Integration in this expression is realized over all values of variable $\tau\in E$. The procedure of calculation (1.1) is rather complicate because the concrete form of the kernel $K(x-\tau)$ in most cases is not known. Therefore, it is important to formulate the following questions which have important value for developing the mathematical physics of a fractal medium:

How to find the function $K(x-\tau)$ for self-similar fractal structure with dimension d_f located in the region $\lambda<\eta<\Lambda$? [1].

How to find the averaged value of the function $f(x)$ distributed over a fractal structure of an arbitrary nature? [1].

These questions are, in turn, closely related with integrals and derivatives associated to net measures discussed in this paper.

The main aim of this paper is to give a framework of calculus on net fractals in one dimensional Euclidean space. In Section 2 we give the definitions of net fractals and net measures supported on net fractals. In Section 3 we discuss integrals and derivatives of functions in net measure on net fractals and give their approximate calculations. In Section 4 we discuss the inverse problem of integrals and give approximate solutions of integral equations in net measure. In Section 5 we give applications of the calculus in some physical systems, such as in diffusion processes and in memory processes.

2. Net Fractals and Net Measures

In this section we give the definitions of net fractals and net measures on net fractals and describe some known results.

Definition 2.1 For any given $T\in(0,\infty)$ and $k\in\{2,3,\cdots\}$, let $E_0=[0,T]$ and let $\{\varphi_j(x)\}_{j=1}^k$ be a family of contractive transformations on E_0, i.e.

$$|\varphi_j(x)-\varphi_j(y)|\leqslant c_j|x-y|,\quad \forall x,y\in E_0,\ 0<c_j<1. \tag{2.1}$$

Suppose $\varphi_j(E_0) \cap \varphi_i(E_0) = \varnothing$, $i \neq j$. Then the non-empty compact set

$$E = \bigcap_{n=1}^{\infty} E(n) \tag{2.2}$$

is a net fractal, where

$$E(n) = \bigcup_{j_1, \cdots, j_n} E_{j_1 \cdots j_n}, \quad E_{j_1 \cdots j_n} = \varphi_{j_n} \circ \cdots \circ \varphi_{j_1}(E_0). \tag{2.3}$$

Remark 2.2

(1) A self-similar set is a net fractal. If $\varphi_j(x)$, $j = 1, \cdots, k$ in Definition 2.1 are the self-similar transformations on E_0, i. e.

$$|\varphi_j(x) - \varphi_j(y)| = c_j |x - y|, \quad \forall x, y \in E_0, 0 < c_j < 1,$$

then the non-empty compact set

$$E = \bigcap_{n=1}^{\infty} E(n)$$

is called a *self-similar set*. Moreover, $\dim_{\mathrm{H}} E = s$ and $0 < H^s(E) < \infty$, where s is the unique positive solution of

$$\sum_{i=1}^{k} c_i^s = 1, \tag{2.4}$$

where $\dim_{\mathrm{H}} E$ and $H^s(E)$ denote the Hausdorff dimension and the Hausdorff measure of E respectively (cf. [2]).

(2) A cookie-cutter set is a net fractal. Let $E_1, E_2, \cdots, E_k$ are disjoint compact subsets of E_0. A *cookie-cutter map* is a mapping $S: \bigcup_{j=1}^{k} E_j \to E_0$ with the properties that

(i) $S|_{E_j}$ is 1—1 mapping onto E_0;

(ii) S is $C^{1+\alpha}$ differentiable, i. e. differentiable with a Hölder continuous derivative S' satisfying $|S'(x) - S'(y)| < c|x - y|^{\alpha}$ for some $c > 0$, and $|S'(x)| > 1$ for all $x \in \bigcup_{j=1}^{k} E_j$.

The *cookie-cutter set* associated to S is the set

$$E = \left\{ x \in \bigcup_{j=1}^{k} E_j : S^n(x) \in \bigcup_{j=1}^{k} E_j, \ \forall n \geqslant 0 \right\}.$$

If let $S|_{E_j} = \varphi_j^{-1}$, then $|\varphi_j'(x)| < 1$, $\forall x \in E_0$. For natural number n, let

$$E_{j_1j_2\cdots j_n}=\varphi_{j_n}\circ\cdots\circ\varphi_{j_1}(E_0),\quad E(n)=\bigcup\nolimits_{j_i\in I}E_{j_1j_2\cdots j_n},\quad I=\{1,2,\cdots,k\}.$$

It is obvious that

$$E_T=\bigcap\nolimits_{n\geqslant 1}E(n)=\bigcap\nolimits_{n\geqslant 1}\bigcup\nolimits_{j_i\in I}E_{j_1j_2\cdots j_n}.$$

Moreover, the Hausdorff dimension of cookie-cutter set E_T is the unique $\beta\in\mathbb{R}$ with $P(\beta f)=0$, where $P(f)$ denotes the pressure of f (cf. [3]).

Lemma 2.3[4] *Let E be a net fractal generated by a family of contraction transformations $\{\varphi_j(x)\}_1^k$ on $E_0=[0,T]$. Then for any normalized probability vector $\mathbf{p}=(p_1,\cdots,p_k)$ satisfying $0\leqslant p_i\leqslant 1$ for all i and $\sum_{i=1}^k p_i=1$, there exists a unique Borel probability measure μ (that is with $\mu(E_0)=1$) such that*

$$\mu(A)=\sum_{i=1}^k p_i\mu(\varphi_i^{-1}(A))\tag{2.5}$$

for all Borel sets A, and

$$\int g(x)\mathrm{d}\mu(x)=\sum_{i=1}^k p_i\int g(\varphi_i(x))\mathrm{d}\mu(x)\tag{2.6}$$

for all continuous $g: E_0\to\mathbb{R}$. Moreover, $\mathrm{supp}(\mu)=E$.

We call the measure satisfying (2.5) a *net measure* on E and $\{p_j\}$ is called the weights. If $\{\varphi_j(x)\}_1^k$ is a family of similarity transformations, then μ is called a *self-similar measure* (cf. [4,5]).

Lemma 2.4 *Let E be a net fractal generated by a family of contraction transformations $\{\varphi_j(x)\}_{j=1}^k$ on E_0. For any given normalized probability vector $\mathbf{p}=(p_1,p_2,\cdots,p_k)$, let the measure μ_n on $E(n)$ be a probability measure defined by*

$$\begin{gathered}\mathrm{d}\mu_n(\tau)=m_n(\tau)\mathrm{d}\tau,\\ m_n(\tau)=\sum_{j_1,\cdots,j_n\in I}p_{j_1}\cdots p_{j_n}\frac{\chi_{E_{j_1\cdots j_n}}(\tau)}{|E_{j_1\cdots j_n}|},\quad I=\{1,\cdots,k\},\end{gathered}\tag{2.7}$$

where

$$\chi_{E_{j_1\cdots j_n}}(\tau)=\begin{cases}1, & \tau\in E_{j_1\cdots j_n}\\ 0, & \text{otherwise}\end{cases}$$

and $|E_{j_1\cdots j_n}|$ *denotes the diameter of the set* $E_{j_1\cdots j_n}$ *in* (2.3). *Then for any continuous function* $f(x)$ *on* $\mathbb{R}$, *there exists a measure* μ *satisfying* $\int_E d\mu(\tau)=1$, $\mathrm{supp}(\mu)=E$ *such that*

$$\lim_{n\to\infty}\int_{\mathbb{R}} f(x)d\mu_n(x)=\int_{\mathbb{R}} f(x)d\mu(x)$$

i.e.

$$\mu_n \to \mu \quad (\textit{weakly converge}).$$

Moreover, for any $A\subset E_0$,

$$\int_A d\mu(\tau)=\sum_{i=1}^{k} p_j\int_A d\mu\circ\varphi_j^{-1}(\tau)$$

i.e.

$$\mu(\cdot)=\sum_{j=1}^{k} p_j\mu\circ\varphi_j^{-1}(\cdot). \tag{2.8}$$

Proof Following the same steps in Refs. [6, 7] we immediately obtain these results.

For any continuous complex function $g(t)$, we can write $g(t)=u(t)+iv(t)$, where $u(t)$, $v(t)$ are continuous real functions. It follows from Lemma 2.4 that

$$\lim_{n\to\infty}\int_0^T u(x)d\mu_n=\int_0^T u(x)d\mu,\quad \lim_{n\to\infty}\int_0^T v(x)d\mu_n=\int_0^T v(x)d\mu$$

and thus $\lim_{n\to\infty}\int_0^T g\,d\mu_n=\int_0^T g\,d\mu$. In particular, if $g(t)=e^{-pt}$, $p\in\mathbb{C}$ ($\mathbb{C}$ denotes the set of complex numbers), then

$$\lim_{n\to\infty}\int_0^T e^{-pt}d\mu_n(t)=\int_0^T e^{-pt}d\mu(t). \tag{2.9}$$

3. Integrals and Derivatives on Net Fractals

In this section we always assume that E is a net fractal generated by a family of contraction transformations $\{\varphi_j(x)\}_{j=1}^k$ on $E_0=[0, T]$ and $\boldsymbol{p}=(p_1, p_2, \cdots, p_k)$ is any given normalized probability vector. From Lemma 2.4 there exists the net measure μ satisfying (2.8). We replace this measure μ by μ_p. Let $\mu_{\bar{p}}$ denote the net measure associated to a normalized probability vector $\bar{\boldsymbol{p}}=(\bar{p}_1, \bar{p}_2, \cdots, \bar{p}_k)$, and let $v=\ln p_1/\ln|\xi_1|$ and $\bar{v}=\ln\bar{p}_1/\ln|\xi_1|$, where $\xi_1=\varphi'_1(0)$. If $v+\bar{v}=1$ we say that the measures μ_p and $\mu_{\bar{p}}$ are mutual conjugate.

Let $f(x)$ be a function on E_0. Define the integral $I^{\mu_p}f$ in measure μ_p by

$$I^{\mu_p}f(x)=\int_0^x f(t)\mathrm{d}\mu_p(t), \quad x\in[0, T]. \tag{3.1}$$

In order to define a derivative of $f(x)$ at x we require that the derivative is a linear operator representing the inverse operator of I^{μ_p}. Inspiring by the notion of fractional integrals and fractional derivatives [8] we define the derivative of $f(x)$ at $x\in E$ by

$$D^{\mu_{\bar{p}}}f(x)=\frac{\mathrm{d}}{\mathrm{d}x}\int_0^x f(t)\mathrm{d}\mu_{\bar{p}}(t) \tag{3.2}$$

if it exists, where μ_p and $\mu_{\bar{p}}$ are mutual conjugate.

Theorem 3.1 *Let $f(x)$ be a continuous function on E_0 and $\varphi_j(x)$ be monotonic increasing and $C^{1+\alpha}$ differentiable. Then*

(i) *If $\varphi_1(0)=0$ and $\varphi_k(T)=T$, then*

$$\int_0^x f(t)\mathrm{d}\mu_p(t) \backsimeq \frac{A_v}{\Gamma(v)}\int_0^x \frac{f(t)\mathrm{d}t}{(x-t)^{1-v}}. \tag{3.3}$$

(ii) *If $\varphi_1(0)=b_1\neq 0$ and $\varphi_k(T)=T$, then*

$$\int_0^x f(t)\mathrm{d}\mu_p(t) \backsimeq \frac{A'_v}{\Gamma(v)}\int_0^x \frac{f(t)\eta(x-t-t_1)\mathrm{d}t}{(x-t-t_1)^{1-v}}, \tag{3.4}$$

where $\Gamma(v)$ is the Gamma function, A_v and A'_v are constants, $v=\ln p_1/\ln\xi_1$, $0<v<1$, $t_1=b_1/(1-\xi_1)$, and

$$\eta(x)=\begin{cases}1, & x>0,\\ 0, & x<0.\end{cases}$$

Further,

$$A_v \backsimeq v\Gamma(v)/T^v \quad \text{and} \quad A'_v \backsimeq v\Gamma(v)/(T-t_1)^v. \tag{3.5}$$

Proof Let

$$F_n(x) = \int_0^x m_n(x-t) f(t) \mathrm{d}t. \tag{3.6}$$

Performing Laplace transform on both sides of (3.6), from the product theorem of Laplace transform, we have

$$\mathscr{F}_n(p) = M_n(p) F(p), \quad p \in \mathbb{C}, \tag{3.7}$$

where

$$\mathscr{F}_n(p) = \int_0^\infty \exp(-pt) F_n(t) \mathrm{d}t,$$

$$M_n(p) = \int_0^\infty \exp(-pt) m_n(t) \mathrm{d}t, \quad F(p) = \int_0^\infty \exp(-pt) F(t) \mathrm{d}t.$$

Let

$$M(p) = \int_0^\infty \exp(-pt) \mathrm{d}\mu_p(t).$$

Noting that (2.7) and (2.9) and the supports of μ_n and μ_p are contained in E_0 we have that $M_n(p) = \int_0^T \exp(-pt) \mathrm{d}\mu_n(t)$ and so $\lim_{n\to\infty} M_n(p) = M(p)$. Therefore, it follows from (3.7) that

$$\mathscr{F}(p) = M(p) F(p),$$

where $\mathscr{F}(p) = \lim_{n\to\infty} \mathscr{F}_n(p)$. Acting the Laplace transform on both sides of (2.8) we have

$$M(p) = \sum_{j=1}^k p_j \int_0^T \mathrm{e}^{-p\varphi_j(\tau)} \mathrm{d}\mu_p(\tau). \tag{3.8}$$

If $\varphi_j(x) \in C^{1+\alpha}$, let $\tilde{\varphi}_j(x) = \varphi_j(x) - b_j$, then $\tilde{\varphi}_j(0) = 0$, $\tilde{\varphi}_j(x) \in C^{1+\alpha}$ and so

$$\tilde{\varphi}_j(x) = \xi_j x + O(x^{1+\alpha}) \quad \text{as } x \to 0,$$

where $\xi_j = \varphi_j'(0)$ and $b_j = \varphi_j(0)$. Thus

$$\int_0^T e^{-p\varphi_j(\tau)} d\mu_p(\tau) = e^{-b_j p} \int_0^T \exp(-p\tilde{\varphi}_j(\tau)) d\mu_p(\tau) = e^{-b_j p} M(\xi_j p) + o(1) \tag{3.9}$$

as Re $p \to +\infty$ (cf. Refs. [9~13]). It follows from (3.8) and (3.9) that

$$M(p) = \sum_{j=1}^{k} p_j e^{-b_j p} M(\xi_j p) + o(1) \tag{3.10}$$

as Re $p \to +\infty$. If $b_1 = 0$, it follows from (3.10) that

$$M(p) = p_1 M(\xi_1 p) + o(1) \quad (\text{as Re } p \to +\infty).$$

Note that $M(0) = \int_0^T d\mu_p(\tau) = 1$ and the solution of the function equation $\overline{M}(p) = p_1 \overline{M}(\xi_1 p)$ is $\overline{M}(p) = Ap^{-v}$, where A is a constant and $v = \ln p_1 / \ln \xi_1$. Since $\overline{M}(0) \neq 1$, thus we choose $\tilde{M}(p)$ as an approximation of $M(p)$:

$$M(p) \backsimeq \tilde{M}(p) = A_v p^{-v} [1 - \exp(-p^v / A_v)] \tag{3.11}$$

as Re $p \to +\infty$ and so

$$\mathscr{F}(p) \backsimeq A_v p^{-v} [1 - \exp(-p^v / A_v)] F(p). \tag{3.12}$$

Applying the theory of complex analysis we can obtain the inverse Laplace transform $\mathscr{F}^{-1}[\tilde{M}(p)]$ of $\tilde{M}(p)$:

$$\begin{aligned} \tilde{m}(t) &= \mathscr{F}^{-1}[\tilde{M}(p)] = \frac{1}{2\pi i} \int_{a-i\infty}^{a+i\infty} e^{pt} \tilde{M}(p) dp \quad (a > 0) \\ &= \frac{1}{2\pi i} \int_{a-i\infty}^{a+i\infty} e^{pt} A_v p^{-v} (1 - \exp(-p^v / A_v)) dp = \frac{A_v \Gamma(1-v) \sin v\pi}{\pi} t^{v-1} \end{aligned} \tag{3.13}$$

if and only if $0 < v < 1$ (see [10, 11]). That is, $d\mu_p(t) \backsimeq \tilde{m}(t) dt$, where

$$\tilde{m}(t) = \frac{A_v}{\Gamma(v)} t^{v-1}. \tag{3.14}$$

It follows from $\int_0^T d\mu_p(t) = 1$ and (3.12)~(3.14) that

$$A_v \backsimeq \frac{v\Gamma(v)}{T^v}, \tag{3.15}$$

$$\widetilde{m}(t) \backsimeq \frac{v}{T^v} t^{v-1}, \tag{3.16}$$

$$\mathscr{F}(x) = \int_0^x f(t)\mathrm{d}\mu_p(t) \backsimeq \frac{v}{T^v}\int_0^x (x-t)^{v-1} f(t)\mathrm{d}t \backsimeq \frac{A_v}{\Gamma(v)}\int_0^x \frac{f(t)\mathrm{d}t}{(x-t)^{1-v}}. \tag{3.17}$$

If $b_1 \neq 0$, since $M(p)$ is bounded and $b_j > b_1 > 0$ for $j = 2, \cdots, k$, we have

$$\sum_{j=1}^{k} p_j \mathrm{e}^{-(b_j - b_1)p} M(\xi_j p) \backsimeq p_1 M(\xi_1 p) + o(1) \quad (\text{as } \operatorname{Re} p \to +\infty). \tag{3.18}$$

It follows from (3.10) and (3.18) that

$$M(p) = \mathrm{e}^{-b_1 p} \sum_{j=1}^{k} p_j \mathrm{e}^{-(b_j - b_1)p} M(\xi_j p) + o(1) \backsimeq p_1 \mathrm{e}^{-b_1 p} M(\xi_1 p) + o(1) \quad (\text{as } \operatorname{Re} p \to +\infty). \tag{3.19}$$

Note that the equation

$$\overline{M}(p) = p_1 \mathrm{e}^{-b_1 p} \overline{M}(\xi_1 p) \tag{3.20}$$

has a unique solution of form

$$\overline{M}(p) = A'_v p^{-v} \exp\left(-\frac{b_1}{1-\xi_1} p\right), \tag{3.21}$$

where $v = \ln p_1 / \ln \xi_1$ and A'_v is a constant. Noting (3.19)~(3.21) and $M(0) = 1$ we can choose

$$\widetilde{M}(p) = A'_v p^{-v} \exp(-t_1 p)(1 - \exp(-p^v / A'_v)) \tag{3.22}$$

as $\operatorname{Re} p \to +\infty$, where $t_1 = b_1/(1-\xi_1)$. It follows from (3.13) and (3.22) that

$$\widetilde{m}(t) = \frac{A'_v}{\Gamma(v)} (t - t_1)^{v-1} \eta(t - t_1)$$

if and only if $0 < v < 1$ and so $\mathrm{d}\mu_p(t) \backsimeq \widetilde{m}(t)\mathrm{d}t$. Therefore

$$A'_v \backsimeq \frac{v\Gamma(v)}{(T-t_1)^v}, \tag{3.23}$$

$$\widetilde{m}(t) \backsimeq \frac{v}{(T-t_1)^v}(t-t_1)^{v-1}\eta(t-t_1),$$

$$\mathscr{F}(x) = \int_0^x f(t)\mathrm{d}\mu_p(t) \backsimeq \frac{A'_v}{\Gamma(v)}\int_0^x \frac{f(t)\eta(x-t-t_1)\mathrm{d}t}{(x-t-t_1)^{1-v}}, \tag{3.24}$$

where $\Gamma(v)$ is the Gamma function, $v = \ln p_1/\ln \xi_1$, $0 < v < 1$, and $t_1 = b_1/(1-\xi_1)$. Let

$$L_r[a, b] = \{f: f \text{ is a Lebesgue measurable function on } [a, b] \text{ and } \int_a^b |f(t)|^r \mathrm{d}t < \infty\}$$

and

$$\mathrm{AC}([0, T]) = \{f(x): f(x) \text{ is an absolutely continuous function on } [0, T]\}.$$

$f(x)$ is absolutely continuous if for any given $\varepsilon > 0$, there exist $\delta > 0$ such that, for any finite disjoint closed intervals $[a_k, b_k] \subset [0, T]$, $k = 1, \cdots, n$,

$$\sum_{k=1}^n |f(b_k) - f(a_k)| < \varepsilon$$

if $\sum_{k=1}^n |b_k - a_k| < \delta$.

Lemma 3.2 [8, p. 35] *Assume* $\varphi_1(0) = 0$ *and* $\varphi_k(T) = T$. *If* $f(x) \in \mathrm{AC}([0, T])$, *then*

$$\frac{1}{\Gamma(1-v)}\frac{\mathrm{d}}{\mathrm{d}x}\int_0^x \frac{f(t)\mathrm{d}t}{(x-t)^v}$$

exists almost everywhere for $0 < v < 1$. *Moreover*,

$$\frac{1}{\Gamma(1-v)}\frac{\mathrm{d}}{\mathrm{d}x}\int_0^x \frac{f(t)\mathrm{d}t}{(x-t)^v} \in L_r(0, T),\ 1 \leqslant r < 1/v, \text{ and}$$

$$\frac{1}{\Gamma(1-v)}\frac{\mathrm{d}}{\mathrm{d}x}\int_0^x \frac{f(t)\mathrm{d}t}{(x-t)^v} = \frac{1}{\Gamma(1-v)}\left[\frac{f(0)}{x^v} + \int_0^x \frac{f'(t)\mathrm{d}t}{(x-t)^v}\right].$$

Theorem 3.3 *Let* $f(x) \in \mathrm{AC}([0, T])$ *and* $\varphi_j(x)$ *be monotonic increasing and* $\mathrm{C}^{1+\alpha}$ *differentiable. Suppose the measures* μ_p *and* $\mu_{\bar{p}}$ *are* *mutual conjugate.*

(i) *If* $\varphi_1(0)=0$ *and* $\varphi_k(T)=T$, *then*

$$D^{\mu_{\bar{p}}} f(x) = \frac{\mathrm{d}}{\mathrm{d}x}\int_0^x f(t)\mathrm{d}\mu_{\bar{p}}(t) \backsimeq \frac{A_{\bar{v}}}{\Gamma(1-v)}\frac{\mathrm{d}}{\mathrm{d}x}\int_0^x \frac{f(t)\mathrm{d}t}{(x-t)^v} \tag{3.25}$$

and $D^{\mu_{\bar{p}}} f(x)$ *exists almost everywhere for* $0<v<1$. *Moreover,*

$$D^{\mu_{\bar{p}}} f(x) \backsimeq \frac{A_{\bar{v}}}{\Gamma(1-v)}\left[\frac{f(0)}{x^v}+\int_0^x \frac{f'(t)\mathrm{d}t}{(x-t)^v}\right]. \tag{3.26}$$

(ii) *If* $\varphi_1(0)=b_1\neq 0$ *and* $\varphi_k(T)=T$, *then*

$$D^{\mu_{\bar{p}}} f(x) = \frac{\mathrm{d}}{\mathrm{d}x}\int_0^x f(t)\mathrm{d}\mu_{\bar{p}}(t) \backsimeq \frac{A'_{\bar{v}}}{\Gamma(1-v)}\frac{\mathrm{d}}{\mathrm{d}x}\int_0^x \frac{f(t)\eta(x-t-t_1)\mathrm{d}t}{(x-t-t_1)^v} \tag{3.27}$$

and $D^{\mu_{\bar{p}}} f(x)$ *exists almost everywhere for* $0<v<1$, *where* $A_{\bar{v}} \backsimeq \bar{v}\Gamma(\bar{v})/T^{\bar{v}}$, $A'_{\bar{v}} \backsimeq \bar{v}\Gamma(\bar{v})/(T-t_1)^{\bar{v}}$, $v+\bar{v}=1$, $v=\ln p_1/\ln\xi_1$, $0<v<1$, *and* $t_1=b_1/(1-\xi_1)$.

Proof It follows from (3.3) that

$$\int_0^x f(t)\mathrm{d}\mu_{\bar{p}}(t) \backsimeq \frac{A_{\bar{v}}}{\Gamma(\bar{v})}\int_0^x \frac{f(t)\mathrm{d}t}{(x-t)^{1-\bar{v}}}.$$

Note that $\bar{v}=1-v$. Thus (3.25) is right. Similarly, we can obtain (3.27).

It follows from Lemma 3.2 and (3.25) that (3.26) holds.

Theorem 3.4 *Assume* μ_p *and* $\mu_{\bar{p}}$ *are mutual conjugate*, $\varphi_1(0)=0$ *and* $\varphi_k(T)=T$. *Let* $f(x): E_0 \to \mathbb{R}$ *be continuous and absolutely integrable, and*

$$F(x)=\int_0^x f(t)\mathrm{d}\mu_p(t),\ x\in(0,T). \tag{3.28}$$

Then

$$D^{\mu_{\bar{p}}} F(x) \backsimeq A_v A_{\bar{v}} f(x),\ x\in(0,T). \tag{3.29}$$

Proof It follows from (3.3) that

$$F(x) \backsimeq \frac{A_v}{\Gamma(v)}\int_0^x \frac{f(t)\mathrm{d}t}{(x-t)^{1-v}}. \tag{3.30}$$

Replacing x and t by t and s in (3.30) respectively, multiplying both sides of (3.30) by $(x-t)^{-v}$ and integrating we have

$$\frac{A_v}{\Gamma(v)}\int_0^x \frac{\mathrm{d}t}{(x-t)^v}\int_0^t \frac{f(s)\mathrm{d}s}{(t-s)^{1-v}} \backsimeq \int_0^x \frac{F(t)\mathrm{d}t}{(x-t)^v}. \tag{3.31}$$

Interchanging the order of integration in the left-hand we obtain

$$\frac{A_v}{\Gamma(v)}\int_0^x f(s)\mathrm{d}s\int_s^x \frac{\mathrm{d}t}{(x-t)^v(t-s)^{1-v}} \backsimeq \int_0^x \frac{F(t)\mathrm{d}t}{(x-t)^v}.$$

Let $t=s+\tau(t-s)$. From the definition of Beta functions and the relation of Beta functions and Gamma functions we have that

$$\int_s^x (x-t)^{-v}(t-s)^{v-1}\mathrm{d}t=\int_0^1 \tau^{v-1}(1-\tau)^{-v}\mathrm{d}\tau=\mathrm{B}(v,1-v)=\Gamma(v)\Gamma(1-v).$$

Therefore

$$A_v\int_0^x f(s)\mathrm{d}s \backsimeq \frac{1}{\Gamma(1-v)}\int_0^x \frac{F(t)\mathrm{d}t}{(x-t)^v}. \tag{3.32}$$

Differentiating both the sides of (3.32) and using (3.25) and (3.2) we have

$$A_v f(x) \backsimeq \frac{1}{\Gamma(1-v)}\frac{\mathrm{d}}{\mathrm{d}x}\int_0^x \frac{F(t)\mathrm{d}t}{(x-t)^v}=\frac{1}{A_{\bar{v}}}\int_0^x F(t)\mathrm{d}\mu_{\bar{p}}(t)=\frac{1}{A_{\bar{v}}}D^{\mu_{\bar{p}}}F(x).$$

Therefore, (3.29) is right.

Theorem 3.4 shows that the derivative of integral is approximately equal to integrand function besides a constant factor difference.

Remark 3.5 The expression

$$I_0^v f(x)=\frac{1}{\Gamma(v)}\int_0^x \frac{f(t)\mathrm{d}t}{(x-t)^{1-v}}=\frac{x^v}{\Gamma(v)}\int_0^1 \frac{f(ux)}{(1-u)^{1-v}}\mathrm{d}u \tag{3.33}$$

is called the Riemann-Liouville fractional integral of order v and

$$D_0^v f(x)=\frac{1}{\Gamma(1-v)}\frac{\mathrm{d}}{\mathrm{d}x}\int_0^x \frac{f(t)\mathrm{d}t}{(x-t)^v}=\frac{1}{\Gamma(1-v)}\frac{\mathrm{d}}{\mathrm{d}x}\left[x^{1-v}\int_0^1 \frac{f(ux)}{(1-u)^v}\mathrm{d}u\right] \tag{3.34}$$

is called the Riemann-Liouville fractional derivative of order v (cf. [8]). Therefore, (3.3), (3.4), (3.25) and (3.27) can write respectively as

$$I^{\mu_p} f(x) = \int_0^x f(t) \mathrm{d}\mu_p(t) \cong A_v I_0^v f(x), \tag{3.35}$$

$$D^{\mu_{\bar{p}}} f(x) = \frac{\mathrm{d}}{\mathrm{d}x} \int_0^x f(t) \mathrm{d}\mu_{\bar{p}}(t) \cong A_{\bar{v}} D_0^v f(x), \tag{3.36}$$

$$I^{\mu_p} f(x) = \int_0^x f(t) \mathrm{d}\mu_p(t) \cong \frac{A'_v (x-t_1)^v}{\Gamma(v)} \int_0^1 \frac{f(u(x-t_1))}{(1-u)^{1-v}} \mathrm{d}u = A'_v I_0^v f(x-t_1), \tag{3.37}$$

$$D^{\mu_{\bar{p}}} f(x) = \frac{\mathrm{d}}{\mathrm{d}x} \int_0^x f(t) \mathrm{d}\mu_{\bar{p}}(t) \cong \frac{A'_{\bar{v}}}{\Gamma(1-v)} \frac{\mathrm{d}}{\mathrm{d}x} \left[(x-t_1)^{1-v} \int_0^1 \frac{f(u(x-t_1))}{(1-u)^v} \mathrm{d}u \right]$$
$$= A'_{\bar{v}} D_0^v f(x-t_1). \tag{3.38}$$

4. Inverse Problem

In order to discuss the solvability of integral equation $\int_0^x f(t) \mathrm{d}\mu_p(t) = F(x)$, $x > 0$, we give some known results for the Abel integral equation. The integral equation

$$\frac{1}{\Gamma(v)} \int_0^x \frac{g(t)\mathrm{d}t}{(x-t)^{1-v}} = G(x), \quad x > 0, \tag{4.1}$$

where $0 < v < 1$, is called *Abel's equation*. Equation (4.1) has a unique solution [8]:

$$g(x) = \frac{1}{\Gamma(1-v)} \frac{\mathrm{d}}{\mathrm{d}x} \int_0^x \frac{G(t)\mathrm{d}t}{(x-t)^v}. \tag{4.2}$$

Lemma 4.1 [8, p. 31] *Abel equation* (4.1) *with* $0<v<1$ *is solvable in* $L_1(0, T)$ *if and only if*

$$G_{1-v}(x) \in AC([0, T]) \quad \text{and} \quad G_{1-v}(0) = 0, \tag{4.3}$$

where

$$G_{1-v}(x) = \frac{1}{\Gamma(1-v)} \int_0^x \frac{G(t)\mathrm{d}t}{(x-t)^v}. \tag{4.4}$$

These conditions being satisfied the equation has a unique solution given by

(4.2).

Corollary 4.2 [8, p.32] *If* $G(x) \in AC([0, T])$, *then Abel's equation* (4.1) *with* $0 < v < 1$ *is solvable in* $L_1(0, T)$ *and its solution* (4.2) *may herein be represented in the form*

$$g(x) = \frac{1}{\Gamma(1-v)}\left[\frac{G(0)}{x^v} + \int_0^x \frac{G'(s)\,ds}{(x-s)^v}\right]. \qquad (4.5)$$

Theorem 4.3 *Let* $\varphi_1(0) = 0$ *and* $\varphi_k(T) = T$. *If*

$$\overline{F}_{1-v}(x) \in AC([0, T]) \quad \text{and} \quad \overline{F}_{1-v}(0) = 0, \qquad (4.6)$$

where

$$\overline{F}_{1-v}(x) = \frac{1}{A_v\Gamma(1-v)}\int_0^x \frac{F(t)\,dt}{(x-t)^v}, \quad F(x) = \int_0^x f(t)\,d\mu_p(t), \qquad (4.7)$$

then the equation $F(x) = \int_0^x f(t)\,d\mu_p(t)$ *is solvable in* $L_1(0, T)$ *and its unique solution is given by*

$$f(x) \backsimeq \frac{1}{A_v A_{\bar{v}}}\frac{d}{dx}\int_0^x F(t)\,d\mu_{\bar{p}}(t) = \frac{1}{A_v A_{\bar{v}}} D^{\mu_{\bar{p}}} F(x), \qquad (4.8)$$

where μ_p *and* $\mu_{\bar{p}}$ *are mutual conjugate.*

Proof It follows from Lemma 4.1, (3.25) and (3.2) that

$$f(x) = \frac{1}{A_v\Gamma(1-v)}\frac{d}{dx}\int_0^x \frac{F(t)\,dt}{(x-t)^v} \backsimeq \frac{1}{A_v A_{\bar{v}}}\frac{d}{dx}\int_0^x F(t)\,d\mu_{\bar{p}}(t) = \frac{1}{A_v A_{\bar{v}}} D^{\mu_{\bar{p}}} F(x)$$

is the unique solution of the equation

$$\frac{1}{\Gamma(v)}\int_0^x \frac{f(t)\,dt}{(x-t)^{1-v}} = \frac{F(x)}{A_v} \quad \text{or} \quad F(x) = \int_0^x f(t)\,d\mu_p(t).$$

5. Applications of Integrals and Derivatives

5.1 In memory processes

When we describe a structure of the evolution of a physical system far from thermodynamic equilibrium, it has been found that the medium exhibits memory. The existence of memory means that if at time τ a force $f(\tau)$ acts on the system, then there arises a flux J whose magnitude at time $t > \tau$ is given through a memory function $m(\tau)$ by the equation

$$J(t)=\int_0^t m(t-\tau)f(\tau)\mathrm{d}\tau.$$

(see Ref. [14] and references therein). The set with "residual" memory, i. e. the support of $m(t)$, is called the *memory set*.

If the memory set is a net set generated by a family of contractive mappings $\{\varphi_j(x)\}_{j=1}^k$ on $[0, T]$ and μ_p is the net measure associated to a normalized probability vector $\boldsymbol{p}=(p_1, \cdots, p_k)$, then

$$J(t)=\int_0^t f(\tau)\mathrm{d}\mu_p(\tau). \tag{5.1}$$

Assume that $\varphi_1(0)=0$ and $\varphi_k(T)=T$. Then, it follows from (3. 3) and (3. 35) that

$$J(t)\backsimeq A_v(\Gamma(v))^{-1}\int_0^t \frac{f(\tau)}{(t-\tau)^{1-v}}\mathrm{d}\tau=A_v I_0^v f(t), \tag{5.2}$$

where $v=\ln p_1/\ln \xi_1$. Let

$$\overline{J}_{1-v}(x)=\frac{1}{A_v\Gamma(1-v)}\int_0^x \frac{J(t)\mathrm{d}t}{(x-t)^v}. \tag{5.3}$$

If $\overline{J}_{1-v}(t)\in \mathrm{AC}([0, T])$ and $\overline{J}_{1-v}(0)=0$, then it follows from Theorem 4. 3 that the integral equation (5. 1) is solvable in $L_1(0, T)$ and its solution is

$$f(t)\backsimeq\frac{1}{A_vA_{\bar{v}}}D^{\mu_{\bar{p}}}J(t)=\frac{1}{A_vA_{\bar{v}}}\frac{\mathrm{d}}{\mathrm{d}t}\int_0^t J(\tau)\mathrm{d}\mu_{\bar{p}}(\tau), \tag{5.4}$$

where $\mu_{\bar{p}}$ is the net measure associated to a normalized probability vector $\overline{\boldsymbol{p}}=(\bar{p}_1, \cdots, \bar{p}_k)$ and μ_p and $\mu_{\bar{p}}$ are mutual conjugate. It follows from (5. 4) and (3. 25) that

$$f(t)\backsimeq\frac{1}{A_v\Gamma(1-v)}\frac{\mathrm{d}}{\mathrm{d}t}\int_0^t \frac{J(\tau)\mathrm{d}\tau}{(t-\tau)^v}. \tag{5.5}$$

The above formula (5. 5) gives the approximate solution of the integral equation

$$J(t)=\int_0^t f(\tau)\mathrm{d}\mu_p(\tau).$$

5.2 In diffusion processes

In Ref. [15], Giona and Roman discussed the fractional diffusion equation for transport phenomena in random isotropic and homogeneous fractal structures media. Since they dealt with stationary stochastic processes, they had a restriction for the diffusion kernel. That is, for the radial probability current $i(r, t)$ leaving the spherical region of radius r at time t and the average probability density $P(r, t)$, they dealt with an integral relation of the form

$$\int_0^t i(r, \tau)\mathrm{d}\tau=r^{d_f-1}\int_0^t K(t, \tau)P(r, \tau)\mathrm{d}\tau=r^{d_f-1}\int_0^t K(t-\tau)P(r, \tau)\mathrm{d}\tau \tag{5.6}$$

and with the constitutive local equation on fractals

$$i(r, t)=-Br^{d_f-1}r^{-\theta'}\left(\frac{\partial P(r, t)}{\partial r}+\frac{\mathcal{K}}{r}P(r, t)\right), \tag{5.7}$$

where $K(t, \tau)=K(t-\tau)$ is the diffusion kernel, d_f is the fractal dimension, $B>0$ is the diffusion coefficient, the parameter $\theta'\geqslant 0$, and $\mathcal{K}$ remains to be determined. In Refs. [15, 16] the asymptotic solution of the fractional diffusion equations (5. 6) and (5. 7) on some fractals is given.

Now we discuss asymptotic solutions of (5. 6) and (5. 7) on net fractals using the results in Sections 3 and 4.

We say the support of $K(t)$ the diffusion set. If the diffusion set is a net fractal generated by a family of contractive mappings $\{\varphi_j(x)\}_{j=1}^k$ on $[0, T]$ and μ_p is the net measure associated to a normalized probability vector $\boldsymbol{p}=(p_1, \cdots, p_k)$, then

$$j(r, t)\equiv\int_0^t i(r, \tau)\mathrm{d}\tau=r^{d_f-1}\int_0^t K(t-\tau)P(r, \tau)\mathrm{d}\tau=r^{d_f-1}\int_0^t P(r, \tau)\mathrm{d}\mu_p(\tau). \tag{5.8}$$

Therefore, when $\varphi_1(0)=0$ and $\varphi_k(T)=T$, we have

$$j(r,t) \backsim \frac{A_v}{\Gamma(v)} r^{d_f-1} \int_0^t \frac{P(r,\tau)\mathrm{d}\tau}{(t-\tau)^{1-v}}, \tag{5.9}$$

$$i(r,t) = r^{d_f-1} \frac{\mathrm{d}}{\mathrm{d}t} \int_0^t P(r,\tau)\mathrm{d}\mu_p(\tau) \backsim \frac{A_v}{\Gamma(v)} r^{d_f-1} \frac{\mathrm{d}}{\mathrm{d}t} \int_0^t \frac{P(r,\tau)\mathrm{d}\tau}{(t-\tau)^{1-v}}, \tag{5.10}$$

where $v=\ln p_1/\ln \xi_1$. Combining (5.7) with (5.10) we obtain the fractional diffusion equation on the net fractal:

$$\frac{\partial^\gamma P(r,t)}{\partial t^\gamma} = -Br^{-\theta'} \left(\frac{\partial P(r,t)}{\partial r} + \frac{\mathscr{K}}{r} P(r,t) \right), \tag{5.11}$$

where

$$\frac{\partial^\gamma P(r,t)}{\partial t^\gamma} = \frac{1}{\Gamma(1-\gamma)} \frac{\mathrm{d}}{\mathrm{d}t} \int_0^t \frac{P(r,\tau)\mathrm{d}\tau}{(t-\tau)^\gamma}, \quad \gamma = 1-v.$$

Replacing $P(r,t)$ in (5.9) and (5.10) by the solution $P(r,t)$ of (5.11) we can obtain the approximate estimates of $j(r,t)$ and $i(r,t)$.

On the other hand, assume $j(r,t)$ given. If

$$\bar{j}_{1-v}(r,t) = \frac{1}{A_v\Gamma(1-v)} \int_0^t \frac{j(r,\tau)\mathrm{d}\tau}{(x-\tau)^v} \in \mathrm{AC}([0,T]) \quad \text{and } \bar{j}_{1-v}(r,0)=0,$$

then the equation

$$j(r,t) = r^{d_f-1} \int_0^t P(r,\tau)\mathrm{d}\mu_p(\tau) \tag{5.12}$$

is solvable in $L_1([0,T])$ and its solution is

$$P(r,t) \backsim \frac{1}{A_v A_{\bar{v}} r^{d_f-1}} \frac{\mathrm{d}}{\mathrm{d}t} \int_0^t j(r,\tau)\mathrm{d}\mu_{\bar{p}}(\tau) \backsim \frac{1}{A_v r^{d_f-1}\Gamma(1-v)} \frac{\mathrm{d}}{\mathrm{d}t} \int_0^t \frac{j(r,\tau)}{(t-\tau)^v}\mathrm{d}\tau, \tag{5.13}$$

where $\mu_{\bar{p}}$ is the net measure associated to a normalized probability vector $\bar{\boldsymbol{p}} = (\bar{p}_1, \cdots, \bar{p}_k)$ and μ_p and $\mu_{\bar{p}}$ are mutual conjugate. This is the approximate solution of fractional integral equation (5.9).

Remark The results in this paper are valid in incomplete statistical physics. In incomplete statistical physics the state normalized probability vector $\boldsymbol{p}=(p_1, \cdots, p_k)$ is incomplete, i. e. $\sum_{j=1}^{k} p_j < 1$ (cf. [17]). Because for any $\boldsymbol{p}=(p_1, \cdots, p_k)$ with $\sum_{j=1}^{k} p_j < 1$ there exists a unique real number q such that $\sum_{j=1}^{k} p_j^q = 1$, thus for the normalized probability vector $\boldsymbol{p}^q=(p_1^q, \cdots, p_k^q)$ and a family of contraction transformations $\{\varphi_j(x)\}_{j=1}^{k}$, there exists a unique net measure μ_{p^q}. Therefore, replacing $\boldsymbol{p}=(p_1, \cdots, p_k)$ in this paper by $\boldsymbol{p}^q=(p_1^q, \cdots, p_k^q)$ the results in the paper are still right.

Acknowledgements

We would like to express our thanks to the Editor Prof. H.-O. Peitgen and the referee for having spent a lot of their invaluable time in preparing this revised version of our submission for publication.

◇ References ◇

[1] Nigmatullin R R, Nivanen L, Le Mehaute A. Time arrows and fractal geometry (Editions HERMES), in preparation.

[2] Falconer K J. Fractal geometry: mathematics foundations and applications. Chichester: John Wiley and Son; 1990.

[3] Bedford T. Applications of dynamical systems theory to fractal: a study of cookie-cutter Cantor set. Preceeding of the Séminaire de Mathématiques Superiéurs, Fractal Geometry and Analysis, Université de Montréal, NAIO ASI Series, Amsterdam, 1989.

[4] Falconer K J. Techniques in fractal geometry. New York: John Wiley and Sons; 1997.

[5] Hutchinson J E. Fractals and self-similarity. Indiana Univ Math J 1981; 30:713-47.

[6] Ren F-Y, Yu Z-G, Su F. Phys Lett A 1996; 219: 59-69.

[7] Yu Z-G, Ren F-Y. J Phys A: Math Gen 1997; 30: 5569-77.

[8] Samko S G, Kilbas A A, Marichev O I. Fractional integrals and derivatives, theory and application. New York: Gordon and Breach Science Publisher; 1993.

[9] Qiu W-Y, Lü J. Phys Lett A 2000; 272: 353-8.

[10] Ren F-Y, Qiu W-Y, Liang J-R, Wang X-T. Determination of memory function and flux on fractals, in press.

[11] Ren F-Y, Wang X-T, Liang J-R. Determination of diffusion kernel on fractals, in press.

[12] Ren F-Y, Yu Z-G, Zhou J. The Southeast Asian Bull Math 1999; 23: 497 - 505.
[13] Ren F-Y, Yu Z-G. Progr Natural Sci 1997; 7: 422 - 8.
[14] Nigmatullin R R. Theor Matem Fiz 1992; 90: 354.
[15] Giona M, Roman H E. Physica A 1992; 185: 87.
[16] Ren F-Y, Liang J-R, Wang X-T. Phys Lett A 1999; 252: 141 - 50.
[17] Wang Q A. Chaos, Solitons and Fractals 2001; 12: 1431 - 7.

Universality of Stretched Gaussian Asymptotic Behaviour for the Fractional Fokker-Planck Equation in External Force Fields*

Fu-Yao Ren[1], Jin-Rong Liang[2], Wei-Yuan Qiu[1] and Yun Xu[1]

[1] *Department of Mathematics, Fudan University, Shanghai 200433, People's Republic of China*

[2] *Department of Mathematics, East China Normal University, Shanghai 200062, People's Republic of China*

Abstract: We introduce a heterogeneous fractional Fokker-Planck equation (HFFPE) on heterogeneous fractal structure media describing systems involving external force fields. The HFFPE is shown to obey the generalized Einstein relation, and its stationary solution is the Boltzmann distribution. It is proven that the asymptotic shape of its solution is a stretched Gaussian and that its solution can be expressed in the form of a function of a dimensionless similarity variable for constant and generic potentials with polar singularity at origin.

1. Introduction

In recent years, much attention has been paid to physical systems driven by transport mechanisms other than ordinary diffusion, in particular, diffusion in a disordered crystalline medium. This leads to many anomalous physical properties [1]. In the physics of complex systems, anomalous transport

* Originally published in *J. Phys. A: Math. Gen.*, Vol. 36, (2003), 7533 - 7543.

properties and their description have attracted considerable interest. Applications have been found in a wide field ranging from physics and chemistry to biology and medicine [1 ~ 5]. Anomalous diffusion in one dimension is characterized by the occurrence of a mean square displacement of the form

$$X^2 = \langle X^2 \rangle(t) = \langle\langle \Delta x \rangle^2 \rangle(t) = \frac{2K_\gamma}{\Gamma(1+\gamma)} t^\gamma \tag{1.1}$$

which deviates from the linear Brownian dependence on time [3]. In equation (1.1), the anomalous diffusion coefficient K_γ is introduced, which has the dimension $[K_\gamma] = \mathrm{cm}^2 \mathrm{s}^{-\gamma}$.

Recently in the literature, in order to describe diffusion processes in disordered media, some authors have proposed an extension of the Fokker-Planck (FP) equation, which is called the fractional Fokker-Planck equation (FFPE) [6~13].

An FFPE describing anomalous transport close to thermal equilibrium has been presented recently [6]. Since it describes subdiffusion in the force-free case, it involves a strong, i. e. slowly decaying, memory. In their original framework [9], Metzler *et al* gave a seminal framework and investigated the anomalous diffusion and relaxation involving external fields with the one-dimensional FFPE for one variable

$$\dot{W}(x, t) = {}_0D_t^{1-\gamma} L_{\mathrm{FP}} W \tag{1.2}$$

with respect to its physical properties. Here, $W(x, t)$ is the probability density function (pdf) at position x at time t, and the FP operator

$$L_{\mathrm{FP}} = \frac{\partial}{\partial x}\left(\frac{V'(x)}{m\eta_\gamma} + K_\gamma \frac{\partial}{\partial x}\right), \tag{1.3}$$

with the external potential $V(x)$ [15], contains the anomalous diffusion constant K_γ and the anomalous friction coefficient η_γ with the dimension $[\eta_\gamma] = \mathrm{s}^{\gamma-2}$. Herein, m denotes the mass of the diffusion particle, and

$${}_0D_t^{1-\gamma} W = \frac{1}{\Gamma(\gamma)} \frac{\partial}{\partial x} \int_0^t \mathrm{d}\tau \frac{W(x, \tau)}{(t-\tau)^{1-\gamma}}. \tag{1.4}$$

The interesting part has asymptotic behaviour $\ln W(x, t) \sim -c\xi^u$ where $\xi \equiv x/t^{\alpha/2} \gg 1$ which is expected to be universal. Here, $u = 1/(1-\alpha/2)$ with the anomalous diffusion exponent α which is the order of the fractional derivative [14].

In [30], the FFPE

$$ {}_0D_t^{\alpha} W(x, t) = G x^{-\theta'} L_{\mathrm{FP}} W(x, t) \tag{1.5} $$

has been presented, where $G > 0$ is to be determined, $\theta' \geqslant 0$ is a parameter, $0 < \alpha < 1$, if $\theta' = 0$ then $\alpha = \gamma$ and equation (1.5) reduces to

$$ {}_0D_t^{\alpha} W(x, t) = G L_{\mathrm{FP}} W(x, t) \tag{1.6} $$

which leads to the FFPE (1.2). It is proven that for the FFPE (1.6), its solution has asymptotic behaviour

$$ \ln W(x, t) \sim -c\xi^u \tag{1.7} $$

where

$$ \xi \equiv x/t^{\gamma/2} \gg 1, \quad u = 1/\left(1 - \frac{\gamma}{2}\right) \tag{1.8} $$

and possesses a scaling variable for constant potentials, linear potentials, logarithm potentials and harmonic potentials.

El-Wakil and Zahranit [12] have discussed the fact that anomalous diffusion in a heterogeneous fractal medium in one dimension is characterized by the occurrence of a mean square displacement of the form

$$ X_\theta^2 = \langle\langle \Delta x\rangle^2\rangle \sim x^{-\theta} t^{\gamma}, \quad 0 < \gamma \leqslant 1, \quad \theta = d_{\mathrm{w}} - 2. \tag{1.9} $$

We call θ a heterogeneous exponent and γ is known to be the diffusion exponent.

By a simple scaling consideration as in [17], according to equation (1.9) we require that $x^{-\theta} t^{\gamma} \sim x^2$, i.e. $x \sim t^{\frac{\gamma}{(2+\theta)}}$. Thus, we have that $X_\theta^2 \sim t^{\frac{2}{(2+\theta)}\gamma}$. Hence we can rewrite equation (1.9) as

$$ X_\theta^2 = \langle\langle \Delta x\rangle^2\rangle = \frac{2K_\gamma^\theta}{\Gamma(1+\gamma_\theta)} t^{\gamma\theta} \tag{1.10} $$

where $\gamma\theta=\frac{2\gamma}{2+\theta}$. In equation (1.10) the anomalous diffusion coefficient K_γ^θ is introduced, which has the dimension $[K_\gamma^\theta]=\mathrm{cm}^2\mathrm{s}^{-2\gamma/(2+\theta)}$. In equation (1.9), $d_w>2$ is the anomalous diffusion exponent [1].

It is easy to see that for $\theta\to 0$, equation (1.10) reduces to equation (1.1) and K_γ^θ to K_γ.

So we naturally ask the following question. Does the FFPE with multi-parameters α, θ, θ', γ and μ

$$ {}_0D_t^\alpha W(x,t)=Gx^{-\theta'}L_{\mathrm{FP}}^\mu W(x,t),\quad \mu>0 \tag{1.11}$$

still possess the stretched Gaussian and a scaling variable for constant and generic potentials where $L_{\mathrm{FP}}^\mu=\frac{\partial}{\partial x}\left(\frac{V'(x)}{m\eta_\gamma}+K_\gamma^\theta\frac{\partial}{\partial x}x^{-\mu}\right)$?

The main purpose of this paper is to solve this problem. By using the heuristic argument of Giona and Roman [17], we introduce an FFPE on heterogeneous fractal structures which can lead to the FFPE (1.5). It is proven that the asymptotic shape of its solution is a stretched Gaussian and that its solution can be expressed in the form of a function of a dimensionless similarity variable for constant and generic potentials.

2. The Derivation of the FFPE with External Potentials

In this section, we derive the fractional diffusion equation involving the external potential $V(x)$ by using the heuristic argument of Giona and Roman [17].

The relationship between the total flux of probability current $S(x,t)$ from time $t=0$ to time t and the average probability density $W(x,t)$, considered as the input and the output of the fractal system [18] (cf [17]), should satisfy the following equation:

$$\int_0^t S(x,\tau)\mathrm{d}\tau=x^{d_f-1}\int_0^t K(t-\tau)W(x,\tau)\mathrm{d}\tau \tag{2.1}$$

where d_f is the fractal dimension of the system considered. This is a conservation equation containing an explicit reference to the history of the diffusion process on a fractal structure. Since we are dealing with stationary processes, we expect $K(t, \tau)$ to be a function of difference $t-\tau$ only, i. e. $K(t, \tau)=K(t-\tau)$. $K(t, \tau)$ is the diffusion kernel. We assume that diffusion sets are underlying fractals (underlying fractals denote self-similar sets in [19] or net fractals in [20, 21]) and the diffusion kernel on the underlying fractal should behave as

$$K(t-\tau)=\frac{A_\alpha}{(t-\tau)^\alpha} \tag{2.2}$$

with $0<\alpha<1$, where α is a diffusion exponent and A_α is a constant that can be determined [21].

On the other hand, on the above fractional structure we propose that the probability current $S(x, t)$ satisfies the following structure equation:

$$S(x, t)=Bx^{d_f-1}x^{-\theta'}\frac{\partial}{\partial x}\left(\frac{V'(x)}{m\eta_\gamma}+K_\gamma^\theta\frac{\partial}{\partial x}x^{-\mu}\right)W(x, t), \quad \mu>0, \tag{2.3}$$

i. e.

$$S(x, t)=Bx^{d_f-1}x^{-\theta'}L_{FP}^\mu W(x, t)$$

where $B>0$ is to be determined, θ' is still a parameter and $L_{FP}^\mu=\frac{\partial}{\partial x}\left(\frac{V'(x)}{m\eta_\gamma}+K_\gamma^\theta\frac{\partial}{\partial x}x^{-\mu}\right)$ is the FP operator.

From equations (2.1), (2.2) and (2.3), we have

$$_0D_t^\alpha W(x, t)=Gx^{-\theta'}L_{FP}^\mu W(x, t) \tag{2.4}$$

where

$$G=B/\Gamma(1-\alpha)A_\alpha>0. \tag{2.5}$$

We call equation (2.4) a heterogeneous fractional Fokker-Planck equation (HFFPE).

It is easy to see that for $\mu\to 0$, L_{FP}^μ reduces to the L_{FP} and equation (2.4) to FFPE (1.5).

Furthermore, we suppose that the pdf $W(x, t)$ satisfies the following normalization conditions:

$$\int_0^{\infty} \mathrm{d}x \cdot x^{d_f - 1} W(x, t) = 1 \tag{2.6}$$

and the extraction of moments $\langle (\Delta X)^n \rangle$ is defined by

$$\langle (\Delta X)^n \rangle = \int_0^{\infty} \mathrm{d}x \cdot x^{d_f - 1} x^n W(x, t). \tag{2.7}$$

Remark 1 We know that the integral equation

$$\frac{1}{\Gamma(v)} \int_0^t \frac{g(\tau)\mathrm{d}\tau}{(t-\tau)^{1-v}} = G(t), \quad t > 0 \tag{2.8}$$

where $0 < v < 1$, is called Abel's equation and for any summable function $g(t)$ it has a unique solution [25]

$$g(t) = \frac{1}{\Gamma(1-v)} \frac{\mathrm{d}}{\mathrm{d}t} \int_0^t \frac{G(\tau)\mathrm{d}\tau}{(t-\tau)^{v}}. \tag{2.9}$$

Conversely, if equation (2.9) holds, then equation (2.8) is satisfied for $G(t) \in I_0^{\alpha}(L_1)$, i.e. there exists a summable function $h(t)$ such that

$$G(t) = \frac{1}{\Gamma(v)} \int_0^t \frac{h(\tau)\mathrm{d}\tau}{(t-\tau)^{1-v}}, \quad t > 0. \tag{2.10}$$

Thus, from equation (2.4) we have

$$W(x, t) = \frac{1}{\Gamma(\alpha)} \int_0^t G x^{-\theta'} \frac{L_{\mathrm{FP}}^{\mu} W(x, \tau)}{(t-\tau)^{1-\alpha}} \mathrm{d}\tau \tag{2.11}$$

for $W(x, t) \in I_0^{\alpha}(L_1)$. Hence

$$\dot{W}(x, t) = \frac{1}{\Gamma(\alpha)} \frac{\partial}{\partial t} \int_0^t G x^{-\theta'} \frac{L_{\mathrm{FP}}^{\mu} W(x, \tau)}{(t-\tau)^{1-\alpha}} \mathrm{d}\tau,$$

i.e.

$$\dot{W}(x, t) = G x^{-\theta'} {}_0D_t^{1-\alpha} L_{\mathrm{FP}}^{\mu} W(x, t) \tag{2.12}$$

where

$${}_0D_t^{1-\alpha} W = \frac{1}{\Gamma(\alpha)} \frac{\partial}{\partial t} \int_0^t \frac{W(x, \tau)\mathrm{d}\tau}{(t-\tau)^{1-\alpha}}. \tag{2.13}$$

Especially, when $G=1$, $\theta'=0$ and $\alpha=\gamma$

$$\dot{W}(x, t)={}_0D_t^{1-\alpha}L_{\mathrm{FP}}^{\mu}W. \tag{2.14}$$

Equation (2.14) is just the one-dimensional FFPE (1.2). The derivation of equation (2.14) is different from those in [9, 26]. This shows that the asymptotic equation (2.3) is reasonable. Conversely, if $W(x, 0)=0$ and equation (2.12) holds, then equation (2.4) is satisfied if $Gx^{-\theta'}L_{\mathrm{FP}}^{\mu}W(x, t)$ is a summable function.

If in the presence of an external nonlinear and time-independent field the stationary state is reached, then S must be constant. Thus, if $S=0$ for any x, it vanishes for all x [15], and the stationary solution is given by $L_{\mathrm{FP}}^{\mu}W(x, t)=0$, i. e. $V'(x)W_{\mathrm{st}}/(m\eta_{\gamma})+K_{\gamma}^{\theta}\mathrm{d}(x^{-\mu}W_{\mathrm{st}})/\mathrm{d}x=C$ (constant) from which the exponential result

$$W_{\mathrm{st}}(x)=A^{\delta}x^{\mu}\left\{C'\exp\left\{-\left[\int x^{\mu}V'(x)\mathrm{d}x/(m\eta_{\gamma}K_{\gamma}^{\theta})\right]\right\}+C\left[\frac{m\eta_{\gamma}}{A^{\delta}x^{\mu}V'(x)}\right.\right.$$
$$\left.\left.-\mathrm{e}^{-\int x^{\mu}V'(x)\mathrm{d}x/(m\eta_{\gamma}K_{\gamma}^{\theta})}\int \mathrm{e}^{\int x^{\mu}V'(x)\mathrm{d}x/(m\eta_{\gamma}K_{\gamma}^{\theta})}\mathrm{d}\left(\frac{m\eta_{\gamma}}{A^{\delta}x^{\mu}V'(x)}\right)\right]\right\} \tag{2.15}$$

can be inferred, where $\delta=\mu/(\mu+1)<1$ and C' is a constant. Requiring, in analogy with the standard case, that W_{st} is given by the generalized Boltzmann distribution in a heterogeneous fractal medium, i. e.

$$W_{\mathrm{st}}\propto x^{\mu}\exp\left\{-\int x^{\mu}V'(x)\mathrm{d}x/(k_{\mathrm{B}}T)\right\}, \tag{2.16}$$

C must be equal to zero. Thus, the generalized Einstein-Stokes-Smoluchowski relation

$$K_{\gamma}^{\theta}=\frac{k_{\mathrm{B}}T}{m\eta_{\gamma}} \tag{2.17}$$

is readily recovered. Thus, the FFPE (2.4) obeys some generalized fluctuation-dissipation theorem.

It is easy to see that for $\mu\to 0$ the generalized Boltzmann distribution (2.16) and Einstein relation (2.17) reduce to the well-known Boltzmann distribution $W_{\mathrm{st}}\propto\exp\{-V(x)/(k_{\mathrm{B}}T)\}$ and Einstein relation $K_{\gamma}=$

$k_B T/(m\eta_\gamma)$, respectively.

We can anticipate the relation between exponents α, γ, θ, μ and θ' by simple scaling considerations. From equation (2. 4) we see that $t^\alpha \sim x^{2+\theta'+\mu}$, and according to equation (1. 9) we require that $t^\gamma \sim x^{2+\theta}$. Thus

$$\alpha = \gamma\left(\frac{2+\mu+\theta'}{2+\theta}\right). \tag{2.18}$$

Corresponding to the value $\theta' = 0$, the new lower bound $\alpha \geqslant \gamma$ follows from equation (2. 18). From the upper bound $\alpha < 1$, equation (2. 18) follows the upper bound for θ':

$$\theta' < \frac{(2+\theta)}{\gamma} - (\alpha - \mu). \tag{2.19}$$

3. Solutions and the Properties of the HFFPE

In this section we give the solutions of the FFPE (2. 4).

It can be shown that the Laplace transform of the fractional derivative ${}_0D_t^\alpha W(x, t)$ is

$$L\left[{}_0D_t^\alpha W(x, t)\right] = s^\alpha W(x, s).$$

Using this result, from equation (2. 4) we obtain

$$s^\alpha W(x, s) = -Gx^{-\theta'}\frac{\partial}{\partial x}\left[\frac{V'(x)}{m\eta_\gamma}W(x, s) + K_\gamma^\theta \frac{\partial x^{-\mu} W(x, s)}{\partial x}\right] \tag{3.1}$$

where $W(x, s)$ denotes the Laplace transform of $W(x, t)$.

Case I: *Constant potential*. First we consider the constant potential $V(x) =$ constant, leading to the force-free case. In this case, equation (3. 1) reduces to

$$x^2 \frac{\partial^2 W(x, s)}{\partial x^2} - 2\mu x \frac{\partial W(x, s)}{\partial x} - \left[qx^{2+\mu+\theta'} - \mu(\mu+1)\right]W(x, s) = 0 \tag{3.2}$$

where $q = s^\alpha / G_\gamma$, $G_\gamma = GK_\gamma^\theta$. In order to solve equation (3. 2), it is convenient to perform the transformation

$$y=A(s)x^{v}, \quad W(x, s)=y^{\delta}Z(y) \tag{3.3}$$

to cast equation (3.2) into the second-order Bessel equation as

$$y^{2}\frac{d^{2}Z}{dy^{2}}+y\frac{dZ}{dy}-(\lambda^{2}+y^{2})Z=0 \tag{3.4}$$

with parameter λ^2 under the following conditions:

$$v=\frac{2+\mu+\theta'}{2},\ A(s)=\frac{q^{1/2}}{v},\ \delta=\frac{1}{2v}(1+2\mu) \tag{3.5}$$

where

$$\lambda^{2}=\frac{2\delta\mu}{v}-\delta\left(\delta-\frac{1}{v}\right)-\frac{\mu(\mu+1)}{v^{2}}=1/4v^{2}. \tag{3.6}$$

The solution of this equation, satisfying the summability condition $\lim_{x\to+\infty}W(x, t)=0$, i.e. $W(x, s)=0$ $(x\to+\infty)$, is given by $Z(y)=C(s)K_{\lambda}(y)$ since $y\to+\infty$,

$$K_{\lambda}(y)=e^{-y}\left(\frac{\pi}{2y}\right)^{1/2}\left[1+O\left(\frac{1}{y}\right)\right] \tag{3.7}$$

in the domain $|\arg y|<\frac{3}{2}\pi$, where K_λ is the modified Bessel function of second-order and $C(s)$ is to be determined. So the solution of equation (3.2) is given by

$$W(x, s)=C(s)y^{\delta}K_{\lambda}(y), \quad y=A(s)x^{v} \tag{3.8}$$

where

$$C(s)=G's^{(\alpha d_{f}/2v)-1} \quad G'=v^{1-d_{f}/v}/[C_{\lambda}(GK_{\gamma}^{\theta})^{d_{f}/2v}] \tag{3.9}$$

ensures the normalization condition (2.6) of $W(x, t)$, i.e. $\int_0^{\infty}dx\cdot x^{d_f-1}W(x, s)=1/s$, if and only if $0<\lambda<1$. $C_\lambda=\int_0^{\infty}dy\cdot y^{\frac{d_f}{v}+\delta-1}K_\lambda(y)$ is a constant since [28]

$$\int_0^{\infty}dy\cdot y^{v}K_{\lambda}(ay)=2^{v-1}a^{-v-1}\Gamma\left(\frac{1+v+\lambda}{2}\right)\Gamma\left(\frac{1+v-\lambda}{2}\right).$$

It is easy to see from equations (1.10) and (2.7) (when $n=2$) that

$$\alpha=\left(\frac{2+\mu+\theta'}{2+\theta}\right)\gamma,\ B=\frac{A_{\alpha}\Gamma(1-\alpha)}{v^{2}}\left(2\frac{C_{\lambda}}{C'_{\lambda}}\right)^{v}(K_{\gamma}^{\theta})^{v-1} \tag{3.10}$$

since $\int_{0}^{\infty}\mathrm{d}x\cdot x^{d_{\mathrm{f}}-1}x^{2}W(x,s)=2K_{\gamma}^{\theta}/s^{2\gamma/(2+\theta)+1}$ corresponding to the Laplace transform of $\langle\langle\Delta x\rangle^{2}\rangle(t)$. $C'_{\lambda}=\int_{0}^{\infty}\mathrm{d}y\cdot y^{\frac{d_{\mathrm{f}}+2}{v}+\delta-1}K_{\lambda}(y)$.

Let us now discuss the asymptotic behaviour of $W(x,t)$ as predicted by equation (3.8). We have

$$W(x,s)\approx G''s^{-(1-\alpha d_{\mathrm{f}}/2v)}\frac{1}{(xs^{\alpha/2v})^{K}}\exp\{-(xs^{\alpha/2v}/G''')^{v}\} \tag{3.11}$$

for $|xs^{\alpha/2v}|\gg 1$, where $G''=\sqrt{\frac{\pi}{2}}v^{1-(d_{\mathrm{f}}+\kappa)/v}/C_{\lambda}(GK_{\gamma}^{\theta})^{(d_{\mathrm{f}}+\kappa)/2v}>0$, $G'''=(v\sqrt{GK_{\gamma}^{\theta}})^{1/v}$, $\kappa=v\left(\frac{1}{2}-\delta\right)$.

Using the same method as in [17], we expect that

$$W(x,t)\sim t^{-(\alpha d_{\mathrm{f}}/2v)}(x/X_{\theta})^{\delta'}\exp\{-\mathrm{const}\cdot(x/X_{\theta})^{u'}\} \tag{3.12}$$

when $x/X_{\theta}\sim x/t^{\gamma/(2+\theta)}\gg 1$ and $t\to+\infty$. The Laplace transform of equation (3.12) can be evaluated by applying the method of steepest descent and the result compared with equation (3.11). This yields

$$u'=v/\left(1-\frac{\alpha}{2}\right) \tag{3.13}$$

and

$$\delta'=u'\left[\frac{1}{2}(\alpha d_{\mathrm{f}}/v-1)-\kappa\right]. \tag{3.14}$$

It is interesting that if $\mu=\theta=\theta'=0$, we have $u'=1/\left(1-\frac{\gamma}{2}\right)$ and $\delta'=u'\left[\frac{1}{2}(\gamma d_{\mathrm{f}}-1)\right]$, which are the same of those of FFPE [30].

Using the inversion theorem of Laplace transform, from equation (3.11) we have (see the appendix)

$$W(x, t) \approx \frac{G''}{\pi} t^{-\alpha d_f/2v} z^{-\kappa} f(z) \tag{3.15}$$

where

$$f(z) = \sum_{n=0}^{\infty} C_n(\beta) z^{nv}, \ z = x/t^{\alpha/2v}, \ \beta = \alpha(d_f - \kappa)/2v - 1.$$

In particular, for $z \ll 1$, $W(x, t) \approx \frac{G'' C_0(\beta)}{\pi} t^{-\alpha d_f/2v} z^{-\kappa}$; thus, if $\kappa > 0$, $W(x, t)$ diverges on the origin.

Case II: Generic potentials. Now we discuss the general case. In this case, equation (3.1) becomes

$$x^2 \frac{\partial^2 W(x, s)}{\partial x^2} + \left(\frac{V'(x) x^{1+\mu}}{m_\gamma^\theta} - 2\mu\right) x \frac{\partial W(x, s)}{\partial x} - \left[q x^{2+\mu+\theta'} - \mu(\mu+1) - \frac{V''(x) x^{2+\mu}}{m_\gamma^\theta}\right] W(x, s) = 0 \tag{3.16}$$

where $q = s^\alpha / G K_\gamma^\theta$, $m_\gamma^\theta = m\eta_\gamma K_\gamma^\theta$.

We assume that the generic external potential at the origin is given by

$$V(x) = b_p x^{-p} \quad (p \neq 0) \quad \text{and} \quad V(x) = b_p \ln x \quad (p = 0) \tag{3.17}$$

where $b_p \neq 0$.

In order to solve equation (3.16), it is convenient to perform the transform $y = A(s) x^v$, $W(x, s) = y^\delta Z(y)$ to cast equation (3.16) into the second-order Bessel equation as

$$y^2 \frac{d^2 Z}{dy^2} + y \frac{dZ}{dy} - (\lambda_g^2 + y^2) Z = 0 \tag{3.18}$$

with parameter λ_g^2 under the following conditions:

$$v = \frac{2+\mu+\theta'}{2}, \ A = q^{1/2}/v = s^{\alpha/2}/v\sqrt{G K_\gamma^\theta}, \tag{3.19}$$

$$\delta_g = \frac{1+2\mu}{v} - \frac{1}{2v m_\gamma^\theta} V'\left(\left(\frac{y}{A}\right)^{\frac{1}{v}}\right)\left(\frac{y}{A}\right)^{\frac{1+\mu}{v}} \tag{3.20}$$

where

$$\lambda_g^2 = \frac{2\mu\delta_g}{v} - \frac{\mu(\mu+1)}{v^2} - \delta_g\left(\delta_g - \frac{1}{v}\right) - \frac{V''\left(\left(\frac{y}{A}\right)^{\frac{1}{v}}\right)\left(\frac{y}{A}\right)^{\frac{2+\mu}{v}}}{V^2 m_r^\theta}$$

$$- \frac{\delta_g}{v m_\gamma^\theta} V'\left(\left(\frac{y}{A}\right)^{\frac{1}{v}}\right)\left(\frac{y}{A}\right)^{\frac{1+\mu}{v}}. \quad (3.21)$$

Noting for $\mathrm{Re}(s^{\alpha/2})/y \gg 1$, $y/|A| \ll 1$, if $\mu - p > 0$, and for $\mathrm{Re}(s^{\alpha/2})/y \ll 1$, $y/|A| \gg 1$, if $\mu - p < 0$, from equation (3.17), we have $\left(\frac{y}{A}\right)^{\frac{1+\mu}{v}} V'\left(\left(\frac{y}{A}\right)^{\frac{1}{v}}\right) \ll 1$ and $\left(\frac{y}{A}\right)^{\frac{2+\mu}{v}} V''\left(\left(\frac{y}{A}\right)^{\frac{1}{v}}\right) \ll 1$. Then from equations (3.20) and (3.21), we have

$$\delta_g \approx \frac{1+2\mu}{2v} = \delta,\ \lambda_g \approx \frac{1}{4v^2} = \lambda. \quad (3.22)$$

Thus, equation (3.18) can be replaced by

$$y^2 \frac{\mathrm{d}^2 Z}{\mathrm{d}y^2} + y\frac{\mathrm{d}Z}{\mathrm{d}y} - (\lambda^2 + y^2)Z = 0. \quad (3.4)$$

Hence, if $\mu \neq p$, we have

$$w(x,t) \sim t^{-(\alpha d_f/2v)}\left(\frac{x}{X_\theta}\right)^{\delta'} \exp\left\{-\text{const} \cdot \left(\frac{x}{X_\theta}\right)^{u'}\right\} \quad (3.23)$$

where $x/X_\theta \sim x/t^{\gamma/(2+\theta)} \gg 1$ and $t \to +\infty$,

$$u' = v\Big/\left(1 - \frac{\alpha}{2}\right), \quad (3.24)$$

$$\delta' = u'\left[\frac{1}{2}(\alpha d_f/v - 1) - \kappa\right], \quad (3.25)$$

$$\kappa = v\left(\frac{1}{2} - \delta\right), \quad (3.26)$$

and

$$W(x,t) \approx \frac{G''}{\pi} t^{-\alpha d_f/2v} z^{-\kappa} f(z) \quad (3.27)$$

where

$$f(z)=\sum_{n=0}^{\infty}C_n(\beta)z^{nv},\ z=x/t^{\alpha/2v},\ \beta=\alpha(d_f-\kappa)-1. \tag{3.28}$$

If $\mu=p\neq 0$, for $\mathrm{Re}(s^{\alpha/2}/y)\gg 1$, since

$$\left(\frac{y}{A}\right)^{\frac{1+\mu}{v}}V'\left(\left(\frac{y}{A}\right)^{\frac{1}{v}}\right)\approx -pb_p,\ \left(\frac{y}{A}\right)^{\frac{2+\mu}{v}}V''\left(\left(\frac{y}{A}\right)^{\frac{1}{v}}\right)\approx p(p+1)b_p,$$

we have

$$\delta_g\approx\frac{1+2\mu}{2v}+\frac{pb_p}{2vm_\gamma^\theta}=\delta_p, \tag{3.29}$$

$$\lambda_g\approx\frac{1}{4v^2}\left(1+\frac{pb_p}{m_\gamma^\theta}\right)^2=\lambda_p^2. \tag{3.30}$$

Similarly, if $\mu=p=0$, for $\mathrm{Re}(s^{\alpha/2}/y)\gg 1$, since

$$\left(\frac{y}{A}\right)^{\frac{1+\mu}{v}}V'\left(\left(\frac{y}{A}\right)^{\frac{1}{v}}\right)=b_0,\quad \left(\frac{y}{A}\right)^{\frac{2+\mu}{v}}V''\left(\left(\frac{y}{A}\right)^{\frac{1}{v}}\right)=-b_0,$$

we have

$$\delta_g\approx\frac{1}{2v}-\frac{b_0}{2vm_\gamma^\theta}=\delta_0, \tag{3.31}$$

$$\lambda_g\approx\frac{1}{4v^2}\left(1-\frac{b_0}{m_\gamma^\theta}\right)^2=\lambda_0^2. \tag{3.32}$$

In this case, equation (3.18) can be replaced by

$$y^2\frac{\mathrm{d}^2Z}{\mathrm{d}y^2}+y\frac{\mathrm{d}Z}{\mathrm{d}y}-(\lambda_p^2+y^2)Z=0. \tag{3.33}$$

As in the discussion of case I, we have

$$W(x,t)\sim t^{-(\alpha d_f/2v)}(x/X_\theta)^{\delta'}\exp\{-\mathrm{const}\cdot(x/X_\theta)^{u'}\} \tag{3.34}$$

where $x/X_\theta\sim x/t^{\gamma/(2+\theta)}\gg 1$ and $t\to+\infty$,

$$u'=v/\left(1-\frac{\alpha}{2}\right), \tag{3.35}$$

$$\delta'=u'\left[\frac{1}{2}(\alpha d_f/v-1)-\kappa_p\right], \tag{3.36}$$

$$\kappa_p = v\left(\frac{1}{2} - \delta_p\right), \tag{3.37}$$

and

$$W(x, t) \approx \frac{G''}{\pi} t^{-\alpha d_f/2v} z^{-\kappa_p} f(z) \tag{3.38}$$

where

$$f(z) = \sum_{n=0}^{\infty} C_n(\beta) z^{nv}, \quad z = x/t^{\alpha/2v}, \quad \beta = \alpha(d_f - \kappa_p) - 1. \tag{3.39}$$

Remark 2 It is worth pointing out that, for $\theta \to 0$, since equation (1.10) reduces to equation (1.1) and $K_\gamma^\theta \to K_\gamma$, with the above discussion and results, the solution of FFPE (2.4) with respect to equation (1.1) has the asymptotic behaviour

$$W_0(x, t) \sim t^{-(\alpha d_f/2v)} (x/X_\theta)^{\delta'} \exp\{-\text{const} \cdot (x/X_\theta)^{u'}\} \tag{3.40}$$

when $x/X \sim x/t^{\gamma/2} \gg 1$ and $t \to +\infty$. It possesses a scaling variable, i. e.

$$W_0(x, t) \approx \frac{G_0''}{\pi} t^{-\alpha d_f/2v} z^{-\kappa_0} f_0(z), \tag{3.41}$$

$$f_0(z) = \sum_{n=0}^{\infty} C_n(\beta_0) z^{nv}, \quad z = x/t^{\alpha/2v}, \quad \beta_0 = \alpha(d_f - \kappa_0)/2v - 1$$

where

$$\alpha = v\gamma, \quad v = (2 + \mu + \theta')/2, \tag{3.42}$$

and

$$u' = v/\left(1 - \frac{\alpha}{2}\right), \quad \delta' = u'\left[\frac{1}{2}(d_s - 1) - \kappa_0\right]. \tag{3.43}$$

For the constant potentials and the generic potentials of equation (3.17) with $u \neq p$,

$$\kappa_0 = v\left(\frac{1}{2} - \delta\right), \quad \delta = \frac{1 + 2\mu}{2v}, \tag{3.44}$$

and for the generic potentials of equation (3.17) with $u = p \neq 0$,

$$\kappa_0 = v\left(\frac{1}{2} - \delta_p\right) \tag{3.45}$$

where δ_p is given by equations (3.29) or (3.31) for $\mu = p \neq 0$ and $\mu = p = 0$, respectively.

$d_s = \gamma d_f$ is called the fraction or spectral dimension [22, 23].

4. Conclusion and Discussion

In order to describe anomalous diffusion processes involving external potential fields on heterogeneous fractal structures, we introduce an HFFPE. Its solution possesses the following properties. The necessary and sufficient condition of its stationary solution, which is the generalized Boltzmann distribution, is that the generalization of the Einstein relation is also preferred as the Stokes-Einstein-Smoluchowski relation holds

$$K_\gamma^\theta = k_B T / m\eta_\gamma$$

for the anomalous coefficients K_γ and η_γ. For $\theta \to 0$ and $\theta' = 0$ this result is consistent with that in [9].

There exists an intrinsic relationship $\alpha = \gamma\left(\frac{2+\mu+\theta'}{2+\theta}\right)$ between γ, α, θ, μ and θ'. γ, θ and α are structural parameters of the underlying fractal structure and α can be determined explicitly; see [18~20]. If $\theta' = 0$ and $\mu = \theta$ then $\alpha = \gamma$.

The solution of the FFPE has asymptotic behaviour $\ln W(x, t) \sim -c\xi^u$ where $\xi \equiv x/t^{\gamma/2} \gg 1$, $u = v/\left(1 - \frac{\alpha}{2}\right)$, $v = (2 + \mu + \theta')/2$, and possesses a scaling variable for constant and generic potentials.

If θ tends to zero, we obtain the corresponding results for the FFPE on homogeneous fractal structures.

Acknowledgment

This work was supported by NSFC and by the Shanghai Priority Academic Discipline (Project No. 10271031).

Appendix

Using the inversion theorem of Laplace transform, from equation (3.11) we have

$$W(x,t)=\frac{G_0}{\pi}\int_0^\infty \tau^\beta \exp\{-[t\tau+G_2\tau^\alpha\cos(\alpha\pi)]\}\sin[G_2\tau^\alpha\sin(\alpha\pi)-\beta\pi]\mathrm{d}\tau$$

$$=\frac{G_0}{\pi}t^{-\beta-1}\int_0^\infty y\exp\{-[y+G_2(y/t)^\alpha\cos(\alpha\pi)]\}\sin[G_2(y/t)^\alpha\sin(\alpha\pi)-\beta\pi]\mathrm{d}y \tag{A.1}$$

where

$$G_2=\frac{x^{\theta'+1}}{(\theta'+1)GK_\gamma^\theta}.$$

Using the expressions of power series of entire-functions e^z, $\cos z$ and $\sin z$ and the integral representation of the Γ function, a simple integration of equation (A.1) yields the following expression with the dimensionless similarity variable $z=x/t^{\alpha/2v}$:

$$W(x,t)=\frac{G''}{\pi}t^{-\alpha d_f/2v}z^{-\kappa}f(z) \tag{A.2}$$

where

$$f(z)=\sum_{m=0}^{\infty}C_m(\beta)z^{mv},$$

$$C_0=a'\Gamma(\gamma d_f/2)>0,\quad C_m(\beta)=a'b_m+b'a_{m-1},$$

$$a_m(\beta)=\sum_{k,\,n\geqslant 0,\,2k+1+n=m}(-1)^{k+n}\frac{a^{2k+1}b^n}{(2k+1)!\,n!}C^{m+1}\Gamma\left(\beta+1+\frac{\alpha}{2}(2k+1+n)\right),$$

$$b_m(\beta)=\sum_{k,\,n\geqslant 0,\,2k+n=m}(-1)^{k+n}\frac{a^{2k}b^n}{(2k)!\,n!}C^m\Gamma\left(\beta+1+\frac{\alpha}{2}(2k+n)\right), \tag{A.3}$$

$$a'=-\sin(\beta\pi)>0,\quad b'=\cos(\beta\pi),\quad a=\sin\left(\frac{\alpha}{2}\pi\right)/G''',\quad b=\cos\left(\frac{\alpha}{2}\pi\right)/G'''.$$

◇ References ◇

[1] Havlin S and Avraham D B 1987 *Adv. Phys.* **36** 695

[2] Isichenko M B 1992 *Rev. Mod. Phys.* **64** 961

[3] Bouchaud J P and Georges A 1990 *Phys. Rep.* **195** 127

[4] Blumen A, Klafter J and Zumofen G 1986 *Optical Spectroscopy of Glasses* ed I Zschokke (Dordrecht: Kluwer)

[5] Losa G A and Weibl E R 1993 *Fractals in Biology and Medicine* (Basel: Birkhauser)

[6] Metzler R, Barkai E and Klafter J (unpublished)

[7] Metzler R, Klafter J and Sokolov L 1998 *Phys. Rev.* E **58** 1621

[8] Metzler R, Glöckle W G and Nonnenmacher T F 1997 *Fractals* **5** 597

[9] Metzler R, Barkai E and Klafter J 1999 *Phys. Rev. Lett.* **82** 3563

[10] El-Wakil S A, Elhanbaly A and Zahran M A 2001 *Chaos Solitons Fractals* **12** 1035

[11] Jumarie G 2001 *Chaos Solitons Fractals* **12** 1873

[12] El-Wakil S A and Zahran M A 2001 *Chaos Solitons Fractals* **12** 1929

[13] Jesperson S, Metzler R and Fogedby H C 1999 *Phys. Rev.* E **59** 2736

[14] Metzler R and Klafter J 2000 *Phys. Rep.* **339** 1

[15] Risken H 1989 *The Fokker-Planck Equation* (Berlin: Springer)

[16] Schneider W R and Wyss W 1989 *J. Math. Phys.* **30** 134

[17] Giona M and Roman H E 1992 *Physica* A **185** 87

[18] Le Mehaute A 1984 *J. Stat. Phys.* **36** 665

[19] Ren F-Y, Yu Z-G and Su F 1996 *Phys. Lett.* A **219** 59

[20] Qiu W-Y and Lü J 2000 *Phys. Lett.* A **272** 353

[21] Ren F-Y, Wang X-T and Liang J-R 2001 Determination of diffusion kernel on fractals *J. Phys. A: Math. Gen.* **34** 9815. 25

[22] Alexander S and Orbach R 1982 *J. Physique Lett.* **43** 2625

[23] Rammel R and Toulouse G 1983 *J. Physique Lett.* **44** 13

[24] Roman H E and Giona M 1992 *J. Phys. A: Math.* Gen. **25** 2107

[25] Samk S G, Kilba A A and Marichev O L 1993 *Fractional Integrals and Derivatives, Theory and Applications* (London: Gordon and Breach)

[26] Barkai E 2001 *Phys. Rev.* E **63** 046118

[27] Ahlfors L V 1979 *Complex Analysis* (New York: McGraw-Hill)

[28] Gradshteyn I S and Ryzhik I M 1980 *Table of Integrals Series and Products* (London: Academic)

[29] Abramovitz M and Stegun I A 1970 *Handbook of Mathematical Functions* (New York: Dover)

[30] Ren F-Y, Qiu W-Y, Liang J-R and Xu Y 2003 *Preprint*

Answer to an Open Problem Proposed by E Barkai and J Klafter*

Fu-Yao Ren[1], Wei-Yuan Qiu[1], Yun Xu[1] and Jin-Rong Liang[2]

[1] *Department of Mathematics, Fudan University, Shanghai 200433, People's Republic of China*

[2] *Department of Mathematics, East China Normal University, Shanghai 200062, People's Republic of China*

Abstract: In a negative answer to the open problem proposed by E Barkai and J Klafter, it is proved that the theory presented by Amblard *et al* is not consistent with the GER. We suggest applying a theory in terms of a fractional Fokker-Planck equation to model the experiment measured by Amblard *et al*. The result obtained is consistent with the statement by Amblard *et al* (1996 *Phys. Rev. Lett.* 77 4470) that the numerical pre-factors of the power law would be modified by local geometry and are not exact.

1. Introduction

In a recent experiment Amblard *et al* [1] measured anomalous transport properties of magnetic beads embedded in a three-dimensional polymer network. In their paper, Amblard *et al* suggested a theory describing the motion of the beads.

Amblard *et al* found that when the beads are subjected to an external uniform force $\vec{F}$, the drift follows

---

* Originally published in *J. Phys. A: Math. Gen.*, Vol. 37, (2004), 9919 - 9922.

$$\langle x_{\parallel}(t)\rangle_F \sim t^p, \qquad \text{with } p=0.76\pm 0.03, \tag{1}$$

where $x_{\parallel}$ is the component of $\vec{x}$ along $\vec{F}$. In addition, they showed that the response to the external bias is linear. The authors also measured the diffusion of the beads in the absence of the field, and found

$$\langle x_{\parallel}^2(t)\rangle_0 \sim t^q, \qquad \text{with } q=0.73\pm 0.01. \tag{2}$$

These observed anomalous power laws (1) and (2) can be readily explained if we consider the bead deforming the filaments of the cage surrounding it. Indeed, let us consider such a filament with bending constant κ, in a solvent of viscosity η. If s is the internal curvilinear coordinate along the polymer and $r(t, s)$ the transverse deformation of the filament, then the equation of movement for $r(t, s)$ is

$$\eta\frac{\partial r}{\partial t}=\kappa\frac{\partial^4 r}{\partial s^4}+f(t, s), \tag{3}$$

where $f(t, s)$ is the force acting on the filament. It is straightforward to show that the Green function, $F(t, s)$, associated with this equation can be written in the following scaling form:

$$F(t, s)=\frac{1}{\eta s}F\left(\frac{\eta s^4}{\kappa t}\right)\equiv\frac{1}{\eta^{3/4}\kappa^{1/4}t^{1/4}}\widetilde{F}\left(\frac{\eta s^4}{\kappa t}\right). \tag{4}$$

If, for the sake of simplicity, we assume that the magnetic bead applies a constant point force, f at $s=0$, then the displacement of the centre of mass of the bead is given by

$$\langle x\rangle_f \approx \langle r(t, 0)\rangle_f = f\int_0^t F(t-t', 0)\mathrm{d}t' \approx \frac{4}{3}f\frac{\widetilde{F}(0)t^{3/4}}{\eta^{3/4}\kappa^{1/4}}. \tag{5}$$

In the absence of the external force, the bead movement is dominated by the thermal motions of the surrounding filaments to which it is coupled. In this case, the mean-square displacement of the bead is the mean-square displacement of the filament,

$$\langle x^2\rangle_0 = \langle r^2(t, s)\rangle_0 = k_B T\eta\int \mathrm{d}s'\int_0^t F^2(t-t', s-s')\mathrm{d}t' \propto \frac{k_B T t^{3/4}}{\eta^{3/4}\kappa^{1/4}}. \tag{6}$$

It is interesting to note that, when one evaluates the numerical pre-factors in (5) and (6), one obtains a semi-quantitative agreement with known values of the f -actin rigidity [2]. The power laws of the time dependence described by (5) and (6), $t^{3/4}$, are not altered if one considers the bead interacting simultaneously with several filaments of the cage. The details of local geometry should modify numerical pre-factors in these theoretical formulae; undoubtedly, the bead has to push against several filaments simultaneously while the confined geometry must change the effective friction constants involved. In [3], Barkai and Klafter pointed out that although the validity of the generalized Einstein relation (GER) has not been discussed by the authors of [1], their measurements provide a direct opportunity to check this relation. When analysing the corrected results of [1] one finds that to a good approximation (equation (7)), the GER [4]

$$\langle x_{\|}^{2}(t)\rangle_{0}=\frac{2k_{\mathrm{B}}T\langle x_{\|}(t)\rangle_{F}}{|\vec{F}|} \tag{7}$$

is valid. They conclude that the GER is well suited to describe the anomalous transport properties. They could not conclude whether the theory presented by Amblard *et al* is consistent with the GER. It would be interesting to clarify this issue. In section 2, we answer this open problem. In section 3, we give a theory to the GER.

2. Answer to the Above Open Problem

In this section, we discuss the open problem, i. e., whether the theory presented by Amblard *et al* [1] is consistent with the GER. From the second equality of equation (4), we have

$$F(\zeta)=\zeta^{\frac{1}{4}}\widetilde{F}(\zeta), \tag{8}$$

where $\zeta=\eta s^{4}/\kappa t$. Since $\widetilde{F}(0)=\lim_{\zeta\to 0}[F(\zeta)/\zeta^{1/4}]$ and $\widetilde{F}(0)\neq 0$, for any given $\varepsilon>0$ there exists a positive number $\delta=\delta(\varepsilon)$ related with ε such that for all ζ, $|\zeta|<\delta$, we have

$$F(\zeta) \approx \widetilde{F}(0)\zeta^{1/4}. \tag{9}$$

Hence,

$$F(t, s) \approx \frac{\widetilde{F}(0)}{\eta^{3/4}\kappa^{1/4}} t^{-\frac{1}{4}} \tag{10}$$

for all s such that $|s| < \left(\frac{\kappa\delta}{\eta}\right)^{1/4} t^{1/4}$. From (6) and (10), a simple calculation yields

$$\langle x^2 \rangle_0 \approx 2k_B T \frac{s\widetilde{F}^2(0)}{\eta^{1/2}\kappa^{1/2}} t^{1/2} \tag{11}$$

for $|s| < \left(\frac{\kappa\delta}{\eta}\right)^{1/4} t^{1/4}$. This implies that

$$\langle x^2 \rangle_0 < 2k_B T \delta^{\frac{1}{4}} \frac{\widetilde{F}^2(0)}{\eta^{3/4}\kappa^{1/4}} t^{3/4}. \tag{12}$$

From (5) and (12), finally we have

$$\langle x^2 \rangle_0 < \frac{3\widetilde{F}(0)}{4} \delta^{\frac{1}{4}} \left[\frac{2k_B T}{f} \langle x \rangle_f\right] \tag{13}$$

for any given ε. Since $\delta = \delta(\varepsilon) \to 0$ when $\varepsilon \to 0$, then for any given $\rho < 1$ and for sufficiently small ε, we have

$$\langle x^2 \rangle_0 < \rho \frac{2k_B T}{f} \langle x \rangle_f. \tag{14}$$

But we know that $\langle x^2 \rangle_0 = \frac{2k_B T}{f} \langle x \rangle_f$ [3]. This shows that the theory of [1] is not consistent with the GER.

3. GER for FFPE

In [5], in order to describe anomalous systems close to thermal equilibrium based on fractional derivatives, Metzler, Barkai and Klafter presented a one-dimensional fractional Fokker-Planck equation (FFPE):

$$ {}_0D_t^{\alpha} W(x, t) = G L_{\mathrm{FP}} W(x, t), \tag{15} $$

where $W(x, t)$ is the probability density function (pdf) at position x at time t. The L_{FP} operator

$$ L_{\mathrm{FP}} = \frac{\partial}{\partial x}\left(\frac{V'(x)}{m\eta_\gamma} + K_r \frac{\partial}{\partial x}\right) \tag{16} $$

with the external potential $V(x)$ [6] contains the anomalous diffusion constant K_γ, and the anomalous friction coefficient η_γ with the dimension $[\eta_\gamma] = \mathrm{s}^{\gamma-2}$; herein, m denotes the mass of the diffusion particle. And

$$ {}_0D_t^{\alpha} W(x, t) = \frac{1}{\Gamma(1-\alpha)} \frac{\partial}{\partial t} \int_0^t \mathrm{d}\tau \frac{W(x, \tau)}{(t-\tau)^{\alpha}}, \quad 0 < \alpha < 1. \tag{17} $$

Applying Laplace transformation to (15), it becomes

$$ s^{\alpha} W(x, s) = G \frac{\partial}{\partial x}\left[\frac{V'(x)}{m\eta_\gamma} + K_\gamma \frac{\partial}{\partial x}\right] W(x, s). \tag{18} $$

Anomalous diffusion in the homogeneous fractal medium in one dimension is characterized by the occurrence of a mean-square displacement of the form

$$ \langle (\Delta x)^2(t) \rangle_0 = \frac{2K_\gamma}{\Gamma(1+\gamma)} t^{\gamma} \tag{19} $$

when no external driving force is applied to the particle. The extraction of moments $\langle (\Delta x)^n \rangle$ for anomalous diffusion on fractal medium with dimension d_{f} is given by [7]

$$ \langle (\Delta x)^n(t) \rangle = \int_0^{\infty} \mathrm{d}x \cdot x^{d_{\mathrm{f}}-1} x^n W(x, t). \tag{20} $$

When the particle is assumed in a constant force field, say $V(x) = -Fx$, by the moment expression (equation (20)) for order one and from the solution of (18), we have the displacement of the particle

$$ \langle (\Delta x)(t) \rangle_F = \frac{F\Gamma(d_{\mathrm{f}}+1)}{m\eta_\gamma \Gamma(d_{\mathrm{f}}+2)\Gamma(1+\gamma)} t^{\gamma}. \tag{21} $$

Thus, from (19), (21) and the generalized Einstein-Stokes-Smoluchowski relation [5]

$$K_\gamma = \frac{k_B T}{m\eta_\gamma}, \tag{22}$$

we get the GER

$$\langle(\Delta x)(t)\rangle_F = \frac{1}{d_f + 1}\frac{F\langle(\Delta x)^2(t)\rangle_0}{k_B T}. \tag{23}$$

It shows that the GER holds for FFPE (equation (15)), and that the pre-factor of GER is not a universal constant, while it is given by $d_f + 1$, where d_f is the fractal dimension of the fractal structure considered and $1 < d_f < 2$. This just shows that the numerical pre-factors of the power law should be modified by the local geometry as Amblard *et al* pointed out. In particular, when $d_f \to 1$, it reduces to

$$\langle(\Delta x)(t)\rangle_F = \frac{1}{2}\frac{F\langle(\Delta x)^2(t)\rangle_0}{k_B T}. \tag{24}$$

This is just the result of [5].

4. Conclusion

It is proved that the theory presented by Amblard *et al* is not consistent with the GER. The theory in terms of a fractional Fokker-Planck equation (FFPE) can be well modelled for the experiment by Amblard *et al*. The result obtained by FFPE is consistent with the statement by Amblard *et al* [1] that the numerical pre-factors of the power law would be modified by local geometry and are not exact.

Acknowledgments

Project No. 10271031 is supported by NSFC and the Shanghai Priority Academic Discipline.

◇ References ◇

[1] Amblard F, Maggs A C, Yurke B, Pargellis A N and Leibler S 1996 *Phys. Rev. Lett.*

77 4470

[2] Farge E and Maggs A C 1993 *Macromolecules* **26** 5041

[3] Barkai E and Klafter J 1998 *Phys. Rev. Lett.* **81** 1134

[4] Bouchaud J P and Georges A 1990 *Phys. Rep.* **195** 127

[5] Metzler R, Barkai E and Klafter J 1999 *Phys. Rev. Lett.* **82** 3563

[6] Risken H 1989 *The Fokker-Planck Equation* (Berlin: Springer)

[7] Ren F Y, Liang J R, Qiu W Y and Xu Y 2003 *J. Phys. A: Math. Gen.* **36** 7533

Answer to an Open Problem Proposed by R Metzler and J Klafter*

Fu-Yao Ren[1], Jin-Rong Liang[2], Wei-Yuan Qiu[1] and Jian-Bin Xiao[3]

[1] *Department of Mathematics, Fudan University, Shanghai 200433, People's Republic of China*

[2] *Department of Mathematics, East China Normal University, Shanghai 200062, People's Republic of China*

[3] *School of Science, Hangzhou Dianzi University, Hangzhou 310037, People' Republic of China*

Abstract: In a positive answer to the open problem proposed by R Metzler and J Klafter, it is proved that the asymptotic shape of the solution for a wide class of fractional Fokker-Planck equations is a stretched Gaussian for the initial condition being a pulse function in the homogeneous and heterogeneous fractal structures, whose mean square displacement behaves like $\langle(\Delta x)^2(t)\rangle \sim t^\gamma$ and $\langle(\Delta x)^2(t)\rangle \sim x^{-\theta} t^\gamma (0 < \gamma < 1, -\infty < \theta < +\infty)$, respectively.

1. Introduction

Anomalous diffusion in one dimension is characterized by the occurrence of a mean square displacement of the form

$$X_0^2 = \langle(\Delta x)^2(t)\rangle_0 = \frac{2K_\gamma}{\Gamma(1+\gamma)} t^\gamma. \tag{1}$$

* Originally published in *J. Phys. A: Math. Gen.*, Vol. 39, (2006), 4911 - 4919.

The subscript zero in $\langle \cdot \rangle_0$ denotes the case when no external driving force is applied to the particle. When $\gamma \neq 1$ the diffusion is anomalous; the case $0<\gamma<1$ is called slow diffusion or sub-diffusion and $\gamma > 1$ is called super-diffusion.

Recently, in the literature, in order to describe diffusion processes in disordered media, some authors proposed an extension of the Fokker-Planck equation, which is called the fractional Fokker-Planck equation (FFPE) [1~10]. In [1], a fractional Fokker-Planck equation involving external field

$$\frac{\partial W(x, t)}{\partial t} = {}_0D_t^{1-\gamma} L_{\mathrm{FP}} W, \quad 0<\gamma<1, \tag{2}$$

was presented to describe the anomalous transport close to thermal equilibrium, where $W(x, t)$ is the probability density function at position x at time t, and the Fokker-Planck operator

$$L_{\mathrm{FP}} = \frac{\partial}{\partial x}\left(\frac{V'(x)}{m\eta_\gamma} + K_\gamma \frac{\partial}{\partial x}\right) \tag{3}$$

with the external potential $V(x)$ [11] contains the anomalous diffusion constant K_γ and the anomalous friction coefficient η_γ with the dimension $[\eta_\gamma] = \sec^{\gamma-2}$; herein m denotes the mass of the diffusion particle, and

$${}_0D_t^{1-\gamma} W = \frac{1}{\Gamma(\gamma)} \frac{\partial}{\partial t} \int_0^t \mathrm{d}\tau \frac{W(x, \tau)}{(t-\tau)^{1-\gamma}}. \tag{4}$$

The fractional Fokker-Planck equation is shown [5] to be closely related to the Scher-Lax-Montroll (SLM) model [12~15] defined within the context of the continuous time random walk and to the collision models [15~18].

In [9], it is discussed that anomalous diffusion in a heterogeneous fractal medium in one dimension is characterized by the occurrence of a mean square displacement of the form

$$X_\theta^2 = \langle (\Delta x)^2(t) \rangle_0 \sim x^{-\theta} t^\gamma, \quad 0<\gamma<1, \quad \theta = d_{\mathrm{w}} - 2, \tag{5}$$

where $d_{\mathrm{w}} > 2$ is the anomalous diffusion exponent [19]. By simple scaling consideration, equation (5) is equivalent to [20, 21]

$$X_\theta^2 = \langle (\Delta x)^2(t) \rangle_0 = \frac{2K_\gamma^\theta}{\Gamma(1+\gamma_\theta)} t^{\gamma_\theta}, \tag{6}$$

where

$$\gamma_\theta = 2\gamma/(2+\theta), \quad 0 < \gamma_\theta < 1. \tag{7}$$

In equation (6) the anomalous diffusion coefficient K_γ^θ is introduced, which has the dimension $[K_\gamma^\theta] = \mathrm{cm}^2\,\mathrm{sec}^{-2\gamma/(2+\theta)}$. It is easy to see that equation (6) reduces to equation (1) and K_γ^θ to K_γ when $\theta \to 0$. Indeed, θ can be any real number in this paper.

In [22], it is pointed out that the detailed structure of the propagator $W(x, t)$ (i. e. the probability density function) for the initial condition $\lim_{t\to 0+} W(x, t) = \delta(x)$ depends generally on the special shape of the underlying geometry. However, the interesting part of the propagator has the asymptotic behaviour $\ln W(x, t) \sim -c\xi^u$ where $\xi \equiv x/t^{\alpha/2} \gg 1$, which is expected to be universal. Here, $u = 1/(1-\alpha/2)$ with the anomalous diffusion exponent α which is the order of the fractional derivative. Does the stretched Gaussian universality hold? It is an open problem.

It would be interesting to clarify this issue. In section 4, in a positive answer to the open problem proposed by R Metzler and J Klafter [22], it is proved that the propagator has the asymptotic behaviour $\ln W(x, t) \sim -c\xi^u$ for the solution of a wide class of the fractional Fokker-Planck equations (FFPE) defined below, where $\xi \equiv x/t^{\alpha/2} \gg 1$ and $u = 1/(1-\alpha/2)$.

2. FFPE with Variable Coefficient

In this section, we will derive the fractional Fokker-Planck equation with a variable coefficient involving the external potential $V(x)$ by using heuristic argument of Giona and Roman [23].

The relationship between the total flux of probability current $S(x, t)$ from time $t = 0$ to time t and the average probability density $W(x, t)$, considered as the input and output of the fractal system [24], should satisfy the following equation (cf [23]):

$$\int_0^t S(x, \tau)\mathrm{d}\tau = x^{d_f - 1}\int_0^t K(t, \tau) W(x, \tau)\mathrm{d}\tau, \tag{8}$$

where d_f is the fractal dimension of the fractal structure of the system considered. This is a conservation equation containing an explicit reference to the history of the diffusion process on the fractal structure. Since we are dealing with stationary processes, we expect that $K(t, \tau)$ be a function of difference $t-\tau$ only, i. e. $K(t, \tau)=K(t-\tau)$. We say $K(t, \tau)$ the diffusion kernel. We assume that diffusion sets are underlying fractals (underlying fractals denote self-similar sets in [25] or net fractals in [26, 27]), and the diffusion kernel on the underlying fractal should behave as

$$K(t-\tau)=\frac{A_\alpha}{(t-\tau)^\alpha}, \quad 0<\alpha<1, \tag{9}$$

where α is a diffusion exponent and A_α is a constant that can be determined [27].

On the other hand, we propose that the probability current $S(x, t)$ on the above fractional structure satisfies the following structure equation:

$$S(x, t)=Bx^{d_f-1}x^{-\theta'}L_{FP}^{K,\nu,\mu}W(x, t), \tag{10}$$

where $\nu, \mu \in \mathbb{R}$, $B>0$ is to be determined, θ' is a non-negative parameter, and

$$L_{FP}^{K,\nu,\mu}W(x, t)=\frac{\partial}{\partial x}\left(\frac{V'(x)}{m\eta_\gamma}W(x, t)+Kx^{-\nu}\frac{\partial}{\partial x}x^{-\mu}W(x, t)\right), \tag{11}$$

where $K=K_\gamma^\theta$. From equations (8)~(10), we have

$${}_0D_t^\alpha W(x, t)=Gx^{-\theta'}L_{FP}^{K,\nu,\mu}W(x, t), \tag{12}$$

where $G=B/\Gamma(1-\alpha)A_\alpha>0$. We call equation (12) a conjugate fractional Fokker-Planck equation with a variable coefficient.

By the theorem of the conjugate operator of the Riemann-Liouville fractional integrals and derivatives ([28], p 45), it follows from equation (12) that

$$W(x, t)=\frac{1}{\Gamma(\alpha)}\int_0^t Gx^{-\theta'}\frac{L_{FP}^{K,\nu,\mu}W(x, \tau)}{(t-\tau)^{1-\alpha}}d\tau, \tag{13}$$

for $W(x, t)\in I_0^\alpha(L_1)$. Hence, we obtain

$$\frac{\partial W(x,t)}{\partial t}=Gx^{-\theta'}{}_0D_t^{1-\alpha}L_{\mathrm{FP}}^{K,\nu,\mu}W(x,t). \tag{14}$$

This is a fractional Fokker-Planck equation (FFPE). Especially, when $G=1$, $\theta'=\nu=\mu=0$ and $\alpha=\gamma$, we have

$$\frac{\partial W(x,t)}{\partial t}={}_0D_t^{1-\gamma}L_{\mathrm{FP}}W. \tag{15}$$

This is just the fractional Fokker-Planck equation (2). Its derivation is different from that in [1, 29].

Furthermore, we suppose the probability density function $W(x,t)$ satisfying the following normalization condition:

$$\int_0^\infty \mathrm{d}x\cdot x^{d_f-1}W(x,t)=1, \tag{16}$$

and boundary condition

$$\lim_{x\to\infty}W(x,t)=0. \tag{17}$$

The extraction of moments $\langle(\Delta X)^n\rangle$ is defined by

$$\langle(\Delta X)^n\rangle=\int_0^\infty \mathrm{d}x\cdot x^{d_f-1}x^nW(x,t). \tag{18}$$

We can anticipate the relation between exponents α, γ, θ, ν, μ and θ' by simple scaling considerations. From (14) we see that $t^\alpha\sim x^{2+\theta'+\nu+\mu}$, and according to (5) we require that $t^\gamma\sim x^{2+\theta}$. Thus,

$$\alpha=\frac{\gamma(2+\nu+\mu+\theta')}{2+\theta},\quad 0<\frac{\gamma(2+\nu+\mu+\theta')}{2+\theta}<1. \tag{19}$$

3. Fokker-Planck Equations in Frequency Domain

In this section, we respectively give the solutions of the fractional Fokker-Planck equation (14) in the case of constant potentials, linear potentials, harmonic potentials, analytic potentials, logarithm potentials, pole potentials and generic potentials, which lead to force-free, uniform force, the Hookean force directed at the original, nonlinear force, hyperbolic force and mixed force, respectively. Using the Laplace transform of the fractional derivative, it follows from equation (14) that

$$x^2\frac{\partial^2 W(x,s)}{\partial x^2}+\left[x^{\nu+\mu}\frac{xV'(x)}{Km\eta_\gamma}-(\nu+2\mu)\right]x\frac{\partial W(x,s)}{\partial x}$$
$$+\left[\mu(\nu+\mu+1)+\frac{x^{\nu+\mu+2}V''(x)}{Km\eta_\gamma}-\frac{x^{\nu+\mu+\theta'+2}s^{\alpha}}{KG}\right]W(x,s)$$
$$=-\frac{x^{\nu+\mu+\theta'+2}s^{\alpha-1}}{GK}W(x,0). \tag{20}$$

We assume that the external potential at the original has the form

$$V(x)=\sum_{j=1}^{k}b_jx^{-j}+b\ln x+\sum_{n=0}^{\infty}a_nx^n, \tag{21}$$

where b_j, b, a_n can be zero.

In order to solve equation (20), it is convenient to perform the transform

$$y=A(s)x^{v},\quad W(x,s)=y^{\delta}Z(y) \tag{22}$$

to cast equation (20) into the second-order Bessel equation as

$$y^2\frac{\mathrm{d}^2z}{\mathrm{d}y^2}+y\frac{\mathrm{d}z}{\mathrm{d}y}-(\lambda^2+y^2)z(y)=-\frac{y^{2-\delta}}{s}W(x,0), \tag{23}$$

with parameter λ^2 under the following conditions:

$$v=\frac{1}{2}(\nu+\mu+\theta'+2),\quad A(s)=\frac{1}{v}\left(\frac{s^{\alpha}}{KG}\right)^{1/2}, \tag{24}$$

$$\delta=\frac{1}{2v}\left\{\nu+2\mu+1-\frac{1}{Km\eta_\gamma}\left(\frac{y}{A}\right)^{(\nu+\mu)/v}\left[-\sum_{j=1}^{k}jb_j\left(\frac{y}{A}\right)^{-j/v}+b+\sum_{n=1}^{\infty}na_n\left(\frac{y}{A}\right)^{n/v}\right]\right\}, \tag{25}$$

$$\lambda^2=\frac{\delta(1-v\delta)}{v}+\frac{\delta(\nu+2\mu)}{v}-\frac{\mu(\nu+\mu+1)}{v^2}$$
$$-\frac{1}{v^2Km\eta_\gamma}\left(\frac{y}{A}\right)^{(\nu+\mu-k)/v}\sum_{j=1}^{k}j(j+1-\delta v)b_j\left(\frac{y}{A}\right)^{(k-j)/v}$$
$$+\frac{1}{v^2Km\eta_\gamma}\left(\frac{y}{A}\right)^{(\nu+\mu)/v}\left[(1-\delta v)b-\sum_{n=1}^{\infty}n(n-1+\delta v)a_n\left(\frac{y}{A}\right)^{n/v}\right]. \tag{26}$$

When $\operatorname{Re}(s^{\alpha/2})/y \gg 1$, i. e. $y/|A| \ll 1$, we have the asymptotic representations of δ, λ corresponding to different potentials as follows.

Case 1. $b_k \neq 0$, i. e. the origin is a pole of order k. It follows from (25) and (26) that if $\nu+\mu>k$,

$$\delta \approx \delta_0=\frac{\nu+2\mu+1}{2v}, \quad \lambda^2 \approx \lambda_0^2=\frac{(1+\nu)^2}{4v^2}>0, \tag{27}$$

which are independent of the external potentials (21) and if $\nu+\mu=k$,

$$\delta \approx \delta_0=\frac{1}{2v}(k+\mu+1+\hat{b}_k), \quad \hat{b}_k=\frac{kb_k}{Km\eta_\gamma}, \tag{28}$$

$$\lambda^2 \approx \lambda_0^2=\frac{1}{4v^2}(k+1-\mu-\hat{b}_k)^2=\frac{1}{4v^2}(1+\nu-\hat{b}_k)^2>0, \tag{29}$$

which depend only upon the coefficient of the pole of order k.

Case 2. $b_j=0$ $(j=1, 2, \cdots, k)$ but $b \neq 0$, i. e.

$$V(x)=b\ln x+\sum_{n=0}^{\infty} a_n x^n. \tag{30}$$

It follows from (25) and (26) that if $\nu+\mu>0$,

$$\delta \approx \delta_0=\frac{\nu+2\mu+1}{2v}, \quad \lambda^2 \approx \lambda_0^2=\frac{(1+\nu)^2}{4v^2}>0, \tag{31}$$

which are independent of the external potentials (30) and if $\nu+\mu=0$,

$$\delta \approx \delta_0=\frac{1}{2v}(\mu+1-\hat{b}), \quad \hat{b}=\frac{b}{Km\eta_\gamma}, \tag{32}$$

$$\lambda^2 \approx \lambda_0^2=\frac{1}{4v^2}(1-\mu+\hat{b})^2>0, \tag{33}$$

which depend only upon b, i. e. the coefficient of the term of the logarithmic singularity at the origin of the external potentials (30).

Case 3. $b_j=b=a_n=0$ for $1 \leqslant j \leqslant k$ and $n<N$, but $a_N \neq 0$, i. e.

$$V(x)=\sum_{n=N}^{\infty} a_n x^n. \tag{34}$$

It follows from (25) and (26) that if $\nu+\mu+N>0$,

$$\delta \approx \delta_0 = \frac{1}{2v}(\nu + 2\mu + 1), \quad \lambda^2 \approx \lambda_0^2 = \frac{(1+\nu)^2}{4v^2} > 0, \tag{35}$$

which are independent of the external potentials (34) and if $\nu + \mu + N = 0$, i. e. $\nu + \mu = -N$,

$$\delta \approx \delta_0 = \frac{1}{2v}(1 + \nu - N - A_N), \quad A_N = \frac{Na_N}{Km\eta_\gamma}, \tag{36}$$

$$\lambda^2 \approx \lambda_0^2 = \frac{1}{4v^2}(\mu + N - 1 - A_N)^2 = \frac{1}{4v^2}(1 + \nu + A_N)^2 > 0, \tag{37}$$

which depend only upon a_N, the coefficient of the first nonzero term of the external potentials (34).

Thus, equation (22) can be replaced by

$$y^2 \frac{d^2 z}{dy^2} + y \frac{dz}{dy} - (\lambda_0^2 + y^2) z(y) = -\frac{y^{2-\delta_0}}{s} W(x, 0). \tag{38}$$

4. Solutions and Its Properties of FFPE

In the case of $W(x, 0) = \delta(x)$, equations (20) and (38) become

$$x^2 \frac{\partial^2 W(x, s)}{\partial x^2} + \left[x^{\nu+\mu} \frac{xV'(x)}{Km\eta_\gamma} - (\nu + 2\mu)\right] x \frac{\partial W(x, s)}{\partial x} + \left[\mu(\nu + \mu + 1) + \frac{x^{\nu+\mu+2} V''(x)}{Km\eta_\gamma} - \frac{x^{\nu+\mu+\theta'+2} s^\alpha}{KG}\right] W(x, s) = -\frac{x^{\nu+\mu+\theta'+2} s^{\alpha-1}}{GK} \delta(x) \tag{39}$$

and

$$y^2 \frac{d^2 z}{dy^2} + y \frac{dz}{dy} - (\lambda_0^2 + y^2) z(y) = -\frac{y^{2-\delta_0}}{s} \delta\left(\left(\frac{y}{A(s)}\right)^{1/v}\right), \tag{40}$$

respectively.

The Delta function $\delta(x)$ can be approximated by the function

$$\delta_h(x) = \begin{cases} 0, & x < 0 \text{ or } x > h, \\ \dfrac{1}{h}, & 0 < x \leqslant h. \end{cases} \tag{41}$$

Thus, equations (39) and (40) can also be approximated by

$$x^2 \frac{\partial^2 W(x, s)}{\partial x^2} + \left[x^{\nu+\mu} \frac{xV'(x)}{Km\eta_\gamma} - (\nu + 2\mu)\right] x \frac{\partial W(x, s)}{\partial x}$$
$$+ \left[\mu(\nu + \mu + 1) + \frac{x^{\nu+\mu+2} V''(x)}{Km\eta_\gamma} - \frac{x^{\nu+\mu+\theta'+2} s^\alpha}{KG}\right] W(x, s)$$
$$= -\frac{x^{\nu+\mu+\theta'+2} s^{\alpha-1}}{GK} \delta_h(x) \tag{42}$$

and

$$y^2 \frac{\mathrm{d}^2 z}{\mathrm{d}y^2} + y \frac{\mathrm{d}z}{\mathrm{d}y} - (\lambda_0^2 + y^2) z(y) = -\frac{y^{2-\delta_0}}{s} \delta_h \left(\left(\frac{y}{A(s)}\right)^{1/v}\right), \tag{43}$$

where δ_0 and λ_0^2 are respectively defined by (27)~(29); (31)~(33) and (35)~(37).

Using the operator method, we know that

$$Z_h^*(y) = \mathrm{e}^{-\int A_2 \mathrm{d}y} \int \mathrm{e}^{\int (A_2 - A_1) \mathrm{d}y} \left(\int f_h(y) \mathrm{e}^{\int A_1 \mathrm{d}y} \mathrm{d}y\right) \mathrm{d}y \tag{44}$$

is a special solution of equation (43), where

$$A_1 = -\frac{1}{2}\left[\frac{1}{y} - \sqrt{\frac{1}{y^2} + 4\left(1 + \frac{\lambda_0^2}{y^2}\right)}\right], \tag{45}$$

$$A_2 = -\frac{1}{2}\left[\frac{1}{y} + \sqrt{\frac{1}{y^2} + 4\left(1 + \frac{\lambda_0^2}{y^2}\right)}\right], \tag{46}$$

$$f_h(y) = -\frac{y^{-\delta_0}}{s} \delta_h \left(\left(\frac{y}{A(s)}\right)^{1/v}\right). \tag{47}$$

To fit the boundary condition $\lim_{x \to +\infty} W(x, t) = 0$, i.e. $W(x, s) = 0 (x \to +\infty)$, we should take an appropriate integral constant in equation (44) such that $Z_h^*(y) = 0$ as $y \to +\infty$. Thus, by the summability, we get the solution of equation (43),

$$Z_h(y) = C_h(s) K_{\lambda_0}(y) + Z_h^*(y), \tag{48}$$

where

$$K_{\lambda_0}(y)=e^{-y}\left(\frac{\pi}{2y}\right)^{1/2}\left[1+O\left(\frac{1}{y}\right)\right] \tag{49}$$

in the domain $|\arg y|<\frac{3}{2}\pi$ is the modified Bessel function of second order, and $C_h(s)$ is to be determined.

So the solution of equation (42) is given by

$$W_h(x, s)=C_h(s)y^{\delta_0}K_{\lambda_0}(y)+y^{\delta_0}Z_h^*(y), \quad y=A(s)x^v. \tag{50}$$

By the normalization condition of $W(x, t)$, i. e. $\int_0^\infty dx \cdot x^{d_f-1}W(x, s)=\frac{1}{s}$, it is not difficult to see that

$$C_h(s)=G's^{(\alpha d_f/2v)-1}+G''s^{\alpha(d_f/2v)+\alpha}h^{(d_f-1+2v)}[1+O(h^{2v})]. \tag{51}$$

Thus, if $d_f-1-2v>0$, i. e. $\nu+\mu>-(d_f+1-\theta')$, noting that $\lim_{h\to 0}C_h(s)=C(s)$, where

$$C(s)=G's^{(\alpha d_f/2v)-1}, \quad G'=v^{1-d_f/v}/[C_{\lambda_0}(GK_\gamma^\theta)^{d_f/2v}], \tag{52}$$

$C_{\lambda_0}=\int_0^\infty dy \cdot y^{\frac{d_f}{v}+\delta_0-1}K_{\lambda_0}(y)$ is a constant since [30]

$$\int_0^\infty dy \cdot y^\mu K_{\lambda_0}(ay)=2^{\mu-1}a^{-\mu-1}\Gamma\left(\frac{1+\mu+\lambda_0}{2}\right)\Gamma\left(\frac{1+\mu-\lambda_0}{2}\right).$$

The solution of equation (39) is given by

$$W(x, s)=C(s)y^{\delta_0}K_{\lambda_0}(y)+y^{\delta_0}Z^*(y), \quad y=A(s)x^v, \tag{53}$$

where

$$Z^*(y)=e^{-\int A_2 dy}\int e^{\int(A_2-A_1)dy}\left(\int f(y)e^{\int A_1 dy}dy\right)dy, \tag{54}$$

$$f(y)=-\frac{-y^{\delta_0}}{s}\delta\left(\left(\frac{y}{A(s)}\right)^{1/v}\right)=-\frac{y^{-\delta_0}}{s}\delta(x). \tag{55}$$

It is easy to see from equations (6) and (18) (when $=2$) that

$$\alpha=\frac{(2+\theta+\theta')\gamma}{2+\theta}, \quad B=\frac{A_\alpha\Gamma(1-\alpha)}{v^2}\left(\frac{2C_{\lambda_0}}{C'_{\lambda_0}}\right)^v(K_\gamma^\theta)^{v-1}, \tag{56}$$

since $\int_0^\infty \mathrm{d}x \cdot x^{d_f-1} x^2 W(x, s) = 2K_\gamma^\theta / s^{2\gamma/(2+\theta)+1}$ corresponding to the Laplace transform of $\langle (\Delta x)^2 \rangle (t)$, $C'_{\lambda_0} = \int_0^\infty \mathrm{d}y \cdot y^{\frac{d_f+2}{v}+\delta_0-1} K_{\lambda_0}(y)$.

Let us now turn to discuss the asymptotic behaviour of $W(x, t)$ as predicted by equation (53). Noting that $Z^*(y) = 0$ when $x > 0$, it follows from (53) and (49) that

$$W(x, s) \approx G'' s^{-(1-\alpha d_f/2v)} \frac{1}{(xs^{\alpha/2v})^\kappa} \exp\{-(xs^{\alpha/2v}/G''')^v\} \tag{57}$$

for $|xs^{\alpha/2v}| \gg 1$, where $G'' = \sqrt{\frac{\pi}{2}} v^{1-(d_f+\kappa)/v} / C_\lambda (GK_\gamma^\theta)^{(d_f+\kappa)/2v} > 0$, $G''' = (v\sqrt{GK_\gamma^\theta})^{1/v}$, $\kappa = v\left(\frac{1}{2} - \delta_0\right)$.

By the same method of [31, 32], we expected that

$$W(x, t) \sim t^{-(\alpha d_f/2v)} (x/X_\theta)^{\delta'} \exp\{-\text{const} \cdot (x/X_\theta)^{u'}\}, \tag{58}$$

when $\xi \equiv x/X_\theta \sim x/t^{\gamma/(2+\theta)} \gg 1$ and $t \to +\infty$. The Laplace transform of (58) can be evaluated by applying the method of steepest descent, and the result is compared with (57). This yields

$$u' = v\Big/\left(1 - \frac{\alpha}{2}\right) \tag{59}$$

and

$$\delta' = u'\left[\frac{1}{2}(\alpha d_f/v - 1) - \kappa\right]. \tag{60}$$

It is interesting that if $\theta = \theta' = 0$, $\nu = 0$, we have $u' = 1\Big/\left(1 - \frac{\gamma}{2}\right)$ and $\delta' = u'\left[\frac{1}{2}(\gamma d_f - 1)\right]$, which are same as that of FFPE in [33, 34] under the condition $W(x, 0) = 0$.

This shows that if $d_f - 1 - 2v > 0$, i. e. $\nu + \mu > -(d_f + 1 - \theta')$, the stretched Gaussian universality holds for the fractional Fokker-Planck equations (14) as expected by R Metzler and J Klafter.

Remark From the above discussion, we can see that the results (58)~(60) for the asymptotic behaviour of $W(x, t)$ also hold under the initial condition $W(x, 0)=0$.

Acknowledgments

We would like to acknowledge the support of projects No. 10571028 and No. 10371043 by NSFC and by the Shanghai Leading Academic Discipline.

◇ References ◇

[1] Metzler R, Barkai E and Klafter J 1999 *Phys. Rev. Lett.* **82** 3563

[2] Metzler R, Klafter J and Sokolov I M 1998 *Phys. Rev.* E **58** 1621

[3] Metzler R and Klafter J 2001 *Adv. Chem. Phys.* **116** 223

[4] Metzler R, Barkai E and Klafter J 1999 *Europhys. Lett.* **46** 431

[5] Barkai E, Metzler R and Klafter J 2000 *Phys. Rev.* E **61** 132

[6] Metzler R, Glöckle W G and Nonnenmacker T E 1997 *Fractals* **5** 597

[7] El-Wakil S A, Elhanbaly A and Zahran M A 2001 *Chaos Solitons Fractals* **12** 1035

[8] Jumarie G 2001 *Chaos Solitons Fractals* **12** 1873

[9] El-Wakil S A and Zahran M A 2001 *Chaos Solitons Fractals* **12** 1929

[10] Jesperson S, Metzler R and Fogedby H C 1999 *Phys. Rev.* E **59** 2736

[11] Risken H 1989 *The Fokker-Planck Equation* (Berlin: Springer)

[12] Scher H and Lax M 1973 *Phys. Rev.* B **7** 4491

[13] Scher H and Lax M 1973 *Phys. Rev.* B **7** 4502

[14] Scher H and Montroll E W 1975 *Phys. Rev.* B **12** 2455

[15] Barkai E and Fleurov V 1996 *Chem. Phys.* **212** 69

[16] Barkai E and Fleurov V N 1998 *Phys. Rev.* E **58** 1296

[17] Barkai E and Fleurov V 1997 *Phys. Rev.* E **56** 6355

[18] Suzuki N and Biyajima M 2001 *Phys. Rev.* E **65** 016123 - 1

[19] Havlin S and Ben Avrahan D 1987 *Adv. Phys.* **36** 695
Reprint: Havlin S and Ben Avrahan D 2002 *Adv. Phys.* **51** 187

[20] Gefen Y, Aharony A and Alexander S 1983 *Phys. Rev. Lett.* **50** 77

[21] O'Shaughnessy B and Procaccia I 1985 *Phys. Rev.* A **32** 3073

[22] Metzler R and Klafter J 2000 *Phys. Rep.* **339** 1

[23] Giona M and Roman H E 1992 *Physica* A **185** 87

[24] Le Mehaute A 1984 *J. Stat. Phys.* **36** 665

[25] Re F-Y, Yu Z-G and Su F 1996 *Phys. Lett.* A **219** 59

[26] Qiu W-Y and Lü J 2000 *Phys. Lett.* A **272** 353

[27] Ren F-Y, Wang X-T and Liang J-R 2001 *J. Phys. A: Math. Gen.* **34** 9815 - 25

[28] Samk S G, Kilba A A and Marichev O L 1993 *Fractional Integrals and Derivatives, Theory and Applications* (London: Gordon and Breach)

[29] Barkai E 2001 *Phys. Rev.* E **63** 046118

[30] Gradshteyn I S and Ryzhik I M 1980 *Table of Integrals Series and Products* (London: Academic)

[31] Giona M and Roman H E 1992 *Physica* A **185** 87

[32] Roman H E and Giona M 1992 *J. Phys. A: Math. Gen.* **25** 2107

[33] Ren F-Y, Liang J-R, Qiu W-Y and Yun Xu 2003 *J. Phys. A: Math. Gen.* **36** 7533

[34] Ren F-Y, Liang J-R, Qiu W-Y and Yun Xu 2003 *Physica* A **326** 430

Fractional Nonlinear Diffusion Equation and First Passage Time*

Jun Wang[a], Wen-Jun Zhang[b], Jin-Rong Liang[c], Jian-Bin Xiao[d], Fu-Yao Ren[a]

[a] *Department of Mathematics, Fudan University, Shanghai 200433, China*

[b] *College of Math. & Computing Science, Shenzhen University, Shenzhen 518060, China*

[c] *Department of Mathematics, East China Normal University, Shanghai 200062, China*

[d] *School of Science, Hangzhou Dianzi University, Hangzhou 310037, China*

Abstract: We investigate the solutions and the first passage time for anomalous diffusion processes governed by the fractional nonlinear diffusion equation with a space- and time-dependent diffusion coefficient subject to absorbing boundaries and the initial condition. We obtain explicit analytical expression for the probability distribution, the first passage time distribution, the mean first passage time, and the mean squared displacement corresponding to different time-dependent diffusion coefficient. In addition, we compare our results for the first passage time distribution and the mean first passage time with the one obtained by usual linear diffusion equation with time-dependent diffusion coefficient.

Keywords: anomalous diffusion, fractional nonlinear diffusion equation, first passage time

---

* Originally published in *Physica A*, Vol. 387, (2008), 764 - 772.

1. Introduction

Anomalous diffusion is a ubiquitous phenomenon in nature and it appears in many fields of physics, chemistry, and biology. The processes associated with anomalous diffusion are investigated by using the Langevin equation, the Fokker-Planck equation, the generalized Langevin equation [1], the generalized Fokker-Planck equation [2], and fractional equations [3, 4] for the probability density $p(x, t)$.

In connection to these approaches, the investigation of a stochastic process, such as anomalous diffusion, is also associated with the mean first passage time (MFPT). The MFPT is defined as the time T when a process, starting from a given point, reaches a predetermined level for the first time. Examples of the MFPT are the escape time from a random potential, intervals between neural spikes [5] and fatigue failure [6]. In this context, the knowledge of the first passage time (FPT) distribution $F(t)$ is essential for the effective application of probabilistic analysis. The concept of the FPT was originally introduced by Schrödinger when discussing behavior of Brownian particles in external fields [7]. Despite the long history, unfortunately, only in a few cases one has explicit analytical expressions for the FPT distribution, as was pointed out in Refs. [8,9]. Recently, Kwok Sau Fa and Lenzi [10], by using the Green function method, investigated the FPT distribution, the MFPT and the solution related to the following diffusion equation

$$\frac{\partial}{\partial t}p(x, t)=\frac{\partial}{\partial x}\left\{D(t, x)\frac{\partial}{\partial x}p(x, t)\right\} \tag{1}$$

in an interval $[0, L]$ with the diffusion coefficient given by $D(t, x) = D(t)|x|^{-\theta}$, subject to absorbing boundaries, i. e. $p(0, t)=p(L, t)=0$ and the initial condition given by $p(x, 0) = p_0(x)$. They obtained an analytical solution for the probability density and for the FPT distribution in a finite interval $[0, L]$ and a semi-infinite interval.

Due to the broadness of physical application to anomalous diffusions, nonlinear diffusion equations, fractional diffusion equations, and the diffusion equations that contain a mix of nonlinear terms and fractional derivative have been extensively investigated ([11～19] and references therein). A typical nonlinear diffusion equation is

$$\frac{\partial}{\partial t}p(x, t)=D\frac{\partial^2}{\partial x^2}[p(x, t)]^{\tau}, \tag{2}$$

usually referred to as the *porous medium equation* ([20] or [13] and references therein), which has been applied in several situations such as percolation of gases through porous media ($\tau \geqslant 2$) [21], thin saturated regions in porous media ($\tau = 2$) [22], a standard solid-on-solid model for surface growth ($\tau=3$) [20], thin liquid films spreading under gravity ($\tau = 4$) [23] and plasma flows ($\tau < 1$) [24, 25]. Here, the nonlinearity in $p(x, t)$ is known to lead to anomalous diffusion if $\tau \neq 1$ (namely super-diffusion for $\tau < 1$ and sub-diffusion for $\tau > 1$ as $\langle x^2(t)\rangle \propto t^{2/(\tau+1)}$) [26]. In this context, a detailed analysis for solutions, the FPT distribution, the MFPT and the influence of diffusion coefficient associated with fractional nonlinear diffusion equation has not been performed. Thus, it is an open problem.

In this paper, our focus is to analyze the MFPT, the FPT distribution, and the solution related to the following fractional nonlinear diffusion equation:

$$\frac{\partial}{\partial t}p(x, t)=\frac{\partial^{\mu'}}{\partial x^{\mu'}}\left\{D(t, x)[p(x, t)]^{\gamma}\frac{\partial^{\mu}}{\partial x^{\mu}}\{|x|^{-\eta}[p(x, t)]^{v}\}\right\} \tag{3}$$

in an interval $[0, L]$ with the diffusion coefficient given by $D(t, x) = KD(t)|x|^{-\theta}$ and with absorbing boundary $p(L, t)=0$ and the initial condition $p(x, 0)=\delta(x)$, and analyze the mean squared displacement of the processes given by Eq. (3) with different forms of $D(t)$. Where for functions $f(x)$ given in the interval $[a, b]$, the expression

$$\frac{\partial^q}{\partial x^q}f(x)=\frac{1}{\Gamma(n-q)}\frac{\mathrm{d}^n}{\mathrm{d}x^n}\int_a^x\frac{f(y)\mathrm{d}y}{(x-y)^{1-n+q}}$$

is the Riemann-Liouville derivative [27], $a, b, q \in \mathbb{R}$ and q is the order of the operation and n is an integer that satisfies $n-1 \leqslant q < n$.

2. Diffusion Equation and FPT Distribution

We consider a particle diffusing in the interval $[0, L]$, whose dynamics is governed by Eq. (3), subject to absorbing boundary $p(L, t)=0$ and the initial condition given by $p(x, 0)=\delta(x)$. We work with the positive x-axis because we can use symmetry to extend the results to the entire real axis.

Let us investigate time-dependent solutions for Eq. (3) by using similarity methods to ordinary differential equations. In this direction, we restrict our analysis to find a solution that can be expressed as a scaled function of the type

$$p(x, t)=\frac{1}{\Phi(t)}\tilde{p}(z),\ z=\frac{x}{\Phi(t)}. \tag{4}$$

Inserting (4) into Eq. (3) we obtain that

$$-\frac{\Phi'(t)}{\Phi(t)^2}\frac{\mathrm{d}}{\mathrm{d}z}[z\tilde{p}(z)]=KD(t)[\Phi(t)]^{-(\mu'+\theta+\gamma+\mu+\eta+\nu)}\frac{\mathrm{d}^{\mu'}}{\mathrm{d}z^{\mu'}}\left\{z^{-\theta}[\tilde{p}(z)]^{\gamma}\frac{\mathrm{d}^{\mu}}{\mathrm{d}z^{\mu}}\{z^{-\eta}[\tilde{p}(z)]^{\nu}\}\right\}.$$

Choosing

$$[\Phi(t)]^{\xi}\Phi'(t)/[KD(t)]=k,$$

where $\xi=\mu'+\theta+\gamma+\mu+\eta+\nu-2$, we obtain that

$$\Phi(t)=\left[(1+\xi)kK\int_0^t D(\tau)\mathrm{d}\tau\right]^{1/(1+\xi)} \tag{5}$$

with $\Phi(0)=0$ when $1+\xi>0$.

Thus, we have

$$\frac{\mathrm{d}^{\mu'}}{\mathrm{d}z^{\mu'}}\left\{z^{-\theta}[\tilde{p}(z)]^{\gamma}\frac{\mathrm{d}^{\mu}}{\mathrm{d}z^{\mu}}\{z^{-\eta}[\tilde{p}(z)]^{\nu}\}\right\}+k\frac{\mathrm{d}}{\mathrm{d}z}[z\tilde{p}(z)]=0. \tag{6}$$

In order to investigate the spatial behavior, we perform an integration in (6) which leads to

$$\frac{\mathrm{d}^{\mu'-1}}{\mathrm{d}z^{\mu'-1}}\left\{z^{-\theta}[\tilde{p}(z)]^{\gamma}\frac{\mathrm{d}^{\mu}}{\mathrm{d}z^{\mu}}\{z^{-\eta}[\tilde{p}(z)]^{\nu}\}\right\}=-kz\tilde{p}(z), \tag{7}$$

from which we can analyze many different situations. In particular, we start by analyzing the case $\mu'=1$ with θ, γ, μ, η and ν arbitraries after we consider μ' arbitrary and, later on, we consider particular cases of these parameters. Now, we start with $\mu'=1$. In this case, $\xi=\theta+\gamma+\mu+\eta+\nu-1$, written by ξ_1, we obtain

$$z^{-\theta}[\tilde{p}(z)]^{\gamma}\frac{d^{\mu}}{dz^{\mu}}\{z^{-\eta}[\tilde{p}(z)]^{\nu}\}=-kz\tilde{p}(z). \tag{8}$$

For this case, we consider the ansatz

$$\tilde{p}(z)=Nz^{\alpha/\nu}(a+bz)^{\beta/\nu} \tag{9}$$

as a solution to Eq. (8). By using the definition of fractional Riemann-Liouville derivative, more precisely the result [12]

$$\frac{d^{\delta}}{dz^{\delta}}[z^{\alpha}(a+bz)^{\beta}]=a^{\delta}\frac{\Gamma(\alpha+1)}{\Gamma(\alpha+1-\delta)}z^{\alpha-\delta}(a+bz)^{\beta-\delta} \tag{10}$$

for $\delta=\alpha+\beta+1$, we obtain that

$$\alpha/\nu=\frac{(1+\theta+\eta+\mu)(2+\theta+\mu)}{(\gamma-1)(1+\theta+\eta+2\mu)}, \tag{11}$$

$$\beta/\nu=\frac{\mu(2+\theta+\mu)}{(\gamma-1)(1+\theta+\eta+2\mu)}, \tag{12}$$

$$\nu=\frac{(1-\gamma)(1-\eta-\mu)}{2+\theta+\mu}, \tag{13}$$

where $N=[L^{\mu}\Gamma_{\mu}/(-k)[\Phi(t)]^{\mu}]^{1/(1-\gamma-\nu)}$ and $\Gamma_{\mu}=\dfrac{\Gamma(\alpha-\eta+1)}{\Gamma(\alpha-\eta+1-\mu)}$.

By substituting these results found above for the parameters α, β and ν, our solution of Eq. (8) can be written in the form

$$\tilde{p}(z)=\left[\left(\frac{\Gamma_{\mu}}{-k}\right)^{\frac{1}{1-\gamma}}\left(\frac{L}{\Phi(t)}\right)^{\frac{\mu}{1-\gamma}}z^{\frac{1+\theta+\eta+\mu}{\gamma-1}}\left(\frac{L}{\Phi(t)}-z\right)^{\frac{\mu}{\gamma-1}}\right]^{\frac{2+\theta+\mu}{1+\theta+\eta+2\mu}}. \tag{14}$$

The solution $p(x, t)$ of Eq. (3) satisfies the boundary condition $p(L, t)=0$ if and only if $\dfrac{\beta}{\nu}>0$, and the initial condition $p(x, 0)=\delta(x)$ holds if and only if $\alpha/\nu<-1$, and

$$-1 < \alpha/\nu - (1+\alpha/\nu)(\mu' + \theta + \mu + \eta)a/(1+\xi) < 0 \tag{15}$$

for $D(t) = t^{a-1}$, $a > 0$, or

$$-1 < \alpha/\nu - (1+\alpha/\nu)(\mu' + \theta + \mu + \eta)\min(b, q)/(1+\xi) < 0 \tag{16}$$

for $D(t) = \frac{\mathrm{d}}{\mathrm{d}t}\left[\frac{(1+at^b)^c}{(1+gt^q)^h}\right]$, b, c, q, $h > 0$ (cf. Ref. [10]), respectively.

In this case, the corresponding mean squared displacement is

$$\langle x^2 \rangle = \int_0^L \mathrm{d}x \cdot x^2 \cdot p(x, t)$$

$$= L^{2+u}\mathrm{B}(3+\alpha/\nu, 1+\beta/\nu)\left[\frac{\Gamma(\alpha-\eta+1)}{(-k)\Gamma(\alpha-\eta+1-\mu)}\right]^{1/(1-\gamma-\nu)}[\Phi(t)]^{-u}, \tag{17}$$

where $\mathrm{B}(\cdot, \cdot)$ is the Beta function, $\Phi(t) = \left[(1+\xi)kK\int_0^t D(\tau)\mathrm{d}\tau\right]^{1/(1+\xi)}$, $1+\xi_1 > 0$, and $u = 1+\alpha/\nu < 0$.

By using Eq. (14) and the expression $F(t)$ [2,8] of FPT distribution and of MFPT as follows

$$F(t) = -\frac{\mathrm{d}}{\mathrm{d}t}\int_0^L \mathrm{d}x p(x, t), \quad T = \int_0^\infty \mathrm{d}t \int_0^L \mathrm{d}x p(x, t),$$

we obtain that

$$F(t) = -L^u \mathrm{B}(1+\alpha/\nu, 1+\beta/\nu)\left[\frac{\Gamma_\mu}{(-k)}\right]^{1/(1-\gamma-\nu)}\frac{\mathrm{d}}{\mathrm{d}t}[\Phi(t)]^{-u} \quad \text{and} \tag{18}$$

$$T = L^u \mathrm{B}(1+\alpha/\nu, 1+\beta/\nu)\left[\frac{\Gamma_\mu}{(-k)}\right]^{1/(1-\gamma-\nu)} I, \tag{19}$$

respectively, where $I = \int_0^\infty \mathrm{d}t[\Phi(t)]^{-u}$.

In the following we give the properties of the mean squared displacement, the FPT and the MFPT for different diffusion coefficient in time.

Case Ⅰ: $D(t) = t^{a-1}$, $a > 0$. We have

$$\langle x^2 \rangle \sim t^\sigma. \tag{20}$$

Thus for $\sigma = a|u|/(1+\xi_1) < 1$, $=1$, >1, we have a sub-, normal- or supper-diffusive behavior.

In this case, the FPT

$$F(t)=C_1 t^{\sigma-1}, \tag{21}$$

where

$$C_1=-\sigma L^u \mathrm{B}(1+\alpha/\nu, 1+\beta/\nu)\left[\frac{\Gamma_\mu}{(-k)}\right]^{1/(1-\gamma-\nu)}\left[\frac{kK(1+\xi)}{a}\right]^{-\frac{u}{1+\xi_1}},$$

and the corresponding MFPT

$$T=\infty.$$

The MFPT is infinite as well as in the case of linear diffusive equation [10].

Case II: $D(t)=\frac{\mathrm{d}}{\mathrm{d}t}\left[\frac{(1+at^b)^c}{(1+gt^q)^h}\right]$, $b, c, q, h>0$ (cf. Ref. [10]). In this case,

$$\Phi(t)=\left\{(1+\xi_1)kK\left[\frac{(1+at^b)^c}{(1+gt^q)^h}-1\right]\right\}^{1/(1+\xi_1)}. \tag{22}$$

Inserting this to Eq. (15), we obtain that

$$\langle x^2\rangle \sim \left[\frac{(1+at^b)^c}{(1+gt^q)^h}-1\right]^{|u|/(1+\xi_1)} \sim \begin{cases} t^{\min(b, q)|u|/(1+\xi_1)}, & t\ll 1, \\ t^{[(bc-qh)-\min(b, q)]|u|/(1+\xi_1)}, & t\gg 1, \end{cases}$$

where $bc-hq-\min(b, q)\geqslant 0$. The diffusion exponent of $\langle x^2\rangle$ is different for

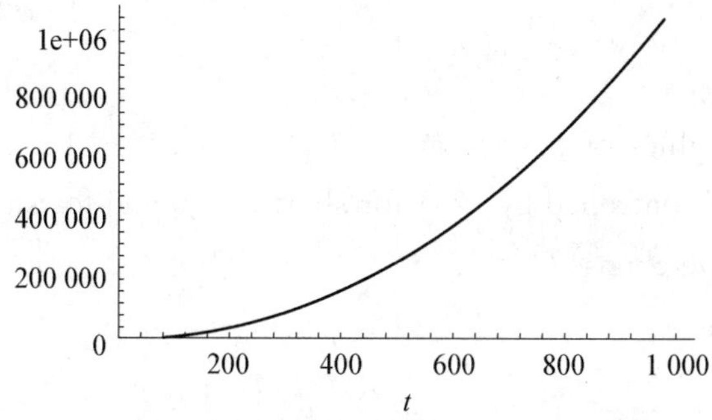

Fig. 1 Plot of $F(t)$ controlled by (23) for $t\geqslant 1$ with the diffusion coefficient $D(x, t)=x^{-\theta}\mathrm{d}[(1+at^b)^c/(1+gt^q)^h]/\mathrm{d}t$. The corresponding parameters are $a=0.4$, $b=2.3$, $\theta=-1$, $c=0.68(2+\theta)$, $g=7.7\times10^{-8}$, $q=0.7$, $h=0.165(2+\theta)$, $\gamma=0$, $\eta=1$, $\mu=-1/3$ and $\mu'=1$.

small t and for large time t, these behavior seem to be verified in turbulent processes [28, 29].

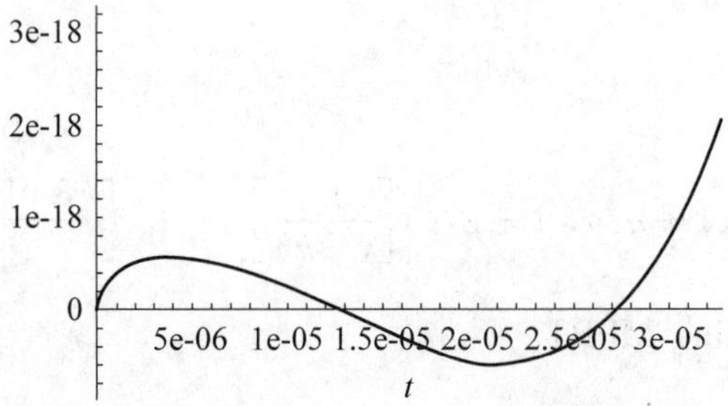

Fig. 2 Plot of $F(t)$ controlled by (23) for $t \ll 1$ with the same $D(x, t)$ and the parameters as in Fig. 1. For example, $5e-06 = 5 \times 10^{-6}$.

In this case,

$$F(t) = C_2 \left(\frac{abct^{b-1}}{1+at^b} - \frac{gqht^{q-1}}{1+gt^q} \right) \frac{(1+at^b)^c}{(1+gt^q)^h} \left[\frac{(1+at^b)^c}{(1+gt^q)^h} - 1 \right]^{\left(\frac{|u|}{1+\xi_1}-1\right)}, \quad (23)$$

where

$$C_2 = -L^u \mathrm{B}(1+\alpha/\nu, 1+\beta/\nu) \frac{u}{1+\xi_1} \left[\frac{\Gamma_\mu}{(-k)} \right]^{1/(1-\gamma-\nu)} [kK(1+\xi_1)]^{-\frac{u}{1+\xi_1}}.$$

For simplicity, suppose that all parameters a, b, c, g, h and q are positive. In this case, the FPT distribution exhibits two power laws

$$F(t) \sim \begin{cases} t^{\min(b,q)|u|/(1+\xi_1)-1}, & t \ll 1, \\ t^{[(bc-qh)-\min(b,q)]|u|/(1+\xi_1)+\min(b,q)-1}, & t \gg 1. \end{cases} \quad (24)$$

For the typical values of parameters in (23) and (24), Figs. 1 and 2 show behavior of $F(t)$ controlled by (23) for short $t \geqslant 0$ and for short time $t \ll 1$. For simplicity, we choose $C_2 = 1$.

Furthermore

$$T = C_3 \int_0^\infty \left[\frac{(1+at^b)^c}{(1+gt^q)^h} - 1 \right]^{|u|/(1+\xi_1)} \mathrm{d}t, \quad (25)$$

where

$$C_3 = L^u \mathrm{B}(1+\alpha/\nu,\ 1+\beta/\nu)\left[\frac{\Gamma_\mu}{(-k)}\right]^{1/(1-\gamma-\nu)}\left[kK(1+\xi_1)\right]^{-\frac{u}{1+\xi_1}}.$$

Moreover, the integral in the right side of (25) is finite, consequently, T is finite, if and only if

$$1-|u|[(bc-hq)-\min(b,\ q)]/(1+\xi_1)<0,$$

and otherwise, $T=\infty$.

It is interesting that the MFPT T may be finite iff $1-|u|[(bc-hq)-\min(b,\ q)]/(1+\xi_1)<0$.

In the rest of this section we consider the case $\mu'\neq 1$. In this case, $\xi=\mu'+\theta+\gamma+\mu+\eta+\nu-2$. By applying the same procedure used above, from (7), (9) and (10) we obtain that

$$\alpha/\nu=\frac{(\theta+\eta+\mu+\mu')(\eta+\mu+\mu')}{(\nu-1)(\theta+\eta+2\mu+2\mu'-1)}, \tag{26}$$

$$\beta/\nu=\frac{(\mu+\mu'-1)(\eta+\mu+\mu')}{(\nu-1)(\theta+\eta+2\mu+2\mu'-1)}, \tag{27}$$

$$\gamma=\frac{(\nu-1)(\theta+\mu+\mu'-1)}{(\eta+\mu+\mu')}, \tag{28}$$

$$N=\left(\frac{a^{\mu+\mu'-1}}{-k}\Gamma_{\mu,\ \mu'-1}\right)^{1/(1-\gamma-\nu)}, \tag{29}$$

$$p(x,\ t)=\frac{1}{\Phi(t)}\left\{\left[\frac{\Gamma_{\mu,\ \mu'-1}}{-k}\frac{L}{\Phi(t)}\right]^{\frac{\mu+\mu'-1}{\nu-1}}\left[\frac{x}{\Phi(t)}\right]^{\frac{\theta+\eta+\mu+\mu'}{1-\nu}}\left[\frac{L-x}{\Phi(t)}\right]^{\frac{\mu+\mu'-1}{1-\nu}}\right\}^{\frac{\eta+\mu+\mu'}{1-\theta-\eta-2\mu-2\mu'}}, \tag{30}$$

where

$$\Gamma_{\mu,\ \mu'-1}=\frac{\Gamma(\alpha-\eta+1)}{\Gamma(\alpha-\eta+1-\mu)}\frac{\Gamma(\alpha(1+\gamma/\nu)-\theta-\eta-\mu+1)}{\Gamma(\alpha(1+\gamma/\nu)-\theta-\eta-\mu-\mu'+2)}.$$

The $p(x,\ t)$ is the solution of Eq. (7) and satisfies $p(L,\ t)=0$ iff $\beta/\nu>0$ and the initial condition $p(x,\ 0)=\delta(x)$ iff $u=1+\alpha/\nu<0$.

Similarly, for $D(t)=t^{a-1}$, $a>0$, we have

$$\langle x^2\rangle\sim t^{a|u|/(1+\xi)},\quad F(t)=C_4 t^{\sigma-1}, \tag{31}$$

and the corresponding MFPT $T=\infty$, where

$$C_4=-\sigma L^u \mathrm{B}(1+\alpha/\nu,\ 1+\beta/\nu)\left[\frac{\Gamma_{\mu,\ \mu'-1}}{(-k)}\right]^{1/(1-\gamma-\nu)}\left[\frac{kK(1+\xi)}{a}\right]^{-\frac{u}{1+\xi}}.$$

For $D(t)=\frac{\mathrm{d}}{\mathrm{d}t}\left[\frac{(1+at^b)^c}{(1+gt^q)^h}\right]$, $b,\ c,\ q,\ h>0$, we have

$$\langle x^2\rangle \sim \begin{cases} t^{\min(b,\ q)\,|u|/(1+\xi)}, & t\ll 1,\\ t^{[(bc-qh)-\min(b,\ q)]\,|u|/(1+\xi)}, & t\gg 1, \end{cases} \tag{32}$$

$$F(t)=C_5\left(\frac{abct^{b-1}}{1+at^b}-\frac{gqht^{q-1}}{1+gt^q}\right)\frac{(1+at^b)^c}{(1+gt^q)^h}\left[\frac{(1+at^b)^c}{(1+gt^q)^h}-1\right]^{\left(\frac{|u|}{1+\xi}-1\right)}, \tag{33}$$

where

$$C_5=-L^u \mathrm{B}(1+\alpha/\nu,\ 1+\beta/\nu)\frac{u}{1+\xi}\left[\frac{\Gamma_{\mu,\ \mu'-1}}{(-k)}\right]^{1/(1-\gamma-\nu)}[kK(1+\xi)]^{-\frac{u}{1+\xi}}.$$

For simplicity, suppose that all parameters a, b, c, g, h and q are positive. In this case, the FPTD exhibits two power laws

$$F(t)\sim \begin{cases} t^{\min(b,\ q)\,|u|/(1+\xi)-1}, & t\ll 1,\\ t^{[(bc-qh)-\min(b,\ q)]\,|u|/(1+\xi)+\min(b,\ q)-1}, & t\gg 1, \end{cases} \tag{34}$$

where $bc-hq\geqslant 0$. Furthermore,

$$T=C_6\int_0^\infty\left[\frac{(1+at^b)^c}{(1+gt^q)^h}-1\right]^{|u|/(1+\xi)}\mathrm{d}t, \tag{35}$$

where

$$C_6=L^u \mathrm{B}(1+\alpha/\nu,\ 1+\beta/\nu)\left[\frac{\Gamma_{\mu,\ \mu'-1}}{(-k)}\right]^{1/(1-\gamma-\nu)}[kK(1+\xi)]^{-\frac{u}{1+\xi}}.$$

The MFPT, T, may be finite iff $1-|u|[(bc-hq)-\min(b,\ q)]/(1+\xi)<0$.

Remark In the case $\mu'\neq 1$, from (23) to (25) we can obtain that α, β and ν also satisfy

$$\alpha/\nu=\frac{(\mu'+\theta+\eta+\mu)(2\mu'+\theta+\mu)}{(\gamma-1)(2\mu'-1+\theta+\eta+2\mu)}, \tag{36}$$

$$\beta/\nu=\frac{(\mu+\mu'-1)(2\mu'+\theta+\mu)}{(\gamma-1)(2\mu'-1+\theta+\eta+2\mu)}, \tag{37}$$

$$\nu=\frac{(\gamma-1)(\eta+\mu-1)}{2\mu'+\theta+\mu}. \tag{38}$$

The results (36)~(38) coincide with (11)~(13) when $\mu'=1$.

Now, let us extend the above results to semi-infinite interval. In this direction, the boundary condition and initial condition is given by $p(\infty, t)=0$ and $p(x, 0)=\delta(x)$, respectively. By using these considerations, we can show that the probability distribution is given by

$$p(x, t)=\frac{1}{\Phi(t)}\left\{\left[\frac{\Gamma_{\mu,\mu'-1}}{-k}\frac{1}{\Phi(t)}\right]^{\frac{\mu+\mu'-1}{\nu-1}}\left[\frac{x}{\Phi(t)}\right]^{\frac{\theta+\eta+\mu+\mu'}{1-\nu}}\left[\frac{1+x}{\Phi(t)}\right]^{\frac{\mu+\mu'-1}{1-\nu}}\right\}^{\frac{\eta+\mu+\mu'}{1-\theta-\eta-2\mu-2\mu'}}. \tag{39}$$

The function $p(x, t)$ is the solution of Eq. (3) and satisfies $p(\infty, t)=0$ iff $\beta/\nu<0$ and the initial condition $p(x, 0)=\delta(x)$ iff $u=1+\alpha/\nu<0$. From this expression, we can respectively obtain the FPT distribution and MFPT related to this process for $D(t)=t^{a-1}$, $a>0$, as follows:

$$F(t)=C'_4 t^{\sigma-1} \quad \text{and} \quad T=\infty, \tag{40}$$

where

$$C'_4=-\sigma \mathrm{B}(1+\alpha/\nu, 1+\beta/\nu)\left[\frac{\Gamma_{\mu,\mu'-1}}{(-k)}\right]^{1/(1-\gamma-\nu)}\left[\frac{kK(1+\xi)}{a}\right]^{-\frac{u}{1+\xi}}.$$

For $D(t)=\frac{\mathrm{d}}{\mathrm{d}t}\left[\frac{(1+at^b)^c}{(1+gt^q)^h}\right]$, $b, c, q, h>0$,

$$F(t)=C'_5\left(\frac{abct^{b-1}}{1+at^b}-\frac{gqht^{q-1}}{1+gt^q}\right)\frac{(1+at^b)^c}{(1+gt^q)^h}\left[\frac{(1+at^b)^c}{(1+gt^q)^h}-1\right]^{\left(\frac{|u|}{1+\xi}-1\right)} \quad \text{and} \tag{41}$$

$$T=C'_6\int_0^{\infty}\left[\frac{(1+at^b)^c}{(1+gt^q)^h}-1\right]^{|u|/(1+\xi)}\mathrm{d}t,$$

where

$$C'_5=-\mathrm{B}(1+\alpha/\nu, 1+\beta/\nu)\frac{u}{1+\xi}\left[\frac{\Gamma_{\mu,\mu'-1}}{(-k)}\right]^{1/(1-\gamma-\nu)}[kK(1+\xi)]^{-\frac{u}{1+\xi}},$$

$$C_6'=\mathrm{B}(1+\alpha/\nu,\ 1+\beta/\nu)\left[\frac{\Gamma_{\mu,\ \mu'-1}}{(-k)}\right]^{1/(1-\gamma-\nu)}[kK(1+\xi)]^{-\frac{u}{1+\xi}}.$$

The MFPT, T, may be finite iff $1-|u|[(bc-hq)-\min(b,\ q)]/(1+\xi)<0$.

3. Summary and Conclusions

In summary, we have investigated some solutions of Eq. (3) and the FPT for a system governed by the fractional nonlinear diffusion equation whose diffusion coefficient is space- and/or time-dependent. We have obtained an analytic solution for the probability density and the FPT distribution in a finite interval $[0,\ L]$ and a semi-infinite interval with absorbing boundary at L or at infinity. In deed, the determination of analytic solutions for a fractional nonlinear diffusion equation and, consequently, to the corresponding mean squared displacement and the FPT distribution is important to the study of diffusion processes due to the fact that the system can be analyzed concisely. In this direction, the study of diffusion processes with the diffusion coefficient $D(x,\ t)=D(t)|x|^{-\theta}$ is significant to both theoretical and experimental physics due to the fact pointed out as in [17] that it can be useful in describing diverse physical processes such as diffusion in systems with porous media [30~33]. We should mention that despite the singular nature of the diffusion coefficient at $x=0$, all the quantities obtained in this work are well-behaved.

Acknowledgments

This work is supported in part by the Science and Technology Commission of Shanghai Municipality (No. 06ZR14029), by the NNSF (No. 10571121), the NSFC (No. 10571028), the ESTF (No. 204063) and by the ZJNSF (No. Y604569).

◇ References ◇

[1] R. Muralidhar, D. Ramkrishna, H. Nakanishi, D. Jacobs, Physica A 167 (1990) 539; K. G. Wang, L. K. Dong, X. F. Wu, F. W. Zhu, T. Ko, Physica A 203 (1994) 53;

K. G. Wang, M. Tokuyama, Physica A 265 (1999) 341.
[2] H. Risken, The Fokker-Planck Equation, Springer, Berlin, 1996;
C. W. Gardiner, Handbook of Stochastic Methods, Springer, Berlin, 1997.
[3] R. Metzler, J. Klafter, Phys. Rep. 339 (2000) 1.
[4] R. Metzler, J. Klafter, J. Phys. A 37 (2004) R161.
[5] H. C. Tuckwell, Introduction to Theoretical Neurobiology, vol. 2, Cambridge University Press, Cambridge, 1988.
[6] Y. K. Lin, G. Q. Cai, Probabilistic Structural Dynamics, McGraw-Hill, New York, 1995.
[7] E. Schrödinger, Phys. Z. 16 (1915) 289.
[8] G. Rangarajan, M. Ding, Phys. Lett. A 273 (2000) 322.
[9] T. Verechtchaguina, I. M. Sokolov, L. Schinmansky-Geier, Phys. Rev. E 73 (2006) 031108.
[10] Kwok Sau Fa, E. K. Lenzi, Phys. Rev. E 71 (2005) 012101.
[11] C. Tsallis, D. J. Bukman, Phys. Rev. E 54 (1996) R2197.
[12] M. Bologna, C. Tsallis, Pl Grigolini, Phys. Rev. E 62 (2) (2000) 2213.
[13] C. Tsallis, E. K. Lenzi, Chem. Phys. 284 (2002) 341.
[14] E. K. Lenzi, L. C. Malacarne, R. S. Mendes, I. T. Pedron, Physica A 319 (2003) 245.
[15] E. K. Lenzi, G. A. Mendsa, R. S. Mendes, L. R. da Silva, L. S. Luccna, Phys. Rev. E 67 (2003) 051109 - 1.
[16] E. K. Lenzi, R. S. Mendes, Kwok Sau Fa, J. Math. Phys. 45 (9) (2004) 3444.
[17] Kwok Sau Fa, E. K. Lenzi, Phys. Rev. E 72 (2005) 011107.
[18] G. A. Mendsa, E. K. Lenzi, R. S. Mendes, L. R. da Silva, Physica A 271 (2005) 346.
[19] F. -Y. Ren, J. -R. Liang, W. -Y. Qiu, Jian-Bin Xiao, Physica A 3857 (2007) 80.
[20] H. Spohn, J. Phys. I France 3 (1993) 69.
[21] M. Muskat, The Flow of Homogeneous Fluid through Porous Media, McGraw-Hill, New York, 1937.
[22] P. Y. Polubarinova-Kochina, Theory of Ground Water Movement, Princeton University Press, Princeton, NJ, 1962.
[23] J. Buckmaster, J. Fluid Mech. 81 (1984) 735.
[24] R. Rosenau, Phys. Rev. Lett. 74 (1995) 1056.
[25] A. Compte, D. Jou, Y. Katayama, J. Phys. A 30 (1997) 1023.
[26] E. K. Lenzi, C. Anteneodo, L. Borland, Phys. Rev. E. 63 (2001) 051109.
[27] I. Podlubny, Fractional Differential Equations, Academic Press, San Diego, CA, 1999.

[28] I. M. Sokolov, J. Klafter, A. Blumen, Phys. Rev. E 61 (2000) 2717.
[29] G. Boffetta, I. M. Sokolov, Phys. Rev. Lett. 88 (2002) 094501.
[30] B. O'Shaughnessy, I. Procaccia, Phys. Rev. Lett. 54 (1985) 455.
[31] L. F. Richardson, Proc. R. Soc. Lond. Ser. A 110 (1926) 709;
A. N. Kolmogorov, C. R. (Dokl.) Acad. Sci. URSS 30 (1941) 301.
[32] A. A. Vedenov, Rev. Plasma Phys. 3 (1967) 229.
[33] Fujisaka, S. Grossmann, S. Thomac, Z. Naturforsch. Tcil A 40 (1985) 867;
E. N. Glass, J. P. Krisch, Classical Quantum Gravity 17 (2000) 2611;
J. P. Krisch, E. N. Glass, Gen. Relativ. Gravitation 33 (2001) 1449;
P. Hanggi, H. Thomas, Phys. Rep. 88 (1982) 207;
P. Hanggi, Helv. Phys. Acta 51 (1978) 183.

Generalized Einstein Relation and the Metzler-Klafter Conjecture in a Composite-subdiffusive Regime*

Pan Hua, Wei-Yuan Qiu, Fu-Yao Ren, Long-Jin Lv

Department of Mathematics, Fudan University, Shanghai 200433, China

Abstract: In this paper, a generalized diffusion model driven by the composite-subdiffusive fractional Brownian motion (FBM) is employed. Based on this stochastic process, we derive a fractional Fokker-Planck equation (FFPE) and obtain its solution. It is proved that the Generalized Einstein Relation (GER) and the Metzler and Klafter conjecture on the asymptotic behavior of stretched Gaussian hold the FFPE in a composite-subdiffusive regime.

Keywords: fractional Brownian motion, fractional Fokker-Plank equation (FFPE), generalized Einstein relation (GER), asymptotic behavior

1. Introduction

Diffusion is a fundamental transport process of matter and energy in various physical, chemical and biological systems. Two types of anomalous diffusion obviously are characterized as superdiffusion and subdiffusion. For a description of these two diffusion regimes a number of effective models and methods have been suggested [1]. The continuous-time random walk (CTRW) model of Scher and Montroll [2] leading to strongly subdiffusion

* Originally published in *Physica A*, Vol. 390, (2011), 2920 – 2925.

behavior, provides a basis for understanding photoconductivity in strongly disordered and glassy semiconductors. The Lévy flight model [3], leading to superdiffusion, describes transport processes in heterogeneous rocks [4], self-diffusion in micelle systems [5], reactions and transport in polymer systems [6], and the stochastic description of financial market indices [7]. For these two cases, the so-called fractional differential equations in coordinate and time spaces are applied as an effective approach [8]. So searching the diffusion character is meaningful.

The Generalized Einstein Relation (GER) relates the fluctuations of the test particle position in the absence of an external field to its behavior under the influence of a constant force field F, i. e. $\langle X^2(t)\rangle_0 \sim \langle X(t)\rangle_F$. It connects the first moment in the presence of a uniform force field with the second moment in the absence of the force. Recent experimental results corroborate the validity of the relation above in polymeric systems in the subdiffusive regime [9~11]. The relation can be used also for gaseous ions in an electrostatic field [12], for disordered semiconductors (implications for device performance) [13] and to ions in molecular gases [14].

In order to better understanding the application of subdiffusive processes, a generalization of GER in the CTRW-subdiffusion was mentioned in Ref. [15], which was obtained for time averaged mean-square displacement differs from the ensemble average.

In Ref. [11], it introduces a particle motion under the combined influence. The FFPE is shown to obey GER and its stationary solution is the Boltzmann distribution. Then in Ref. [16], Metzler and Klafter point out that the detailed structure of the propagator $p(x, t)$ (i. e. the probability density function (PDF)) for the initial condition $\lim_{t\to 0} p(x, t) = \delta(x)$, depends generally on the special shape of the underlying geometry. However, the interesting part of the propagator has the asymptotic behavior $\ln p(x, t) \sim C\xi^u$ where $\xi \equiv x/t^{\alpha/2} \gg 1$, which is expected to be universal.

In Ref. [17], it demonstrates that the anomalous diffusion with the asymptotic behavior is corresponded to a fractional partial differential equation which is uniquely determined by the fractal Hausdorff dimension and the anomalous diffusion exponent.

In Refs. [18~20], it is proved that the propagator has the asymptotic behavior $\ln p(x, t) \sim -C\xi^u$ for the solution of a wide class of the FFPE, where $\xi = x/t^{\alpha/2} \gg 1$, $u = 1/(1-\alpha/2)$ and $p(x, t)$ is a stretched Gaussian distribution.

In this paper, we introduce a composite-subdiffusive fractional Brownian motion model

$$X_\alpha(t) = X(S_\alpha(t)), \quad 0 < \alpha < 1, \tag{1}$$

where the parent process $X(\tau)$ is given by the Itô stochastic differential equation

$$\mathrm{d}X(\tau) = F(X(\tau))(\mathrm{d}\tau)^{2H} + K\,\mathrm{d}b(\tau, H). \tag{2}$$

Here, $b(\tau, H)$ is defined by $\mathrm{d}b(\tau, H) = w(\tau)(\mathrm{d}\tau)^H$, $0 < H < 1$, $w(\tau)$ is a Gaussian white noise with mean value 0 and variance 1, $F(x)$ denotes the external force and the constant K denotes the anomalous diffusion coefficient. The subordinator $S_\alpha(t)$ is the inverse-time α-stable subordinator, $\alpha \in (0, 1)$, defined by

$$S_\alpha(t) = \inf\{\tau > 0 : U(\tau) > t\} \tag{3}$$

where $U(\tau)$ denotes a strictly increasing α-stable Lévy motion. For every jump of $U(\tau)$ there is a corresponding flat period of its inverse $S_\alpha(t)$. These heavy-tailed flat periods of $S_\alpha(t)$ represent long waiting times in which the subdiffusive particle gets immobilized in the trap. The sense of Model (1) can be presented by the figure in Ref. [21]. It will time randomized. When $\alpha \uparrow 1$, $S_\alpha(t)$ reduces to the "objective time" t.

Do the GER and the Metzler-Klafter conjecture hold in composite-subdiffusive regime? It is an open problem.

It would be interesting to clarify this issue. In Section 2 we derive the FFPE governing the PDF of a composite-subdiffusive process $X_\alpha(t)$ and its solution. Then, we discuss the diffusion character, GER and the asymptotic behavior of $X_\alpha(t)$ in a composite-subdiffusive regime.

2. The FFPE of Composite-subdiffusive Processes $X_\alpha(t)$ and Its Solution

The main idea of this section is to find the FFPE governing the PDF of the composite-subdiffusive processes $X_\alpha(t)$ and its solution.

Theorem 1 *Let $S_\alpha(t)$ be the inverse-time α-stable subordinator* (3) *and $X(\tau)$ be the solution of the Itô stochastic differential equation* (2). *Assume that the process of the stochastic time $S_\alpha(t)$ and the process $X(\tau)$ are independent. Then, the PDF of $X_\alpha(t)$ denoted by $p(x, t)$ satisfies*

$$\frac{\partial}{\partial t}p(x, t)=\Gamma(1+2H)\left[-\frac{\partial}{\partial x}F(x)+\frac{1}{2}K^2\frac{\partial^2}{\partial x^2}\right]{}_0D_t^{1-2H\alpha}p(x, t)$$

$$\left(0<H\leqslant\frac{1}{2}\right), \tag{4}$$

$$\frac{\partial}{\partial t}p(x, t)=\Gamma(2H)\left[-\frac{\partial}{\partial x}F(x)+\frac{1}{2}K^2\frac{\partial^2}{\partial x^2}\right]{}_0D_t^{1-(2H-1)\alpha}p(x, t)$$

$$\left(\frac{1}{2}<H\leqslant 1\right), \tag{5}$$

where the operator ${}_0\mathrm{D}_t^{1-2H\alpha}$ and ${}_0\mathrm{D}_t^{1-(2H-1)\alpha}$ are the fractional derivative of the Riemann-Liouville type. In particular, when $\alpha\uparrow 1$, we have

$$\frac{\partial}{\partial t}p(x, t)=\Gamma(1+2H)\left[-\frac{\partial}{\partial x}F(x)+\frac{1}{2}K^2\frac{\partial^2}{\partial x^2}\right]{}_0D_t^{1-2H}p(x, t)$$

$$\left(0<H\leqslant\frac{1}{2}\right), \tag{6}$$

$$\frac{\partial}{\partial t}p(x, t)=\Gamma(2H)\left[-\frac{\partial}{\partial x}F(x)+\frac{1}{2}K^2\frac{\partial^2}{\partial x^2}\right]{}_0D_t^{2-2H}p(x, t)$$

$$\left(\frac{1}{2}<H\leqslant 1\right). \tag{7}$$

Proof From the total probability formula and the independence of $X(\tau)$ and $S_\alpha(t)$, we get the Laplace transform

$$\hat{p}(x, k)=\int_0^\infty \mathrm{e}^{-kt}p(x, t)\mathrm{d}t=\int_0^\infty f(x, \tau)\hat{g}(\tau, k)\mathrm{d}\tau.$$

Here, $f(x, \tau)$ and $g(\tau, t)$ are the PDFs of $X(\tau)$ and $S_\alpha(t)$, respectively, and $\hat{g}(\tau, k)=\int^{\infty} e^{-kt} g(\tau, t)dt$; see Ref. [21]. Since the process $X(\tau)$ is given by Eq. (2), using the Kramers-Moyal forward expansion of fractional order [22] and the Modified Riemann-Liouville derivative [23, 24], we have when $0 < 2H \leqslant 1$,

$$\lim_{\Delta\tau\to 0}\frac{f(x, \tau+\Delta\tau)-f(x, \tau)}{(\Delta\tau)^{2H}}=\frac{f^{(2H)}(x, \tau)}{\Gamma(2H+1)}$$
$$=\left[-\frac{\partial}{\partial x}F(x)+\frac{1}{2}K^2\frac{\partial^2}{\partial x^2}\right]f(x, \tau), \tag{8}$$

where the fractional derivative of Modified Riemann-Liouville type of order $2H$ is defined by

$$f^{(2H)}(x, \tau)=\frac{1}{\Gamma(1-2H)}\frac{\partial}{\partial\tau}\int_0^\tau(\tau-\xi)^{-2H}[f(x, \xi)-f(x, 0)]d\xi.$$

Consequently, in the Laplace space, we obtain

$$k^{2H}\hat{f}(x, k)-k^{2H-1}f(x, 0)=\Gamma(1+2H)\left[-\frac{\partial}{\partial x}F(x)+\frac{1}{2}K^2\frac{\partial^2}{\partial x^2}\right]\hat{f}(x, k). \tag{9}$$

Using the result in Ref. [21], we have

$$\hat{p}(x, k)=k^{\alpha-1}\hat{f}(x, k^\alpha).$$

Changing the variable $k\to k^\alpha$ in Eq. (9), we have

$$k\hat{p}(x, k)-p(x, 0)=\Gamma(1+2H)\left[-\frac{\partial}{\partial x}F(x)+\frac{1}{2}K^2\frac{\partial^2}{\partial x^2}\right]k^{1-2H\alpha}\hat{p}(x, k). \tag{10}$$

Finally, inverting the Laplace transform in Eq. (10), we obtain Eq. (4).

When $1 < 2H \leqslant 2$, then $0 < 2H-1 \leqslant 1$. In this case, from Eq. (8) and Lemma 3.2 [20], we have

$$\frac{\partial}{\partial\tau}f(x, \tau)=\left[-\frac{\partial}{\partial x}F(x)+\frac{1}{2}K^2\frac{\partial^2}{\partial x^2}\right]\frac{\partial}{\partial\tau}\int_0^\tau f(x, \tau')(d\tau')^{2H-1}$$

$$=\left[-\frac{\partial}{\partial x}F(x)+\frac{1}{2}K^2\frac{\partial^2}{\partial x^2}\right](2H-1)\frac{\partial}{\partial\tau}\int_0^\tau(\tau-\tau')^{2H-2}f(x,\tau')\mathrm{d}\tau'$$

$$=\Gamma(2H)\left[-\frac{\partial}{\partial x}F(x)+\frac{1}{2}K^2\frac{\partial^2}{\partial x^2}\right]{}_0D_t^{2-2H}f(x,\tau). \tag{11}$$

Here, ${}_0D_t^{2-2H}$ is also the Riemann-Liouville type. Using the same procedure as before, we have Eq. (5). The proof is completed.

When $H=\frac{1}{2}$, the result is same with the conclusion in Refs. [21,25].

Now, we are going to find the solution of the obtained FFPE.

Theorem 2 *Let $S_\alpha(t)$ be the inverse-time α-stable subordinator* (3) *and $X(\tau)$ be the solution of the Itô stochastic differential equation* (2). *Assume that the processes $S_\alpha(t)$ and $X(\tau)$ are independent. Let $f(x,\tau)$ and $g(\tau,t)$ to be the PDF of $X(\tau)$ and $S_\alpha(t)$. Then, the PDF of the composite-subdiffusive processes $X_\alpha(t)$ is given by*

$$p(x,t)=\int_0^\infty f(x,\tau)g(\tau,t)\mathrm{d}\tau, \tag{12}$$

where

$$f(x,\tau)=\frac{1}{\sqrt{2\pi}K\tau^H}\mathrm{e}^{-\frac{[x-x_0-F(x)\tau^{2H}]^2}{2K^2\tau^{2H}}}, \tag{13}$$

$$g(\tau,t)=\frac{1}{t^\alpha}H_{11}^{10}\left[\frac{\tau}{t^\alpha}\,\middle|\,\begin{smallmatrix}(1-\alpha,\ \alpha)\\(0,\ 1)\end{smallmatrix}\right]. \tag{14}$$

Proof Since the random change of time $S_\alpha(t)$ and the process $X(\tau)$ are independent, according to the total probability formula [21], we have

$$p(x,t)=\int_0^\infty f(x,\tau)g(\tau,t)\mathrm{d}\tau.$$

And if $f(x,\tau)$ and $g(\tau,t)$ have achieved, we can know the formula describing $p(x,t)$.

On the one hand, integrating Eq. (2) and noting $w(\tau)\sim N(0,1)$, we get

$$\mathbb{E}(X(\tau))=X(0)+F(x)\tau^{2H},\ \mathbb{D}(X(\tau))=K^2\tau^{2H}.$$

Thus, when $X(0)=x_0$ we have Eq. (13) by the normal distribution probability density formula.

On the other hand, from [21], we know that the Laplace transform of $g(\tau, t)$ has the relation

$$\hat{g}(\tau, k)=k^{\alpha-1}\mathrm{e}^{-\tau k^{\alpha}}. \tag{15}$$

Inverting the Laplace transform of Eq. (15) and exploring the properties of H-Fox functions [26,27], we have Eq. (14). So we can obtain the formula of $p(x, t)$ expressed by $f(x, \tau)$ and $g(\tau, t)$.

3. Generalized Einstein Relation

In this section, we prove that the Generalized Einstein Relation (GER) holds for a composite-subdiffusive system.

From Eq. (13), for $F(x)=0$, we have

$$f_0(x, \tau)=\frac{1}{\sqrt{2\pi}K\tau^H}\mathrm{e}^{-\frac{(x-x_0)^2}{2K^2\tau^{2H}}}. \tag{16}$$

Using Eq. (16) and the relation $\hat{p}(x, k)=\int_0^{\infty}f(x, \tau)k^{\alpha-1}\mathrm{e}^{-\tau k^{\alpha}}\mathrm{d}\tau$ which is proved specifically in Ref. [21], we have

$$\begin{aligned}\hat{p}_0(x, k)&=\int_0^{\infty}\frac{1}{\sqrt{2\pi}K\tau^H}\mathrm{e}^{-\frac{(x-x_0)^2}{2K^2\tau^{2H}}}k^{\alpha-1}\mathrm{e}^{-\tau k^{\alpha}}\mathrm{d}\tau\\&=\frac{k^{\alpha-1}}{\sqrt{2\pi}}\int_0^{\infty}\frac{1}{K\tau^H}\mathrm{e}^{-\frac{(x-x_0)^2}{2K^2\tau^{2H}}}\mathrm{e}^{-\tau k^{\alpha}}\mathrm{d}\tau.\end{aligned}$$

Then the Laplace transform of $\langle X_{\alpha}^2(t)\rangle_0$ comes to

$$\begin{aligned}L[\langle X_{\alpha}^2(t)\rangle_0]&=\int_{-\infty}^{+\infty}x^2\hat{p}_0(x, k)\mathrm{d}x\\&=k^{\alpha-1}\int_0^{\infty}\mathrm{e}^{-\tau k^{\alpha}}\left[\int_{-\infty}^{+\infty}\frac{1}{\sqrt{2\pi}K\tau^H}x^2\mathrm{e}^{-\frac{(x-x_0)^2}{2K^2\tau^{2H}}}\mathrm{d}x\right]\mathrm{d}\tau\end{aligned}$$

$$=k^{\alpha-1}\int_0^{\infty} e^{-\tau k^{\alpha}} K^2 \tau^{2H} d\tau$$
$$=K^2 \frac{\Gamma(2H+1)}{k^{1+2H\alpha}}.$$

Inverting the Laplace transform in the last formula, we have the desired relation

$$\langle X_{\alpha}^2(t)\rangle_0 = K^2 \frac{\Gamma(2H+1)}{\Gamma(2H\alpha+1)} t^{2H\alpha}. \tag{17}$$

Here, for $0 < 2H\alpha < 1$, $2H\alpha = 1$, $2H\alpha > 1$, the dynamic is subdiffusion, normal diffusion and superdiffusion, respectively. Especially, when $\alpha \in (0, 1)$, $H \in \left(0, \frac{1}{2}\right)$, the diffusion is subdiffusion.

Now, let us begin our discussion. As we know, under the influence of a constant external force F, Eq. (13) yields

$$f_F(x, \tau) = \frac{1}{\sqrt{2\pi} K \tau^H} e^{-\frac{(x-x_0-F\tau^{2H})^2}{2K^2\tau^{2H}}}.$$

Noting that

$$\hat{p}_F(x, k) = \int_0^{\infty} f_F(x, \tau) k^{\alpha-1} e^{-\tau k^{\alpha}} d\tau$$

and $\langle X_{\alpha}(t)\rangle_F = \int_{-\infty}^{+\infty} x p_F(x, t) dx$, thus, in the Laplace space, it comes to

$$L[\langle X_{\alpha}(t)\rangle_F] = \int_{-\infty}^{+\infty} x \hat{p}_F(x, k) dx$$
$$= \frac{x_0}{k} + F \frac{\Gamma(2H+1)}{k^{2H\alpha+1}}.$$

The inverse Laplace transform of it is

$$\langle X_{\alpha}(t)\rangle_F = x_0 + F \frac{\Gamma(2H+1)}{\Gamma(2H\alpha+1)} t^{2H\alpha}. \tag{18}$$

It is

$$\langle X_{\alpha}(t)\rangle_F = F \frac{\Gamma(2H+1)}{\Gamma(2H\alpha+1)} t^{2H\alpha}, \tag{19}$$

when $X(0)=x_0=0$.

Comparing Eqs. (17) and (19), we have

$$\langle X_\alpha^2(t)\rangle_0=\frac{K^2}{F}\langle X_\alpha(t)\rangle_F. \tag{20}$$

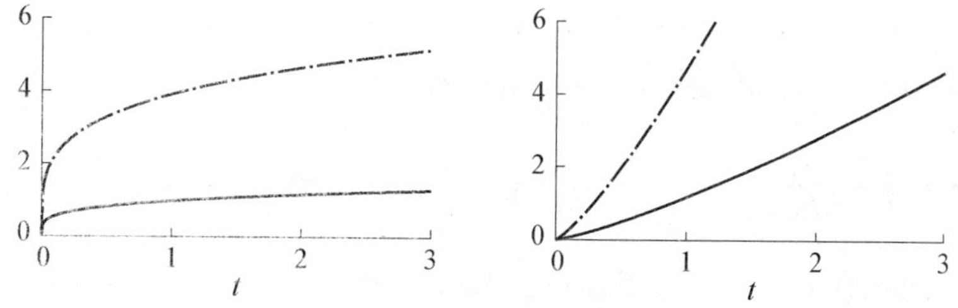

Fig. 1 The figure is in the condition of $H=\frac{1}{4}$, $\alpha=\frac{1}{2}$ and $H=\frac{3}{4}$, $\alpha=\frac{5}{6}$, respectively. The full lines represent the value of $\langle X_\alpha^2(t)\rangle_0$ and the dashed lines demonstrate the value of $\langle X_\alpha(t)\rangle_F$ with variable "t". We can see that its telescopic transform with same transformation trends between mean-square displacement by free force and the mean value by force for both of the numerical.

This shows that the GER is satisfied in composite-subdiffusive regime which is a fractional Brownian motion model in Eq. (1).

The relationship between mean-square displacement by free force and the mean value by force can be obtained by Fig. 1 obviously. It is a telescopic transform in fact.

4. The Metzler-Klafter Conjecture

In this section, it is proved that the propagator has the asymptotic behavior $\ln p(x,t)\sim -C\xi^u$ for the solution of Eqs. (4) and (5). That is, $p(x,t)$ is a stretched Gaussian distribution which has heavy tail and high peak.

Following the same procedures showed in Ref. [18], for Eq. (4), that is when $0<H\leqslant\frac{1}{2}$, the Laplace transform is

$$x^2 \frac{\partial^2}{\partial x^2}\hat{p}(x, k) - \frac{2F(x)}{K^2}x^2 \frac{\partial}{\partial x}\hat{p}(x, k) - \frac{2\left(F'(x) + \frac{k^{2H\alpha}}{\Gamma(2H+1)}\right)}{K^2}x^2\hat{p}(x, k)$$

$$= -\frac{2k^{2H\alpha-1}}{\Gamma(2H+1)K^2}x^2 p(x, 0).$$

Put $y = A(k)x^v$, $\hat{p}(x, k) = y^\delta Z(y)$, then the transform of Eq. (21) arrives into the second-order Bessel equation

$$y^2 \frac{d^2 Z}{dy^2} + y\frac{dZ}{dy} - (\lambda^2 + y^2)Z(y) = -\frac{y^{2-\delta}}{k}p(x, 0)$$

with parameters under the following conditions:

$$v=1, \quad \delta = \frac{1}{2}, \quad \lambda = \frac{1}{2},$$

$$A(k) = \frac{\sqrt{2}k^{H\alpha}}{\Gamma(2H+1)K}.$$

Using the solution of the Bessel equation, we have

$$\hat{p}(x, k) \approx \frac{k^{H\alpha-1}}{\sqrt{2\Gamma(2H+1)}K}\exp\left(-\frac{\sqrt{2}k^{H\alpha}}{\sqrt{\Gamma(1+2H)}K}x\right). \tag{21}$$

By the same method of Refs. [28,29], we expected that

$$p(x, t) \sim t^{-2H\alpha/2}(x/X_\theta)^{\delta'}\exp\{-\text{const}\cdot(x/X_\theta)^{u'}\} \tag{22}$$

when $\xi = x/X_\theta \sim x/t^{H\alpha} \gg 1$ and $t \to +\infty$.

The Laplace transform of Eq. (22) can be evaluated by applying the method of steepest descent, and compared it with Eq. (21), we obtain $u' = 1/(1-H\alpha)$ and $\delta' = \frac{1}{2}u'(2H\alpha - 1)$.

With the same procedures as before, when $\frac{1}{2} < H \leqslant 1$, we obtain that the Laplace transform of Eq. (5) is

$$\hat{p}(x, k) \approx \frac{k^{(2H-1)\alpha/2-1}}{K\sqrt{2\Gamma(2H)}}\exp\left(-\frac{\sqrt{2}k^{(2H-1)\alpha/2}}{\sqrt{\Gamma(2H)}K}x\right). \tag{23}$$

The corresponding equation is

$$p(x, t) \sim t^{-(2H-1)\alpha/2}(x/X_\theta)^{\delta'}\exp\{-\text{const}\cdot(x/X_\theta)^{u'}\} \tag{24}$$

with $u' = 1/(1-(2H-1)\alpha/2)$ and $\delta' = \frac{1}{2}u'((2H-1)\alpha-1)$ when $\xi = x/X_\theta \sim x/t^{(2H-1)\alpha/2} \gg 1$ and $t \to +\infty$.

Especially, when $\alpha \uparrow 1$, we get if $0 < H \leqslant \frac{1}{2}$.

$$f(x, t) \sim t^{-H}(x/X_\theta)^{\delta'}\exp\{-\text{const}\cdot(x/X_\theta)^{u'}\} \tag{25}$$

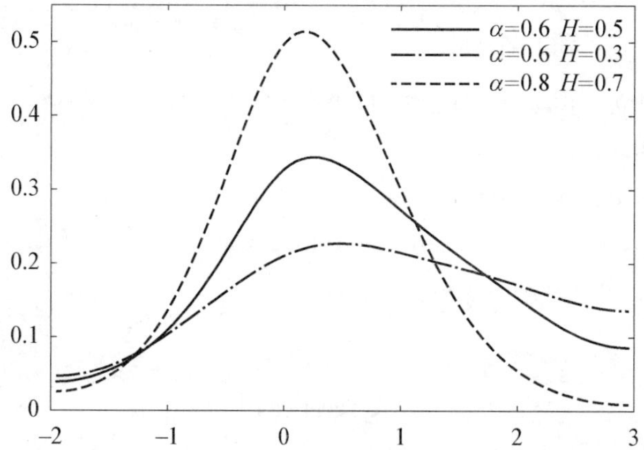

Fig. 2 The figure is in the condition of different α, H values and it satisfies stretched Gaussian distribution generally.

when $\xi = x/X_\theta \sim x/t^H \gg 1$ and $t \to +\infty$, where $u' = 1/(1-H)$ and $\delta' = \frac{1}{2}u'(2H-1)$ if $\frac{1}{2} < H \leqslant 1$.

$$f(x, t) \sim t^{-(2H-1)/2}(x/X_\theta)^{\delta'}\exp\{-\text{const}\cdot(x/X_\theta)^{u'}\} \tag{26}$$

when $\xi = x/X_\theta \sim x/t^{(2H-1)/2} \gg 1$ and $t \to +\infty$, where $u' = 1/(1-(2H-1)/2)$ and $\delta' = u'(H-1)$.

These show that the asymptotic behavior of $f(x, \tau)$ and $p(x, t)$ satisfy stretched Gaussian distribution, consequently, the Metzler and Klafter conjecture holds in a composite-subdiffusive regime. The result can be obtained by Fig. 2 for different α, H.

5. Conclusion

In this paper, we prove that the GER and the asymptotic behavior of stretched Gaussian hold in a composite-subdiffusive regime. We build the relation between FFPE and the subordinated process driven by fractional Brownian motion. Then we find the PDF of this subordinated process is rightly the solution of this FFPE. With the help of this result, the solution of the FFPE can be easily obtained. We expect that the results obtained here may be useful to the decision of the anomalous diffusion systems where fractional diffusion equations play an important role.

Acknowledgments

This work is supported by the NSFC (Nos 10871047 and 10771071). The authors are grateful to the anonymous referees for useful comments and suggestions.

◇ References ◇

[1] S. A. Trigger, G. Heijst, P. Schram, J. Phys. : Conf. Ser. 11 (2005) 37.
[2] H. Scher, E. W. Montroll, Phys. Rev. B 12 (1975) 2455.
[3] J. Klafter, M. F. Schlezinger, G. Zumofen, Phys. Today 49 (1996) 33.
[4] J. Klafter, A. Blumen, G. Zumofen, M. F. Shlesinger, Physica A 168 (1990) 637.
[5] A. Ott, J. P. Bouchaud, D. Langevin, W. Urbach, Phys. Rev. Lett. 65 (1994) 2201.
[6] I. M. Sokolov, J. Mayand, A. Blumen, Phys. Rev. Lett. 79 (1997) 857.
[7] P. Gopikrishnan, V. Plerou, L. Amaral, M. Meyer, H. E. Stanley, Phys. Rev. E 60 (1999) 5305.
[8] B. J. West, M. Bologna, P. Grigolini, Physics of Fractal Operators, Springer-Verlag, New York, 2003.
[9] F. Amblard, A. C. Maggs, B. Yurke, A. N. Pargellis, S. Leibler, Phys. Rev. Lett. 77 (1996) 4470.
[10] E. Barkai, J. Klafter, Phys. Rev. Lett. 81 (1998) 1134.
[11] R. Metzler, E. Barkai, J. Klafter, Phys. Rev. Lett. 82 (1999) 3563.

[12] R. E. Robson, J. Phys. B: At. Mol. Phys. 9 (1976) 337.
[13] Y. Roichman, N. Tessler, Appl. Phys. Lett. 80 (2002) 1948.
[14] R. Y. Pai, H. W. Ellis, G. R. Akridge, E. W. McDaniel, Phys. Rev. A 12 (1975) 5.
[15] Y. He, S. Burov, R. Metzler, E. Barkai, Phys. Rev. Lett. 101 (2008) 058101.
[16] R. Metzler, J. Klafter, Phys. Rep. 339 (2000) 1.
[17] R. Metzler, W. G. Glockle, T. F. Nonnenmacher, Physica A 211 (1994) 13 - 24.
[18] F. Y. Ren, J. R. Liang, W. Y. Qiu, J. B. Xiao, J. Phys. A: Math. Gen. 39 (2006) 4911.
[19] F. Y. Ren, Y. Xu, W. Y. Qiu, J. R. Liang, Chaos Solitons Fractals 24 (2005) 273 - 278.
[20] F. Y. Ren, J. R. Liang, W. Y. Qiu, J. B. Xiao, Physica A 373 (2007) 165 - 173.
[21] M. Magdziarz, A. Weron, Phys. Rev. E 75 (2007) 016708.
[22] H. Risken, The Fokker-Planck Equation, Springer-Verlag, New York, 1984.
[23] G. Jumarie, Comput. Math. Appl. 51 (2006) 1367.
[24] G. Jumarie, Insurance Math. Econom. 42 (2008) 271.
[25] M. Magdziarz, Stochastic Process. Appl. 119 (2009) 3238.
[26] F. Mainardi, G. Pagnini, R. K. Saxena, J. Comput. Appl. Math. 178 (2005) 321.
[27] A. M. Mathai, R. K. Saxena, The *H*-function with Application in Statistics and Other Disciplines, Wiley Eastern, New Delhi, 1978.
[28] M. Giona, H. E. Roman, Physica A 185 (1992) 87.
[29] H. E. Roman, M. Giona, J. Phys. A: Math. Gen. 25 (1992) 2107.

Continuous Time Black-Scholes Equation with Transaction Costs in Subdiffusive Fractional Brownian Motion Regime*

Jun Wang[a], Jin-Rong Liang[b], Long-Jin Lv[a],
Wei-Yuan Qiu[a], Fu-Yao Ren[a]

[a] *Department of Mathematics, Fudan University, Shanghai 200433, China*

[b] *Department of Mathematics, East China Normal University, Shanghai, 200241, China*

Abstract: In this paper, we study the problem of continuous time option pricing with transaction costs by using the homogeneous subdiffusive fractional Brownian motion (HFBM) $Z(t) = X(S_\alpha(t))$, $0 < \alpha < 1$, here $dX(\tau) = \mu X(\tau)(d\tau)^{2H} + \sigma X(\tau) dB_H(\tau)$, as a model of asset prices, which captures the subdiffusive characteristic of financial markets. We find the corresponding subdiffusive Black-Scholes equation and the Black-Scholes formula for the fair prices of European option, the turnover and transaction costs of replicating strategies. We also give the total transaction costs.

Keywords: subdiffusion, Black-Scholes formula, fractional Black-Scholes equation, transaction costs

1. Introduction

The classical and still most popular model of the market is the Black-

---

* Originally published in *Physica A*, Vol. 391, (2012), 750 - 759.

Scholes model based on the diffusion process call geometric Brownian motion (GBM) [1,2]

$$dX(\tau)=\mu X(\tau)d\tau+\sigma X(\tau)dB(\tau),\quad X(0)=X_0 \tag{1}$$

where μ, σ are constants, and $B(\tau)$ is the Brownian motion. In the presence of transaction costs (TC), Leland [3] first examined option replication in a discrete time setting, and pose a modified replicating strategy, which depends upon the level of transaction costs and upon the revision interval, as well as upon the option to be replicated and the environment. Since then, a lot of authors study this problem, but all in a discrete time setting [4~11].

The option pricing theory as developed by Black-Scholes [1, 2] rests on an arbitrage argument: by continuously adjusting a portfolio consisting of a stock and a risk-free bond, an investor can exactly replicate the returns to any option on the stock. It leads us naturally to pose the following question.

In the presence of transaction costs, is there an alternative replicating strategy depending upon the level of transaction costs and a technique leading to the Black-Scholes equation in a continuous time setting? Does the perfect replication incur an infinite amount of transaction costs?

The Black-Scholes (BS) model is based on the diffusion process called geometric Brownian motion (GBM). However, the empirical studies show that many characteristic properties of markets cannot be captured by the BS model, such as: long-range correlations, heavy-tailed and skewed marginal distributions, lack of scale invariance, periods of constant values, etc. Therefore, in recent years one observes many generalizations of the BS model based on the ideas and methods known from statistical and quantum physics [12].

In this paper, we deal with the asset price exhibiting subdiffusive dynamics, $Z(t)=X(S_\alpha(t))$, $0<\alpha<1$, in which the price of an asset $X(\tau)$ follows a stochastic differential equation

$$dX(\tau)=\mu X(\tau)(d\tau)^{2H}+\sigma X(\tau)dB_H(\tau), \tag{2}$$

where $B_H(\tau)$ is the fractional Brownian motion (FBM) with Hurst exponent $H\in(0, 1)$, and $S_\alpha(t)$ is the inverse α-stable subordinator defined as below

$$S_\alpha(t) = \inf\{\tau > 0 : U_\alpha(\tau) > t\}, \tag{3}$$

$U_\alpha(\tau)$ is a strictly increasing α-stable Lévy process [13, 14] with Laplace transform given by $E(e^{-uU_\alpha(\tau)}) = e^{-\tau u^\alpha}$. When $\alpha \uparrow 1$, $S_\alpha(t)$ reduces to the "objective time" t. Here, we apply the subdiffusive mechanism of trapping events in order to describe financial data exhibiting periods of constant values as in Ref. [15].

From the Appendix, we can express the model (2) into the following form

$$\frac{dx(\tau)}{d\tau} = \int_0^\tau \mu X(s) 2H(2H-1)(\tau - s)^{2H-2} ds + \sigma X(\tau)\xi(\tau) \tag{4}$$

where $\xi(\tau)$ is the fractional Gaussian noise, heuristically $\xi(\tau) = dB_H(\tau)/d\tau$. The FBM has two unique properties: self-similarity and stationary increments [16]. The autocorrelation function of fractional Gaussian noise is the memory kernel $K(\tau)$ $(\tau > 0)$.

$$K(\tau) = 2\mathrm{Cov}(\xi(0), \ \xi(\tau)) = 2H(2H-1)\tau^{2H-2}, \tag{5}$$

which is rightly the fractional operator in (4). This model was first proposed by Kou to simulate the fluctuation of the distance between a fluorescein-tyrosine pair within a single protein complex [17]. Eq. (4) can be converted to an equation for the time correlation function $C_x(\tau) = E(X(0)X(\tau))$ of $X(\tau)$ by multiplying X(0) and taking expectation, yields

$$\frac{\partial C_x(\tau)}{\partial \tau} = \int_0^\tau \mu K(\tau - s) C_x(s) ds + \sigma E[\xi(\tau) X(\tau) X(0)]. \tag{6}$$

The last term $E[\xi(\tau)X(\tau)X(0)] = 0$ for $\xi(\tau)$ is orthogonal to $X(\tau)$ in the phase space [18]. The Laplace transform of Eq. (6) gives

$$\hat{C}_x(k) = C_x(0)\frac{k^{2H-1}}{k^{2H} - a}, \quad a = \Gamma(2H+1)\mu.$$

Thus, inverting the Laplace transform, we have

$$C_x(\tau) = C_x(0) E_{2H,1}(a\tau^{2H}), \tag{7}$$

where $E_{\alpha,\beta}(t)$ is the Mittag-Leffler function [19]. So, by computing the covariance of the real data, then using $C_x(\tau)$ to approximate it, we can get the value of H. From the correlation function (7), we can get the model is long-dependent for $1/2 < H < 1$. Noting that, the self-similarity, stationary increments and long dependence are all important properties in financial market.

This paper is organized as follows. In Section 2, by using a delta hedging strategy initiated by Leland [3], we deduce the Black-Scholes equation with transaction costs in continuous time setting for the asset price $Z(t) = X(S_\alpha(t))$, $0<\alpha<1$, where $X(\tau)$ follows (2) and $S_\alpha(t)$ is defined in (3). In the Section 3, we obtain the corresponding Black-Scholes formula. In Section 4, we estimate turnover and transaction costs of replicating strategies. In Section 5, we give the total transaction costs.

2. Continuous Time Black-Scholes Equation with TC

Theorem 1 (*Continuous Time BS Equation with Transaction Costs in Subdiffusive Regime*) *Let $Z(t)=X(S_\alpha(t))$ be the subdiffusive HFBM, here the parent process $X(\tau)$ is defined in* (2), *and $S_\alpha(t)$, $0 < \alpha < 1$, is the inverse α-stable subordinator defined in* (3). *Assume that the process $S_\alpha(t)$ and $B_H(\tau)$ are independent. Then, the price $V(t, z)$ of a derivative on the subdiffusive FBM stock price $Z(t)$ with transaction cost parameter k is given by the Black-Scholes equation*

$$\frac{\partial V(t, z)}{\partial t}+rz\frac{\partial V(t, z)}{\partial z}+\frac{\tilde{\sigma}^2_{H,\alpha}(t)}{2}z^2\frac{\partial^2 V(t, z)}{\partial z^2}-rV(t, z)=0, \quad (8)$$

where the modified volatility $\tilde{\sigma}_{H,\alpha}(t)$ is defined, respectively, by

$$\tilde{\sigma}^2_{H,\alpha}(t)=2Ht^{2H-1}\left(\frac{t^{\alpha-1}}{\Gamma(\alpha)}\right)^{2H}\left(1+\frac{k}{2}\left|\frac{2\mu}{\sigma^2}-1\right|\operatorname{sign}(\Gamma)\right)\sigma^2 \quad (9)$$

and $\Gamma=\dfrac{\partial^2 V(t, z)}{\partial z^2}$.

Proof Let $V(t, z)$ be the option price driven by the stock asset $Z = Z(t)=X(S_\alpha(t))$. By the Taylor's formula, we have

$$V(t+\Delta t, z+\Delta z)=V(t, z)+\frac{\partial V(t, z)}{\partial z}\Delta z+\frac{\partial V(t, z)}{\partial t}(\Delta t)+\frac{1}{2}\frac{\partial^2 V(t, z)}{\partial z^2}(\Delta z)^2$$
$$+\frac{\partial^2 V(t, z)}{\partial z\partial t}\Delta z(\Delta t)+\frac{1}{2}\frac{\partial^2 V(t, z)}{\partial t^2}(\Delta t)^2+o(\Delta t). \tag{10}$$

Noting (3), it is easy to see

$$\Delta z=\mu z(\Delta S_\alpha(t))^{2H}+\sigma z\Delta B_H(S_\alpha(t))$$
$$=\mu z\left(\frac{t^{\alpha-1}}{\Gamma(\alpha)}\Delta t\right)^{2H}+\sigma z\Delta B_H(S_\alpha(t))+o((\Delta t)^{2H}),$$
$$(\Delta z)^2=\sigma^2 z^2\left(\frac{t^{\alpha-1}}{\Gamma(\alpha)}^{2H}\right)(\Delta t)^{2H}+o((\Delta t)^{2H}),$$
$$\Delta z(\Delta t)=o((\Delta t)^{2H}).$$

Thus, by Lemma 1 in Appendix, the process followed by the function $V(t, z)$ of z and t is

$$dV(z, t)=\left[\left(\frac{t^{\alpha-1}}{\Gamma(\alpha)}\right)^{2H}\left(\mu z\frac{\partial V(z, t)}{\partial z}+\frac{1}{2}\sigma^2 z^2\frac{\partial^2 V(z, t)}{\partial z^2}\right)2Ht^{2H-1}+\frac{\partial V(z, t)}{\partial t}\right]dt$$
$$+\sigma z\frac{\partial V(z, t)}{\partial z}dB_H(S_\alpha(t)). \tag{11}$$

We consider a replicating portfolio as in Ref. [11] with $\frac{\partial V(t, z)}{\partial z}$ units of underlying asset and $\delta_2(t)$ units of the riskless bond. The value of the portfolio at current time t is $\Pi(t, z)=\frac{\partial V(t, z)}{\partial z}z+\delta_2(t)D(t)$. In addition, we assume that the riskless bound price dynamics satisfies $dD(t)=rD(t)dt$, where r is a constant.

The change in the value at current time t of the portfolio is

$$d\Pi(t, z)=\frac{\partial V(t, z)}{\partial z}dz+\delta_2(t)dD(t)-\frac{k}{2}\left|\frac{\partial^2 V(t, z)}{\partial z^2}dz\right|z, \tag{12}$$

where, $dD(t)$ is the change in the riskless bond price, $\frac{\partial^2 V(t, z)}{\partial z^2}dz$ is the change in the number of units of asset held in the portfolio, and

$\frac{k}{2}\left|\frac{\partial^2 V(t, z)}{\partial z^2}\mathrm{d}z\right|z$ is the transaction costs. Since the value of the option must equal the value of the replicating portfolio $\Pi(t, z)$ to reduce (but not to avoid) arbitrage opportunities and be consistent with economic equilibrium, therefore

$$V(t, z)=\frac{\partial V(t, z)}{\partial z}z+\delta_2(t)D(t). \tag{13}$$

Thus, from (11)~(13), we have

$$\begin{aligned}\delta_2(t)\mathrm{d}D(t)&=\mathrm{d}V(t, z)-\frac{\partial V(t, z)}{\partial z}\mathrm{d}z+\frac{k}{2}\left|\frac{\partial^2 V(t, z)}{\partial z^2}\mathrm{d}z\right|z\\&=\left[\frac{\partial V(t, z)}{\partial t}+\sigma^2 H\left(\frac{t^{\alpha-1}}{\Gamma(\alpha)}\right)^{2H}t^{2H-1}z^2\frac{\partial^2 V(t, z)}{\partial z^2}\right]\mathrm{d}t\\&\quad+\frac{k}{2}\left|\frac{\partial^2 V(t, z)}{\partial z^2}\mathrm{d}z\right|z.\end{aligned} \tag{14}$$

At the same time, from (13), we have

$$\delta_2(t)\mathrm{d}D(t)=r\delta_2(t)D(t)\mathrm{d}t=r\left(V(t, z)-z\frac{\partial V(t, z)}{\partial z}\right)\mathrm{d}t. \tag{15}$$

Thus, we obtain

$$\begin{aligned}&\left[\frac{\partial V(t, z)}{\partial t}+\sigma^2 H\left(\frac{t^{\alpha-1}}{\Gamma(\alpha)}\right)^{2H}t^{2H-1}z^2\frac{\partial^2 V(t, z)}{\partial z^2}+rz\frac{\partial V(t, z)}{\partial z}-rV(t, z)\right]\mathrm{d}t\\&\qquad+\frac{k}{2}z\left|\frac{\partial^2 V(t, z)}{\partial z^2}\mathrm{d}z\right|=0.\end{aligned} \tag{16}$$

On the other side, from Eq. (2) and Taylor's formula, we have

$$\begin{aligned}\mathrm{d}z(t)&=\mu z(\mathrm{d}S_\alpha(t))^{2H}-\frac{\sigma^2}{2}z(\mathrm{d}S_\alpha(t))^{2H}+\sigma z\mathrm{d}B_H(S_\alpha(t))\\&=\left(\mu-\frac{\sigma^2}{2}\right)z\left(\frac{t^{\alpha-1}}{\Gamma(\alpha)}\right)^{2H}(\mathrm{d}t)^{2H}+o((\mathrm{d}t)^{\min(1, 2H)}).\end{aligned}$$

Consequently, by Lemma 1 in Appendix we have

$$\mathrm{d}z=\left(\mu-\frac{\sigma^2}{2}\right)z\left(\frac{t^{\alpha-1}}{\Gamma(\alpha)}\right)^{2H}2Ht^{2H-1}\mathrm{d}t+o((\mathrm{d}t)^{\min(1, 2H)}). \tag{17}$$

Thus, from (16) and (17), we obtain

$$\frac{\partial V(t,z)}{\partial t}+\frac{\sigma^2}{2}\left(\frac{t^{\alpha-1}}{\Gamma(\alpha)}\right)^{2H}2Ht^{2H-1}z^2\left(\frac{\partial^2 V(t,z)}{\partial z^2}+\frac{k}{2}\left|\left(\frac{2\mu}{\sigma^2}-1\right)\frac{\partial^2 V(t,z)}{\partial z^2}\right|\right)$$
$$+rz\frac{\partial V(t,z)}{\partial z}-rV(t,z)=0. \tag{18}$$

Since $\Gamma=\dfrac{\partial^2 V(t,z)}{\partial z^2}$ represents the degree of mishedging of the portfolio, it is not surprising to observe that Γ is involved in the transaction cost term. Noting $\left|\dfrac{\partial^2 V(t,z)}{\partial z^2}\right|=\operatorname{sign}(\Gamma)\dfrac{\partial^2 V(t,z)}{\partial z^2}$, thus we immediately get Black-Scholes equation (8) from Eq. (18), so the proof is completed.

In particular, letting $\alpha\uparrow 1$ in the last equation, we obtain

Corollary 1 (*BS Equation with Transaction Costs in Real Time*) *Let the price $X(t)$ satisfy Eq.* (2), *then the option price $V(t,x)$ with transaction costs is given by the Black-Scholes equation*

$$\frac{\partial V(t,x)}{\partial t}+rx\frac{\partial V(t,x)}{\partial x}+\frac{\tilde{\sigma}_H^2(t)}{2}x^2\frac{\partial^2 V(t,x)}{\partial x^2}-rV(t,x)=0, \tag{19}$$

where the modified volatility $\tilde{\sigma}_H(t)$ is defined by

$$\tilde{\sigma}_H^2(t)=2Ht^{2H-1}\left(1+\frac{k}{2}\left|\frac{2\mu}{\sigma^2}-1\right|\operatorname{sign}(\Gamma)\right)\sigma^2. \tag{20}$$

Remark 1 Eq. (19) resembles the Black-Scholes equation obtained by Wang [11] in discrete time setting.

If $k=0$, i. e. , with no transaction costs, we obtain

Corollary 2 (*BS Equation with No Transaction Costs in Subdiffusive Regime*)

$$\frac{\partial V(t,z)}{\partial t}+\frac{\sigma^2}{2}\left(\frac{t^{\alpha-1}}{\Gamma(\alpha)}\right)^{2H}2Ht^{2H-1}z^2\frac{\partial^2 V(t,z)}{\partial z^2}+rz\frac{\partial V(t,z)}{\partial z}-rV(t,z)=0. \tag{21}$$

Furthermore, if $H=\dfrac{1}{2}$ and $\alpha\uparrow 1$, from (21) *we have the celebrated Black-Scholes equation* [1,2]

$$\frac{\partial V(t,z)}{\partial t}+rz\frac{\partial V(t,z)}{\partial z}+\frac{\sigma^2}{2}z^2\frac{\partial^2 V(t,z)}{\partial z^2}-rV(t,z)=0. \tag{22}$$

3. Continuous Time Black-Scholes Formula with TC

It is known that Γ is always positive for the simple European call and put option in the absence of transaction costs. If we postulate Γ in the presence of transaction costs is always positive, then

$$\tilde{\sigma}^2_{H,\alpha}(t)=2Ht^{2H-1}\left(\frac{t^{\alpha-1}}{\Gamma(\alpha)}\right)^{2H}\left(1+\frac{k}{2}\left|\frac{2\mu}{\sigma^2}-1\right|\right)\sigma^2. \tag{23}$$

Consequently, Eq. (9) becomes linear under such assumption so the Black-Scholes formula become applicable except that the modifies volatility $\tilde{\sigma}_{H,\alpha}(t)$ should be used as the volatility parameter.

Moreover, using the results in Ref. [11] and Eq. (9), we obtain

Theorem 2 (*Instantaneous BS Formula with Transaction Costs in Subdiffusive Regime*) *Let $z=Z(t)=X(S_\alpha(t))$ as the subdiffusive model of asset price, here the parent process $X(\tau)$ is defined by (2) with initial value $X(0)>0$, where $S_\alpha(t)$, $0<\alpha<1$, is the inverse α-stable subordinator defined in (3). Assume that the process $S_\alpha(t)$ and $B_H(\tau)$ are independent. Then, we obtain the option price $\hat{C}_\alpha(t,z)$ and $\hat{P}_\alpha(t,z)$, of a European call option and a put option with time to maturity T and strike price K are given, respectively, by*

$$\hat{C}_\alpha(t,z)=zN(\hat{d}_1(t))-Ke^{-r(T-t)}N(\hat{d}_2(t)), \tag{24}$$

$$\hat{P}_\alpha(t,z)=Ke^{-r(T-t)}N(-\hat{d}_2(t))-zN(-\hat{d}_1(t)), \tag{25}$$

and the Put-Call Parity relationship

$$\hat{C}_\alpha(t,z)-\hat{P}_\alpha(t,z)=z-Ke^{-r(T-t)} \tag{26}$$

where, the function $N(x)$ is the cumulative probability distribution function for a standard normal distribution, and

$$\hat{d}_1(t)=\frac{\ln(z/K)+\left(r+\frac{1}{2}\hat{\sigma}^2_{H,\alpha}(t)\right)(T-t)}{\hat{\sigma}_{H,\alpha}\sqrt{T-t}},$$

$$\hat{d}_2(t)=\frac{\ln(z/K)+\left(r-\frac{1}{2}\hat{\sigma}^2_{H,\alpha}(t)\right)(T-t)}{\hat{\sigma}_{H,\alpha}(t)\sqrt{T-t}}, \tag{27}$$

here $\hat{\sigma}^2_{H,\alpha}(t)=\frac{\int_t^T \tilde{\sigma}^2_{H,\alpha}(\tau)\mathrm{d}\tau}{T-t}$ *is given by*

$$\hat{\sigma}^2_{H,\alpha}(t)=\frac{\sigma^2}{\alpha[\Gamma(\alpha)]^{2H}}\left(1+\frac{k}{2}\left|\frac{2\mu}{\sigma^2}-1\right|\right)\left(\frac{T^{2H\alpha}-t^{2H\alpha}}{T-t}\right). \tag{28}$$

Corollary 3 (*BS Formula with Transaction Costs in Subdiffusive Regime*) *Let* $z=Z(t)=X(S_\alpha(t))$ *be the same subdiffusive model of asset price defined in Theorem* 2, *where the process* $S_\alpha(t)$ *and* $B_H(\tau)$ *are independent. Then, we obtain the option price* $\hat{C}_\alpha(K, T)$ *and* $\hat{P}_\alpha(K, T)$, *of a European call option and a put option with time to maturity* T *and strike price* K *are given by*

$$\hat{C}_\alpha(K, T)=X_0N(\hat{d}_1)-K\mathrm{e}^{-rT}N(\hat{d}_2), \tag{29}$$

$$\hat{P}_\alpha(K, T)=K\mathrm{e}^{-rT}N(-\hat{d}_2)-X_0N(-\hat{d}_1), \tag{30}$$

and the Put-Call Parity relationship

$$\hat{C}_\alpha(K, T)-\hat{P}_\alpha(K, T)=X_0-K\mathrm{e}^{-rT} \tag{31}$$

where, the function $N(x)$ *is the cumulative probability distribution function for a standard normal distribution, and*

$$\hat{d}_1=\frac{\ln(X_0/K)+\left(r+\frac{1}{2}\hat{\sigma}^2_{H,\alpha}\right)T}{\hat{\sigma}_{H,\alpha}\sqrt{T}},\quad \hat{d}_2=\frac{\ln(X_0/K)+\left(r-\frac{1}{2}\hat{\sigma}^2_{H,\alpha}\right)T}{\hat{\sigma}_{H,\alpha}\sqrt{T}}, \tag{32}$$

here $\hat{\sigma}^2_{H,\alpha}$ *is given by*

$$\hat{\sigma}^2_{H,\alpha}=\frac{\sigma^2}{\alpha[\Gamma(\alpha)]^{2H}}\left(1+\frac{k}{2}\left|\frac{2\mu}{\sigma^2}-1\right|\right)T^{2H\alpha-1}. \tag{33}$$

Clearly, letting $H=\frac{1}{2}$, $k=0$ and $\alpha \uparrow 1$ in Corollary 3, we have the classical Black-Scholes formula [1, 2]. We give two figures of the prices of European call options in the subdiffusive BS model with TC for different parameters (see Figs. 1 and 2). At the same time, it is easy to see that $\hat{\sigma}_{H,\alpha}$ is just the implied volatility of the classical Black-Scholes model. From Eq. (33), we have

$$\sigma=\left(1-\frac{k}{2}\right)^{-1/2}\sqrt{\alpha[\Gamma(\alpha)]^{2H}T^{1-2H\alpha}\hat{\sigma}_{H,\alpha}^{2}-k\mu} \quad \text{for } 2\mu>\sigma^{2},$$

$$\sigma=\left(1+\frac{k}{2}\right)^{-1/2}\sqrt{\alpha[\Gamma(\alpha)]^{2H}T^{1-2H\alpha}\hat{\sigma}_{H,\alpha}^{2}+k\mu} \quad \text{for } 2\mu<\sigma^{2}.$$

Therefore, the implied volatility of our model keep the basic property of that of the classical BS model. In detail, if the asset price following the classical BS model presents volatility smiles, then the implied volatility of our model also smile. For example, we have the implied volatility of the asset price following the classical BS model as follows:

K	80	85	90	95	100	105	110	115	120
Three months	0.1935	0.1761	0.1665	0.1597	0.1545	0.1506	0.1479	0.1472	0.1506
Six months	0.2018	0.1920	0.1834	0.1758	0.1692	0.1634	0.1584	0.1544	0.1514

According to these data, we give two figures of the implied volatility in the subdiffusive BS model with TC (see Figs. 3 and 4).

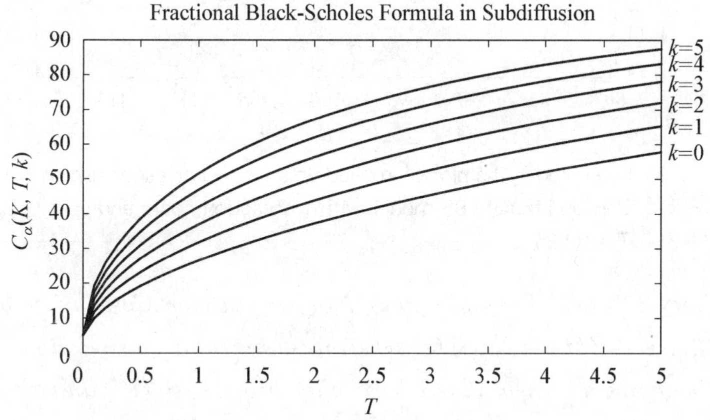

Fig. 1 Presented are prices of European call options corresponding to the subdiffusive BS model with transaction costs for different parameter k, where $X_0 = 100$, $K = 95$, $r = 0.1$, $\mu = 0.2$, $\sigma = 0.5$, $\alpha = 0.8$ and $H = 0.6$.

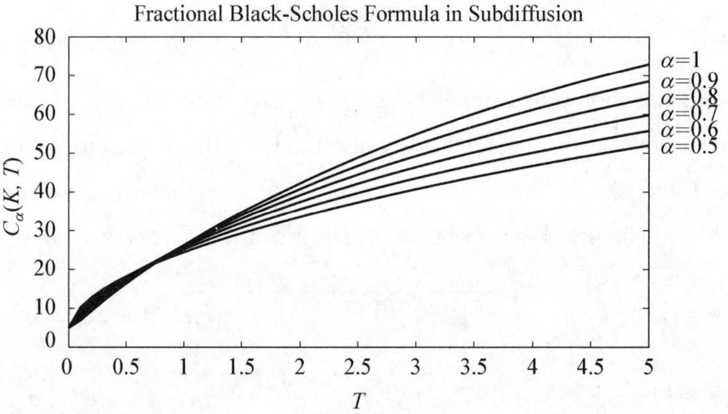

Fig. 2 Presented are prices of European call options corresponding to the subdiffusive BS model without transaction costs for different parameter α, where $X_0 = 100$, $K = 95$, $r = 0.1$, $\mu = 0.2$, $\sigma = 0.5$, $H = 0.8$ and $k = 0$.

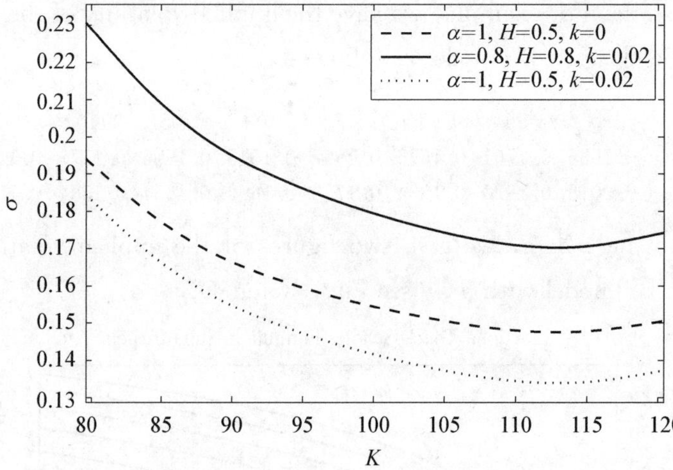

Fig. 3 Presented is the plot of implied volatility σ corresponding to K in the subdiffusive BS model with transaction costs and $\mu = 0.2$, $T = 0.25$.

Corollary 4 *(BS Formula with No Transaction Costs in Subdiffusive Regime) Let* $z = Z(t) = X(S_\alpha(t))$ *be the same subdiffusive model of asset price defined in Theorem 2, where the process* $S_\alpha(t)$ *and* $B_H(\tau)$ *are independent. Then, we obtain the option price* $C_\alpha(K, T)$ *and* $P_\alpha(K, T)$, *of a European call option and a put option with time to maturity* T *and* *strike price K are given by*

$$C_\alpha(K, T) = X_0 N(d_1) - K e^{-rT} N(d_2), \tag{34}$$

$$P_\alpha(K, T) = K e^{-rT} N(-d_2) - X_0 N(-d_1), \tag{35}$$

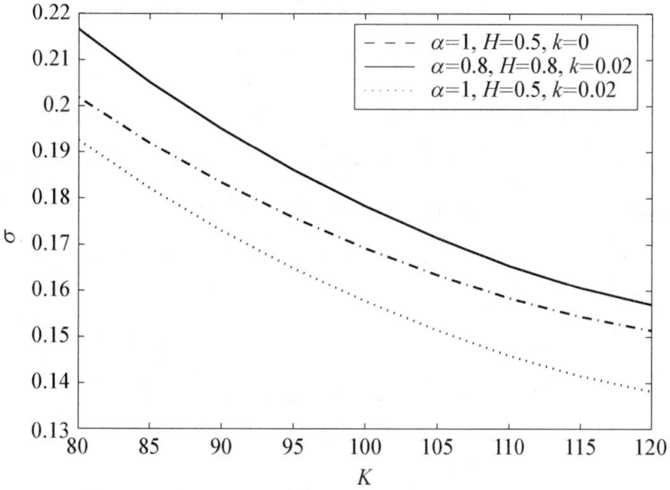

Fig. 4 Presented is the plot of implied volatility σ corresponding to K in the subdiffusive BS model with transaction costs and $\mu = 0.2$, $T = 0.5$.

and the Put-Call Parity relationship

$$C_\alpha(K, T) - P_\alpha(K, T) = X_0 - K e^{-rT} \tag{36}$$

where, the function $N(x)$ *is the cumulative probability distribution function for a standard normal distribution, and*

$$d_1 = \frac{\ln(X_0/K) + \left(r + \frac{1}{2}\sigma_{H,\alpha}^2\right)T}{\sigma_{H,\alpha}\sqrt{T}}, \quad d_2 = \frac{\ln(X_0/K) + \left(r - \frac{1}{2}\sigma_{H,\alpha}^2\right)T}{\sigma_{H,\alpha}\sqrt{T}}, \tag{37}$$

here $\sigma_{H,\alpha}^2 = \frac{\sigma^2}{\alpha[\Gamma(\alpha)]^{2H}} T^{2H\alpha - 1}$, $0 < H \leqslant 1$.

In particular, letting $\alpha \uparrow 1$ *in Corollaries 3 and 4, we can obtain corresponding Black-Scholes formulas in real time with transaction costs and no transaction costs respectively.*

4. Estimating Turnover and Transaction Costs of Replicating Strategies

It follows directly that the difference between the two initial option prices in subdiffusive regime

$$Z_\alpha = \hat{C}_\alpha(K, T) - C_\alpha(K, T) \tag{38}$$

is a valid measure of the total transaction costs associated with the replicating strategies. Since the volatility is adjusted according to the transaction cost rate k, this parameter as well as the environmental parameters (r, σ^2) and the option parameters (K, T) will importantly affect the total transaction cost, Z_α.

For the total transaction costs, Z_α, we have the following theorem:

Theorem 3 *(Transaction Costs in Subdiffusive Regime) When* $k \ll 1$, *we have*

$$Z_\alpha = \frac{1}{4} k K e^{-rT} N'(d_2) \frac{\sigma T^{\alpha H}}{\sqrt{\alpha}[\Gamma(\alpha)]^{\alpha H}} \left| \frac{2\mu}{\sigma^2} - 1 \right|, \tag{39}$$

where $d_2 = \dfrac{\ln(X_0/K) + rT}{\sigma_H \sqrt{T}} - \dfrac{1}{2}\sigma_H \sqrt{T}$, $\sigma_H^2 = \dfrac{\sigma^2}{\alpha[\Gamma(\alpha)]^{2H}} T^{2\alpha H - 1}$, $0 < H \leqslant 1$.

Proof From Eqs. (29) and (34), we have

$$\begin{aligned} Z_\alpha &= X_0[N(\hat{d}_1) - N(d_1)] - K e^{-rT}[N(\hat{d}_2) - N(d_2)] \\ &= X_0 N'(d_1)\Delta d_1 - K e^{-rT} N'(d_2)\Delta d_2 + o(k), \end{aligned}$$

where

$$\hat{d}_1 = \frac{\ln(X_0/K) + rT}{\hat{\sigma}_H \sqrt{T}} + \frac{1}{2}\hat{\sigma}_H \sqrt{T}, \quad \hat{d}_2 = \frac{\ln(X_0/K) + rT}{\hat{\sigma}_H \sqrt{T}} - \frac{1}{2}\hat{\sigma}_H \sqrt{T},$$

$$d_1 = \frac{\ln(X_0/K) + rT}{\sigma_H \sqrt{T}} + \frac{1}{2}\sigma_H \sqrt{T}, \quad d_2 = \frac{\ln(X_0/K) + rT}{\sigma_H \sqrt{T}} - \frac{1}{2}\sigma_H \sqrt{T},$$

and the modified volatility $\hat{\sigma}_H$ is defined by

$$\hat{\sigma}_H = \sigma_H \left(1 + \frac{k}{2}\left|\frac{2\mu}{\sigma^2} - 1\right|\right)^{1/2}, \quad 0 < H \leqslant 1.$$

When $k \ll 1$, it is easy to see $\Delta d_1 = \hat{d}_1 - d_1 = -\frac{k}{4}\left|\frac{2\mu}{\sigma^2} - 1\right| d_2$ and $\Delta d_2 = \hat{d}_2 - d_2 = -\frac{k}{4}\left|\frac{2\mu}{\sigma^2} - 1\right| d_1$.

Thus we have

$$\begin{aligned} Z_\alpha &= \frac{1}{4}k[K e^{-rT} N'(d_2) d_1 - X_0 N'(d_1) d_2]\left|\frac{2\mu}{\sigma^2} - 1\right| \\ &= \frac{1}{4}kK\sigma e^{-rT} N'(d_2)\left|\frac{2\mu}{\sigma^2} - 1\right| \frac{T^{H\alpha}}{\sqrt{\alpha}\,[\Gamma(\alpha)]^{H\alpha}}. \end{aligned}$$

The proof is completed.

In particular, letting $\alpha \uparrow 1$, we have the following corollary:

Corollary 5 *(Transaction Costs in Real Time) Let* $Z = \hat{C}(K, T) - C(K, T)$, *when* $k \ll 1$, *then, we have*

$$Z = \frac{1}{4}kK e^{-rT}\sigma N'(d_2) T^H \left|\frac{2\mu}{\sigma^2} - 1\right|, \quad 0 < H \leqslant 1, \tag{40}$$

where $d_2 = \frac{\ln(X_0/K) + rT}{\sigma T^H} - \frac{1}{2}\sigma T^H$.

Once Z has been computed, round trip turnover estimates follows immediately:

$$\text{Turnover} = Z/(kX_0). \tag{41}$$

Note that turnover will depend upon k, as well as the environmental parameters (r, σ^2) and the option parameters (K, T).

Proposition I *When* $k \ll 1$, *expected turnover is given by*

$$\frac{1}{4}\frac{K}{X_0} e^{-rT}\sigma N'(d_2) T^H \left|\frac{2\mu}{\sigma^2} - 1\right|, \quad 0 < H \leqslant 1, \tag{42}$$

Proposition II *When* $k \ll 1$, *transaction costs are roughly proportional to rate*, k.

This proposition follows directly from (40), noting that d_2 does not depend on k.

Proposition III *When $k \ll 1$, transaction costs will increase with the striking price, K, when $K < K^*$ and decrease with K when $K > K^*$, where $K^* = X_0 e^{\left(r+\frac{1}{2}\sigma^2 T^{2H-1}\right) T}$.*

Proof It follows directly from that

$$\frac{\partial Z}{\partial K} = \frac{1}{4\sigma^2 T^{2H}} k N'(d_2)\sigma T^H \left| \frac{2\mu}{\sigma^2} - 1 \right| \left[\left(r + \frac{1}{2}\sigma^2 T^{2H-1} \right) T - \ln(K/X_0) \right]. \tag{43}$$

It follows immediately from Proposition III that turnover is greatest when the striking price is K^*, i. e., an option whose present value of striking price is slightly in money.

In particular, when $H = \frac{1}{2}$, i. e., for geometric Brownian motion (GBM) model, Eq. (1), we have

Proposition IV *When $k \ll 1$, transaction costs will increase with the maturity, T, when $T < T^*$ and decrease with T when $T > T^*$, where $T^* = \frac{1}{2a}(b + \sqrt{b^2 + 4ac})$, where*

$$a = \frac{1}{2}\left(\frac{1}{2}\sigma^2 + r\right), \quad b = \sigma^2, \quad c = \frac{1}{2}(\ln(X_0/K))^2.$$

Proof It follows directly from that

$$\frac{\partial Z}{\partial T} = \frac{1}{4} k K \sigma \left| \frac{2\mu}{\sigma^2} - 1 \right| \frac{\partial}{\partial T}[T^{1/2} e^{-rT} N'(d_2)] = -\frac{e^{-rT} N'(d_2)}{\sigma^2 T^{3/2}} (aT^2 - bT - c). \tag{44}$$

It follows immediately from Proposition IV that turnover is greatest when the maturity is T^*, i. e., an option whose present value of maturity is slightly in money.

Remark 2 Propositions I～III are first obtained by Leland [3] but in a discrete time setting.

Remark 3 It follows directly from Put-Call Parity relation that above

Propositions I～IV are valid for transaction costs $Z=\hat{P}(K, T)-P(K, T)$.

Remark 4 The Propositions I～III are valid in subdiffusive regime.

5. Total Transaction Costs

From (12), the total transaction costs up to time t is given by

$$T_c(t)=\int_0^t \frac{k}{2}\left|z^2 \frac{\partial^2}{\partial z^2}\hat{C}_\alpha(t', z)\frac{\mathrm{d}z}{z}\right|. \tag{45}$$

Recall $\frac{\partial^2}{\partial z^2}\hat{C}_\alpha(t', z)z^2=zN'(\hat{d}_1)/\hat{\sigma}_{H,\alpha}(t')\sqrt{T-t'}$, $t<T$, [3], where

$$\hat{d}_1(t')=\frac{\ln(z/K\mathrm{e}^{-rT})}{\hat{\sigma}_{H,\alpha}(t)\sqrt{T-t}}+\frac{1}{2}\hat{\sigma}_{H,\alpha}(t')\sqrt{T-t'}, \quad N'(\hat{d}_1)=\frac{\mathrm{e}^{-\frac{1}{2}\hat{d}_1^2}}{\sqrt{2\pi}}. \tag{46}$$

Noting that $\mathrm{d}z/z$ is given by (17), thus, $T_c(t)$ can be approximately written in the following form:

$$T_c(t)=\frac{k}{2}\int_0^t \frac{zN'(\hat{d}_1(t'))}{\hat{\sigma}_{H,\alpha}(t')\sqrt{T-t'}}\left|\mu-\frac{\sigma^2}{2}\right|\left(\frac{t^{\alpha-1}}{\Gamma(\alpha)}\right)^{2H} 2Ht'^{2H\alpha-1}\mathrm{d}t' <+\infty, \quad t\leqslant T, \tag{47}$$

where $\hat{\sigma}_{H,\alpha}(t')$ is defined by (28).

This shows that the total transaction costs is finite for $t\leqslant T$.

6. Conclusions

Using the strategy of Leland [3], this paper has developed a technique for replicating option returns in a continuous time setting in the presence of transaction costs and obtain the corresponding Black-Scholes equations, Black-Scholes formulas and total transaction costs with transaction costs both in subdiffusive regime and in real time.

Acknowledgments

The authors would like to thank the referee for his/her kind suggestions, and also thank Prof. Wenbin Chen. This work was supported by the National

Natural Science Foundation of China (No. 10871047, No. 10771071 and No. 11001057).

Appendix

Lemma 1 *Let $f(x)$ denote a continuous function, then the solution of the equation*

$$dy = f(t)(dt)^\alpha, \quad t \geqslant 0, \ y(0) = 0 \tag{48}$$

is given by the equality

$$y = \int_0^t f(t)(dt)^\alpha = \begin{cases} \alpha\int_0^t (t-\tau)^{\alpha-1} f(\tau)d\tau, & 0 < \alpha \leqslant 1, \ [20] \\ \left(\frac{\alpha}{n}\int_0^t (t-\tau)^{\alpha/n-1} f^{1/n}(\tau)d\tau\right)^n, & n-1 < \alpha \leqslant n. \end{cases} \tag{49}$$

In particular, for $f(t) \equiv 1$ we have

$$y = \int_0^t (dt)^\alpha = t^\alpha, \quad \alpha > 0. \tag{50}$$

Proof Assume that $0 < \alpha \leqslant 1$. Let us consider the fractional differential equation of Eq. (48)

$$y^{(\alpha)}(t) = {}_0D_t^\alpha y(t) = f(t).$$

On the one hand of side, by using Theorem 2.4 [9], from (48) we have

$$y(t) = {}_0I_t^\alpha f(t) = \frac{1}{\Gamma(\alpha)}\int_0^t (t-\tau)^{\alpha-1} f(\tau)d\tau, \quad 0 < \alpha \leqslant 1 \tag{51}$$

for $y(t) \in I_0^\alpha(L_1)$, which denote the space of function $f(x)$, represented by the left-sided functional integral of order α of a function: $f = I_{0+}^\alpha \varphi$, $\varphi \in L_1(a, b)$.

On the other hand of side, we write Eq. (48) in the differential form

$$d^\alpha y = f(t)(dt)^\alpha.$$

By using the relation [14]

$$d^\alpha y = \Gamma(\alpha+1)dy$$

from (48) for $0 < \alpha < 1$, we have

$$y(t)=\frac{1}{\Gamma(\alpha+1)}\int_0^t f(t)(\mathrm{d}t)^\alpha. \tag{53}$$

Combining (51) and (53), and noting that $\Gamma(\alpha+1)=\alpha\Gamma(\alpha)$, the first part of lemma holds for $0<\alpha\leqslant 1$.

For $n-1<\alpha\leqslant n$, let $\mathrm{d}x=(\mathrm{d}y)^{1/n}=f^{1/n}(t)(\mathrm{d}t)^{\alpha/n}$, by the result obtained above and relation (48), we have

$$X(\tau)=\int_0^t f^{1/n}(t)(\mathrm{d}t)^{\alpha/n}=\frac{\alpha}{n}\int_0^t (t-\tau)^{\alpha/n-1}f^{1/n}(t)\mathrm{d}t$$

$$X(\tau)=\int_0^y(\mathrm{d}y)^{\frac{1}{n}}=y^{\frac{1}{n}}.$$

Hence, the second part of lemma holds. Thus the proof is completed.

◇ References ◇

[1] F. Black, M. S. Scholes, The pricing of options and corporate liabilities, J. Polit. Econ. 81 (1973) 637 - 659.

[2] R. C. Merton, Theory of rational option pricing, Bell J. Econ. 4 (1973) 141 - 183.

[3] H. E. Leland, Option pricing and replication with transaction costs, J. Financ. 40 (1985) 1283 - 1301.

[4] S. Perrakis, J. Lefoll, Option pricing and replication with transaction costs and dividends, Research Paper No. 8, International Center for Financial Asset Management and Engineering.

[5] Phelim P. Boyle, Ton Vorst, Option replication in discrete time with transaction costs, The Journal of Finance 47 (1) (1992) 271 - 293.

[6] S. D. Hodges, A. Neuberger, Optimal replication of contingent claims under transaction costs, The Review of Futures Markets 8 (1989) 222 - 239.

[7] Q. Shen, Bid-ask prices for call option with transaction costs part I: discrete time case, Working Paper, Finance Department, The Wharton School, University of Pennsylvania, PA, 1990.

[8] R. C. Merton, Continuous Time Finance, Basil Blackwell Ltd. , Oxford, 1990, Chapter 14, Section 14. 2.

[9] P. P. Boyle, T. Vorst, Option replication in discret time with transaction costs, J. of Finance 47 (1992) 271 - 293.

[10] M. Monoyios, Option pricing with transaction costs using a Markov chain approximation, Journal of Economic Dynamics and Control 28 (2004) 889 – 913.

[11] Xiao-Tian Wang, Scaling and long range dependence in option pricing, IV: pricing European options with transaction costs under the multifractional Black-Scholes model, Physica A 389 (2010) 438 – 444.

[12] R. N. Mantegna, H. E. Stanley, An Introduction to Econophysics—Correlation and Complexity in Finance, Cambridge University Press, Cambridge, 2000.

[13] K.-I. Sato, Processes and Infinitely Divisible Distributions, Cambridge University Press, Cambridge, 1999.

[14] I. M. Sokolov, Lévy flights from a continuous-time processs, Phys. Rev. E 63 (2000) 011104.

[15] M. Magdziarz, Black-Scholes formula in subdiffusive regime, Journal of Statistical Physics 136 (2009) 553 – 564.

[16] P. Embrechts, M. Maejima, Selfsimilar Processes, Princeton University Press, Princeton, 2002.

[17] S. C. Kou, X. Sunney Xie, Generalized Langevin equation with fractional Gaussian noise: subdiffusion within a single protein molecule, Phys. Rev. Lett. 93 (2004) 180603.

[18] B. J. Berne, J. P. Boon, S. A. Rice, J. Chem. Phys. 45 (1966) 1086.

[19] I. Podlubny, Fractional Differential Equations: An Introduction to Fractional Derivatives, Fractional Differential Equations, to Methods of Their Solution and Some of Their Applications, Academic Press, 1999.

[20] Guy Jumarie, Stock exchange fractional dynamics defined as fractional exponential growth driven by (usual) Gaussian white noise. application to fractional Black-Scholes equations, Insurance: Mathematics and Economics 42 (2008) 271 – 287.

Heterogeneous Memorized Continuous Time Random Walks in an External Force Fields*

Jun Wang, Ji Zhou, Long-Jin Lv, Wei-Yuan Qiu, Fu-Yao Ren

Abstract: In this paper, we study the anomalous diffusion of a particle in an external force field whose motion is governed by nonrenewal continuous time random walks with correlated memorized waiting times, which involves Riemann-Liouville fractional derivative or Riemann-Liouville fractional integral. We show that the mean squared displacement of the test particle X_x which is dependent on its location x of the form (El-Wakil and Zahran, Chaos Solitons Fractals, 12, 1929 - 1935, 2001)

$$\langle \mathbb{X}_x^2 \rangle(t) = \langle (\Delta X_x(t))^2 \rangle_0 \sim |x|^{-\theta} t^{\gamma}, \quad 0 < \gamma < 1, \theta = d_w - 2, \tag{1}$$

where $d_w > 2$ is the anomalous exponent, the diffusion exponent γ is dependent on the model parameters. We obtain the Fokker-Planck-type dynamic equations, and their stationary solutions are of the Boltzmann-Gibbs form. These processes obey a generalized Einstein-Stokes-Smoluchowski relation and the second Einstein relation. We observe that the asymptotic behavior of waiting times and subordinations are of stretched Gaussian distributions. We also discuss the time averaged in the case of an harmonic potential, and show

* Originally published in *J. Stat. Phys*, Vol. 156, No. 6, (2014), 1111 - 1124.

J. Wang, W.-Y. Qiu, F.-Y. Ren

Department of Mathematics, Fudan University, Shanghai 200433, China

J. Zhou

Department of Mathematics, Sichuan Normal University, Chengdu 610066, China

L.-J. Lv

Ningbo Institute of Technology, Zhejiang University, Ningbo 315100, China

that the process exhibits aging and ergodicity breaking.

Keywords: Fokker-Planck equation, generalized Einstein relation, stretched Gaussian distribution, ergodicity breaking

1. Introduction

In connection with the growing interest in the physics of complex systems, anomalous diffusion and their description have received considerable interest. Unbiased and biased anomalous diffusion are well established phenomena Refs. [1~8] found in many different systems. In homogeneous fractal medium in one dimension the anomalous diffusion is characterized by the occurrence of a mean square displacement (MSD) of a test particle with some kind of thermal bath can behave

$$\langle \mathbb{X}^2 \rangle_0(t) = \langle (\Delta x)^2 \rangle_0(t) \sim t^{\gamma}, \quad \text{and} \quad \langle x(t) \rangle_0 = 0. \tag{2}$$

The subscript zero in $\langle \cdots \rangle_0$ denotes the case when no external driving force is applied to the particle. When $\gamma=1$ is called normal diffusion, $\gamma < 1$ is called slow diffusion or subdiffusion, and $\gamma > 1$ is called enhanced diffusion. A particular important class are subdiffusion processes, characterized in terms of the MSD of the form

$$\langle \mathbb{X}^2 \rangle(t) = \langle (\Delta x)^2 \rangle_0(t) = \frac{2K_\gamma}{\Gamma(1+\gamma)} t^{\gamma}, \quad 0<\gamma<1 \tag{3}$$

which deviates from the linear Brownian dependence on time Ref. [5]. Subdiffusion has been observed in a large variety of systems, from physics and chemistry to biology and medicine Refs. [9~20].

In the heterogeneous fractal medium in the one dimension, the anomalous diffusion characterized by the occurrence of an MSD of the test particle X_x which is dependent on its location x of the form [21]

$$\langle \mathbb{X}_x^2 \rangle(t) = \langle (\Delta X_x(t))^2 \rangle_0 \sim |x|^{-\theta} t^{\gamma}, \quad 0<\gamma<1, \quad \theta = d_{\mathrm{w}} - 2, \tag{4}$$

where $d_{\mathrm{w}} > 2$.

A most widely used mathematical models effecting subdiffusion of the form of Eq. (2) is the continuous time random walk (CTRW) theory which was originally introduced by Montroll and Weiss in their seminal paper of 1965 Ref. [8]. Since that time the CTRW was proved to be a useful tool for description of systems out of equilibrium, especially of anomalous diffusion phenomena characterized by nonlinear time dependence of MSD Refs. [5, 6, 14] and references therein.

In a CTRW process after each jumps, a new pair of waiting time and jump lengths are drawn from the associated distributions, independent of the previous values. This independence of the waiting time and jump lengths given rise to a renewal process is not always justified. As soon as the random walk has some form of memory, even a short one, the variable become non-independent. Examples are found such as bacterial motion Ref. [19] or the movement ecology of animal motion [22]. In 2012, M. Magdziarz et al authors Refs. [30~37] introduced and comprehensively studied the correlated continuous time random walks (CCTRW) which is represented in the form of the integral $t(s)=\int_0^s |L_\alpha(u)|\,du$.

However, in fractal medium and in financial market dynamics Refs. [23, 24], single trajectories in which there is a directional memory Ref. [25], or in astrophysics [26], or chaotic and turbulent flows Ref. [27], etc., the processes are with memory. Recently, these facts impel one to introduce the correlated continuous random walks with time averaged waiting time [28], the memorized continuous time random walks (MCTRW) and the derivative memorized continuous time random walks (DMCTRW) [29] and studied their anomalous diffusion in the presence of an external force $F(x)$ with operational time. In all these Refs. [30 ~ 37] the anomalous diffusion discussed are homogeneous and the waiting time is produced by CTRW with no memory.

In this paper, we study the dynamics of two alternative model of subdiffusion with intrinsic memory, namely, heterogeneous memorized CTRW (HMCTRW) processes and heterogeneous derivative memorized CTRW (HDMCTRW). Our starting point of the HMCTRW is letting $t_x(s)$ defined as follows

$$D_{0+}^{1-\mu} t_x(s) = |x|^{\theta/\gamma} |L_\alpha(s)|, \tag{5}$$

here x as a parameter, $-\infty < x < +\infty$, and $0 < \mu < 1$, where $D_{0+}^{1-\mu} t_x(s)$ denotes the Riemann-Liouville fractional derivative [38] defined by

$$D_{0+}^{1-\mu} f(s) = \frac{1}{\Gamma(\mu)} \frac{\mathrm{d}}{\mathrm{d}s} \int_0^s \frac{f(u)\mathrm{d}u}{(s-u)^{1-\mu}},\ 0 < \mu < 1. \tag{6}$$

To solve this equation using the Theorem 2. 4 [38, p. 45] we obtain the processes $t_x(s)$ dependent the parameters x, θ, μ, and α,

$$t_x(s) = |x|^{\theta/\gamma} t(s),\quad t(s) = \frac{1}{\Gamma(1-\mu)} \int_0^s \frac{|L_\alpha(u)|\mathrm{d}u}{(s-u)^\mu}, \tag{7}$$

for $t(s) \in I_{0+}^{1-\mu}(L_1)$ [38], where $1/\gamma$ is the self-similar index of process $t(s)$ with $\gamma = \alpha/[1 + \alpha(1-\mu)]$. The corresponding operational time process $s_x(t)$ is inverse to $t_x(s)$ represented as

$$s_x(t) = \inf\{s \geqslant 0 : t_x(s) > t\}. \tag{8}$$

Finally, one assembles the jump lengths $x(\tau) = B(\tau)$ and waiting times $\tau = s_x(t)$ to obtain the subordination $X_x(t) = B[s_x(t)]$ to be called heterogeneous CTRW with memory (HMCTRW) which reduces to CCTRW [35] when μ and θ tends to zero.

Similarly, let $t_x(s)$ is defined as follows

$$I_{0+}^{\mu} t_x(s) = |x|^{\theta/\gamma} |L_\alpha(s)|, \tag{9}$$

here x as a parameter, $-\infty < x < +\infty$, and $0 < \mu < 1$, where $I_{0+}^{\mu} t(s)$ denotes the Riemann-Liouville fractional derivative [38] defined by

$$I_{0+}^{\mu} f(s) = \frac{1}{\Gamma(\mu)} \int_0^s \frac{f(u)\mathrm{d}u}{(s-u)^{1-\mu}},\quad 0 < \mu < 1, \tag{10}$$

where μ to be called memory exponent. We solve out $t_x(s)$ as

$$t_x(s) = |x|^{\theta/\gamma} t(s),\quad t(s) = \frac{1}{\Gamma(1-\mu)} \frac{\mathrm{d}}{\mathrm{d}s} \int_0^s \frac{|L_\alpha(u)|\mathrm{d}u}{(s-u)^\mu}, \tag{11}$$

for $t(s) \in I_{0+}^{1-\mu}(L_1)$ [38], where $1/\gamma$ is the self-similar index of process $t(s)$ with $\gamma = \alpha/(1-\alpha\mu) > 0$, thus yielding the processes $s_x(t)$, which is inverse to $t_x(s)$. Finally, one assembles the jump lengths $x(\tau) = B(\tau)$ and waiting times $\tau = s_x(t)$ to obtain the subordination $X_x(t) = B[s_x(t)]$, which is called heterogeneous derivative CTRW with memory (HDMCTRW), which also reduces to CTRW when μ and θ tends to zero [39].

Where $B(s)$ is the Brownian motion with mean zero and $\langle B^2(1)\rangle = 2D$, and $L_\alpha(s)$ is the α-stable symmetric *Lévy* process with $L_\alpha(0) = 0$ and Laplace transform given by

$$\langle \exp[-kL_\alpha(s)]\rangle = \exp\left(-s\frac{1}{2}|k|^\alpha\right), \quad 0 < \alpha \leqslant 2. \tag{12}$$

When memory set is a fractal, the exact estimates of corresponding memory exponent μ on the fractal can be obtained Ref. [39].

We derive the HMCTRW and HDMCTRW, and apply these results to describe the anomalous diffusion dynamics in the presence of an external force $F(x)$. We obtain the corresponding Fokker-Planck equation and show that their stationary solutions are the Boltzmann distribution and that these processes satisfies generalized Einstein-Stokes-Smoluchowski relation and generalized Einstein relation. We show that the asymptotic behavior of waiting time and subordination are stretched Gaussian, namely, heavy tailed. Additionally, we discuss phenomena of ergodicity breaking of single particle tracking as well as aging.

2. Asymptotic Behavior of HMCTRW and HDMCTRW

From Figs. 1 and 2., we can see, because of the memory effect, there is correction between two consecutive waiting times for CTRW with memory. But, no relation for CTRW without memory. From the figures, we also can find that the time for a particle stopping at one point is almost the same between two consecutive jumps for the CTRW with memory. But, the time is changing remarkably for the CTRW without memory.

It is not difficult to verify that

$$\langle \exp\{ikt_x(s)\}\rangle = \exp\left\{-s^{\alpha/\gamma}\frac{1}{2}|A_x k|^{\alpha}\right\},\quad s_x(t) = |x|^{-\theta}s(t), \tag{13}$$

where $s(t) = \inf\{s \geqslant 0:\ t(s) > t\}$, and

$$\hat{g}_x(s, k) = \frac{\alpha}{2\gamma}A_x^{\alpha}s^{\frac{\alpha}{\gamma}-1}k^{\alpha-1}e^{-\frac{1}{2}s^{\alpha/\gamma}(A_x k)^{\alpha}}. \tag{14}$$

where, $g_x(s, t)$ is the probability density function (PDF) of process $s_x(t)$ and $\hat{g}_x(s, k)$ its Laplace transform, $A_x = |x|^{\theta/\gamma}\mathrm{B}(1+1/\alpha, 1-\mu)/\Gamma(1-\mu)$ for HMCTRW, and $A_x = |x|^{\theta/\gamma}(1-\mu+1/\alpha)\mathrm{B}(1+1/\alpha, 1-\mu)/\Gamma(1-\mu)$ for HDMCTRW, respectively; and that

$$g_x(s, t) \sim |x|^{\theta}t^{-\gamma}(s/S_0)^{\delta'}\exp\{-\mathrm{const}\cdot(s/S_0)^{u'}\}, \tag{15}$$

when $\xi \equiv s/S_0 \sim s/t^{\gamma/2} \gg 1$ and $t \to \infty$, where $S_0^2 = ((\Delta s)^2(t))_0 \sim t^{\gamma}$, $u' = \alpha/[\gamma(1-\alpha)]$, $\delta' = u'(\gamma - 1/2 - \kappa)$, $\kappa = 1 - \alpha/\gamma$.

Proposition 1 *Equation (15) shows that both HMCTRW and HDMCTRW processes can be of stretched Gaussian with heavy-tailed waiting times, of Gaussian, and of stretched Gaussian with thin-tailed waiting times with respect to ξ when $u' < 2$, $u' = 2$ and $u' > 2$, respectively.*

Proposition 2 *Equation (15) shows that the parameter $|x|^{\theta}$ effects only the amplitude of the PDF of HMCTRW and HDMCTRW processes.*

Proposition 3 *It is easy to see that $B(s_x(t)) = |x|^{-\theta/2}B(s(t))$ in distribution. This shows that the difference of the figures of HMCTRW and of MCTRW or of HDMCTRW and DMCTRW only a factor $|x|^{-\theta/2}$.*

In biased case, an external force $F(x)$ should be included in our model as the usual way. This yields the Langevin equation

$$\dot{x}(\tau) = \frac{F[x(\tau)]}{m\eta} + \mathrm{d}B(\tau)/\mathrm{d}\tau, \tag{16}$$

with particle mass m, and friction coefficient η.

3. MSD in the Force-Free Case

In the unbiased case, $F \equiv 0$, $x = B(\tau)$ and the solution to Eqs. (16) and

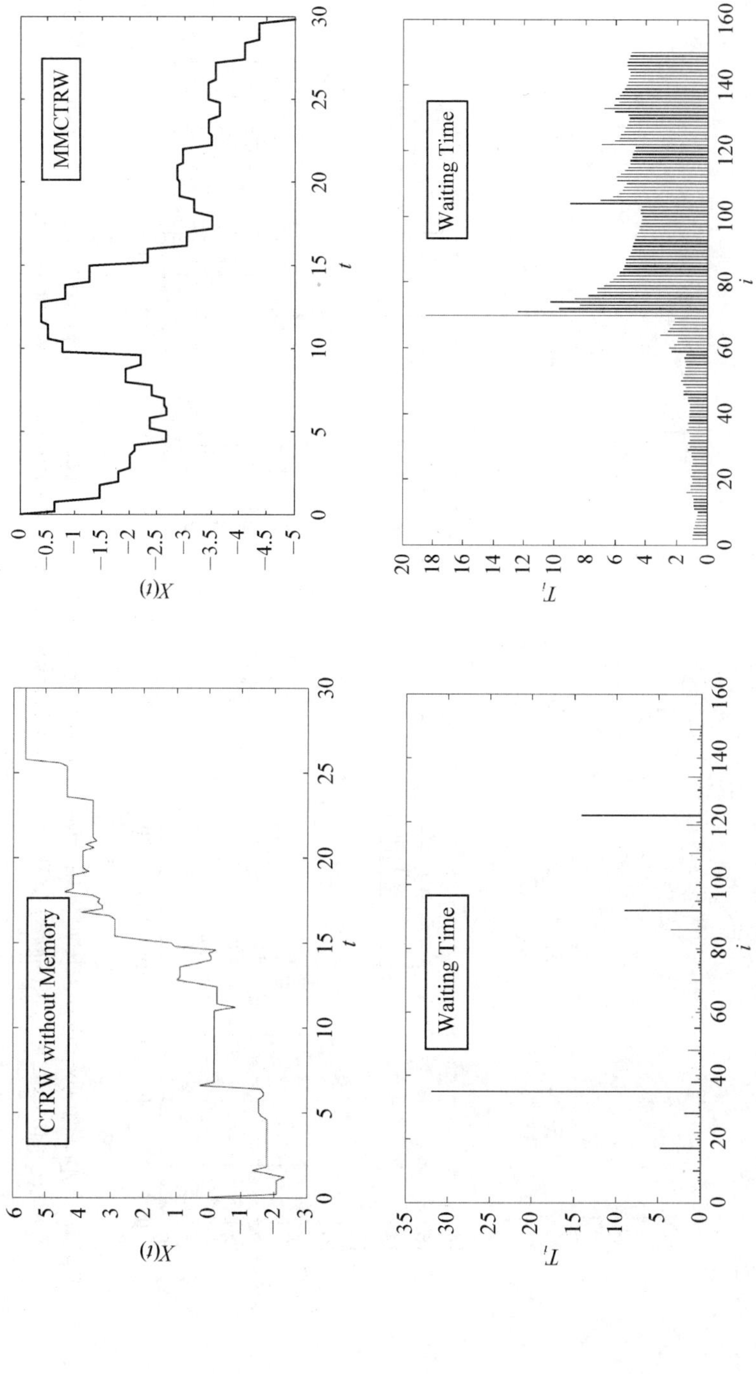

Fig. 1 Typical trajectory of CTRW with power law waiting times [$\alpha = 0.7$ (*Left*)], compared to a trajectory of the MCTRW [$\alpha = 0.7$; $\mu = 0.5$ (*Right*)]. The corresponding waiting times are shown in the two figures *below*

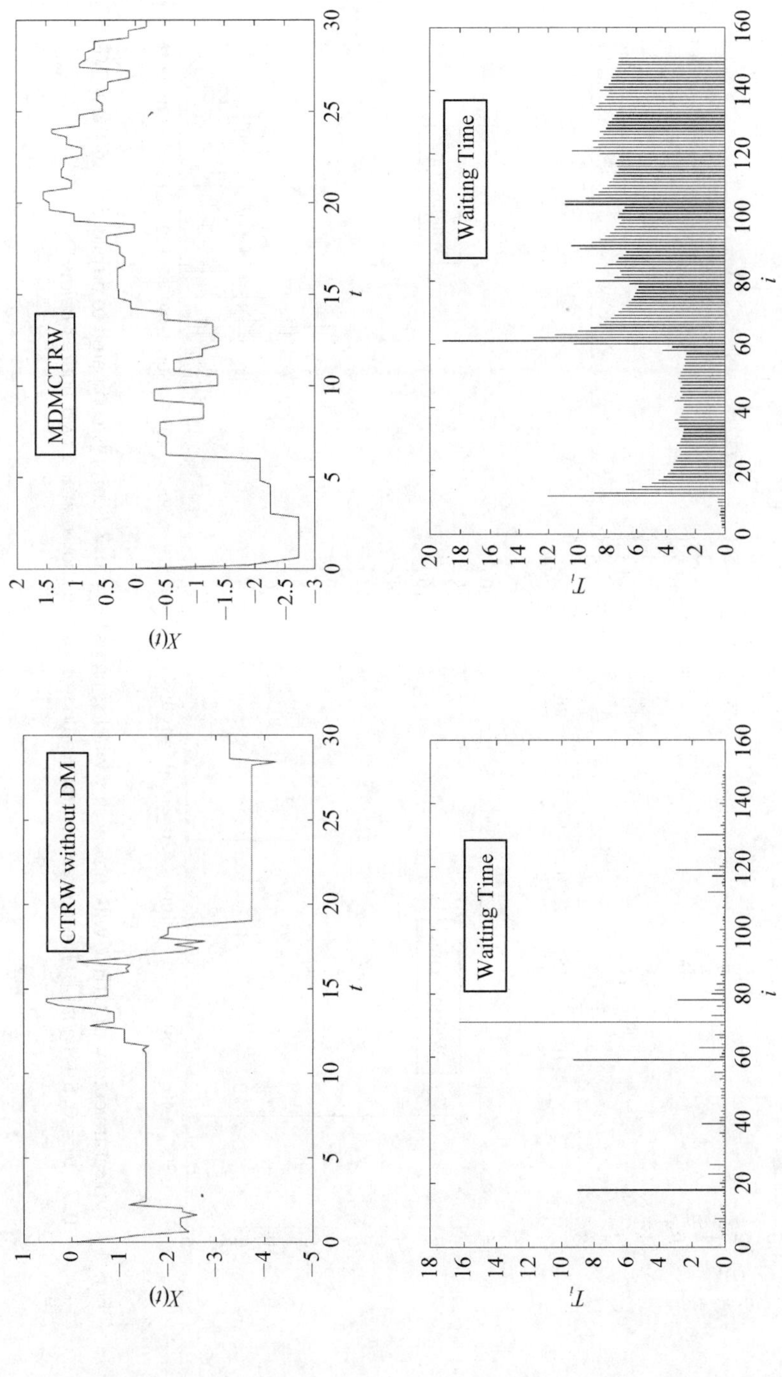

Fig. 2 Typical trajectory of CTRW with power law waiting times [α = 0.7 (*Left*)], compared to a trajectory of the MDMCTRW [α = 0.7; μ = 0.5 (*Right*)]. The corresponding waiting times are shown in the two figures *below*

(5) or (9) has simple form $X_x(t)=B[s_x(t)]$. It is not difficult to verify that $t_x(s)$ is $1/\gamma$-self-similar and $s_x(t)$ a γ-self-similar process with $\gamma=\alpha/[1+\alpha(1-\mu)]$ which reduces to $\alpha/(1+\alpha)$ derived in Ref. [33] when μ tends to zero for HMCTRW, and $\gamma=\alpha/(1-\alpha\mu)$ which reduces to α derived in Ref. [35] when μ tends to zero for HDMCTRW, respectively. Assume that processes $B(\tau)$ and $s(t)$ is independent through the paper. It is not difficult to verify that $s_x(t)=|x|^{-\theta}s(t)$, thus,

$$\langle \mathbb{X}_x^2(t)\rangle_0(t)=2D\langle s(1)\rangle |x|^{-\theta}t^{\gamma}, \tag{17}$$

where $s(t)=\inf\{s\geqslant 0;\ t(s)>t\}$.

Proposition 4 *Equation (17) shows that $X_x(t)$ satisfy heterogeneous anomalous diffusion of kind of Eq. (3).*

Proposition 5 *Equation (17) shows that the diffusion of $X_x(t)$ can be subdiffusion, or normal diffusion, and or superdiffusion when $\gamma<1$, $\gamma=1$, and $\gamma>1$, respectively. In the case of HMCTRW, γ ranges between $\alpha/(1+\alpha)$ and α; so that the diffusion of $X_x(t)$ can be subdiffusion, or normal diffusion, and or superdiffusion. In the case of HDMCTRW, γ ranges between zero and $\alpha/(1-\alpha)$; so that the diffusion of $X(t)$ can be subdiffusion, or normal diffusion, and or superdiffusion.*

4. Fokker-Planck-Type Dynamic Equation for HMCTRW and HDMCTRW Model

Using the total probability formula, the PDF $p(x,t)$ of $X(x,t)=x[s_x(t)]$ in Laplace space k is given by

$$\hat{p}(x,k)=\int_0^{\infty} f(x,s)\hat{g}_x(s,k)\mathrm{d}s=\frac{1}{2}A_x^{\alpha}k^{\alpha-1}\hat{f}_1\left(x,\frac{1}{2}(A_xk)^{\alpha}\right), \tag{18}$$

where $f(x,s)$ be the PDF of $x(s)$, $f_1(x,s)=f(x,s^{\gamma/\alpha})$. Thus, it can be shown that the Fokker-Planck-type dynamic equation is given by

$$\frac{\partial p(x,t)}{\partial t}=\left[\frac{\partial}{\partial x}\left(-\frac{F(x)}{m\eta}+D\frac{\partial}{\partial x}\right)\right]\Phi_t[p(x,t)], \tag{19}$$

where $\Phi_t(.)$ is a double integral operator:

$$\Phi_t f(t) = \frac{1}{2\pi i}\int_{a-i\infty}^{a+i\infty} d\zeta h(x, t, \zeta)\int_0^\infty dt e^{-\frac{1}{A_x}(2\zeta)^{1/\alpha}t} f(t),$$

$$h(x, t, \zeta) = \frac{2^{1/\alpha}}{A_x}\zeta^{\frac{(1-\alpha)}{\alpha}} L_k^{-1}\left[\hat{g}\left(\frac{1}{2}(A_x k)^\alpha - \zeta\right)\right],$$

where $g(s) = \frac{\gamma}{\alpha}s^{\gamma/\alpha-1}$, $a > s_1$ and $\mathrm{Re}\, p > s_2 + a$, s_1 and s_2 are the increasing exponent of $f_1(x, s)$ and $g(s)$, respectively; and L_k^{-1} denotes the inverse of the Laplace transform in variable k.

Proposition 6 *Equation (19) shows that the Fokker-Planck equation is essentially same for HMCTRW and HDMCTRW waiting times and only different from the constant* A_x.

5. Stationary Solution and Generalized Einstein Relations

Since in external force field $F(x) = -V'(x)$, here $V(x)$ is external potential and $S = \left[\frac{-V'(x)}{m\eta} - D\frac{\partial}{\partial x}\right]p(x, t)$ is the probability current. If the stationary state is reached, S must be constant. Thus, if $S = 0$ for any x, from the Eq. (19) it vanishes for all x Ref. [39], and the stationary solution is given by $V'(x)p_{st}(x)/[m\eta] + Dp'_{st}(x) = 0$, hence,

$$p_{st}(x) \propto \exp\left\{-\frac{V(x)}{Dm\eta}\right\}. \tag{20}$$

Comparing this expression with the required Boltzmann distribution $p_{st}(x) \propto \exp\{-V(x/[k_B T])\}$, we recover a generalized Einstein relation, also referred to an Einstein-Stokes-Smoluchowski relation,

$$D = \frac{k_B T}{m\eta}, \tag{21}$$

connecting the noise strength D with the dissipative parameter η via the thermal energy T.

In the presence of a uniform force field, given by $V(x) = -Fx$, a net drift occurs. We calculate the first moment of $X(t)$, obtaining

$$\langle X(t)\rangle_F = \frac{F}{m\eta}\langle s(1)\rangle t^\gamma = \frac{F}{m\eta}\frac{2^{\gamma/\alpha}}{\alpha A^\gamma}\frac{\Gamma(\gamma/\alpha)}{\Gamma(\gamma)}t^\gamma. \tag{22}$$

Comparison of the above result with Eq. (17) indeed leads to the second generalized Einstein relation

$$\langle X(t)\rangle_F = \frac{1}{2}\frac{F\langle X^2(t)\rangle_0}{k_B T} \tag{23}$$

discussed in Refs. [5, 40, 41], connecting the first moment in the presence of a uniform force field with the second moment in absence of the force. Thus,

Proposition 7 *Equations (21) and (23) show that both the HMCTRW and HDMCTRW processes preserve the two fundamental physical properties of diffusion processes, in analogy to regular diffusive CTRW [14, 40~42].*

6. Asymptotic Behavior of the PDF of Subordination

We assume that the external potential at original has the form

$$V(x) = b\ln x + \sum_{n=0}^{\infty} a_n x^n, \tag{24}$$

where b, a_n can be zero, i.e. $F(x) = -V'(x)$. It is proved that

$$p(x,t) \sim |x|^{\theta/2} t^{-\gamma/2} (x^{1+\theta/2}/X_\theta)^\delta \exp\{-\text{const}\cdot (x^{1+\theta/2}/X_\theta)^u\}, \tag{25}$$

when $\xi \equiv x^{1+\theta/2}/X_\theta \sim x^{1+\theta/2}/t^{\gamma/2} \gg 1$ and $t \to \infty$, which reduces to the result of Ref. [37] when θ tends to zero. Here $u = 1/(1-\gamma/2)$, $\delta = u\left[\frac{1}{2}(\gamma-1) - \kappa\right]$, $\kappa = \frac{1}{2} - \delta_0$, while $\delta_0 = \frac{1}{2}\left(1 - \frac{b}{Dm\eta}\right)$ if $b \neq 0$; or $\delta_0 = \frac{1}{2}$ if $b = a_n = 0$ for $n < N$ but $a_N \neq 0$ and $N \geqslant 0$ [43].

Proposition 8 *Equation (25) shows that the PDF $p(x, t)$ of the subordination $X(x, t) = x[s_x(t)]$ corresponding to HMCTRW and HDMCTRW waiting time processes $s_x(t)$ are of heavy-tailed stretched Gaussian Ref. [14], Gaussian, and of thin-tailed stretched Gaussian when $u < 2$, $u = 2$ and $u > 2$, respectively, which equivalent to $\gamma < 1$, $\gamma = 1$, and γ*

>1, *respectively*, *i.e.*, $\mu\alpha<1$, $=1$, >1, *respectively*, *for HMCTRW waiting times*, *and* $\alpha/(1-\alpha\mu)<2$, $=2$, >2, *respectively*, *for HDMCTRW waiting times.*

7. Time Averaged Mean Squared Displacement

7.1 In the unbiased case with $F\equiv 0$

In this case, $X_x(t)=B[s_x(t)]$. We consider the n-th order (TA) moments at point x,

$$\overline{\delta^n(x;\Delta;t_a,T)_0}=\frac{\int_{t_a}^{t_a+T-\Delta}[X_x(t+\Delta)-X_x(t)]^n\mathrm{d}t}{T-\Delta},\tag{26}$$

with process age t_a, the lag time Δ and the measurement time T, which are useful to characterize experiment data Refs. [44~54]. It is not difficult to verify that

$$\langle X_x^n(t)\rangle_0=\langle\zeta_n\rangle t^{nH\gamma},\tag{27}$$

where $H=1/2$, $\zeta_n=|x|^{-nH\theta}B(1)^n s^{nH}(1)$ is a random variable and

$$\langle\zeta_n\rangle=|x|^{-nH\theta}l_n\frac{2^{nH\gamma/\alpha}}{\alpha A_x^{nH\gamma}}\frac{\Gamma(nH\gamma/\alpha)}{\Gamma(nH\gamma)},\tag{28}$$

where $l_n=\langle|B^n(1)|\rangle=(n-1)(n-3)(n-5)\cdots(n-2m+1)(\sqrt{2D})^nJ_n$, $J_n=1$ for $n=2m$ and $J_n=\sqrt{2/\pi}$ for $n=2m+1$, $m\in\mathbb{N}^+$, which recover the result of (7) of Ref. [27] for a random walk on a fractal, approximated by the dynamics scheme of O'Shaughnessy and Procaccia Ref. [44] only different from a constant; and that in the limit $\Delta\ll T$

$$\overline{\delta^n(x;\Delta;t_a,T)_0}=\zeta_n\Lambda_{nH\gamma}(t_a/T)\frac{\Delta}{T^{1-nH\gamma}},\tag{29}$$

$$\langle\overline{\delta^n(x;\Delta;t_a,T)_0}\rangle=\langle X^n(T)\rangle_0\Lambda_{nH\gamma}(t_a/T)\frac{\Delta}{T},\tag{30}$$

$\Lambda_\alpha(z)=(1+z)^\alpha-z^\alpha$, and the ergodicity breaking parameter EB_n of the dimensionless random variable $\xi_n = \frac{\overline{\delta^n(x;\Delta;t_a,T)_0}}{\langle\overline{\delta^n(x;\Delta;t_a,T)_0}\rangle}$ of the n-th order (TA) moment (26),

$$EB_n = \langle\xi_n^2\rangle - 1 = \frac{\ell_{2n}}{\ell_n^2}\frac{\Gamma(2nH\gamma/\alpha)}{\Gamma(2nH\gamma)}\frac{\Gamma^2(nH\gamma)}{\Gamma^2(nH\gamma/\alpha)} - 1, \tag{31}$$

and the PDF of the dimensionless random variable ξ_n

$$\phi_{n,\gamma,H}(x) = \frac{1}{\ell_n}\frac{\alpha A_\rho^{nH\gamma}}{2^{nH\gamma/\alpha}}\frac{\Gamma(nH\gamma)}{\Gamma(nH\gamma/\alpha)}\sqrt{\frac{2}{\pi}}\frac{1}{n^2H}x^{\frac{1}{n}+\frac{1}{nH}-2}e^{-x^{2/n}/2}g(x^{1/nH}),\ x \geqslant 0, \tag{32}$$

where $g(x)$ is the PDF of $s(1)$.

Proposition 9 *Equation (29) shows that* $\overline{\delta^n(x;\Delta;t_a,T)}$ *remains a random variable.*

Proposition 10 *Equation (30) shows that if we know through measurement the ensemble averaged* $\langle X_x^n(T)\rangle_0$ *we can determine the single particle trajectory averaged behavior; conversely, the inverse implication is also true.*

Proposition 11 *Equations (29) and (30) show that the diffusion exponents of* $\langle X_x^n(t)\rangle_0$ *and* $\langle\overline{\delta^n(x;\Delta;t_a,t)}\rangle$ *is* $nH\gamma$ *and* $nH\gamma - 1$*, respectively. This fact suggests one to define the n-th order (TA) moments as follows*

$$\bar{\delta}^n(x;\Delta;t_a,T)_0 = \int_{t_a}^{t_a+T-\Delta}[X(t+\Delta)-X(t)]^n dt. \tag{33}$$

Thus, $\langle X_x^n(t)\rangle_0$ *and* $\langle\bar{\delta}^n(x;\Delta;t_a,t)\rangle$ *have the same diffusion exponents,* $nH\gamma$.

Proposition 12 *Equation (31) shows that the ergodicity breaking parameter* EB_n *of the dimensionless random variable* ξ_n *is independent of the position* x*, the lag time* Δ *and the starting time* t_a *of measurement.*

Interestingly, when $n=2$*, we have*

$$\overline{\delta^2}(x;\Delta;t_a,T)=|x|^{-\theta}B^2(1)s(1)\Lambda_\gamma(t_a/T)\frac{\Delta}{T^{1-\gamma}},\tag{34}$$

and

$$\langle\overline{\delta^2}(x;\Delta;t_a,T)\rangle=2D|x|^{-\theta}\frac{2^{\gamma/\alpha}}{\alpha A^\gamma}\frac{F(\gamma/\alpha)}{\Gamma(\gamma)}\Lambda_\gamma(t_a/T)\frac{\Delta}{T^{1-\gamma}},\tag{35}$$

$$\phi_{\xi_2}(x)=\frac{\alpha A^\gamma}{2^{\gamma/\alpha}}\frac{\Gamma(\gamma)}{\Gamma(\gamma/\alpha)}\frac{1}{4D\sqrt{\pi D}}x^{-\frac{1}{2}}\exp\left(-\frac{x}{4D}\right)g(x),\ x\geqslant 0,\tag{36}$$

and

$$EB=\frac{\langle(\overline{\delta^2}(\Delta;t_a,T))^2\rangle}{(\langle\overline{\delta^2}(\Delta;t_a,T)\rangle)^2}-1=3\alpha\frac{\Gamma^2(\gamma)}{\Gamma(2\gamma)\Gamma(2\gamma/\alpha)}-1.\tag{37}$$

7.2 In the biased case

In a biological cell, the diffusive motion of a tracer particle is spatially confined, or a charge in a disordered semiconductor experiences a drift force. To address such systems we need to determine the TAMSD of subdiffusive particle $X(t)=x[s_x(t)]$ in the presence of an external harmonic potential $V(x)=\lambda x^2/2$ of stiffness λ and discuss the more general case below. To that ehd we consider the following Langevin equation with Brownian motion $B(\tau)$

$$\mathrm{d}x(\tau)=-\lambda' x(\tau)\mathrm{d}\tau+\mathrm{d}B(\tau),\tag{38}$$

where $\lambda'=\lambda/(m\eta)$. Eq. (38) has a unique solution

$$x(\tau)=B(\tau)-\lambda'\int_{-\infty}^{\tau}\mathrm{e}^{-\lambda'(\tau-s)}B(s)\mathrm{d}s.\tag{39}$$

Using the self-similarity of $B(\tau)$ and $s_x(t)$ from Eq. (38) we have

$$X(t)=\zeta_1|x|^{-H\theta}s^H(1)t^{H\gamma}-\lambda\Big\{\zeta_{-1}\sum_{n=0}^{\infty}a_n|x|^{-n\theta}t^{n\gamma}$$

$$+\zeta_1\sum_{n=0}^{\infty}b_n|x|^{-(n+H+1)\theta}t^{\gamma(n+H+1)}\Big\},\tag{40}$$

in distribution, where $\zeta_{-1}=B(-1)$, $\zeta_1=B(1)$, $a_n=a'_n s^n(1)$, $a'_n=\lambda'^{H+1}\Gamma(H$

$+1)\dfrac{(-\lambda')^n}{n!}$, $b_n=b'_n s^{n+H+1}(1)$, $b'_n=\dfrac{(-\lambda')^n}{n!}\left[\sum_{j=0}^{n}\dfrac{(-j)^n}{j+H+1}C_j^n\right]$ and $C_j^n=\dfrac{n(n-1)(n-2)\cdots(n-j+1)}{j!}$. Thus, for $T\gg 1$

$$\langle X^2(T)\rangle \backsimeq 2D\left(\frac{\lambda}{m\eta}\right)^2 g(T)T^{3\gamma}, \tag{41}$$

where

$$g(T)=\frac{2^{3\gamma/\alpha}}{\alpha}\left[\sum_{n=0}^{\infty}b'_n\left(\frac{2^{\gamma/\alpha}}{\gamma}T\right)^n\right]^2,$$

and in the limit $\Delta\ll T$ the TAMSD

$$\overline{\delta^2}(x;\Delta;t_a,T)\backsimeq \lambda'^2|x|^{-3\theta}B^2(1)s^3(1)h(t_a,T)\frac{\Delta}{T^{1-3\gamma}}, \tag{42}$$

$$\langle\overline{\delta^2}(x;\Delta;t_a,T)\rangle \backsimeq 2D\lambda'^2|x|^{-3\theta}\frac{2^{3\gamma/\alpha}}{\alpha A^{3\gamma}}\frac{\Gamma(3\gamma/\alpha)}{\Gamma(3\gamma)}h(t_a,T)\frac{\Delta}{T^{1-3\gamma}}, \tag{43}$$

$$h(t_a,T,x)=\sum_{n,m=0}^{\infty}b'_n b'_m|x|^{-\theta(n+m+3)}\frac{2^{\gamma(n+m+3)/\alpha}}{\alpha A^{\gamma(n+m+3)}}\frac{\Gamma((n+m+3)\gamma/\alpha)}{\Gamma((n+m+3)\gamma)}\cdot T^{\gamma(n+m)}\Lambda_{\gamma(n+m+3)}(t_a/T),$$

where $A=B(1+1/\alpha,1-\mu)/\Gamma(1-\mu)$ for HMCTRW, and $A=(1-\mu+1/\alpha)B(1+1/\alpha,1-\mu)/\Gamma(1-\mu)$ for HDMCTRW, respectively. Thus, it yields that the PDF $\varphi_\xi(x)$ and the ergodicity breaking parameter EB of the dimensionless random variable $\xi=\dfrac{\overline{\delta^2}(\Delta;t_a,T)}{\langle\overline{\delta^2}(\Delta;t_a,T)\rangle}$,

$$\phi_{\xi_2}(x)=\frac{\alpha A^{3\gamma}}{2^{3\gamma/\alpha}}\frac{\Gamma(3\gamma)}{\Gamma(3\gamma/\alpha)}\frac{x^{-7/6}}{12D\sqrt{\pi D}}\exp\left(-\frac{x}{4D}\right)g(x^{\frac{1}{3}}),\ x\geqslant 0, \tag{44}$$

$$EB=\langle\xi^2\rangle-1=3\alpha 2^{6\gamma(1/\alpha-1)}\frac{\Gamma(3\gamma/\alpha+1/2)}{\Gamma(3\gamma+1/2)}\frac{\Gamma(3\gamma)}{\Gamma(3\gamma/\alpha)}-1. \tag{45}$$

Proposition 13 *These show that $\overline{\delta^2}(x;\Delta;t_a,T)$ remains a random variable; if we know through measurement the ensemble averaged $\langle\mathbb{X}^2(T)\rangle$ we can determine the single particle trajectory averaged behavior;*

conversely, the inverse implication is also true; and that the ergodicity breaking parameter EB of the dimensionless random variable ξ is independent of the lag time Δ and the starting time t_a of measurement.

8. Conclusion

In this paper, we studied a model of anomalous diffusion in external force fields, which originates from a heterogeneous CTRW process with memorized waiting times. These HMCTRW and HDMCTRW models was rephrased in terms of a Langevin equation for position x and one of Riemann-Liouville fractional derivative or Riemann-Liouville fractional integral, equivalent to the definition within the framework of subordination, the random time change using the operational time of the system. We show that the mean squared displacement is of the form $\mathbb{X}_x^2 = \langle (\Delta X_x)^2(t) \rangle_0 \sim |x|^{-\theta} t^{\gamma}$, $0 < \gamma < 1$, $\theta = d_w - 2$, where $d_w > 2$ is the anomalous exponent, $\gamma = \alpha/[1+\alpha(1-\mu)]$ for memorized continuous-time random walks (HMCTRW), and $\gamma = \alpha/(1-\alpha\mu)$ for derivative memorized continuous-time random walks (HDMCTRW). We obtained the shape of the corresponding Fokker-Planck-type dynamic equation. We showed that the stationary solution of the model equals the Boltzmann distribution, and that the model satisfies the generalized Einstein relations in consistency with the fluctuation-dissipation relation. We showed that the asymptotic behavior of the waiting time and the PDF of subordination $X(t)$ are stretched Gaussian distribution, respectively. We showed that the diffusion of the subordination can be subdiffusion, or normal diffusion, and or superdiffusion depending on the parameters α and μ, and that our model displays weak ergodicity-breaking.

We believe that the introduced two models are an interesting alternative in the modeling of the complex systems showing anomalous diffusion under the influence of external force fields. It relaxes the renewal property of standard CTRW and CCTRW theory and thus of interest in all those cases, where the non-renewal behavior is relevant: such as in financial market dynamics, in movement ecology, to search processes and human motion patterns, also various physical processes as turbulent flows or complex systems with heterogeneities.

There are numerous open lines to be pursued for HMCTRW and HDMCTRW processes. Thus, we may ask for the scaling limits or for the numerical simulation with memoralized waiting times.

Acknowledgments This work is supported by the National Natural Foundation of China (Grant Nos. 11371266, 10871047, 11001057, 11271074), and ZJNSF No. Q13A010034.

◇ References ◇

[1] Klafter, J., Shlesinger, M. F., Zumofen, G.: Beyond brownian motion. Phys. Today 49(2), 33 - 39 (1996)

[2] Shlesinger, M. F., Zaslavsky, G. M., Frisch, U. (eds.): Lévy Flights and Related Topics in Physics. Springer, Berlin (1994)

[3] Weiss, G. H.: Aspects and Applications of Random Walk. North-Holland, Amsterdam (1994)

[4] Scher, H., Shlesinger, M. F., Bendler, J. T.: Time-scale invariance in transport and relaxation. Phys. Today 44(1), 26 - 34 (1991)

[5] Bouchaud, J. P., Georges, A.: Anomalous diffusion in disordered media: statistical mechanisms, models and physical applications. Phys. Rep. 195, 127 - 293 (1990)

[6] Haus, J. W., Kehr, K. W.: Diffusion in regular and disordered lattices. Phys. Rep. 150, 263 - 406 (1987)

[7] Havlin, S., Ben-Avrahan, D.: Diffusion in disordered media. Adv. Phys. 36, 695 - 798 (1987). Reprint in Adv. Phys. 51, 187 - 292 (2002)

[8] Montroll, E. W., Weiss, G. H.: Random walks on lattices. II. J. Math. Phys. 6, 167 - 181 (1965)

[9] Isichenko, M. B.: Percolation, statistical topography, and transport in random media. Rev. Mod. Phys. 64, 961 - 1043 (1992)

[10] Blumen, A., Klafter, J., Zumofen, G.: Models for reaction dynamics. In: Zschokke, I. (ed.)Optical Spectres Copy of Glasses. Reidel, Dordrecht (1986)

[11] Losa, G. A., Nonnenmacher, T. F., Weibl, E. R.: Fractals in Biology and Medicine. Birkhäuser Verlag, Basel (1993)

[12] Scher, H., Montroll, E. W.: Anomalous transit-time dispersion in amorphous solids.

Phys. Rev. B 12, 2455 - 2477 (1975)

[13] Solomon, T. H., Weeks, E. R., Swinney, H. L.: Observation of anomalous diffusion and Lévy flights in a two-dimensional rotating flow. Phys. Rev. Lett. 71, 3975 - 3978 (1993)

[14] Metzler, R., Klafter, J.: The random walk's guide to anomalous diffusion: a fractional dynamics approach. Phys. Rep. 339, 1 - 77 (2000)

[15] Chechkin, A. V., Gonchar, Y. V., Szydlowski, M.: Fractional kinetics for relaxation and superdiffusion in a magnetic field. Phys. Plasmas 9, 78 - 88 (2002)

[16] Scher, H., et al.: The dynamical foundation of fractal stream chemistry: the origin of extremely long retention times. Geophys. Res. Lett. 29, 1061 (2002)

[17] Berkowitz, B., Cortis, A., Dentz, M., Scher, H.: Modeling non-Fickian transport in geological formations as a continuous time random walk. Rev. Geophys. 44, RG2003 (2006)

[18] Klemm, A., Metzler, R., Kimmich, R.: Diffusion on random-site percolation clusters: theory and NMR microscopy experiments with model objects. Phys. Rev. E 65, 021112 (2002)

[19] Rogers, S. S., van der Walle, C., Waigh, T. A.: Microrheology of bacterial biofilms in vitro: *Staphylococcus aureus* and *Pseudomonas aeruginosa*. Langmuir 24, 13549 - 13555 (2008)

[20] Golding, I., Cox, E. C.: Physical nature of bacterial cytoplasm. Phys. Rev. Lett. 96, 098102 (2006)

[21] El-Wakil, S. A., Zahran, M. A.: Fractional representation of Fokker - Planck equation. Chaos Solitons Fractals 12, 1929 - 1935 (2001)

[22] Nathan, R., et al.: A movement ecology paradigm for unifying organismal movement research. Proc. Natl. Acad. Sci. USA 105, 19052 - 19059 (2008)

[23] Scalas, E.: The application of continuous-time random walks in finance and economics. Phys. A 362, 225 - 239 (2006)

[24] La Spada, G., Farmer, J. D., Lillo, F.: The non-random walk of stock prices: the long-term correlation between signs and sizes. Eur. Phys. J. B 64, 607 - 614 (2008)

[25] Le Borgne, T., Dentz, M., Carrera, J.: Lagrangian statistical model for transport in highly heterogeneous velocity fields. Phys. Rev. Lett. 101, 090601 (2008)

[26] Barkana, R.: On correlated random walks and 21-cm fluctuations during cosmic reionization. Mon. Not. R. Astron. Soc. 376, 1784 - 1792 (2007)

[27] Manneville, P.: Instabilities Chaos and Turbulence. Imperial College Press, London (2010)

[28] Lv, L. J. , Ren, F. Y. , Wang, J. , Xiao, J. B. : Correlated continuous-time random walks with time averaged waiting time. submitted to J. Math. Phys.

[29] Ren, F. Y. , Wang, J. , Lv, L. J. , Pan H. , Qiu, W. Y. : Anomalous diffusion in memorized continuous time random walks in an external force fields. Preprint

[30] Magdziarz, M. , Weron, A. : Fractional Langevin equation with α-stable noise. A link to fractional ARIMA time series. Stud. Math. 181(1), 47 - 60 (2007)

[31] Meerschaert, M. M. , Nane, E. , Xiao, Y. : Correlated continuous time random walks. Stat. Probab. Lett. 79, 1194 - 1202 (2009)

[32] Chechkin, A. V. , Hofmann, M. , Sokolov, I. M. : Continuous-time ramdom walk with correlated waiting times. Phys. Rev. E 80, 031112 (2009)

[33] Tejedor, V. , Metzler, R. : Anomalous diffusion in correlated continuous time random walks. J. Phys. A 43, 082002 (2010)

[34] Magdziartz, M. , Metzler, R. , Szczotka, W. , Zebrowski, P. : Correlated continuous time random walks -scaling limits and Langevin picture. J. Stat. mech. P04010(2012)

[35] Magdziartz, M. , Metzler, R. , Szczotka, W. , Zebrowski, P. : Correlated continuous time random walks in external force fields. Phys. Rev. 85, 051103 (2012)

[36] Eliazar, I. I. , Shlesinger, M. F. : Fractional motions. Phys. Rep. 527(2), 101 - 130 (2013)

[37] Magdziarz, M. , Szczotka, W. , Zebrowski, P. : Asymptotic behaviour of random walks with correlated temporal structure. Proc. R. Soc. A 469, 20130419 (2013)

[38] Samko, S. G. , Kilbas, A. A. , Marichev, O. I. : Fractional Integrals and Derivatives: Theory and Applications. Gorden and Breach, Yverdon (1993)

[39] Weron, A. , Magdziarz, M. , Weron, K. : Modeling of subdiffusion in space-time-dependent force fields beyond the fractional Fokker-Planck equation. Phys. Rev. E 77, 036704 (2008)

[40] Metzler, R. , Barkai, E. , Klafter, J. : Deriving fractional Fokker-Planck equations from a generalised master equation. Europhys. Lett. 46(4), 431 - 436 (1999)

[41] Metzler, R. , Barkai, E. , Klafter, J. : Anomalous diffusion and relaxation close to thermal equilibrium: A fractional Fokker-Planck equation approach. Phys. Rev. Lett. 82, 3563 - 3567(1999)

[42] Barkai, E. , Metzler, R. , Klafter, J. : From continuous time random walks to the fractional Fokker-Planck equation. Phys. Rev. E 16, 132 - 138 (2000)

[43] Ren, F. Y. , Liang, J. R. , Qiu, W. Y. , Xiao, J. B. : Answer to an open problem proposed by R Metzler and J Klafter. J. Phys. A Math. Gen. 39, 4911 - 4919 (2006)

[44] O'Shaughnessy, B. , Procaccia, I. I. : Analytical solutions for diffusion on fractal

objects. Phys. Rev. Lett. 54, 455 - 458 (1985)

[45] Cherstvy, A. G., Chechkin, A. V., Metzler, R.: Anomalous diffusion and ergodicity breaking in heterogeneous diffusion process. New J. Phys. 15, 083039 (2013)

[46] Lubelski, A., Sokolov, A. M., Klafter, J.: Nonergodicity mimics inhomogeneity in single particle tracking. Phys. Rev. Lett. 100, 250602 (2008)

[47] He, Y., Burov, S., Metzler, R., Barkai, E.: Random time-scale invariant diffusion and transport coefficients. Phys. Rev. Lett. 101, 058101 (2008)

[48] Barkai, E., Garini, Y., Metzler, R.: Strange kinetics of single molecules in living cells. Phys. Today 65(8), 29 - 35 (2012)

[49] Bel, G., Barkai, E.: Weak ergodicity breaking in the continuous-time random walk. Phys. Rev. Lett. 94, 240602 (2005)

[50] Margolin, G., Barkai, E.: Nonergodicity of blinking nanocrystals and other Lé-walk processes. Phys. Rev. Lett. 94, 080601 (2005)

[51] Burov, S., Joen, J. H., Metzler, R., Barkai, E.: Single particle tracking in systems showing anomalous diffusion: the role of weak ergodicity breaking. Phys. Chem. Chem. Phys. 13, 1800 - 1812 (2011)

[52] Jeon, J. H., Metzler, R.: Inequivalence of time and ensemble averages in ergodic systems: exponential versus power-law relaxation in confinement. Phys. Rev. E. 85, 021147, 039904 (2012)

[53] Froemberg, D., Barkai, E.: Time-averaged Einstein relation and fluctuating diffusivities for the Lévy walk. Phys. Rev. E. 87, 030104(R) (2013)

[54] Jeon, J. H., et al.: In vivo anomalous diffusion and weak ergodicity breaking of lipid granules. Phys. Rev. Lett. 106, 048103 (2011)

Effect of Different Waiting Time Processes with Memory to Anomalous Diffusion Dynamics in an External Force Field*

Fu-Yao Ren[a], Jun Wang[a], Long-Jin Lv[b],
Hua Pan[a], Wei-Yuan Qiu[a]

[a] *Department of Mathematics, Fudan University, Shanghai 200433, China*

[b] *Ningbo Institute of Technology, Zhejiang University, Ningbo 315100, China*

Abstract: In this paper, we study the anomalous diffusion of a particle in an external force field whose motion is governed by nonrenewal continuous time random walks with memory. In our models, the waiting time involves Riemann-Liouville fractional derivative or Riemann-Liouville fractional integral. We obtain the systematic observation on the mean squared displacement, the Fokker-Planck-type dynamic equations and their stationary solutions. These processes obey a generalized Einstein-Stokes-Smoluchowski relation, and observe the second Einstein relation. The asymptotic behavior of waiting times and subordinations are of stretched Gaussian distributions. We also discuss the time averaged in the case of an external force field, and show that the process exhibits aging and ergodicity breaking.

Keywords: Fokker-Planck equation, generalized Einstein relation, stretched Gaussian distribution, ergodicity breaking

---

* Originally published in *Physica A*, Vol. 417, (2015), 202 - 214.

1. Introduction

Over recent years, anomalous diffusion and their description have received considerable interest due to the growing interest in the physics of complex system. Biased and unbiased anomalous diffusion are well established phenomena [1~8] found in many different systems. For unbiased processes, the mean square displacement of a test particle with some kind of thermal bath can behave as

$$\langle X^2\rangle_0(t)=\langle(\Delta x)^2\rangle_0(t)\sim t^{\gamma},\quad \text{and}\quad \langle x(t)\rangle_0=0. \tag{1}$$

The subscript zero denotes the case when no external driving force is applied to the particle. When $\gamma=1$ is called normal diffusion, $\gamma<1$ is called slow diffusion or subdiffusion, and $\gamma>1$ is called enhanced diffusion. A particular important class are subdiffusion processes, characterized in terms of the mean square displacement (MSD):

$$\langle X^2\rangle(t)=\langle(\Delta x)^2\rangle_0(t)=\frac{2K_\gamma}{\Gamma(1+\gamma)}t^{\gamma},\quad 0<\gamma<1 \tag{2}$$

which deviates from the linear Brownian dependence on time, Ref. [5]. Subdiffusion has been observed in a large variety of systems, from physics and chemistry to biology and medicine, see Refs. [9~20].

The continuous time random walk (CTRW), which was originally introduced by Montroll and Weiss [8], is a most widely used mathematical models effecting subdiffusion of the form of Eq. (2). CTRW was proved to be a useful tool for description of systems out of equilibrium, especially of anomalous diffusion phenomena characterized by nonlinear time dependence of mean squared displacement (MSD), see Refs. [5, 6, 14] and references therein. In CTRW process, after each jumps, a new pair of waiting time and jump lengths is drawn from the associated distributions, independent of the previous values. This independence of the waiting time and jump lengths given rise to a renewal process is not always justified.

As soon as the random walk has some form of memory, even a short one, the variable becomes non-independent. Examples are found such as bacterial motion [19] or the movement ecology of animal motion [21]. In financial market dynamics [22, 23], single trajectories in which there is a directional memory [24], or in astrophysics [25], or chaotic and turbulent flows [26], these processes are with memory. The correlated continuous time random walks (CCTRW) was introduced and comprehensively studied in Refs. [27~31] where the time is represented in the form of the integral $t(s) = \int_0^s |L_\alpha(u)| \, du$. Here, $L_\alpha(u)$ is the α-stable nondecreasing *Lévy* process with $L_\alpha(0)=0$ and Laplace transform given by

$$\langle \exp[-kL_\alpha(s)] \rangle = \exp\left(-s\frac{1}{2}|k|^\alpha\right), \quad 0<\alpha\leqslant 2. \tag{3}$$

However, the waiting time still has no memory in their model.

In this paper, we mainly study the dynamics of two classes of CTRW with intrinsic memory and their basic properties, and the position x always satisfy Langevin equation

$$\dot{x}(s) = \mathrm{d}B(s)/\mathrm{d}s, \tag{4}$$

where $\mathrm{d}B(s)/\mathrm{d}s$ represents standard white Brownian Gaussian noise, $B(s)$ is the Brownian motion with zero mean and $\langle B^2(1) \rangle = 2D$. The time $t(s)$ is defined differently in our two classes, and the corresponding operational time process $s(t)$ is inverse to $t(s)$, so can be represented as

$$s(t) = \inf\{s \geqslant 0: t(s) > t\}. \tag{5}$$

Finally, we assemble the jump lengths $x(\tau)=B(\tau)$ and waiting times $\tau = s(t)$ to obtain the subordination $X(t)=B[s(t)]$, which are our models.

The first class contains the memorized CTRW (MCTRW) process and the derivative memorized CTRW (DMCTRW) process. Our starting point of MCTRW is letting the time t defined as

$$D_{0+}^{1-\mu}\, t(s) = L_\alpha(s), \quad 0<\mu<1, \tag{6}$$

and the time t in DMCTRW is defined by

$$I_{0+}^{\mu} t(s)=L_{\alpha}(s), \quad 0<\mu<1. \tag{7}$$

Here, $D_{0+}^{1-\mu} t(s)$ and $I_{0+}^{\mu} t(s)$ with $0<\mu<1$ denote the Riemann - Liouville fractional derivative and integral [32] with the definition as follows:

$$D_{0+}^{1-\mu} t(s)=\frac{1}{\Gamma(\mu)} \frac{\mathrm{d}}{\mathrm{d}s} \int_{0}^{s} \frac{t(u)\mathrm{d}u}{(s-u)^{1-\mu}}, \quad I_{0+}^{\mu} t(s)=\frac{1}{\Gamma(\mu)} \int_{0}^{s} \frac{t(u)\mathrm{d}u}{(s-u)^{1-\mu}}. \tag{8}$$

When memory set is a fractal, the exact estimates of corresponding memory exponent μ on the fractal can be obtained, Ref. [33].

The second class includes the mean memorized CTRW (MMCTRW) and the mean derivative memorized CTRW (MDMCTRW), where the time t is defined by

$$t(s)=\frac{1}{s^{1-\mu}} \frac{1}{\Gamma(1-\mu)} \int_{0}^{s} \frac{|L_{\alpha}(u)|\,\mathrm{d}u}{(s-u)^{\mu}}, \tag{9}$$

$$t(s)=s^{\mu} \frac{1}{\Gamma(1-\mu)} \frac{\mathrm{d}}{\mathrm{d}s} \int_{0}^{s} \frac{|L_{\alpha}(u)|\,\mathrm{d}u}{(s-u)^{\mu}}, \tag{10}$$

respectively. When $\mu \to 0$, $t(s)$ reduces to $t(s)=s^{-1}\int_{0}^{s}|L_{\alpha}(u)|\,\mathrm{d}u$ and $t(s)=|L_{\alpha}(s)|$ correspondingly.

In biased case, an external force $F(x)$ should be added in our models as the usual way, and this means

$$\dot{x}(s)=\frac{F[x(s)]}{m\eta}+\Gamma(s), \tag{11}$$

with particle mass m and friction coefficient η.

2. The Fundamental Properties of CTRW with Memory

To solve the system of equations (4) and (6), by using the Theorem 2. 4 [32, p. 45] we obtain the process $t(s)$:

$$t(s)=\frac{1}{\Gamma(1-\mu)} \int_{0}^{s} \frac{|L_{\alpha}(u)|\,\mathrm{d}u}{(s-u)^{\mu}}, \tag{12}$$

for $t(s) \in I_{0+}^{1-\mu}(L_1)$. This Abel's equation expresses the relation between the force $|L_\alpha(u)|$ and the flux $t(s)$ of a system with memory function $\Gamma(1-\mu)^{-1}u^{-\mu}$. Thus, MCTRW reduces to CCTRW when μ tends to zero. As for DMCTRW, we obtain the process $t(s)$ from Eq. (7),

$$t(s)=\frac{1}{\Gamma(1-\mu)}\frac{\mathrm{d}}{\mathrm{d}s}\int_0^s \frac{|L_\alpha(u)|\,\mathrm{d}u}{(s-u)^\mu}. \tag{13}$$

This means that $t(s)$ is the rate of output (or flux)

$$\int_0^x \frac{|L_\alpha(u)|}{\Gamma(1-\mu)(s-u)^\alpha}\mathrm{d}u$$

with the input (or, force) $|L_\alpha(u)|$ and the memory function $\Gamma(1-\mu)^{-1}u^{-\mu}$. When $\mu\to 0$, DMCTRW reduces to CTRW.

From Figs. 1 and 2, we can see, because of the memory effect, there is correction between two consecutive waiting times for CTRW with memory. But, no such relation for CTRW without memory. From the figures, we also can find that the time for a particle stopping at one point is almost the same between two consecutive jumps for the CTRW with memory. But, the time is changing remarkably for the CTRW without memory.

Since $L_\alpha(s)$ is $1/\alpha$-self-similar, it is not difficult to verify that $t(s)$ is $1/\gamma$-self-similar and $s(t)$ a γ-self-similar process with $\gamma=\alpha/[1+\alpha(1-\mu)]$ for MCTRW and $\gamma=\alpha/(1-\alpha\mu)$ for DMCTRW, respectively.

Proposition I *This shows that the waiting times for MCTRW and DMCTRW processes have same memory function $s^{-\mu}$, but they have different self-similarity.*

Using Eqs. (3), (12) and (13), we obtain

$$\begin{aligned}\langle \exp\{\mathrm{i}kt(s)\}\rangle &= \langle \exp\{\mathrm{i}Aks^{1/\gamma}|L_\alpha(1)|\}\rangle \\ &= \langle \exp\{\mathrm{i}Ak(s^{\alpha/\gamma})^{1/\alpha}|L_\alpha(1)|\}\rangle \\ &= \exp\left\{-s^{\alpha/\gamma}\frac{1}{2}|Ak|^\alpha\right\},\end{aligned} \tag{14}$$

where $A=\mathrm{B}(1+1/\alpha, 1-\mu)/\Gamma(1-\mu)$ for MCTRW, and $A=(1-\mu+1/\alpha)\mathrm{B}(1+1/\alpha, 1-\mu)/\Gamma(1-\mu)$ for DMCTRW, respectively; $B(.,.)$ is the beta function.

Now, let us pass on to the main problem of this section. Firstly, we establish the relation between the probability density function (PDF) $g(s, t)$ of $s(t)$ and the PDF $u(t, s)$ of $t(s)$. Since $t(s)$ is $1/\gamma$ self-similar and $s(t)$ is γ self-similar, their PDF fulfills the scaling relation

$$u(t, s)=\frac{1}{s^{1/\gamma}}u\left(\frac{t}{s^{1/\gamma}}\right), \quad g(s, t)=\frac{1}{t^{\gamma}}g\left(\frac{s}{t^{\gamma}}\right), \tag{15}$$

where $u(t)=u(t, 1)$, $g(t)=g(1, t)$.

Consider the property $\Pr(s(t)\leqslant s)=\Pr(t(s)\geqslant t)$, and using the same method as in Ref. [34], we know

$$g(s, t)=\frac{t}{s\gamma}u(t, s). \tag{16}$$

Thus,

$$g(\tau, t)=\frac{t}{\gamma}\frac{1}{\tau^{1/\gamma+1}}u\left(\frac{t}{s^{1/\gamma}}\right), \quad u(t, s)=\frac{s\gamma}{t^{\gamma+1}}g\left(\frac{s}{t^{\gamma}}\right). \tag{17}$$

From Eq. (14), we get that $\langle e^{-kt(\tau)}\rangle=\int_0^{\infty}e^{-kt}u(t, \tau)dt=e^{-\frac{1}{2}s^{\alpha/\gamma}(Ak)^{\alpha}}$. Thus, we obtain

$$\begin{aligned}\hat{g}(s, k)&=\int_0^{\infty}e^{-kt}g(s, t)dt\\&=\int_0^{\infty}e^{-kt}\frac{t}{s\gamma}u(t, s)dt\\&=-\frac{1}{s\gamma}\frac{d}{dk}\int_0^{\infty}e^{-kt}u(t, s)dt\\&=\frac{\alpha}{2\gamma}A^{\alpha}s^{\frac{\alpha}{\gamma}-1}k^{\alpha-1}e^{-\frac{1}{2}s^{\alpha/\gamma}(Ak)^{\alpha}},\end{aligned} \tag{18}$$

and that

$$\int_0^{\infty}e^{-kt}\frac{s\gamma}{t^{\gamma+1}}g\left(\frac{s}{t^{\gamma}}\right)dt=e^{-\frac{1}{2}s^{\alpha/\gamma}(Ak)^{\alpha}}. \tag{19}$$

Eq. (18) can be rewritten as

$$\hat{g}(s, k)=\frac{\alpha}{2\gamma}A^{\alpha}s^{\frac{\alpha}{\gamma}-1}k^{\alpha-1}e^{-\frac{1}{2}s^{\alpha/\gamma}(Ak)^{\alpha}}$$

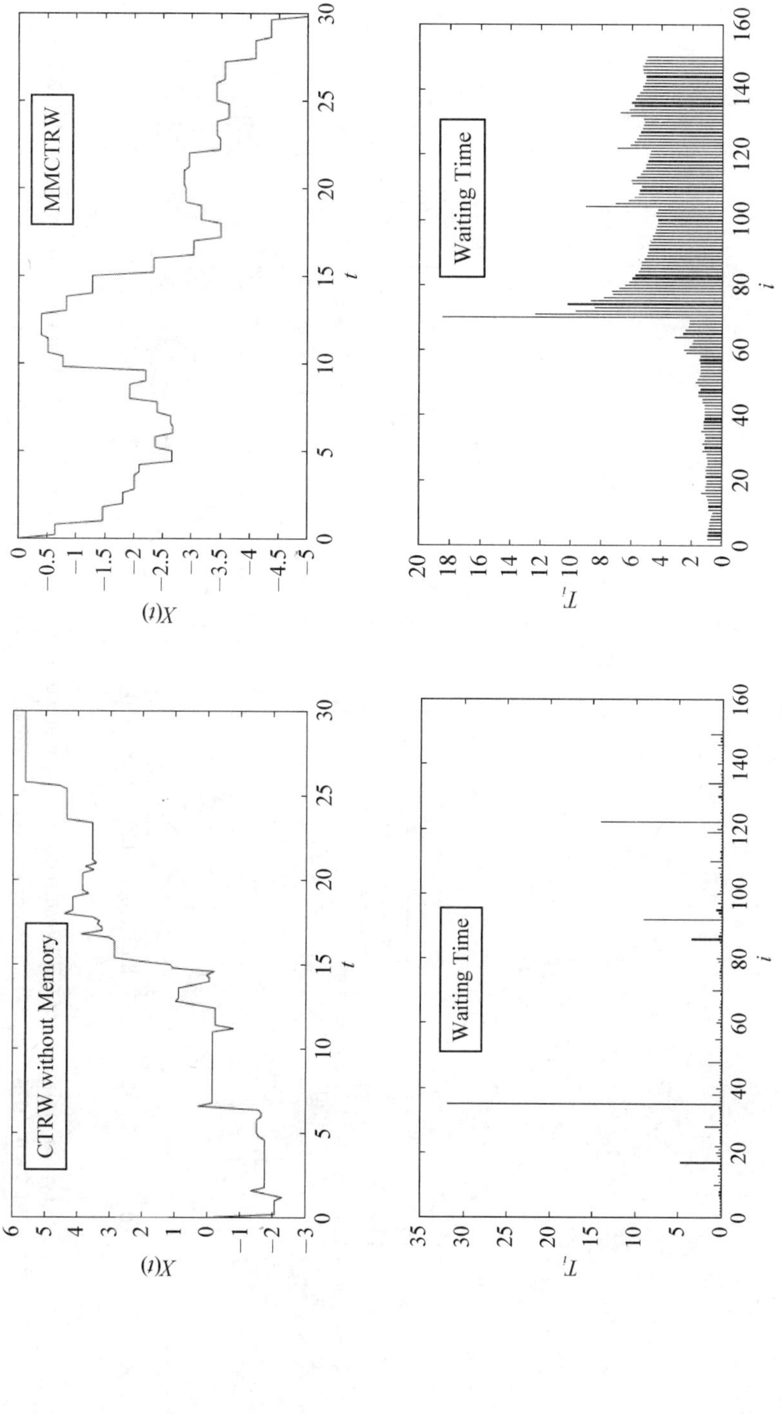

Fig. 1 Typical trajectory of CTRW with power law waiting times ($\alpha = 0.7$ (Left)), compared to a trajectory of the MCTRW ($\alpha = 0.7$; $\mu = 0.5$ (Right)). The corresponding waiting times are shown in the two figures below.

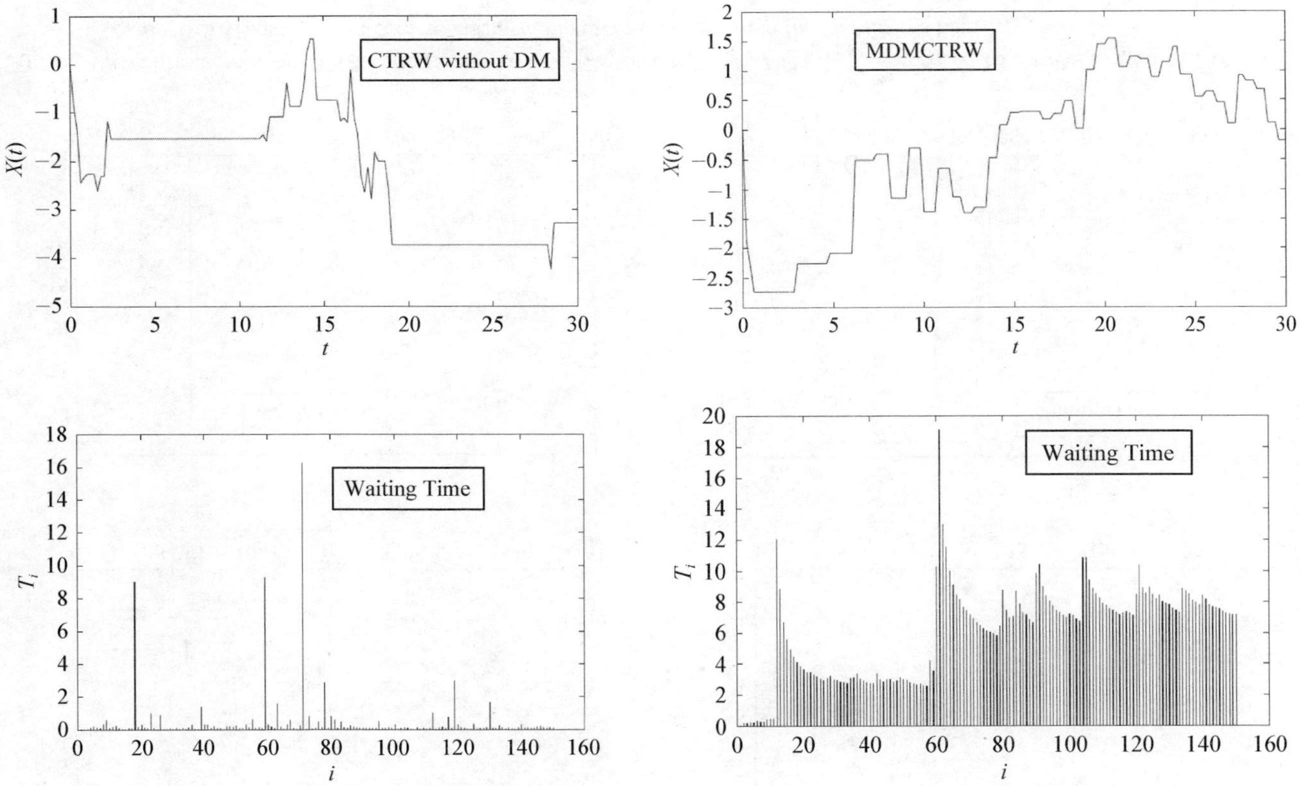

Fig. 2 Typical trajectory of CTRW with power law waiting times ($\alpha = 0.7$ (Left)), compared to a trajectory of the DMCTRW ($\alpha = 0.7$; $\mu = 0.5$ (Right)). The corresponding waiting times are shown in the two figures below.

$$= \frac{\alpha}{2\gamma} A^{\alpha} k^{-(1-\beta/2v)} \frac{1}{(sk^{\beta/2v})^{\kappa}} \exp\{-\text{const} \cdot (sk^{\beta/2v})^{v}\}, \tag{20}$$

where $\beta/2v=\gamma$, $v=\alpha/\gamma$, $\kappa=1-\alpha/\gamma$. Using the result of Ref. [35], we have that

$$g(s,t) \sim t^{-\gamma}(s/X_0)^{\delta'} \exp\{-\text{const} \cdot (s/X_0)^{u'}\}, \tag{21}$$

when $\xi \equiv s/X_0 \sim s/t^{\gamma/2} \gg 1$ and $t \to \infty$, where

$$u' = \alpha/[\gamma(1-\alpha)], \quad \delta' = u'(\gamma - 1/2 - \kappa). \tag{22}$$

Proposition II *This shows that the waiting times of MCTRW and DMCTRW processes are stretched Gaussian with different stretched Gaussian exponent $u'=(1-\alpha\mu+\alpha)/(1-\alpha)$ for MCTRW, and $u'=(1-\alpha\mu)/(1-\alpha)$ for DMCTRW.*

From Eq. (19), for any real number $p > 0$ we have

$$\int_0^{\infty} s^{p-1}\left[\int_0^{\infty} \mathrm{e}^{-kt} \frac{\gamma s}{t^{\gamma+1}} g\left(\frac{s}{t^{\gamma}}\right) \mathrm{d}t\right] \mathrm{d}s = \int_0^{\infty} s^{p-1} \mathrm{e}^{-\frac{1}{2}s^{\alpha/\gamma}(Ak)^{\alpha}} \mathrm{d}s. \tag{23}$$

Thus, we obtain

$$\langle s'^{p} \rangle =: \int_0^{\infty} s'^{p} g(s') \mathrm{d}s' = \frac{2^{p\gamma/\alpha}}{\alpha A^{p\gamma}} \frac{\Gamma(p\gamma/\alpha)}{\Gamma(p\gamma)}. \tag{24}$$

Hence, for any positive number p we have

$$\langle s^{p}(t) \rangle =: \int_0^{\infty} s^{p} g(s,t) \mathrm{d}s = \frac{2^{p\gamma/\alpha}}{\alpha A^{p\gamma}} \frac{\Gamma(p\gamma/\alpha)}{\Gamma(p\gamma)} t^{p\gamma}. \tag{25}$$

Proposition III *This shows that the waiting times of MCTRW and DMCTRW processes have any p-order moments for $p > 0$.*

It is not difficult to verify that $t(s)$ is $1/\alpha$-self-similar and $s(t)$ be α-self-similar process for MMCTRW and MDMCTRW, and similarly

$$\langle \exp\{ikt(s)\} \rangle = \exp\left\{-s\frac{1}{2}|Ak|^{\alpha}\right\}, \tag{26}$$

where $A = \mathrm{B}(1+1/\alpha, 1-\mu)/[\alpha(1-\mu)]$ for MMCTRW, and $A=(1-\mu+1/\alpha)\mathrm{B}(1+1/\alpha, 1-\mu)/[\alpha(1-\mu)]$ for MDMCTRW.

Proposition IV *This shows that waiting time $s(t)$ processes for MMCTRW and MDMCTRW has same self-similarity exponent α.*

Similarly as for MCTRW and DMCTRW, we get $\langle e^{-kt(\tau)}\rangle = e^{-\frac{1}{2}s(Ak)^\alpha}$ and

$$\hat{g}(s,k)=\frac{1}{2}A^\alpha k^{\alpha-1}e^{-\frac{1}{2}s(Ak)^\alpha}, \tag{27}$$

$$\int_0^\infty e^{-kt}\frac{s\alpha}{t^{\alpha+1}}g\left(\frac{s}{t^\alpha}\right)dt=e^{-\frac{1}{2}s(Ak)^\alpha}. \tag{28}$$

Rewrite Eq. (27) as

$$\hat{g}(s,k)=\frac{\alpha}{2\alpha}A^\alpha k^{-(1-\alpha)}\frac{1}{(sk^{\beta/2v})^\kappa}\exp\{-\text{const}\cdot(sk^{\beta/2v})^v\}, \tag{29}$$

where $\beta/2v=\alpha$, $v=1$, $\kappa=(1-\alpha)/\alpha$. Using the result in Ref. [36], we have that

$$\bar{g}(s,t)\sim t^{-\alpha}(s/X_\alpha)^{\delta'}\exp\{-\text{const}\cdot(s/X_\alpha)^{u'}\}, \tag{30}$$

when $\xi\equiv s/X_\alpha\sim s/t^{\alpha/2}\gg 1$ and $t\to\infty$, where

$$u'=1/\left(1-\frac{\alpha}{2}\right),\quad \delta'=u'\left[\frac{1}{2}(\alpha-1)-\kappa\right]. \tag{31}$$

Proposition V *This shows that the waiting times of MMCTRW and MDMCTRW are same stretched Gaussian with same stretched Gaussian exponent $u'=1/\left(1-\frac{\alpha}{2}\right)$, which can be heavy-tailed, Gaussian-tailed, and thin-tailed with respect to ξ for $0<u'<1$, $u'=2$, and $u'>2$, respectively.*

From Eq. (28), for any real number $p>0$ we have

$$\int_0^\infty s^{p-1}\left[\int_0^\infty e^{-kt}\frac{\alpha s}{t^{\alpha+1}}\bar{g}\left(\frac{s}{t^\alpha}\right)dt\right]ds=\int_0^\infty s^{p-1}e^{-\frac{1}{2}s(Ak)^\alpha}ds. \tag{32}$$

Thus, for any positive number p we have

$$\langle s^p(t)\rangle =: \int_0^\infty s^p g(s,t)ds=\frac{2^p}{\alpha A^{p\alpha}}\frac{\Gamma(p)}{\Gamma(p\alpha)}t^{p\alpha}. \tag{33}$$

Proposition VI *This shows that the waiting times of MMCTRW and*

MDMCTRW processes also have any p-order moments, $p > 0$.

In the unbiased case, that is external force free, and $x(s) = B(s)$. Assume that processes $B(s)$ and $s(t)$ is independent through the paper. Thus, the MSD of $X(t)=B(s(t))$ is given by

$$\langle X^2(t)\rangle_0 = 2D\langle s(1)\rangle t^\gamma = 2D\langle s(t)\rangle, \tag{34}$$

where $\gamma = \alpha/[1+\alpha(1-\mu)]$ for MCTRW, and $\gamma = \alpha/(1-\alpha\mu)$ for DMCTRW, respectively. For MMCTRW and MDMCTRW, we have

$$\langle X^2(t)\rangle_0 = 2D\langle s(1)\rangle t^\alpha = 2D\langle s(t)\rangle. \tag{35}$$

Proposition VII *Eqs.* (34) *and* (35) *show that the ensemble-averaged MSD is proportional to* $\langle s(t)\rangle$, *and* $X(t)$ *satisfies anomalous diffusion of the kind of Eq.* (1) *with different diffusion exponent* γ *for MCTRW and DMCTRW, and* α *for MMCTRW and MDMCTRW.*

3. Fokker-Planck-type Dynamic Equation

In biased case, $X(t) = x(s(t))$, where $x(s)$ is given by Eq. (11). Firstly, we consider MCTRW and DMCTRW, so $t(s)$ satisfies Eqs. (12) and (13) respectively. Using the total probability formula, we get that the PDF $p(x, t)$ of $X(t)$ can be given by

$$p(x, t) = \int_0^\infty f(x, s) g(s, t) \mathrm{d}s,$$

where $f(x, s)$ is the PDF of $x(s)$. Consequently, in the Laplace space k, the above formula and Eq. (18) yield

$$\begin{aligned}\hat{p}(x, k) &= \int_0^\infty \mathrm{e}^{-kt} p(x, t)\mathrm{d}t \\ &= \int_0^\infty f(x, s)\hat{g}(s, k)\mathrm{d}s \\ &= \frac{A}{2}(Ak)^{\alpha-1}\hat{f}_1\left(x, \frac{1}{2}(Ak)^\alpha\right),\end{aligned} \tag{36}$$

where $f_1(x, s) = f(x, s^{\gamma/\alpha})$. Since the process $x(s)$ is given by the *Itô* stochastic differential equation (11), its PDF $f(x, s)$ obeys the ordinary Fokker-Planck equation [14]

$$\frac{\partial f(x, s)}{\partial s}=\frac{\partial}{\partial x}\left[\frac{V'(x)}{m\eta}+D\frac{\partial}{\partial x}\right]f(x, s). \tag{37}$$

Thus,

$$\frac{\partial f_1(x, s')}{\partial s'}=\frac{\partial}{\partial x}\left[\frac{V'(x)}{m\eta}+D\frac{\partial}{\partial x}\right]\frac{\gamma}{\alpha}s^{\gamma/\alpha-1}f_1(x, s'), \tag{38}$$

and in the Laplace space p yields

$$p\hat{f}_1(x, p)-f_1(x, 0)=L_{\mathrm{FP}}(x)\frac{1}{2\pi \mathrm{i}}\int_{a-\mathrm{i}\infty}^{a+\mathrm{i}\infty}\hat{f}_1(x, q)\hat{g}(p-q)\mathrm{d}q, \tag{39}$$

where $g(s)=\frac{\gamma}{\alpha}s^{\gamma/\alpha-1}$, $a>s_1$ and $\mathrm{Re}\, p>S_2+a$, $L_{\mathrm{FP}}(x)=\left[-\frac{\partial}{\partial x}\frac{F(x)}{m\eta}+D\frac{\partial^2}{\partial x^2}\right]$. Here s_1 and s_2 are the increasing exponent of $f_1(x, \tau)$ and $g(\tau)$, respectively.

Letting $p=\frac{1}{2}(k\gamma)^{\alpha}$ and noting that $p(x, 0)=f(x, 0)=f_1(x, 0)$, we infer that in the Laplace space $\hat{p}(x, k)$ satisfies the equation

$$\begin{aligned}&k\hat{p}(x, k)-p(x, 0)\\&=L_{\mathrm{FP}}\frac{1}{2\pi \mathrm{i}}\int_{a-\mathrm{i}\infty}^{a+\mathrm{i}\infty}\hat{p}\left(x, \frac{1}{A}(2q)^{\frac{1}{\alpha}}\right)\frac{2}{A}(2q)^{\frac{(1-\alpha)}{\alpha}}\hat{g}\left(\frac{1}{2}(Ak)^{\alpha}-q\right)\mathrm{d}q.\end{aligned} \tag{40}$$

Thus, we obtain the Fokker-Planck-type dynamic equation:

$$\frac{\partial p(x, t)}{\partial t}=\Phi_t\left[\frac{\partial}{\partial x}\left(-\frac{F(x)}{m\eta}+D\frac{\partial}{\partial x}\right)\right]p(x, t), \tag{41}$$

where $\Phi_t(.)$ is a double integral operator:

$$\Phi_t f(t)=\frac{1}{2\pi \mathrm{i}}\int_{a-\mathrm{i}\infty}^{a+\mathrm{i}\infty}\mathrm{d}q h(t, q)\int_0^{\infty}\mathrm{d}t\, \mathrm{e}^{-\frac{1}{\gamma}(2q)^{1/\alpha}t}f(t), \tag{42}$$

$$h(t, q)=\frac{2^{1/\alpha}}{A}q^{\frac{1-\alpha}{\alpha}}L_k^{-1}\left[\hat{g}\left(\frac{1}{2}(Ak)^{\alpha}-q\right)\right],$$

where L_k^{-1} denotes the inverse of the Laplace transform in variable k.

Proposition VIII *This shows that since the self-similarity exponent* γ *of MCTRW and of DMCTRW different from that one of* α*-stable Lévy motion* $L_\alpha(s)$, *the operator* $\Phi_t f(t)$ *of the Fokker-Planck equation becomes a complex double integral operator.*

Now we derive the Fokker-Planck equation for MMCTRW and MDMCTRW, where t is defined by (9) and (10) correspondingly. Similarly, the PDF $p(x, t)$ of the subordinated $X(t)=x(s(t))$ in Laplace space k is given by

$$\hat{p}(x, k)=\frac{1}{2}A^\alpha k^{\alpha-1}\hat{f}\left(x, \frac{1}{2}(Ak)^\alpha\right), \tag{43}$$

where $f(x, s)$ be the PDF of $x(s)$, and $f(x, s)$ obeys the ordinary differential equation [14]

$$\frac{\partial f(x, s)}{\partial s}=L_{\mathrm{FP}}(x)f(x, s). \tag{44}$$

Thus, in Laplace space $p=\frac{1}{2}(Ak)^\alpha$, we have

$$\frac{1}{2}(Ak)^\alpha\hat{f}\left(x, \frac{1}{2}(Ak)^\alpha\right)-f(x, 0)=L_{\mathrm{FP}}(x)\hat{f}\left(x, \frac{1}{2}(Ak)^\alpha\right). \tag{45}$$

Substituting Eq. (43) into Eq. (45), it leads to

$$k\hat{p}(x, k)-f(x, 0)=\frac{2}{A^\alpha}L_{\mathrm{FP}}(x)k^{1-\alpha}\hat{p}(x, k). \tag{46}$$

Thus, noting that $p(x, 0)=f(x, 0)$ and the Laplace transform of the Riemann-Liouville fractional derivative and Riemann-Liouville integral: $\mathscr{L}[D_{0+}^\alpha f(t)]=s^\alpha\hat{f}(k)$ for $0<\alpha<1$, and $\mathscr{L}[I_0^\alpha f(t)]=s^{-\alpha}\hat{f}(k)$ for $\alpha>0$, we obtain that

$$\frac{\partial p(x, t)}{\partial t}=\frac{2}{A^\alpha}\,{}_0D_t^{1-\alpha}\left[-\frac{\partial}{\partial x}\frac{F(x)}{m\eta}+D\frac{\partial^2}{\partial x^2}\right]p(x, t), \quad 0<\alpha<1; \tag{47}$$

$$\frac{\partial p(x, t)}{\partial t}=\frac{2}{A^\alpha}\left[-\frac{\partial}{\partial x}\frac{F(x)}{m\eta}+D\frac{\partial^2}{\partial x^2}\right]p(x, t), \quad \alpha=1; \tag{48}$$

$$\frac{\partial p(x,t)}{\partial t}=\frac{2}{A^{\alpha}}\,{}_0I_t^{\alpha-1}\left[-\frac{\partial}{\partial x}\frac{F(x)}{m\eta}+D\frac{\partial^2}{\partial x^2}\right]p(x,t),\quad 1<\alpha<2. \tag{49}$$

Proposition IX *These show that the Fokker-Planck equations both for MMCTRW and MDMCTRW are in analogy with regular diffusive CTRW* [14, 33, 34].

4. Stationary Solution and Generalized Einstein Relations

Since in external force field $F(x)=-V'(x)$, here $V(x)$ is external potential and $S=\left[\frac{-V'(x)}{m\eta}-D\frac{\partial}{\partial x}\right]p(x,t)$ is the probability current. If the stationary state is reached, S must be constant. Thus, if $S=0$ for any x, from Eq. (44) it vanishes for all x [37], and the stationary solution is given by $V'(x)p_{\mathrm{st}}(x)/(m\eta)+Dp'_{\mathrm{st}}(x)=0$. Hence,

$$p_{\mathrm{st}}(x)\propto\exp\left\{-\frac{V(x)}{Dm\eta}\right\}. \tag{50}$$

Comparing this expression with the required Boltzmann distribution $p_{\mathrm{st}}(x)\propto\exp\{-V(x/[k_{\mathrm{B}}T])\}$, we recover a generalized Einstein relation, also referred to an Einstein-Stokes-Smoluchowski relation,

$$D=\frac{k_{\mathrm{B}}T}{m\eta}, \tag{51}$$

connecting the noise strength D with the dissipative parameter η via the thermal energy T.

For the case of a uniform force field, given by $V(x)=-Fx$, a net drift occurs. We calculate the first moment of $X(t)$, obtaining

$$\langle X(t)\rangle_F=\frac{F}{m\eta}\langle s(1)\rangle t^{\gamma} \tag{52}$$

for MCTRW and DMCTRW, and

$$\langle X(t)\rangle_F=\frac{F}{m\eta}\langle s(1)\rangle t^{\alpha} \tag{53}$$

for MMCTRW and MDMCTRW. Consider (34) and (35), we recover the relation

$$\langle X(t)\rangle_F = \frac{1}{2}\frac{F\langle X^2(t)\rangle_0}{k_B T}, \tag{54}$$

connecting the first moment in the presence of a uniform force field with the second moment in absence of the force. Relation (54) is the generalized Einstein relation discussed in Refs. [5, 37].

Proposition X *Our two classes of CTRW processes preserve the two fundamental physical properties of diffusion processes, in analogy to regular diffusive CTRW.*

5. Asymptotic Behavior of the PDF of Subordination

We remember that

$$\frac{\partial f(x, s)}{\partial s} = \frac{\partial}{\partial x}\left[\frac{V'(x)}{m\eta} + D\frac{\partial}{\partial x}\right] f(x, s), \quad f(x, 0) = \delta(x). \tag{55}$$

We assume that the external potential at original has the form

$$V(x) = b\ln x + \sum_{n=0}^{\infty} a_n x^n, \tag{56}$$

where b, a_n can be zero. By the results of Ref. [35], we obtain

$$\hat{f}(x, k) \sim k^{-\frac{1}{2}} \frac{1}{(xk^{1/2})^{\kappa}} \exp\{-\text{const} \cdot xk^{1/2}\} \tag{57}$$

for $|x/k^{1/2}| \gg 1$, where $\kappa = \frac{1}{2} - \delta_0$. Thus, for $f_1(x, s) = f(x, s^{\gamma/\alpha})$, we have

$$\hat{f}_1(x, k) \sim k^{-\frac{1}{2}} \frac{1}{(xk^{1/2})^{\kappa}} \exp\{-\text{const} \cdot xk^{1/2}\} \tag{58}$$

when $\zeta := \frac{x}{x_0} \sim x/s^{\gamma/2\alpha} \gg 1$ and $s \to \infty$. Consequently, using Eq. (36), we obtain

$$\hat{P}(x, k) \sim k^{-\left(1-\frac{\gamma}{2}\right)} \frac{1}{(xk^{\gamma/2})^{\kappa}} \exp\{-\text{const} \cdot xk^{\gamma/2}\}. \tag{59}$$

By the same method of Refs. [38~41], we expected that

$$p(x, t) \sim t^{-\gamma/2}(x/X_0)^{\delta'} \exp\{-\text{const} \cdot (x/X_0)^{u'}\}, \tag{60}$$

when $\xi \equiv x/X_0 \sim x/t^{\gamma/2} \gg 1$ and $t \to \infty$. The Laplace transform of (60) can be evaluated by applying the method of steepest descent, and the result is compared with (59). This yields

$$u' = 1/(1-\gamma/2), \quad \delta' = u'\left[\frac{1}{2}(\gamma-1)-\kappa\right], \tag{61}$$

where $\kappa = \frac{1}{2} - \delta_0$, while $\delta_0 = \frac{1}{2}\left(K+1+\frac{Kb_K}{Dm\eta}\right)$ if $b_K \neq 0$; or $\delta_0 = \frac{1}{2}\left(1-\frac{b}{Dm\eta}\right)$ if $b_j = 0$ $(j=1, 2, \cdots, m)$ but $b \neq 0$; or $\delta_0 = \frac{1}{2}$ if $b_j = b = a_n$ for $1 \leqslant j \leqslant K$ and $n < N$ but $a_N \neq 0$ and $N \geqslant 0$ [35].

Proposition XI *Thus, the stretched Gaussian university holds for the PDF of the subordination $X(t)$ as expected by R. Metzler and J. Klafter [14]. Moreover, the PDF of $X(t)$ can be heavy-tailed, Gaussian-tailed, and thin-tailed with respect to ξ for $0 < u' = 1/(1-\gamma/2) < 2$, $u' = 2$, or $u' > 2$, respectively, where $\gamma = \alpha/[1+\alpha(1-\mu)]$ for MCTRW and $\gamma = \alpha/(1-\alpha\mu)$ for DMCTRW.*

Next, we treat the asymptotic behavior in MMCTRW and MDCTRW, and now assume

$$V(x) = \sum_{j=1}^{K} x^{-j} + b\ln x + \sum_{n=0}^{\infty} a_n x^n, \tag{62}$$

where b_j, b, and a_n can be zero. By the results of Ref. [36], we obtain

$$\hat{f}(x, k) = G'k^{\alpha/2-1} y^{\delta_0} K_{\lambda_0}(y), \quad y = x\sqrt{k^{\alpha}/D}, \tag{63}$$

where $K_{\lambda_0}(y)$ is the modified Bessel function of second kind, $G' = 1/[C_{\lambda_0} \cdot \sqrt{K_\alpha}]$, $C_{\lambda_0} = \int_0^{\infty} y^{\delta_0} K_{\lambda_0}(y)\mathrm{d}y$ is a constant [35], $\lambda_0^2 = \frac{1}{4}(1+\hat{b})^2 > 0$, $\delta_0 =$

$\frac{1}{2}(1-\hat{b})$, and $\hat{b}=b/(Dm\eta)$.

Noting that

$$K_{\lambda_0}(y)=\mathrm{e}^{-y}\left(\frac{\pi}{2y}\right)\left[1+O\left(\frac{1}{y}\right)\right] \tag{64}$$

in the domain $|\arg y|<\frac{3\pi}{2}$, we obtain

$$\hat{f}(x,k)\sim G''k^{-(1-\frac{1}{2}\alpha)}\frac{1}{(xk^{\alpha/2})^{\kappa}}\exp\{-\text{const}\cdot xk^{\alpha/2}/G'''\} \tag{65}$$

for $|x/k^{\alpha/2}|\gg 1$, where $\kappa=\frac{1}{2}-\delta_0$, $G''=\sqrt{\pi/2}/C_{\lambda_0}(GK_\alpha)^{(1+K)/2}>0$, $G'''=\sqrt{GK_\alpha}$.

Thus, it follows from Eqs. (65) and (43) that

$$\hat{p}(x,k)\sim k^{-(1-\frac{\alpha}{2})}\frac{1}{(xk^{\alpha/2})^{\kappa}}\exp\{-\text{const}\cdot xk^{\alpha/2}\} \tag{66}$$

for $|x/k^{\alpha/2}|\gg 1$.

By the same method of Refs. [37,42], we expected that

$$p(x,t)\sim t^{-\alpha/2}(x/X_\alpha)^{\delta'}\exp\{-\text{const}\cdot(x/X_\alpha)^{u'}\}, \tag{67}$$

when $\xi\equiv x/X_\alpha\sim x/t^{\alpha/2}\gg 1$ and $t\to\infty$. The Laplace transform of (67) can be evaluated by applying the method of steepest descent, then compare it with (66). This yields

$$u'=1/\left(1-\frac{\alpha}{2}\right),\quad \delta'=u'\left[\frac{1}{2}(\alpha-1)-\kappa\right], \tag{68}$$

where $\kappa=\frac{1}{2}-\delta_0$, while $\delta_0=\frac{1}{2}\left(1-\frac{b}{Dm\eta}\right)$ if $b_j=0$ $(j=1,2,\cdots,K)$ but $b\neq 0$; or $\delta_0=\frac{1}{2}$ if $b_j=b=a_n=0$ for $1\leqslant j\leqslant K$ and $n<N$ but $a_N\neq 0$ and $N\geqslant 0$ [36].

Proposition XII *Thus, PDF of the subordination in MMCTRW and MDMCTRW can be heavy-tailed, Gaussian-tailed, and thin-tailed depending on* *the different value of* $u' = 1/(1-\alpha/2)$. *The stretched Gaussian university holds with respect to* ξ *for the PDF of the subordination* $X(t) = x(s(t))$ *as expected by R. Metzler and J. Klafter* [14] *only when* $u' < 2$.

6. Time Averaged Mean Squared Displacement

We consider the nth order time averaged (TA) moments

$$\overline{\delta^n}(\Delta;\ t_a,\ T)_0 = \frac{\int_{t_a}^{t_a+T-\Delta} [X(t+\Delta) - X(t)]^n \mathrm{d}t}{T-\Delta} \tag{69}$$

with process age t_a, the lag time Δ and the measurement time T, which are useful to characterize experiment data, Refs. [43,44]. When $n=2$, it is just the TA mean square displacement (MSD). The equivalence of time and ensemble MSD is the hallmark of ergodicity. If the time and ensemble averages of diffusion processes are nonidentical, this means that ergodicity is broken. The phenomenon of weak ergodicity breaking for scale-free CTRW processes is observed in Refs. [45, 46]. Moreover, there exist other processes with similar violation of ergodicity, see Refs. [29, 31, 47, 48].

6.1 The unbiased case with $F \equiv 0$

In this case, we note $X(t) = B(s(t))$. Recall the self-similarity with index $H = 1/2$ of $B(\tau)$ and γ of $s(t)$ where $\gamma = \alpha/[1+\alpha(1-\mu)]$ for MCTRW, $\gamma = \alpha/(1-\alpha\mu)$ for DMCTRW, $\gamma = \alpha$ for MMCTRW and MDMCTRW respectively. We know $X(t) = B(1)s^H(1)t^{H\gamma}$ in distribution. Thus,

$$\langle X^n(t)\rangle_0 = \langle \zeta_n \rangle t^{nH\gamma}, \tag{70}$$

where $\zeta_n = B(1)^n s^{nH}(1)$ is a random variable and

$$\langle \zeta_n \rangle = \ell_n \frac{2^{nH\gamma/\alpha}}{\alpha A^{nH\gamma}} \frac{\Gamma(nH\gamma/\alpha)}{\Gamma(nH\gamma)},$$

where $\ell_n = \langle |B^n(1)| \rangle = (n-1)(n-3)(n-5)\cdots(n-2m+1)(\sqrt{2D})^n J_n$, $J_n = 1$ for $n=2m$ and $J_n = \sqrt{2/\pi}$ for $n=2m+1$, $m \in \mathbb{N}^+$, which recover the result of (7) of Ref. [28] for a random walk on a fractal, approximated by the dynamics scheme of O'Shaughnessy and Procaccia [43] only different from a constant; and that in the limit $\Delta \ll T$ the aged TA nth order TA moments

$$\begin{aligned}\overline{\delta^n}(\Delta;\ t_a,\ T)_0 &= \sigma^n \zeta_n \frac{\int_{t_a}^{t_a+T-\Delta} [(t+\Delta)^{H\gamma} - t^{H\gamma}]^n \mathrm{d}t}{T-\Delta} \\ &= \sigma^n \zeta_n \frac{1}{\alpha'(T-\Delta)} \{[(t_a+T)^{\alpha'+1} - (t_a+\Delta)^{\alpha'+1}] \\ &\quad - [(t_a+T-\Delta)^{\alpha'+1} - t_a^{\alpha'+1}]\} \end{aligned} \tag{71}$$

in distribution is obtained, where $\alpha' = nH\gamma$. Hence, in the limit $\Delta \ll T$ we obtain

$$\overline{\delta^n}(\Delta;\ t_a,\ T)_0 = \zeta_n \Lambda_{nH\gamma}(t_a/T) T^{nH\gamma}(\Delta/T), \tag{72}$$

$$\langle \overline{\delta^n}(\Delta;\ t_a,\ T)_0 \rangle = (\Delta/T) \Lambda_{nH\gamma}(t_a/T) \langle X^n(T) \rangle_0, \tag{73}$$

where $\Lambda_\alpha(z) := (1+z)^\alpha - z^\alpha$ is identical to that obtained in Refs. [49,50]. The ergodicity breaking parameter EB_n of the dimensionless random variable $\xi_n = \overline{\delta^n}(\Delta;\ t_a,\ T)_0 / \langle \overline{\delta^n}(\Delta;\ t_a,\ T)_0 \rangle$ has the form as

$$EB_n = \langle \xi_n^2 \rangle - 1 = \frac{l_{2n}}{l_n^2} \frac{\Gamma(2nH\gamma/\alpha)}{\Gamma(2nH\gamma)} \frac{\Gamma^2(nH\gamma)}{\Gamma^2(nH\gamma/\alpha)} - 1, \tag{74}$$

and the PDF of the dimensionless random variable ξ_n given by

$$\phi_{n;\gamma;H}(x) = \frac{1}{l_n} \frac{\alpha A^{nH\gamma}}{2^{nH\gamma/\alpha}} \frac{\Gamma(nH\gamma)}{\Gamma(nH\gamma/\alpha)} \sqrt{\frac{2}{\pi}} \frac{1}{n^2 H} x^{\frac{1}{n}+\frac{1}{nH}-2} \mathrm{e}^{-x^{2/n}/2} g(x^{1/nH}),\ x \geqslant 0, \tag{75}$$

where $g(x)$ is the PDF of $s(1)$.

Interestingly, when $n=2$, we have

$$\overline{\delta^2}(\Delta;\ t_a,\ T) = B^2(1) s(1) \Lambda_\gamma(t_a/T) T^\gamma(\Delta/T), \tag{76}$$

and

$$\langle \overline{\delta^2}(\Delta;\ t_a,\ T) \rangle = 2D \frac{2^{\gamma/\alpha}}{\alpha A^\gamma} \frac{\Gamma(\gamma/\alpha)}{\Gamma(\gamma)} \Lambda_\gamma(t_a/T) T^\gamma(\Delta/T), \tag{77}$$

$$\Phi_{\xi_2}(x)=\frac{\alpha A^{\gamma}}{2^{\gamma/\alpha}}\frac{\Gamma(\gamma)}{\Gamma(\gamma/\alpha)}\frac{1}{4D\sqrt{\pi D}}x^{-\frac{1}{2}}\exp\left(-\frac{x}{4D}\right)g(x),\quad x\geqslant 0,\tag{78}$$

and

$$EB=\frac{\langle(\overline{\delta^2}(\Delta;\ t_a,\ T))^2\rangle}{(\langle\overline{\delta^2}(\Delta;\ t_a,\ T)\rangle)^2}-1=3\frac{\Gamma(2\gamma/\alpha)}{\Gamma(2\gamma)}\frac{\Gamma^2(\gamma)}{\Gamma^2(\gamma/\alpha)}-1.\tag{79}$$

Proposition XIII *Eq. (72) shows that $\overline{\delta^n}(\Delta;\ t_a,\ T)$ remains a random variable. Eq. (73) means that $\langle\overline{\delta^n}(\Delta;\ t_a,\ T)\rangle$ and $\langle X^n(T)\rangle_0$ have different diffusion exponents $nH\gamma-1$, $nH\gamma$, respectively. The behavior of the MSD and the TAMSD may be fundamentally different, so ergodicity is broken. Furthermore, if we know through measurement the ensemble averaged $\langle X^n(T)\rangle_0$ we can determine the single particle trajectory averaged behavior; conversely, the inverse implication is also true.*

Proposition XIV *Eq. (74) shows that the ergodicity parameter EB_n of the dimensionless random variable ξ_n is independent of the lag time Δ and the starting time t_a of measurement.*

6.2 In biased case

We first study the time-averaged first moment. In the presence of a uniform force field, given by $V(x)=-Fx$, a net drift occurs. In this case, $X(t)=x(s(t))=\frac{F}{m\eta}s(t)+B(s(t))$. We calculate its first moment, obtained in the limit $\Delta\ll T$

$$\overline{\delta^1}(\Delta;\ t_a,\ T)_F=\frac{F}{m\eta}s(1)\Lambda_{\gamma}(t_a/T)\frac{\Delta}{T}T^{\gamma}.\tag{80}$$

Thus, combining Eqs. (54), (73) and (80) indeed leads to the time-averaged second generalized Einstein relation

$$\langle\overline{\delta^1}(\Delta;\ t_a,\ T)\rangle_F=\frac{1}{2}\frac{F\langle\overline{\delta^2}(\Delta;\ t_a,\ T)\rangle_0}{k_B T}\tag{81}$$

discussed in Refs. [5, 37], connecting the time-averaged first moment in the presence of a uniform force field with the time-averaged second moment in absence of the force.

Proposition XV *This shows that the second generalized Einstein relation in ensemble average discussed in Refs.* [5, 37] *holds in time-average of a single particle tracking both for our models.*

In a biological cell, the diffusive motion of a tracer particle is spatially confined, or a charge in a disordered semiconductor experiences a drift force. To address such systems we need to determine the TAMSD of subdiffusive particle $X(t)=x(s(t))$ in the presence of an external harmonic potential $U(x)=\lambda x^2/2$ of stiffness λ and discuss the more general case below. To that end we consider the following Langevin equation with Brownian motion $B(\tau)$

$$\mathrm{d}x(\tau)=-\lambda' x(\tau)\mathrm{d}\tau+\mathrm{d}B(\tau), \tag{82}$$

with $\lambda'=\lambda/(m\eta)$, which has a unique solution

$$x(\tau)=B(\tau)-\lambda'\int_{-\infty}^{\tau}\mathrm{e}^{-\lambda'(\tau-s)}B(s)\mathrm{d}s. \tag{83}$$

Using the self-similarity of $B(\tau)$ and $s(t)$ from Eq. (84) we have

$$X(t)=\zeta_1 s^H(1)t^{H\gamma}-\lambda\left\{\zeta_{-1}\sum_{n=0}^{\infty}a_n t^{n\gamma}+\zeta_1\sum_{n=0}^{\infty}b_n t^{\gamma(n+H+1)}\right\}, \tag{84}$$

in distribution, where $\zeta_1=B(1)$, $\zeta_{-1}=B(-1)$, $a_n=a'_n s^n(1)$, $a'_n=\lambda'^{H+1}\Gamma(H+1)\dfrac{(-\lambda')^n}{n!}$, $b_n=b'_n s^{n+H+1}(1)$,

$$b'_n=\frac{(-\lambda')^n}{n!}\left(\sum_{j=0}^{n}\frac{(-j)^n}{j+H+1}\mathrm{C}_j^n\right),\quad \mathrm{C}_j^n=\frac{n(n-1)(n-2)\cdots(n-j+1)}{j!}.$$

Here, $H=1/2$, and $\gamma=\alpha/[1+\alpha(1-\mu)]$ for MCTRW, $\gamma=\alpha/(1-\alpha\mu)$ for DMCTRW, $\gamma=\alpha$ for MMCTRW and MDMCTRW. Thus, for $T\gg 1$

$$\langle X^2(T)\rangle \backsimeq 2D\lambda'^2 g(T)T^{3\gamma}, \tag{85}$$

where

$$g(T)=\frac{2^{3\gamma/\alpha}}{\alpha}\left[\sum_{n=0}^{\infty}b'_n\left(\frac{2^{\gamma/\alpha}}{\gamma}T\right)^n\right]^2,$$

and in the limit $\Delta \ll T$ the TAMSD

$$\overline{\delta^2}(\Delta;\ t_a,\ T) \backsimeq \lambda'^2 B^2(1) s^3(1) h(t_a,\ T) T^{3\gamma}(\Delta/T), \tag{86}$$

$$\langle\overline{\delta^2}(\Delta;\ t_a,\ T)\rangle \backsimeq 2D\lambda'^2 \frac{2^{3\gamma/\alpha}}{\alpha A^{3\gamma}} \frac{\Gamma(3\gamma/\alpha)}{\Gamma(3\gamma)} h(t_a,\ T) T^{3\gamma}(\Delta/T), \tag{87}$$

$$h(t_a,\ T) = \sum_{n,\ m=0}^{\infty} b'_n b'_m \frac{2^{\gamma(n+m+3)/\alpha}}{\alpha A^{\gamma(n+m+3)}} \frac{\Gamma((n+m+3)\gamma/\alpha)}{\Gamma((n+m+3)\gamma)} T^{\gamma(n+m)} \Lambda_{\gamma(n+m+3)}(t_a/T).$$

Thus, it yields the PDF $\phi_{\xi_2}(x)$ and the ergodicity breaking parameter EB of the dimensionless random variable $\xi_2 = \dfrac{\overline{\delta^2}(\Delta;\ t_a,\ T)}{\langle\overline{\delta^2}(\Delta;\ t_a,\ T)\rangle}$

$$\phi_{\xi_2}(x) = \frac{\alpha A^{3\gamma}}{2^{3\gamma/\alpha}} \frac{\Gamma(3\gamma)}{\Gamma(3\gamma/\alpha)} \frac{x^{-7/6}}{12D\sqrt{\pi D}} \exp\left(-\frac{x}{4D}\right) g(x^{\frac{1}{3}}), \quad x \geqslant 0, \tag{88}$$

and

$$EB = 3\alpha 2^{6\gamma(1/\alpha-1)} \frac{\Gamma(3\gamma/\alpha+1/2)}{\Gamma(3\gamma+1/2)} \frac{\Gamma(3\gamma)}{\Gamma(3\gamma/\alpha)} - 1. \tag{89}$$

Proposition XVI *Eq.* (86) *shows that* $\overline{\delta^2}(\Delta;\ t_a,\ T)$ *remains a random variable. Eqs.* (85) *and* (87) *imply that* $\langle\overline{\delta^2}(\Delta;\ t_a,\ T)\rangle$ *and* $\langle X^2(T)\rangle$ *differ from a factor*

$$R(T) = \frac{\langle\overline{\delta^2}(\Delta;\ t_a,\ T)\rangle}{\langle X^2(T)\rangle} = \frac{2^{3\gamma/\alpha}}{\alpha A^{3\gamma}} \frac{\Gamma(3\gamma/\alpha)}{\Gamma(3\gamma)} \frac{h(t_a,\ T)}{g(T)} \left(\frac{\Delta}{T}\right) \tag{90}$$

depending on the lag time Δ, *the starting time* t_a *of measurement and the measurement time* T.

Proposition XVII *Eq.* (89) *shows that the ergodicity parameter EB of the dimensionless random variable* ξ_2 *is independent of the lag time* Δ *and the starting time* t_a *of measurement, and tends to* 2 *when* α *tend to* 1, *i.e.*, $\lim_{\alpha\to 1} EB = 2$.

7. Conclusion

In this paper, we introduce some models of anomalous diffusion in external force fields, which originates from a CTRW process with memorized waiting times. Our models are rephrased in terms of a Langevin equation for position x and the random time change involving Riemann-Liouville fractional derivative or Riemann-Liouville fractional integral. We obtain the shape of the corresponding Fokker-Planck-type dynamic equation. We show that the stationary solution of the model equals the Boltzmann distribution, and that the model satisfies the generalized Einstein relations in consistency with the fluctuation-dissipation relation. We show that the asymptotic behavior of the waiting time and the PDF of subordination $X(t)$ are stretched Gaussian distribution, respectively. The diffusion of the subordination can be subdiffusion, normal diffusion, or superdiffusion depending on the self-similar exponent for MCTRW, DMCTRW, MMCTRW and MDMCTRW, and that our models display weak ergodicity-breaking. Interestingly, the second generalized Einstein relation also holds for the time-averaged first moment of a single particle tracking in our models.

We believe that the introduced models could be seen as an interesting alternative in the modeling of the complex systems showing anomalous diffusion under the influence of external force fields. It relaxes the renewal property of standard CTRW and CCTRW theory, and thus of interest in all those cases where the non-renewal behavior is relevant: such as in financial market dynamics, in movement ecology, to search processes and human motion patterns, also various physical processes as turbulent flows or complex systems with heterogeneities.

Acknowledgments

This work is dedicated to the 120th Birthday of Prof. Jiangong Chen, and is supported by the National Natural Science Foundation of China (Grants No. 11371266, 11001057, 11271074) and ZJNSF (No. Q13A010034). The authors would like to thank the referee for his/her kind suggestions.

◇ References ◇

[1] J. Klafter, M. F. Shlesinger, G. Zumofen, Phys. Today 49 (2) (1996) 33.

[2] M. F. Shlesinger, G. M. Zaslavsky, U. Frisch (Eds.), Lévy Flights and Related Topics in Physics, Springer-Verlag, Berlin, 1994.

[3] G. H. Weiss, Aspects and Applications of Random Walk, North-Holland, Amsterdam, 1994.

[4] H. Scher, M. F. Shlesinger, J. T. Bendler, Phys. Today 44 (1) (1991) 26.

[5] J. P. Bouchaud, A. Georges, Phys. Rep. 195 (1990) 127.

[6] W. Haus, K. W. Kehr, Phys. Rep. 150 (1987) 263.

[7] S. Havlin, D. B. Avrahan, Adv. Phys. 36 (1987) 695. Reprint in Adv. Phys. 51 (2002) 187.

[8] E. W. Montroll, G. G. Weiss, J. Math. Phys. 6 (1965) 167.

[9] M. B. Isichenko, Rev. Modern Phys. 64 (1992) 961.

[10] A. Blumen, J. Klafter, Z. Zumofen, Models for reaction dynamics in glasses, in: I. Zschokke (Ed.), Optical Spectres Copy of Glasses, Reidel, Dordrecht, 1986.

[11] G. A. Losa, E. R. Weibl, Fractals in Biology and Medicine, Birkhäuser, Basel, 1993.

[12] H. Scher, E. W. Montroll, Phys. Rev. B 12 (1975) 2455.

[13] T. H. Solomon, R. E. Weeks, L. H. Swinney, Phys. Rev. Lett. 71 (1993) 3975.

[14] R. Metzler, J. Klafter, Phys. Rep. 339 (2000) 1.

[15] A. V. Chechkin, Y. V. Gonchar, M. Szydlowsky, Phys. Plasmas 9 (2002) 78.

[16] H. Scher, et al., Geophys. Res. Lett. (2002) 1061.

[17] B. Berkowitz, A. Cortis, M. Dentz, H. Scher, Rev. Geophys. 44 (2006) RG2003.

[18] A. Klemm, R. Metzler, R. Kimmich, Phys. Rev. E 65 (2002) 021112.

[19] S. S. Rogers, C. V. D. Wahle, T. A. Waigh, Langmuir 24 (2008) 13549.

[20] I. Golding, E. C. Cox, Phys. Rev. Lett. 96 (2006) 098102.

[21] R. Nathan, et al., Proc. Natl. Acad. Sci. USA 105 (2008) 19052.

[22] E. Scalas, Physica A 362 (2006) 225.

[23] G. La Spada, J. D. Farmer, F. Lillo, Eur. Phys. J. B 64 (2008) 607.

[24] R. Le Borgne, M. Dentz, J. Carrera, Phys. Rev. Lett. 101 (2008) 090601.

[25] R. Barkana, Mon. Not. R. Astron. Soc. 376 (2007) 1784.

[26] P. Manneville, Instabilities, Chaos and Turbulence, Imperial College Press, London, 2010.

[27] M. M. Meerschaert, E. Nane, Y. Xiao, Statist. Probab. Lett. 79 (2009) 1194.

[28] A. V. Chechkin, M. Hofmann, I. M. Sokolov, Phys. Rev. E 80 (2009) 031112.

[29] V. Tejedor, R. Metzler, J. Phys. A 43 (2010) 082002.

[30] M. Magdziartz, R. Metzler, W. Szczotka, P. Zebrowski, J. Stat. Mech. (2012) P04010.

[31] M. Magdziartz, R. Metzler, W. Szczotka, P. Zebrowski, Phys. Rev. E 85 (2012) 051103.

[32] S. G. Samko, A. A. Kilbas, O. I. Marichev, Fractional Integrals and Derivatives, Theory and Applications, Gorden and Breach Science Publishers, Yverdon, ISBN: 2-88124-864-0, 1993.

[33] M. Magdziarz, A. Weron, K. Weron, Phys. Rev. E 77 (2008) 036704.

[34] M. Magdziarz, A. Weron, Phys. Rev. E 75 (2007) 016708.

[35] F. Y. Ren, J. R. Liang, W. Y. Qiu, J. B. Xiao, J. Phys. A: Math. Gen. 39 (2006) 4911.

[36] F. Y. Ren, W. Y. Qiu, J. R. Liang, X. T. Wang, Phys. Lett. A 288 (2001) 79.

[37] H. Risken, The Fokker-Planck Equation, Springer-Verlag, Berlin, 1989.

[38] M. Giona, H. E. Romman, Physica A 185 (1992) 87.

[39] H. E. Roman, M. Giona, J. Phys. A: Math. Gen. 25 (1992) 2107.

[40] R. Metzler, W. G. Glöck, T. F. Nonnenmacher, Physica A 211 (1994) 13.

[41] L. Acedo, S. Bravo Yuste, Phys. Rev. E 57 (1998) 5160.

[42] E. Barkai, V. N. Fleurov, Phys. Rev. E 58 (1998) 1296.

[43] B. O' Shaughnessy, I. Procaccia, Phys. Rev. Lett. 54 (1985) 455 - 458.

[44] J. H. Jeon, et al. , Phys. Rev. Lett. 106 (2011) 048103.

[45] Y. He, S. Burov, R. Metzler, E. Barkai, Phys. Rev. Lett. 101 (2008) 058101.

[46] E. Barkai, Y. Garini, R. Metzler, Phys. Today 65 (8) (2012) 29.

[47] A. G. Cherstvy, A. V. Chechkin, R. Metzler, New J. Phys. 15 (2013) 083039.

[48] J. H. Jeon, A. V. Chechkin, R. Metzler, Phys. Chem. Chem. Phys. 16 (2014) 15811.

[49] J. H. P. Schulz, E. Barkai, R. Metzler, Phys. Rev. Lett. 110 (2013) 020602.

[50] J. H. P. Schulz, E. Barkai, R. Metzler, Phys. Rev. X 4 (2014) 011028.

附录一　任福尧教授论著目录

I. 专著和教材

[1] 应用复分析,复旦大学出版社,1993.

[2] 复解析动力系统,复旦大学出版社,1997.

II. 学术论文

[1] 开拓戈鲁辛和夏尔绳斯基的几个定理,复旦学报(自然科学版),第1期,(1957),117-133.

[2] 关于比霸巴赫函数和列别杰夫—米林函数,科学记录,第4期,(1958),110-112.

[3] On the functions of Bieberbach and of Lebedev-Milin, *Science Record*, Vol. II, No. 4, (1958), 121-125.

[4] 开拓劳宾生和戈鲁辛的几个定理,数学学报,第8卷,第2期,(1958),181-189.

[5] 比霸巴赫函数和列别杰夫—米林函数,复旦学报(自然科学版),第1期,(1958),35-46.

[6] 具有无界型特征的拟似共形映照的存在性定理,复旦大学数学系数学论文集,1962,427-432.

[7] 无界的广义拟共形映照的孤立奇点,复旦学报(自然科学版),第8卷,第2期,(1963),237-242.

[8] 关于射流计数触发器的最大项,射流技术讨论,第1期,(1973),1-44.

[9] 射流反馈振荡器的理论和实验研究,射流技术讨论,第3期,(1973),1-37.

[10] 压力阶跃信号在流体管路中的传输,复旦学报(自然科学版),第2期,(1975),42-59.(和柳兆荣合作)

[11] 以腔室为终端的流体管路的传输特性(I),复旦学报(自然科学版),第1期,(1977),19-30.(和柳兆荣合作)

[12] 以腔室为终端的流体管路的传输特性(II),复旦学报(自然科学版),第2期,(1977),94-

105.（和柳兆荣合作）

[13] 流体瞬态理论在双风机并联运行管系中的应用，复旦学报（自然科学版），第3期，(1977)，85-95.（和柳兆荣合作）

[14] 单叶亚纯函数的系数，复旦学报（自然科学版），第4期，(1978)，93-96.

[15] 在第三项系数限制下的单叶函数的 Bieberbach 猜想，中国科学，数学专辑(I)，(1979)，275-280.

[16] 加强的偏差定理和葛隆斯基不等式，复旦学报（自然科学版），第3期，(1979)，69-75.

[17] 压力阶跃在复杂管系中的传输，力学学报，第4期，(1979)，380-384.（和柳兆荣合作）

[18] 在第三项系数限制下的单叶函数的比霸巴赫猜想，自然杂志，第1期，(1979)，6.

[19] 在第三项系数限制下的单叶函数的比霸巴赫猜想，复旦学报（自然科学版），第2期，(1979)，92-99.

[20] 具有拟共形扩张的单叶函数的菲斯杰拉尔特不等式，自然杂志，第10期，(1979)，598.

[21] Coefficient of inverse of meromorphic univalent functions, *Kexue Tongbao*, No. 4, (1980), 277-280.

[22] 关于单叶函数的局部极大值定理，科学通报，第1期，(1980)，47.

[23] 单叶半纯函数之逆函数的系数，科学通报，第5期，(1980)，193-195.

[24] 关于单叶半纯函数的 Springer 猜想，复旦学报（自然科学版），第2期，(1980)，229-231.

[25] 关于由抽水或灌水引起的隔水层的一维固结问题，上海地质，第2期，(1981)，44-55.

[26] 灌采条件下地下水温度动态，上海地质，第4期，(1981)，33-40.（和姚邦基合作）

[27] 灌采条件下承压水温度纵向变化规律，水文地质工程地质，第6期，(1981)，34-37.（和姚邦基合作）

[28] 关于单叶函数的系数估计，科学通报，第8期，(1981)，509.

[29] Bieberback 函数族和 Grunsky 函数族的另一种充要条件和偏差定理，科学通报，第9期，(1981)，516-519.

[30] Another necessary and sufficient condition and distortion theorems for Bieberbach and Grunsky families of functions, *Kexue Tongbao*, Vol. 26, No. 9, (1981), 781-785.

[31] Hayman 正则性定理和 Fitzgerald 不等式的改进，复旦学报（自然科学版），第1期，(1981)，1-14.

[32] 巴齐列维奇不等式的应用和改进，复旦学报（自然科学版），第4期，(1981)，402-410.

[33] 单复变函数的某些进展，自然科学年鉴，1981，2.7-2.9.（和杨乐、沈燮昌合作）

[34] 具有拟共形扩张的不重叠区域的共形映照，科学通报，第15期，(1981)，905-909.

[35] 具有拟共形扩张的不重叠区域的共形映照的偏差定理，复旦学报（自然科学版），第1期，(1982)，56-70.

[36] The distortion theorems for Bieberbach class and Grunsky class of univalent functions,

Chinese Ann. of Math., Vol. 3, No. 6, (1982), 691 - 703.

[37] 关于具有拟共形扩张的比伯霸赫函数族,数学学报,第 25 卷,第 4 期,(1982),441 - 455.

[38] Stronger distortion theorems of univalent functions and its applications, *Chinese Ann. of Math. B*, Vol. 4, No. 4, (1983), 425 - 441.

[39] Conformal mapping of non-overlapping domain with quasiconformal extension, *Kexue Tongbao*, Special Issue, (1983), 100 - 105.

[40] 关于地下水水位分布形态的若干问题,复旦学报(自然科学版),第 22 卷,第 3 期,(1983),295 - 306.(和张景辉合作)

[41] 管井回灌规律和堵塞情况的判别,工程勘察,第 1 期,(1985),60 - 65.(和姚邦基、孙永福、徐永春合作)

[42] 关于串联式管系共振脉冲射流的瞬态特性,应用数学学报,第 10 卷,第 1 期,(1987),81 - 90.

[43] 关于均匀单管共振射流的瞬态特性分析,纯粹数学与应用数学,总第 3 期,(1987),1 - 12.

[44] 关于 corona (日冕) 问题,数学进展,第 16 卷,第 2 期,(1987),159 - 169.

[45] 虚瓦茨导数和对数导数的一些估计,复旦学报(自然科学版),第 27 卷,第 3 期,(1988),297 - 304.

[46] 关于 Teichmüller 极值拟共形映照的三个 Reich 猜想,复旦学报(自然科学版),第 27 卷,第 4 期,(1988),439 - 445.

[47] 一个 Zygmund 定理的推广及应用,数学研究与评论,第 8 卷,第 3 期,(1988),391 - 400.

[48] A note on a class of p-valently α-convex functions, *C. R. Math. Rep. Acad. Sci. Canada*, Vol. 10, No. 1, (1988), 35 - 40. (with S. Owa)

[49] Extension of a theorem of Carleson-Duren, *J. Math. Research and Exposition*, Vol. 9, No. 1, (1989), 31 - 39. (with Li-Feng Huang)

[50] Some problems of univalent functions and some applications of complex analysis, *Univalent Functions, Fractional Calculus and Their Applications*, Ed. H. M. Srivastava and S. Owa, Ellis Horwood Limited, Chichester, 1989, 229 - 244.

[51] Some inequalities on quasi-subordinate functions, *Bull. Austral. Math. Soc.*, Vol. 43, (1991), 317 - 324. (with S. Owa and S. Fukui)

[52] The Julia sets of the random iteration of rational functions, *Chinese Sci. Bull.*, Vol. 37, No. 12, (1992), 969 - 971. (with Wei-Min Zhou)

[53] A dynamical sysmtem formed by a set of rational functions, in *Current Topics in Analytic Function Theory*, Ed. H. M. Srivastava and S. Owa, World Scientific, Singapore, 1992, 437 - 449. (with Wei-Min Zhou)

[54] An invariant measure for the dynamical system formed by a set of rational functions, in *Current Topics in Analytic Function Theory*, Ed. H. M. Srivastava and S. Owa, World Scientific, Singapore, 1992, 450 - 456. (with Wei-Min Zhou)

[55] 超越函数随机迭代系统的 Julia 集，科学通报，第 38 卷，第 4 期，(1993)，289 - 290.(和周维民合作)

[56] The Julia sets of rational iteration system of transcendental meromorphic functions, *Chinese Sci. Bull.*, Vol. 38, No. 4, (1993), 289 - 290. (with Wei-Min Zhou)

[57] On the iterations of entire algebroid functions, *Sciences in China* (*Ser. A*), Vol. 37, No. 4, (1994), 422 - 431. (with Jian-Yong Qiao)

[58] H^p multipliers on bounded symmetric domains, *Science in China* (*Ser. A*), Vol. 37, No. 3, (1994), 257 - 264. (with Jian-Bin Xiao)

[59] Univalence of holomorphic mappings and growth theorem for close-to-starlike mappings in finitely dimensional Banach space, *Acta Math. Sinica* (*N. S.*) Special Issue, Vol. 10, (1994), 207 - 214. (with Hong-Bin Chen)

[60] Some results on complex analytic dynamics, *J. Fudan Univ. Natur. Sci.*, Vol. 33, No. 2, (1994), 197 - 208.

[61] Iterations of holomorphic self-maps of C^N, *J. Fudan Univ.* (*Natur. Sci.*), Vol. 33 No. 4, (1994), 452 - 462. (with Wen-Jun Zhang)

[62] On iterates of holomorphic maps in strongly pseudoconvex domains, *Proceedings of the Conference on Complex Analysis* (Tianjin, 1992), Int. Press, Cambridge, MA, 1994, 245 - 253, (with Wen-Jun Zhang and Jian-Yong Qiao)

[63] A proof of Taylor's conjecture for packing measure, *Progress in Natural Science*, Vol. 4, No. 1, (1994), 60 - 67. (with You Xu)

[64] The starlikeness of a certain class of integral operators, *Complex Variables Theory Appl.*, Vol. 27, No. 2, (1995), 185 - 191. (with S. Owa, H. M. Srivastava and Wei-Qi Yang)

[65] Random iteration of holomorphic self-maps over bounded domains in C^N. *Chinese Ann. Math. Ser. B*, Vol. 16, No, 1, (1995), 33 - 42. (with Wen-Jun Zhang)

[66] The Hausdorff dimension and measure of the generalized Moran fractals and Fourier series, *Chinese Ann. Math. Ser. B*, Vol. 16, No. 2, (1995), 153 - 162. (with Jin-Rong Liang)

[67] Dynamics on weakly pseudoconvex domains, *Chinese Ann. Math. Ser. B*, Vol. 16, No. 4, (1995), 467 - 476. (with Wen-Jun Zhang)

[68] Quasiconformal extension of biholomorphic mappings of several complex variables, *J. Fudan Univ.* (*Natur. Sci.*), Vol. 34, No. 5, (1995), 545 - 556. (with Jian-Guo

Ma)

[69] Bounded projections and duality on spaces of holomorphic functions in the unit ball of C^n, *Acta Math. Sinica* (*N. S.*), Vol. 11, No. 1, (1995), 29 - 36. (with Jian-Bin Xiao)

[70] On Taylor's conjecture about the packing measures of Cartesian product sets, *Chinese Ann. Math. Ser. B*, Vol. 17, No. 1, (1996), 121 - 126. (with You Xu)

[71] Local fractional Brownian motions and Gaussian noises and application, *J. Fudan Univ.* (*Natur. Sci.*), Vol. 35, No. 4, (1996), 361 - 372. (with Xing-Qiu Zhao, Feng Jiang, Wei-Yuan Qiu, Feng Su, Shao-Xing Qian, Pin-Pin Shen)

[72] A random dynamical system formed by infinitely many functions, *J. Fudan Univ.* (*Natur. Sci.*), Vol. 35, No. 4, (1996), 387 - 392. (with Zhi-Min Gong)

[73] Fractional integral associated to the self-similar set or the generalized self-similar set and its physical interpretation, *Physics Letters A*, Vol. 219, (1996), 59 - 68. (with Zu-Guo Yu, Feng Su)

[74] A negative answer to a problem of Bergweiler, *Complex Variables Theory Appl.*, Vol. 30, No. 4, (1996), 315 - 322. (with Zhi-Min Gong, Wei-Yuan Qiu)

[75] On the Fatou and Julia conjecture for the random dynamical system, *Pure Appl. Math.*, Vol. 12, No. 2, (1996), 23 - 27. (with Zhi-Min Gong)

[76] The study of the Fourier series of functions defined on Moran fractals, *Acta Math. Appl. Sinica*, Vol. 13, No. 2, (1997), 158 - 166. (with Jin-Rong Liang)

[77] The Fourier series expansions of functions defined on *s*-sets, *Chinese Ann. Math. Ser. B*, Vol. 18, No. 2, (1997), 201 - 212. (with Jin-Rong Liang, Wan-She Li, Feng Su)

[78] An affirmative answer to a problem of Baker, *J. Fudan Univ.* (*Natur. Sci.*), Vol. 36, No. 2, (1997), 231 - 233. (with Wan-She Li)

[79] Advances and problems in random dynamical systems, *Adv. in Math.* (*China*), Vol. 26, No. 5, (1997), 385 - 394.

[80] On interior points of the Julia set $J(R)$ for random dynamical system R, *Chinese Ann. Math. Ser. B*, Vol. 18, No. 4, (1997), 503 - 512. (with Zhi-Min Gong)

[81] Fractional integral associated to cookie-cutter set and its physical interpretation, *Progr. Natur. Sci.*, Vol. 7, No. 4, (1997), 422 - 428. (with Zu-Guo Yu)

[82] Fractional integral associated to generalized cookie-cutter set and its physical interpretation, *J. Phys. A: Math. Gen.*, Vol. 30, (1997), 5569 - 5577. (with Zu-Guo Yu, Ji Zhou)

[83] The relationship between the fractional integral and the fractal structure of a memory set, *Physica A*, Vol. 246, (1997), 419 - 429. (with Zu-Guo Yu)

[84] Domains of explosion of the dynamic system generated by $P(\lambda e^z + \mu e^{-z})$, *Southeast Asian Bull. Math.*, Vol. 21, No. 2, (1997), 193 - 201. (with Wan-She Li)

[85] The random iterations of commutable rational functions, *J. Fudan Univ.* (*Natur. Sci.*), Vol. 36, No. 6, (1997), 685 - 690. (with Wan-She Li)

[86] On fractal properties of reflection sequences from a well log and seismic trace, *J. Fudan Univ.* (*Natur. Sci.*), Vol. 36, No. 6, (1997), 691 - 699. (with Xing-Qiu Zhao, Feng Jiang, Wei-Yuan Qiu, Feng Su, Shao-Xing Qiang)

[87] Self-similar measures on the Julia sets, *Progr. Natur. Sci.*, Vol. 8, No. 1, (1998), 24 - 34. (with Ji Zhou, Wei-Yuan Qiu)

[88] Julia sets of the random iteration systems and its subsystems, *Chinese Sci. Bull.*, Vol. 43, No. 4, (1998), 265 - 268. (with Ji Zhou, Wei-Yuan Qiu)

[89] Hausdorff dimensions of random net fractals, *Stochastic Processes Appl.*, Vol. 74, No. 2, (1998), 235 - 250. (with Jin-Rong Liang)

[90] On computation for the multipliers of periodic orbits of polynomial maps, *J. Xiangtan Univ.* (*Natur. Sci.*), Vol. 20, No. 3, (1998), 51 - 58. (with Zhi-Min Gong, Jian Wang)

[91] Dynamics of periodically random orbits, *Progress in Natural Science*, Vol. 9, No. 4, (1999), 248 - 255. (with Ji Zhou, Wei-Yuan Qiu)

[92] On three open questions proposed by Falconer, *Progr. Natur. Sci.*, Vol. 9, No. 3, (1999), 180 - 188. (with Jin-Rong Liang, Xiao-Tian Wang)

[93] The determination of the diffusion kernel on fractals and fractional diffusion equation for transport phenomena in random media, *Physics Letter A*, Vol. 252, (1999), 141 - 150. (with Jin-Rong Liang, Xiao-Tian Wang)

[94] Hausdorff dimension, mean, quadratic variation of infinite self-similar measures, *Bull. Hong Kong Math. Soc.*, Vol. 2, No. 2, (1999), 347 - 355. (with Zu-Guo Yu, Jin-Rong Liang)

[95] Fine approximation of flux and fractional integrals on a (infinite) self-similar set or generalized self-similar set, *Southeast Asian Bull. Math.*, Vol. 23, No. 3, (1999), 497 - 505. (with Zu-Guo Yu, Ji Zhou)

[96] 随机网分形的多重分形分解,数学年刊(A辑),第20卷,第1期,(1999),81 - 90.(和梁金荣、徐先进合作)

[97] Multifractal decompositions of random net fractals, *Chinese J. Contemp. Math.*, Vol. 20, No. 1, (1999), 51 - 60. (with Jin-Rong Liang, Xian-Jin Xu)

[98] Noise and chaotic disturbance on self-similar sets, *Southeast Asian Bull. Math.*, Vol. 23, No. 1, (1999), 145 - 152. (with Zu-Guo Yu)

[99] Self-similar measures on the Julia sets of polynomials, *Progr. Natur. Sci.*, Vol. 10, No. 4, (2000), 266 - 271. (with Ji Zhou, Wei-Yuan Qiu)

[100] Measures and their dimension spectrums for cookie-cutter sets in Rd, *Acta Math. Appl. Sinica*, Vol. 16, No. 1, (2000), 9 - 21. (with Jin-Rong Liang, Zu-Guo Yu)

[101] The non-integer operation associated to random variation sets of the self-similar set, *Physica A*, Vol. 286, (2000), 45 - 55. (with Jin-Rong Liang)

[102] Determination of memory function and flux on fractals, *Physics Letters A*, Vol. 288, No. 2, (2001), 79 - 87. (with Wei-Yuan Qiu, Jin-Rong Liang, Xiao-Tian Wang)

[103] Determination of diffusion kernel on fractals, *J. Phys. A: Math. Gen.*, Vol. 34, No. 46, (2001), 9815 - 9825. (with Xiao-Tian Wang, Jin-Rong Liang)

[104] Option pricing of fractional version of the Black-Scholes model with Hurst exponent H being in $\left(\frac{1}{3}, \frac{1}{2}\right)$, *Chaos, Solitons and Fractals*, Vol. 12, No. 3, (2001), 599 - 608. (with Xiao-Tian Wang, Wei-Yuan Qiu)

[105] A proof for French's empirical formula on option pricing, *Chaos, Solitons and Fractals*, Vol. 12, No. 13, (2001), 2441 - 2453. (with Xiao-Tian Wang, Jin-Rong Liang)

[106] Whitening filter and innovations representation of self-similar process, *Chaos Solitons Fractals*, Vol. 14, No. 7, (2002), 1047 - 1057. (with Xiao-Tian Wang, Wei-Yuan Qiu)

[107] A fractional version of the Merton model, *Chaos Solitons Fractals*, Vol. 15, No. 3, (2003), 455 - 463. (with Xiao-Tian Wang, Xiang-Qian Liang)

[108] An anomalous diffusion model in an external force fields on fractals, *Physics Letters A*, Vol. 312, No. 3 - 4, (2003), 187 - 197. (with Jin-Rong Liang, Wei-Yuan Qiu, Xiao-Tian Wang, Yun Xu, R. R. Nigmatullin)

[109] Integrals and derivatives on net fractals, *Chaos Solitons Fractals*, Vol. 16, No. 1, (2003), 107 - 117. (with Jin-Rong Liang, Xiao-Tian Wang, Wei-Yuan Qiu)

[110] Fractional Fokker-Planck equation on heterogeneous fractal structures in external force fields and its solutions, *Physica A*, Vol. 326, No. 3-4, (2003), 430 - 440. (with Jin-Rong Liang, Wei-Yuan Qiu, Yun Xu)

[111] Universality of stretched Gaussian asymptotic behaviour for the fractional Fokker-Planck equation in external force fields, *J. Phys. A: Math. Gen.*, Vol. 36, No. 27, (2003), 7533 - 7543. (with Jin-Rong Liang, Wei-Yuan Qiu, Yun Xu)

[112] Fractional Giona-Roman equation on heterogeneous fractal structures in external force fields and its solutions, *Proceedings of DETC*. 03, 2003, DETC 2003 / VIB - 48401. (with Jin-Rong Liang, Wei-Yuan Qiu, Yun Xu)

[113] Stretched Gaussian asymptotic behavior for fractional Giona-Roman equation on biased

heterogeneous fractal structure in external force fields, *Nonlinear Dynamics*, Vol. 38, No. 1-4, (2004), 285-294. (with Wei-Yuan Qiu, Yun Xu, Jin-Rong Liang)

[114] Fractional diffusion equations involving external forces in the higher dimensional case, *Chaos Solitons Fractals*, Vol. 21, No. 3, (2004), 679-687. (with Fu Zou, Wei-Yuan Qiu)

[115] Answer to an open problem proposed by E Barkai and J Klafter, *J. Phys. A: Math. Gen.*, Vol. 37, No. 42, (2004), 9919-9922. (with Wei-Yuan Qiu, Yun Xu, Jin-Rong Liang)

[116] Option pricing of a mixed fractional-fractional version of the Black-Scholes model, *Chaos Solitons Fractals*, Vol. 21, No. 5, (2004), 1163-1174. (with Jian-Hong Chen, Wei-Yuan Qiu)

[117] Universality of stretched Gaussian asymptotic diffusion behavior on biased heterogeneous fractal structure in external force fields, *Chaos Solitons Fractals*, Vol. 24, (2005), 273-278. (with Wei-Yuan Qiu, Yun Xu, Jin-Rong Liang)

[118] On some generalization of fractional Brownian motions, *Chaos Solitons Fractals*, Vol. 28, No. 4, (2006), 949-957. (with Xiao-Tian Wang, Xiang-Qian Liang, Shi-Ying Zhang)

[119] Answer to an open problem proposed by R Metzler and J Klafter, *J. Phys. A: Math. Gen.*, Vol. 39, No. 16, (2006), 4911-4919. (with Jin-Rong Liang, Wei-Yuan Qiu, Jian-Bin Xiao)

[120] Asymptotic behavior of a fractional Fokker-Planck-type equation, *Physica A*, Vol. 373, (2007), 165-173. (with Jin-Rong Liang, Wei-Yuan Qiu, Jian-Bin Xiao)

[121] Exact solutions for nonlinear fractional anomalous diffusion equations, *Physica A*, Vol. 385, No. 1, (2007), 80-94. (with Jin-Rong Liang, Wei-Yuan Qiu, Jian-Bin Xiao)

[122] Solutions for multidimensional fractional anomalous diffusion equations, *J. Math. Phys.*, Vol. 49, No,7, (2008), 073302, 9 pp. (with Long-Jin Lü, Jian-Bin Xiao, Lei Gao)

[123] Fractional nonlinear diffusion equation and first passage time, *Physica A*, Vol. 387, No. 4, (2008), 764-772. (with Jun Wang, Wen-Jun Zhang, Jin-Rong Liang, Jian-Bin Xiao)

[124] Solutions for a time-fractional diffusion equation with absorption: influence of different diffusion coefficients and external forces, *J. Phys. A: Math. Theor.*, Vol. 41, No. 4, (2008), 045003, 10 pp. (with Wen-Bin Chen, Jun Wang, Wei-Yuan Qiu)

[125] The solutions to a bi-fractional Black-Scholes-Merton differential equation, *Int. J.*

Pure Appl. Math., Vol. 58, No. 1, (2010), 99 - 112. (with Jin-Rong Liang, Jun Wang, Wen-Jun Zhang, Wei-Yuan Qiu)

[126] Option pricing of a bi-fractional Black-Merton-Scholes model with the Hurst exponent H in $\left[\frac{1}{2}, 1\right]$, *Appl. Math. Letters*, Vol. 23, No. 8, (2010), 859 - 863. (with Jin-Rong Liang, Jun Wang, Wen-Jun Zhang, Wei-Yuan Qiu)

[127] The application of fractional derivatives in stochastic models driven by fractional Brownian motion, *Physica A*, Vol. 389, No. 21, (2010), 4809 - 4818. (with Long-Jin Lü, Wei-Yuan Qiu)

[128] Generalized Einstein relation and the Metzler-Klafter conjecture in a composite-subdiffusive regime. *Physica A*, Vol. 390, (2011), 2920 - 2925. (with Hua Pan, Wei-Yuan Qiu, Long-Jin Lü)

[129] Fractional Fokker-Planck equation and Black-Scholes formula in composite-diffusive regime, *J. Stat. Phys.*, Vol. 146, No. 1, (2012), 205 - 216. (with Jin-Rong Liang, Jun Wang, Long-Jin Lü, Hui Gu, Wei-Yuan Qiu)

[130] Continuous time Black-Scholes equation with transaction costs in subdiffusive fractional Brownian motion regime, *Physica A*, Vol. 391, No. 3, (2012), 750 - 759. (with Jun Wang, Jin-Rong Liang, Long-Jin Lü, Wei-Yuan Qiu)

[131] Fractional Fokker-Planck equation with space and time dependent drift and diffusion, *J. Stat. Phys.*, Vol. 149, No. 4, (2012), 619 - 628. (with Long-Jin Lü, Wei-Yuan Qiu)

[132] Heterogeneous memorized continuous time random walks in an external force fields, *J. Stat. Phys.*, Vol. 156, No. 6, (2014), 1111 - 1124. (with Jun Wang, Ji Zhou, Long-Jin Lü, Wei-Yuan Qiu)

[133] Effect of different waiting time processes with memory to anomalous diffusion dynamics in an external force fields, *Physica A*, Vol. 417, (2015), 202 - 214. (with Jun Wang, Long-Jin Lü, Hua Pan, Wei-Yuan Qiu)

[134] Correlated continuous time random walk with time averaged waiting time, *Physica A*, Vol. 422, (2015), 101 - 106. (with Long-Jin Lü, Jun Wang, Jian-Bin Xiao)

[135] Correlated continuous time random walk and option pricing, *Physica A*, Vol. 447, (2016), 100 - 107. (with Long-Jin Lü, Jian-Bin Xiao, Liang-Zhong Fan)

附录二　任福尧教授指导的学生和博士后

一、博士后

苏　峰,1994—1996 年。

二、博士研究生

1. 黄立丰,1989 年 7 月,日冕问题和有界函数生成的理想。

2. 肖建斌,1989 年 7 月,混合范数空间的乘子及对偶。

3. 邱维元,1991 年 1 月,复解析动力系统中的某些课题。

4. 尹永成,1991 年 7 月,二次有理函数的动力系统。

5. 周维民,1991 年 7 月,随机迭代动力系统。

6. 陈红斌,1992 年 7 月,全纯映射的单叶性准则、近于星象映射以及凸象映射的增长性定理。

7. 马建国,1993 年 7 月,Heisenberg 群上的拟共形映照和多复变全纯映照的拟共形扩张。

8. 陈　正,1993 年 7 月,多复变几何理论中的一些课题。

9. 乔建永,1994 年 1 月,关于复解析动力系统的若干研究。

10. 张文俊,1994 年 1 月,多复变全纯映照的迭代。

11. 李万社,1995 年 7 月,关于复解析动力系统的若干问题。

12. 龚志民,1996 年 7 月,复解析动力系统若干问题的研究。

13. 梁金荣,1997 年 7 月,Hausdorff 维数、多重分形分解及 Fourier 分析。

14. 喻祖国,1997 年 7 月,分形几何与高维 Möbius 群中若干问题的研究。

15. 周　吉,1998 年 7 月,复解析动力学若干问题的研究。

16. 王晓天,2000 年 7 月,分形几何在统计物理中的应用。

17. 吕龙进,2012 年 7 月,分数阶奇异扩散方程的几种解法及其应用。

三、硕士研究生

1. 谭德邻，1981年7月（与何成奇教授联合指导）。
2. 陈纪修，1981年7月（与何成奇教授联合指导）。
3. 王　键，1985年7月（与何成奇教授联合指导）。
4. 方　向，1985年7月（与何成奇教授联合指导）。
5. 尹永成，1988年7月。
6. 周维民，1988年7月。
7. 高福昌，1989年7月。
8. 黄　宏，1989年7月。
9. 许　友，1990年7月。
10. 康　荣，1990年7月。
11. 石丽夏，1991年7月。
12. 王　蕾，1991年7月。
13. 彭岳建，1992年7月。
14. 曹建乔，1993年7月。
15. 沈　洪，1993年7月。
16. 吴少轩，1994年7月。
17. 许　昀，2004年7月。
18. 陈建红，2004年7月。
19. 邹　富，2004年7月。
20. 陈文彬，2008年7月。
21. 张　攀，2008年7月。
22. 潘　华，2012年7月。

图书在版编目(CIP)数据

任福尧数学论文选:汉、英/任福尧著. —上海:复旦大学出版社,2019.10
ISBN 978-7-309-14573-1

Ⅰ.①任… Ⅱ.①任… Ⅲ.①数学-文集-汉、英 Ⅳ.①O1-53

中国版本图书馆 CIP 数据核字(2019)第 179966 号

任福尧数学论文选:汉、英
任福尧 著
责任编辑/陆俊杰

复旦大学出版社有限公司出版发行
上海市国权路 579 号 邮编:200433
网址:fupnet@fudanpress.com http://www.fudanpress.com
门市零售:86-21-65642857 团体订购:86-21-65118853
外埠邮购:86-21-65109143
上海春秋印刷厂

开本 787×960 1/16 印张 33 字数 546 千
2019 年 10 月第 1 版第 1 次印刷

ISBN 978-7-309-14573-1/O·676
定价:80.00 元
